FISH
and
FISHERIES

FISH and FISHERIES

[Anatomy, Physiology, Applied Fisheries, Genetics, Biotechnology, and Fish Legislature]

Santosh Kumar
Formerly Professor, Bioscience Department
Vice Chancellor, Barkatullaha Vishwavidyalaya, Bhopal &
Dr H S Gour University, Sagar

Manju Tembhre
Head, Dept of Life Sciences
Principal, Sant Hirdaram Girls' College
Bairagarh, Bhopal

New Central Book Agency (P) Ltd
LONDON
HYDERABAD ERNAKULAM BHUBANESWAR
NEW DELHI KOLKATA PUNE GUWAHATI

NCBA

REGD OFFICE

8/1 Chintamoni Das Lane, Kolkata 700 009, India
email: ncbapvtltd@eth.net

OVERSEAS

NCBA (UK) Ltd, 149 Park Avenue North, Northampton, NN3 2HY, UK
email: ncbauk@yahoo.com

EXPORT

NCBA Exports Pvt. Ltd, 212 Shahpur Jat, New Delhi 110 049
email: ncbaexports@ncbapvtltd.com

BRANCHES

212 Shahpur Jat
New Delhi 110 049
email: ncbadel@ncbapvtltd.com

Shop Nos. 3 & 4, Vinayak Towers
681/B Budhwar Peth, Appa Balwant Chowk
Pune 411 002
email: ncbapune@ncbapvtltd.com

Shop No. 15, Bharati Towers, Ground Floor
Forest Park, **Bhubaneswar** 751 001
email: ncbabub@ncbapvtltd.com

House No. 3-1-315, 1st Floor
Nimboliadda, Kachiguda, **Hyderabad** 500 027
email: ncbahydb@ncbapvtltd.com

GSS Shopping Complex, 1st Floor
Opposite College Post Office, Convent Road
Ernakulam 682 035
email: ncbaernk@ncbapvtltd.com

Radhamayee Complex (Second Floor)
Opp. SBI, Panbazar, Jaswanta Road
Guwahati 781 001
email: ncbaguwahati@ncbapvtltd.com

FISH AND FISHERIES • Santosh Kumar and Manju Tembhre

First Published: January 2010
Reprinted: July 2011
Reprinted: June 2014

PUBLISHER AND TYPESETTER
New Central Book Agency (P) Ltd
8/1 Chintamoni Das Lane, Kolkata 700 009

PRINTER
New Central Book Agency (P) Ltd
Web-Offset Division, Dhulagarh, Sankrail, Howrah

COVER DESIGNER
Soumen Paul

COVER PRINTER
Biswanath Studio, Kolkata

TECHNICAL EDITORS
Subas Maitra, Dipan Roy

PROJECT TEAM
Prabhat Jas, Pradip Das

ISBN 978 81 7381 536 2

Price: ₹ 850.00

CONTENTS

Part III deals with **Genetics.** In this part the uses of recent techniques of molecular biology and biotechnology of commercial importance to enhance the aquaculture products have been dealt for fisheries students. Special detailed attention has been given on chromosome, chromosome banding, recombinant DNA Technology, fish mitochondria genome, transgenic fishes, gynogenesis, and monosex culture. New technologies such as Polymerase chain reaction and advanced Flow Cytometry have been introduced. Qualitative trait and the pattern of inheritance, and methods for the measurement of quantitative inheritance and binomial distribution in relation to profitable aquaculture have been dealt with thoroughly.

Part IV deals with **Fisheries Laws and Regulations**—both State and Central—for the protection and conservation of fisheries. Brief introduction has been provided for Marine Export Development Authority Act, 1972, and international HACCP export rules. International Seabed Authority and its functions in relation to Seabed nodules and continental shelf have been fully described.

The authors express their heatfelt gratitude to the staff of Biosciences Department of Barkatullah University, Sant Hirdaram Girls' College, and the Chairman, Shaheed Hemu Kalani Educational Society, Bhopal, for their generous help during the preparation of the manuscript. We also extend our heartfelt thanks to Dr Subodh Jain, Reader, Dr Harisingh Gour University, Sagar, and Dr Praveen Tamot, Prof. of Zoology, M.V.M., Bhopal, for comments and suggestions given on the chapters of PCR and Cage Culture, respectively. We express our sincere obligations to Dr Vijay Laxmi Saxena, Prof. of Zoology, Kanpur, for her help in this endeavour. We also wish to acknowledge the help extended by Dr Madhavi Gaur, Ms Shanta Ahirwar, Ms Sandhya Gaur and Ms Anjali Namdeo for their active and valuable help.

1 January 2010

Prof. Santosh Kumar
Dr Manju Tembhre

PART I

ANATOMY AND PHYSIOLOGY

Chapter 1

PETROMYZON (LAMPREY)

The lampreys or lamper eel and hagfishes are the sole existing representative of group 'Agnatha', i.e., 'Jawless fish'. They are characterised by the absence of both jaws, pectoral and pelvic fins or paired fins. They possess a single nostril and have a protocercal tail, pore like gill openings, and a monomeric haemoglobin. The Lampreys are generally active predators, attaching themselves to other animals and extracting blood and the product of tissue cytolysis. They possess teeth on a well developed suctorial disc and tongue. It is generally agreed that the lampreys evolved from an anapsid or anapsid like stock and are related to the ostracoderm of the ordovician, silurian and devonian periods.

Distribution

Lampreys are found in both fresh and marine waters in the Northern and the Southern Hemispheres. The marine water species are *Petromyzon fluviatilis* (smaller), *P. marinus* (largest), *Lampetra ayresii*, *L, tridentata*, and *L. japonica*. Marine species have anadromous migratory habit. Fresh water species are *Tetrapleurodon spadiceus* and *T. genuinis*, which are both restricted to streams and lakes in high altitudes (Hardisty and Potter, 1972; Ganguly, *et al.*, 2008).

Habit and Habitat : Life history of all lampreys includes two distinct stages, the ammocoete larva lives in fresh water, buried in the mud and is microphagous while the adult lamprey has a sucking mouth, and usually lives in the sea, where it feeds on the fishes and turtles. Adult lamprey migrate to the river for the purpose of spawning after which they eventually die.

Shape and size : The adult has an eel-like smooth, slimy body, black on the back and white below. The adult nonparasitic lampreys are generally in the length ranges 100–180 mm while sexually mature parasitic species are between 180 and 800 mm long. The larvae of parasitic forms metamorphose at a smaller size than their non-parasitic derivatives.

External Features

The body comprises a cylindrical head and trunk and laterally flattened tail. The paired fins are absent, but the tail bears a median fin, which is expanded in front as a dorsal fin. In female the anal fin is present while in male copulatory papilla is found. The male has different shape of dorsal fin. The head is characterised by downwardly directed depression called the buccal funnel which is beset with numerous horny teeth. A single median nostril is present on the dorsal side of the head. There lies a yellow spot behind the nostril representing the position of the internal eye. A pair of lateral eyes are well developed. There are seven pairs of small and round gill openings present at the posterolateral side of the head (Fig. 1.1a, b).

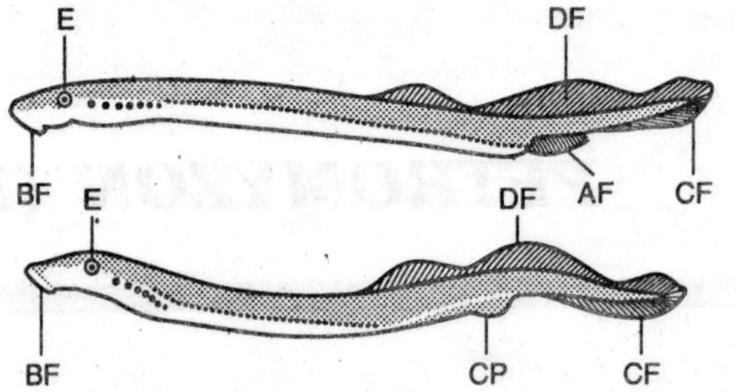

Fig. 1.1a, b : Brook lamprey – *Lampetra pleneri* (a) Female with anal fin. (b) Male with copulatory papilla and different shape of dorsal fin. [AF, anal fin; BF, buccal funnel; CF, caudal fin; E, eye; CP, copulatory papilla.]

Skin : The skin of both ammocoetes and adult lampreys is soft and formed of multilayered epidermis and a dense dermis (Fig. 1.2).

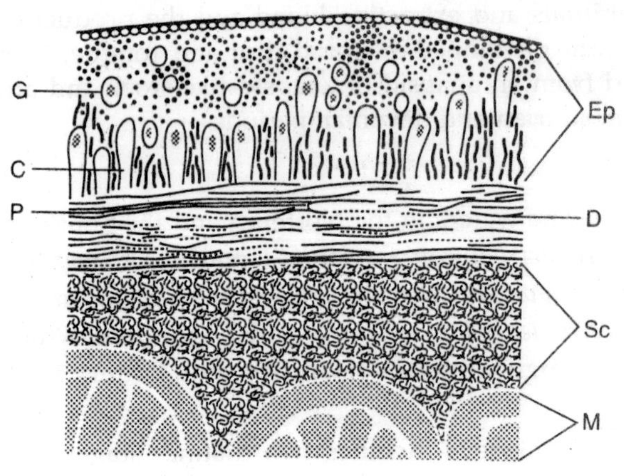

Fig. 1.2 : Section of skin of lamprey. [C, club cell; D, dermis, Ep, epidermis; G, granular gland cells; M, myotomal muscle; P, pigment cells; Sc, subcutaneous connective tissue.]

Both the layers are separated from each other by connective tissue which contains blood capillaries and migratory pigment cells or the chromatophore.

The epidermis consists of three or more layers of mucous cells, provided with large intracellular spaces. The outer layer of cells possesses brush border and stores mucus. Inner club shaped cells rest on the dermal layer and have a cytoplasmic cone. The dermis comprises circularly arranged bundles of collagenous and elastic fibres.

Musculature : The trunk musculature is well developed and consists of a series of myomeres, i.e. there are a series of myotomes separated by myocommas. Each myotome has a w-shape instead of the simple v-shape of *Amphioxus*. The muscle fibres are striated.

Skeleton : The skeleton of lamprey consists of notochord and various cartilaginous structures. The notochord is well developed throughout life as a rod below the nerve cord. It consists of large turgid vacuolated cells, enclosed in a thick fibrous sheath (Fig. 1.3).

It prevents shortening of body when myotomes contract. The notochordal sheath also surrounds the spinal cord and joins the myocommas and thus eventually the subcutaneous connective tissues.

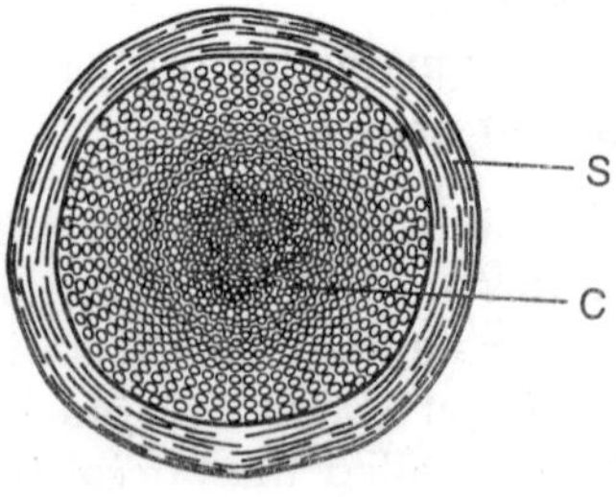

Fig. 1.3 : T-S of notochord of lamprey, showing fibrous sheath and vacuolated cells. [C, cells. S, sheath.]

Within this connective tissue there are present certain irregular cartilaginous thickenings which may be compared with vertebrae. Rods of cartilage extend dorsally and ventrally into the fins (Fig. 1.4).

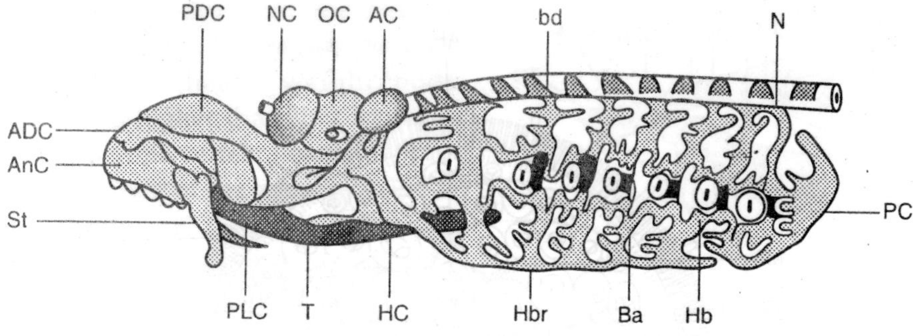

Fig. 1.4 : Skeleton of head and branchial arches of *Petromyzon* from lateral side. AC, auditory capsule; ADC, annular-dorsal cartilage; AnC, annular cartilage; Ba, branchial arch; bd, basidorsal; HC, hyoidcartilage; Hbr, hypobranchial rod; Hb, horizontal bar, N, notochord; NC, nasal capsule; OC, orbital cartilage; PC, pericardial cartilage; PDC, postero-dorsal cartilage; PLC, postero-lateral cartilage; St, styliform cartilage; T, tendon of tongue.

The skull is of a primitive type. The floor is made up of paired parachordals and paired trabeculae. A series of incomplete cartilagenous boxes surrounding the brain and sense organs are attached to the base of skull. The cranium has a floor around the end of the notochord, and in front is a pit containing the pituitary gland.

The skeleton of the branchial organ consists of a system of vertical plates between the gill slits, joined by horizontal bars above and below them. The elastic action of the cartilages produces the movement of inspiration. A backward extension of the branchial basket forms a box surrounding the heart.

Digestive System

The alimentary canal starts from small mouth leading behind into the spacious buccal cavity. The mouth is surrounded by a buccal funnel. The buccal funnel of lamprey is rounded, discoidal and ornamented with various types of replaceable teeth series which are as follows (Fig. 1.5).

1. Lingual lamina— These are multicuspid teeth of the tongue born on lingual lamina.
2. Supraoral (SO)— These are present anterior to the oesophageal opening.
3. Infraoral (IO)— These are present posterior to the oesophageal opening.
4. Circumoral row— This comprises those teeth forming a single continuous or subcontinuous series around the oesophageal opening, outside the supraoral and infraoral plates.
5. Intermediate disc teeth (IT)— Those teeth present between the circumoral and the marginals.
6. Median anterior teeth row (MA) — These are present on midline from supraoral.
7. Anterior field (AF)— Such teeth are present on the anterior area of the buccal disc.
8. Lateral field (LF)— These teeth are present on each side of the disc lying between the anterior and posterior fields.

which the veins pour blood. The heart is innervated by the vagus branch, they penetrate the heart and profusely innervate the various chambers.

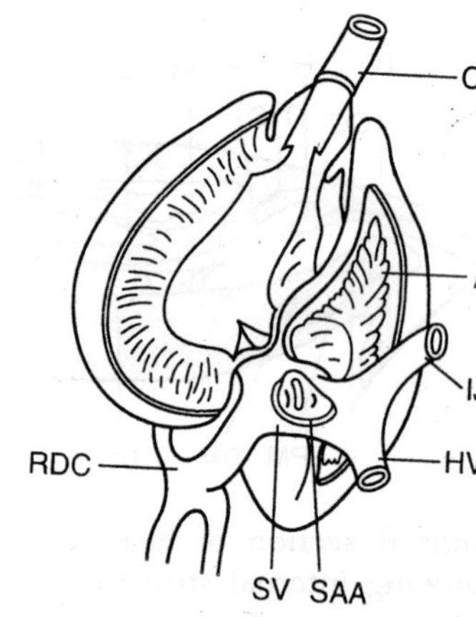

Fig. 1.8 : Diagrammatic L S. of heart of Lamprey. A, atrium; CA, conus arteriosus; HV, hepatic vein; IJ, inferior, jugular; RDC, right ductus Cuvieri; SAA, sinu auricular aperture; SV, sinus venosus; Source; Jollie, 1969.

Presence of chromaffin cells are noticed and they contain monoamines. The cytoplasm of these cells contain a large number of granules with a size of 1000–3000 Å which are surrounded by a membrane about 50 Å thickness. This leads to a thin-walled auricle, lying above the sinus. The auricle passes blood into the ventricle below it, a thick-walled chamber, pumping blood to the body with main force. The sinus venosus receives two ducti Cuvieri in larvae, but in adult lampreys the left ductus Cuvieri is obliterated or disappeared and only right persists (Hatta, 1923).

Arterial system : A large ventral aorta arises from the anterior end of the ventrical, and runs between the gill pouches. The ventral aorta swells at its base to form bulbus arteriosus. From the ventral aorta, eight branchial arteries originate and run to gill pouches. In the gill pouches they break up into capillaries. The blood from the gills is collected through efferent branchial arteries. Each afferent and efferent branchial artery are present in the interbranchial septum and thus supplies blood to posterior hemibranch of one gill and anterior hemibranch of the other. The afferent branchial arteries in turn open into the paired dorsal aortae. Both dorsal aortae run posteriorly and merge together to form a single median dorsal aorta, which gives out several arteries supplying blood to the myotomes (Young, 1962). The gut, kidneys and gonads receive blood supply from special arteries coming out from the unpaired dorsal aorta. Generally, all the arteries (except the renal and efferent branchial artery) possess the valve at their point of origin (Fig. 1.10).

These valves regulate the blood pressure in the arteries. In dorsal aorta the blood flows anteriorly while in the ventral aorta it flows backwardly.

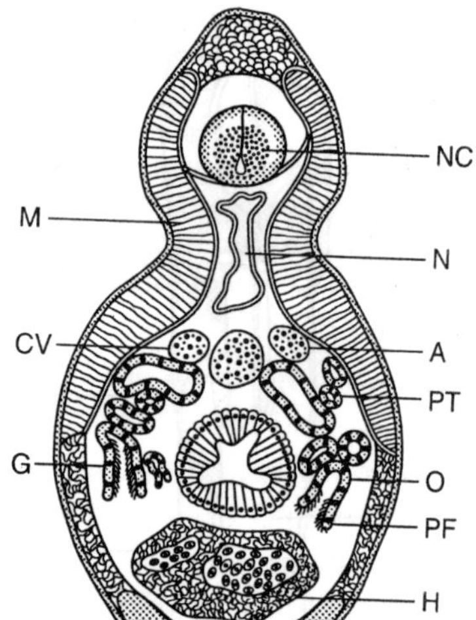

Fig. 1.9 : Section of body of larva of Lampetra showing various organs. A, aorta; CV, cardinal vein; G, glomerulus; H, heart; M, myotome; NC, nerve-cord; N, notochord; O, oesophagus; PF, ciliated funnel of pronephros; PT, twisted pronephric tubule.

Venous system : The venous system is composed of veins and a complex network of venous sinuses. A large caudal vein collects blood from the tail region. On reaching the abdominal cavity the caudal vein divides into two posterior cardinal veins which collect blood from the anterior end of the body and discharge into the heart. In addition to this the median and inferior jugular vein collect blood from the muscles of the buccal funnel and gill pouches. The renal portal vein is absent. The hepatic portal vein drains blood from the gut and pour into the liver by a contractile portal heart. The hepatic vein carries blood from the liver and discharges into the heart. Besides the veins, the branchial sinuses are present in the head region which comprises three longitudinal channels. These are :

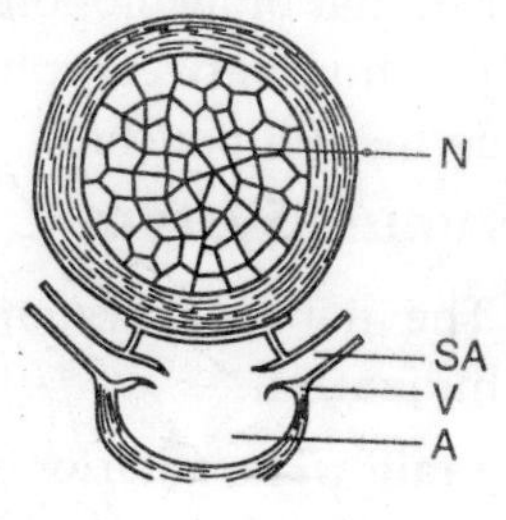

Fig. 1.10 : T.S. of artery of lamprey showing value at the origin of segmental artery. A, aorta; N, notochord; SA, segmental artery, V, valve.

(i) ventral branchial sinus or ventral jugular sinus

(ii) inferior branchial sinus present below the gill pouches

(iii) superior branchial sinus present on the gill pouches.

Blood : The red cells of lamprey are nucleated and spherical in shape of about 710 μ in size. It contains about six times as much haemoglobin as is contained in human red cells. This arrangement immensely increases the oxygen carrying power of blood. The white corpuscles are lymphocytes and polymorphs similar to the higher vertebrates. Hemopoiesis occurs in the kidney and spinal cord. The blood of adult lampreys shows a large Bohr effect, particularly in sea lamprey *(P. marius)* while ammocoete blood has a high affinity for oxygen and weaker Bohr effect (Manwell, 1963; for Bohr effect see blood of fishes).

Haemoglobin of lamprey has been studied and amino acid sequences of two lamprey haemoglobin are known. It had special importance and has a special role in the physiology of the animal due to migratory habit from ocean to fresh waters. Carbon-dioxide is present in three forms, i.e., CO_2, HCO_3 and H_2CO_3. The haemoglobin of the river lamprey, *Lampetra fluviatilis* is also polymorphic (for detail see haemoglobins by Riggs (1972) in : *Biology of Lampreys,* Vol. 2).

Respiratory System

In the lamprey the funnel shaped oral hood and mouth cavity opens behind into the pharyngeal cavity and opens dorsally into an oesophagus and ventrally into a blind ending respiratory tube. There are seven gill pouches (marsipobranch) on either side of the respiratory tube, each opening to exterior via an external branchiopore and connected to the water tube by an internal branchiopore. The respiratory tube and gill pouches are contained within an unjointed cartilaginous framework, the branchial basket. Each gill pouch has shape like a biconvex lens. Its inner wall becomes folded to form numerous gill lamellae while the outer wall is highly thick. The gill pouches are separated from each other by a septum called interbranchial septum. The water enters and leaves the body through gill pouches by a peculiar mechanism because of their specialised mode of attachment. Contraction of the branchial constrictor muscles and the ventral and dorsal diagonal muscles cause a reduction in the volume of the branchial basket, which forces water out of the gill pouches and fill with water as the volume of the branchial basket increases due to elastic recoil. The direction of water flow is regulated by valves and sphincters associated with the external branchiopore, the junction between the water tube and pharyngeal cavity, and the junction between pharyngeal and mouth cavities. The gaseous exchange takes place inside the gill pouches.

Nervous System

The nervous system of lampreys is well developed, and shows high degree of cephalization.

Brain : The anterior end of spinal cord is enlarged into a complicated brain from which special cranial nerves arise. The brain is distinguished into forebrain (prosencephalon), mid-brain (mesencephalon) and hind-brain (rhombencephalon) (Figs. 1.11, 1.12).

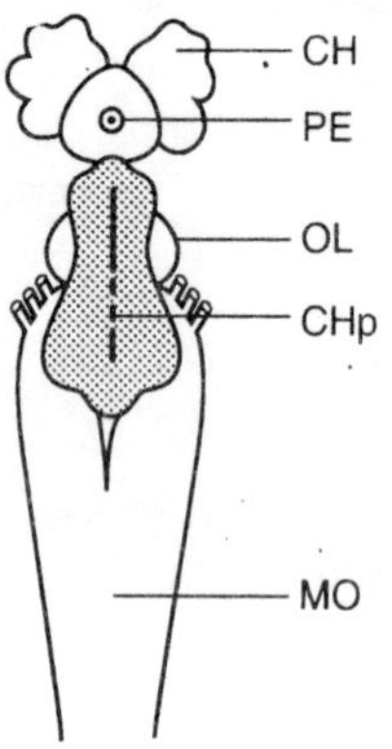

Fig. 1.11 : Brain of *Petromyzon* from dorsal view with choroid plexus. CH, cerebral hemisphere; Chp. choroid plexus; MO medulla oblongata; OL ,optic lobe, PE, pineal eye.

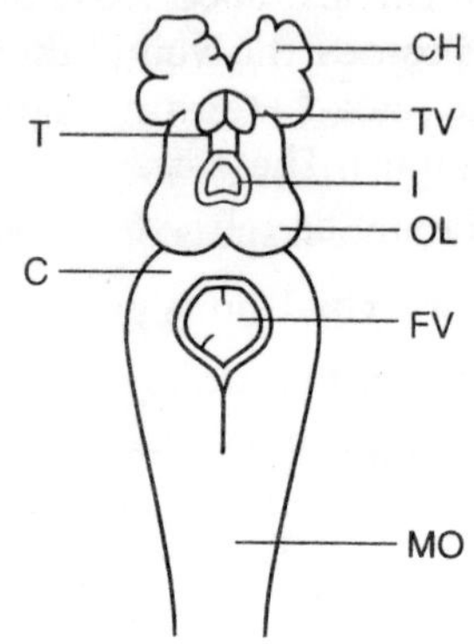

Fig. 1.12 : Dorsal view of brain of *Petromyzon* after removing choroid plexus. C, cerebellum; CH, cerebral hemisphere; FV, fourth ventricle; I, iter; MO, medulla oblongata; OL, optic lobe; T, thalamus; TV, third ventricle.

The forebrain comprises a pair of olfactory bulb and two cerebral hemisphere or telencephalon which are connected through the interventricular foramen with an unpaired

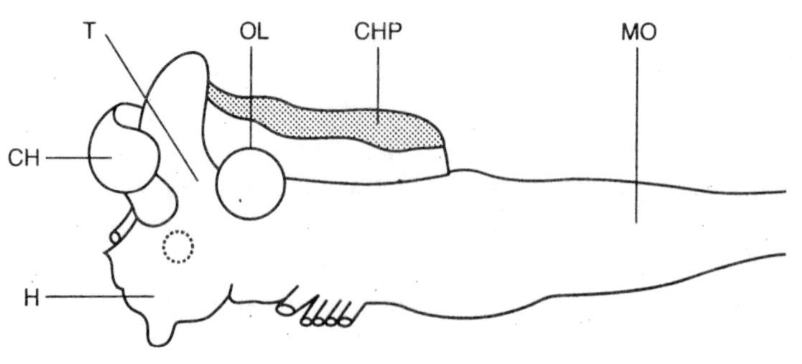

Fig. 1.13 : Lateral view of brain of Petromyzon. CH cerebral hemisphere; CHp, choroid plexus; H, hypothalamus; MO, medulla oblongata; OL, optic lobe; T, thalamus.

ventricle, which leads into two diencephalon. The diencephalon is displaced due to great development of olfactory system. The diencephalon can be distinguished into a dorsal epithalamus, median thalamus, and a ventral hypothalamus (Fig. 1.13).

The mid-brain or optic lobes extend dorsally from posterior commissure to the cerebellum. Ventrally it hardly projects from the general surface of the brain.

The hind-brain comprises an anterior cerebellum and a posterior medulla oblongata. It occupies almost half of the whole brain. The medulla leads behind the spinal cord. The upper surface of the brain is covered by an extensive vascular pad, the choroid plexus or tela choroidea. The roof of the brain is thus non-nervous in these region.

Spinal cord : The spinal cord has a uniform transparent grey colour. It is flattened dorsoventrally, which allows excess of oxygen and metabolites. The spinal cord does not possess blood vessels which is a unique feature.

Cranial nerves : Except the olfactory and optic nerves, all other nerves originate from the lower part of the midand hind-brains. There are 10 pairs of cranial nerves present in lamprey which are as follows :

Nerve Number	Name	Origin	Function/Innervation
(i)	Olfactory	Forebrain	Smell, innervates tip of snout.
(ii)	Optic	Mid-brain	Vision, innervates retina.
(iii)	Oculomotor	Mid-brain	Movement of eye muscles.
(iv)	Trochlear	Cerebellum	Innervates one of the eye muscle.
(v)	Trigeminal	Lateral of medulla	Innervates upper and lower jaws.
(vi)	Abducens. Its existence is doubtful in lamprey.		
(vii)	Facial	From medulla behind the V	Taste buds.
(viii)	Auditory	Medulla	Innervates infraorbital lateral line, and internal ear.
(ix)	Glossopharyngeal	Medulla	Innervates pharyngeal region and first gill.
(x)	Vagus	Medulla	Innervates visceral organs, gills.

The olfactory nerve consists of many nerve fibres. The optic nerves do not form optic chiasma. The trigeminal nerve (V) and facial nerve (VII) are closely associated. The roots of seventh and eighth nerve are closely present. The branches of the glossopharyngeal and vagus nerves are similar to that of fishes.

Spinal nerves : The dorsal and ventral roots of the spinal nerves are separate and is different from that of a gnathostomes in which both the roots are joined together. The ventral root possesses motor nerve fibres which innervate the myotomes. However, the dorsal root comprises sensory fibres coming from myotomes. The hypoglossal is the first spinal nerve.

All the nerves of lamprey are unmyelinated as a result of very slow impulse conduction across them.

The sympathetic system consists of isolated fibres running in both dorsal and ventral roots.

Sense Organs

The various sense organs are well developed. These are as under :

Paired Eyes : The paired eyes of adult lamprey are well developed (Fig. 1.14).

Its characteristic features are non-pigmented cornea, which is made up of two layers containing gelatinous matrix in between them. The eye lens is globular and comprises pigmented granules arranged in three distinct layers of fibres. The retina is about 200 mm thick and consists of both rod and cone cells. The retina also possesses bipolar, amacrine and a few ganglion cells (Fig. 1.15).

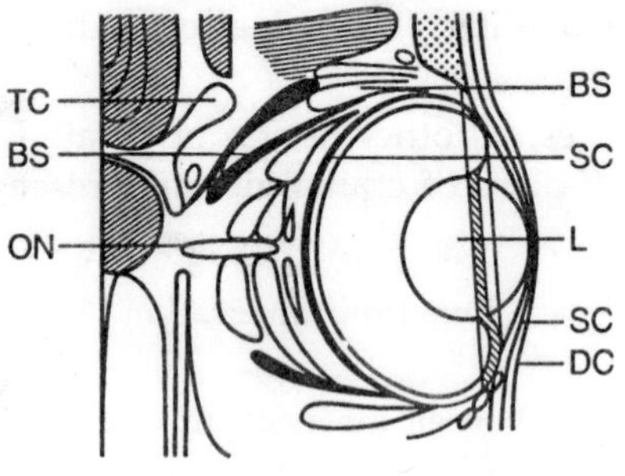

Fig. 1.14 : Vertical section of paired eye of *Lampetra fluviatilis.* BS, blood sinus; DC dermal layer of cornea; L, lens; ON, optic nerve; SC, scleral layer of cornea; TC, cartilage.

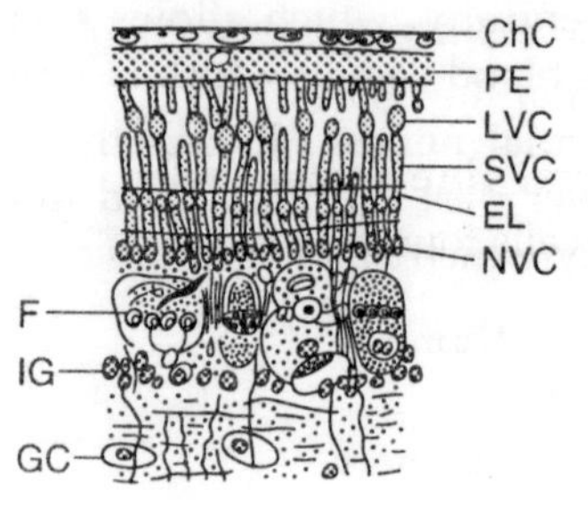

Fig. 1.15 : Transverse section of retina of paired eyes of lamprey. ChC, chorio-capillaries; EL, external limitans; F, fibres connecting visual cells with bipolar cells; GC, ganglion cells; IG; internal germinal layer with bipolar cells, LVC, long visual cells, NVC, layer of nuclear of visual cells; PE, pigmented epithelium; SVC, short visual cells.

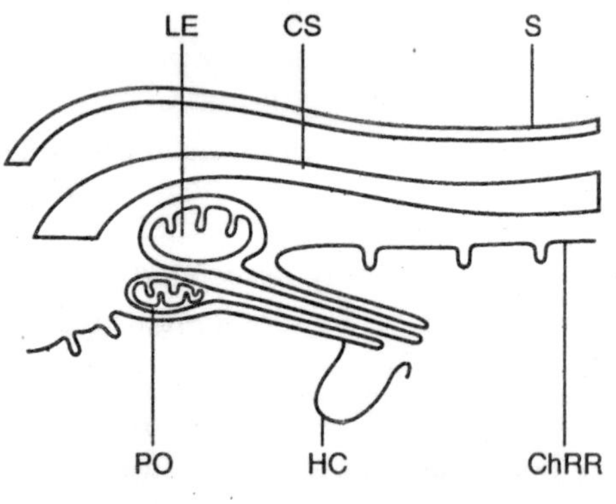

Fig. 1.16 : Diagrammatic sagittal section of pineal eye ChRR, choroid roof of brain; CS, cartilagenous skull; HC, habenular commissure; LE, lumen of epiphysis; PO, parapineal organ; S, skin.

There is an iris which surrounds the spherical pupil. The pupil is capable of little change in diameter. Many species of lamprey are diurnal. In ammocoete larva the paired eyes are buried below the pigmented skin. A special muscle originates from the myotome, which is called 'cornealis muscles,' helps in the accommodation. It inserts on the outer wall of the cornea. Its contraction results in flattening of cornea and also pushing of the lens inward.

Pineal Eye : Like the paired eyes, the pineal eye or epiphysial eye also develops from the diencephalon of the brain. The pineal eye first develops as two equal sacs. One sac is present on the dorsal side and is larger, it is known as *pineal eye* whereas another sac lies ventral to the first one and is called *parapineal organ* (Fig. 1.16).

Morphologically the pineal eye seems to be placed on the right side while para-pineal organ is on the left side. Both organs are formed by invagination of the roof of the brain and lie connected with two stalks. These organs contain narrow lumen and their walls are composed of receptor cells. The retinal layer is made up of sensory, pigmented and supporting cells. Histologically, pineal eye is similar to that of paired eyes.

Inspite of the fact that function of the pineal apparatus is not well understood, it is considered that it has the power of response to change of light and thus helps in initiation of movement. In ammocoetes larvae pineal eye helps in changing colour of body. They become pale at night and dark during daytime. Ablation of pineal eye results in lack of power of changing colour and becomes permanently dark. The pineal eye also regulates secretion of the melanophores expanding substance from the pituitary gland.

Lateral line system : It comprises receptors present in the form of little patches of sensory cells found on the head and trunk (Fig. 1.17).

All the sensory receptors are exposed and are not enclosed in a canal. They are innervated by the branches of V and X cranial nerves (Fig. 1.18).

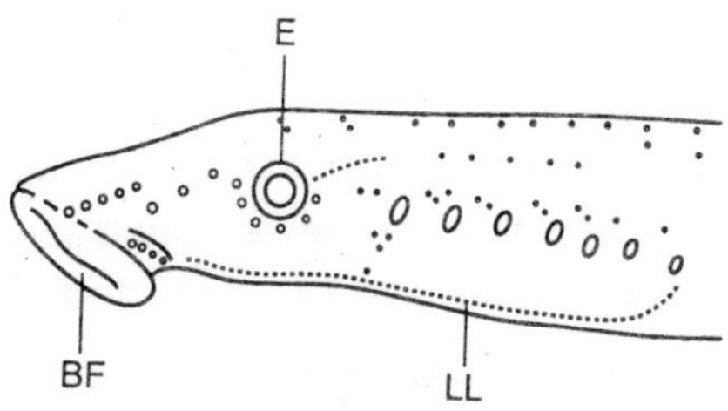

Fig. 1.17 : Lateral sensory organs of the head and body of *Petromyzon*. BF, buccal funnel; E, eye; LL, lateral line.

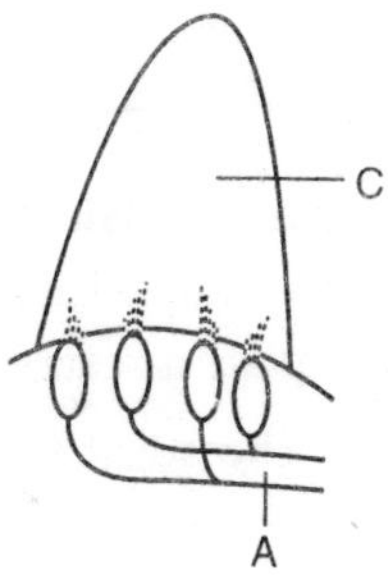

Fig. 1.18 : Lateral line neuro-mast. A, axon; C, cupula.

Lateral line system helps in searching food, in escaping from enemies and in orienting the body while swimming.

Olfactory organs : Cyclostomes are monorhinic, i.e., they have a single, median olfactory organ. The single nostril opens behind into a rounded nasal or olfactory sac through a short duct. In vertebrates the pituitary develops from buccal epithelium and has no connection with the exterior, but in lamprey the lumen of hypophyseal rudiment is continuous with the nasal sac by a duct known as nasohypophyseal tube or nasopalatine canal. This duct runs posteriorly and opens behind into hypophyseal sac. The hypophyseal sac is placed below the brain and notochord and extends between the first pair of branchial pouches. The epithelium of nasal sac is raised into numerous longitudinal folds, which extends radially into the lumen of the nasal capsule forming a number of pockets. In lamprey the glandular accessory olfactory organ is also present.

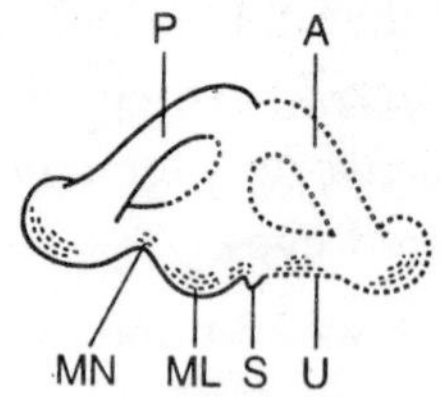

Fig. 1.19 : Auditory labyrinth of lamprey. A, anterior; P, posterior, MN, macula neglecte; ML, macula lagena; S, sacculus; U, utriculus.

The vestibular organ : The labyrinth develops by an inpushing of the wall of the head, and later on becomes closed off from the exterior. Internally it is divisible into chambers, which are provided with patches of sensory hairs similar to macula of higher vertebrates. There are two semicircular canals which open into sac called vestibule (Fig. 1.19).

The vestibule subdivides by a median fold into an anterior and posterior utricular chambers. Below the utricular chambers are present small anterior sacculus and posterior lagina. The canals and sacs are filled with endolymph. The walls of chambers are ciliated and their cilia produce current in the endolymph.

Photoreceptors : Lampreys possess light sensitive cells in the skin and in the eyes. These are abundantly present in the tail and when light falls on them, the animal rapidly moves away. The pigment present is probably a porphyropsin.

Urinogenital System

This system consists of various tubes opening via coelom to the exterior through which both excretory and genital products are carried away. These tubules are modified for the excretory purpose a result of adaption of the freshwater habit.

Kidney : The adult functional kidney is mesonephric and is attached to the dorsal wall of the coelom by peritoneal sheet. The kidney develops from the part between the dorsal scleromyotome and the ventro-lateral plate mesoderm, i.e., nephrostome. It forms a series of segmental funnels opening into a common archinephric duct. The network of blood-vessels surrounds each funnel, forming the glomerulus. The osmotic flow of water into the body is relieved by the pressure of heart beat forcing water out from the glomeruli into the coelomic fluid and then it is removed through the funnels with the help of their cilia. After hatching the tubules become longer and coiled and help in salt reabsorption. The anterior funnels together form the pronephros (Fig. 1.20).

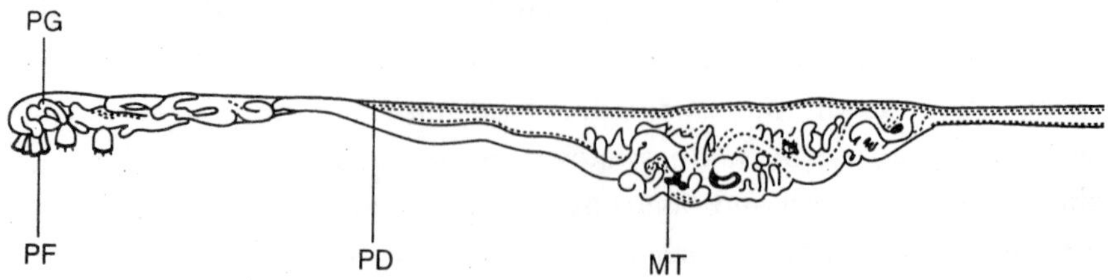

Fig. 1.20 : Kidney of larval Petromyzon. PD, pronephric duct; PF, pronephric funnels; PG, pronephric glomeruli; MT, mesonephric tubule.

As the animal's growth proceeds they are shifted to the posterior and form mesonephros. In adult the pronephric tubules remain in the form of mass of lymphoid tissue. The mesonephros develops in the form of large fold, hanging into the coelom. The mesonephric tubules do not open into coelom but into malpighian capsule, which contains a part of coelom and the glomerulus (Fig. 1.21).

As the animal grows the mesonephros extends at its hind end and forms the adult kidney. Apart from its excretory function the kidney also contains lymphoid tissue and fat and take part in the production and destruction of red and white corpuscles.

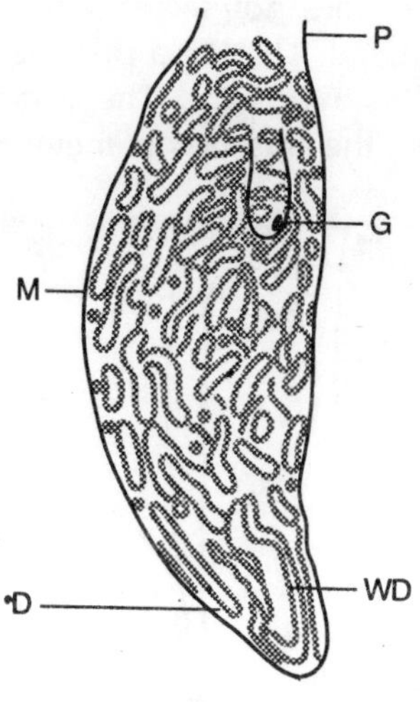

Fig. 1.21 : T.S. of kidney of lampetra. G, glomerulus; M, middle section of tubule; P, proximal region of tubule; D, distal region; WD, Wolffian duct.

The Gonads : The lampreys breed only once at the end of their life-cycle. The sexes are separate in the adult while the early developmental stages of ammocoete larvae possess hermaphrodite gonad containing both oocyte and spermatocyte. At metamorphosis there are no marked difference in the size or degree of development of either ovary or testis in the parasitic and non-parasitic forms. The mature ovary contains ova covered by a single layered follicular epithelium, when ruptured it releases telolecithal egg into coelom. The spermatozoa is about 14 mm in length and 0.5 mm in diameter with rounded head, and pointed tail with no middle piece. The ovary is suspended in coelom by a mesentery, called *mesovarium* and the testis by *mesorchium*. The testis is made-up of follicles containing sperms. On ripening follicles rupture and liberate spermatozoa into the coelom. The gonads extend almost the whole of the body cavity and lack reproductive duct. From the coelom the sex cell are carried away through abdominal pores. These pores are present in the wall of the urinogenital Sinus which opens a few week before spawning. The fertilization is external.

Development

The lamprey has telolecithal eggs having large amount of yolk. The cleavage is holoblastic but unequal and results into the formation of blastula having an upper half of micromeres and a lower half of macromeres. The gastrula is formed by the process of invagination and the blastopore is shifted to posterodorsal position and converted it to anus. In early stages the neural tube does not possess neurocoel and is called a neural rod, later it is transformed into brain and spinal cord. The archenteron leads to blastopore. The archenteron is distinguished into the gut and the liver is developed from the anterior part to the gut.

The Ammocoete larva : After about three weeks, hatching takes place in 7 mm long young transparent ammocoete larva which remains buried in the mud. The mouth is devoid of real sucker and is surrounded by an oral hood (Fig. 1.15a, b, c). The eyes are covered by mucscles and skin. This makes the head less sensitive, but when the light falls on the tail which is provided by photoreceptors the animal quickly moves. The nasal sac is poorly developed. Small food particles are taken through the mouth with water. The pharynx strains the smaller food particles with the help of mucus secreted by endostyle. The endostyle comprises a pair of tubes containing four rows of secretary cells. At the end of life the endostyle forms a large mass, which is made up of secretory and ciliated cells. The endostyle does not secrete any enzyme it only secretes mucus into which food particles are entangled. The larva possesses seven pairs of branchial sac, opening out by gill slits (Figs. 1.22a, b & 1.23). The excretion takes place by the pronephros. After spending about 3 to 4 years, the ammocoete metamorphoses into the adult lamprey.

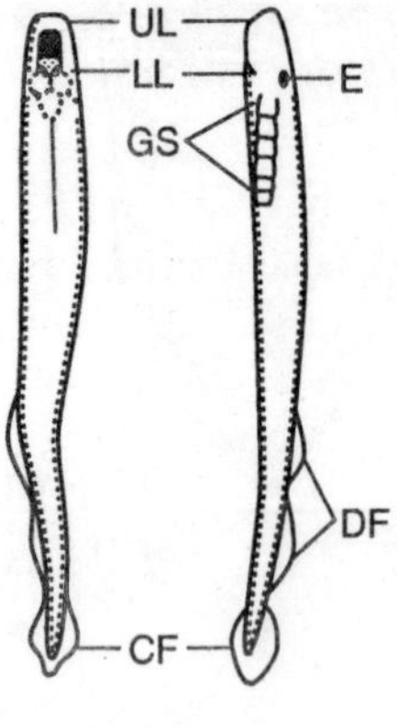

Fig. 1.22a, b : Ammocoete larva of Petromyzon. (a) central view (b) Lateral view. CF, caudal fin; DF, dorsal fin; E, eye; GS, gill-slits, LL, lower lip; UL, upper lip.

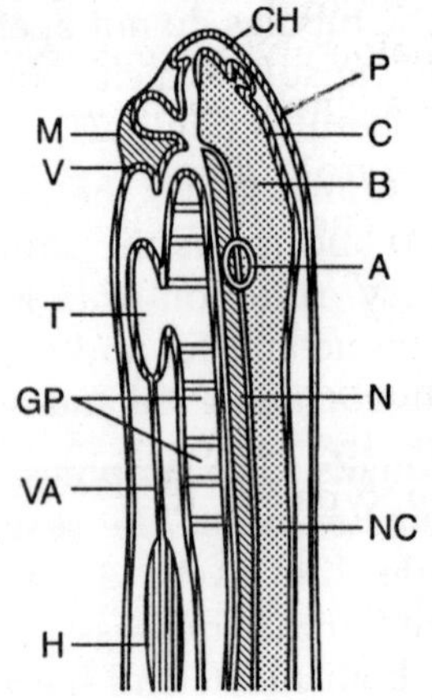

Fig. 1.23 : L-S. of anterior part of ammocoete larva. A, auditory vesicle; B, brain; CH, cerebral hemisphere; C, cerebellum; GP, gill-pouches; H, heart; M, mouth; N, notochord; NC, nerve cord; P, pineal body, T, thyroid invulsion; V, velum; VA, ventral aorta.

Changes during metamorphosis : The larval characters are lost during metamorphosis and it grows into an adult by acquiring following characters :

(1) The mouth becomes surrounded by suctorial buccal funnel with tongue apparatus.

(2) The endostyle is replaced by thyroid gland ventral to the pharynx.

(3) The eyes become open and functional.

(4) The single dorsal fin is divided into two by a median notch.

(5) Velum reduces and surrounds the opening of respiratory tube.

(6) The oesophagus and respiratory tube becomes separated.

(4) The single dorsal fin is divided into two by a median notch.

(5) Velum reduces and surrounds the opening of respiratory tube.

(6) The oesophagus and respiratory tube becomes separated.

(7) The intestine becomes modified.

(8) The pericardial cavity becomes completely detached from the coelom.

(9) The gall bladder disappears.

(10) The pronephros disappears but mesonephros is retained.

(11) The spinal cord becomes compressed.

(12) The body colour changes from yellow brown to adult texture.

The larva migrates from river to the sea and starts its adult life. When the gonads become mature the adult again migrates to freshwater for the purpose of spawning. Thus life cycle is completed into an adult and larval phase.

The Endocrine Organs

1. **The hypophysis :** In lamprey, the pituitary gland or hypophysis is present between diencephalon and nasopharyngeal pouch. It comprises an anterior neurohypophysis and a posterior adenohypophysis. Its cells are arranged in cords.

2. **Thyroid gland :** After metamorphosis in lamprey the thyroid gland develops from the endostyle of the ammocoete larva. After metamorphosis of ammocoete larva, its endostyle forms thyroid gland of the adult lamprey which is diffused gland and consists of follicle cells found to be scattered around the ventral aorta. It secretes thyroxin hormone.

In ammocoete larva topographic study show that larval endostyle is made up of five cells type. Its radiography reveals that iodine accumulates in (type III) and to some extent to type V. It is believed that adult thyroid follicles originate from type III and IV.

3. **Parathyroid gland :** These are very small sized glands and lie diffused in dorsal and ventral parts of pharyngeal pouches.

4. **Adrenal glands :** Proper mammal-like adrenal glands are lacking in lamprey. It is represented by a series of structures. The adrenocortical or inter-renal tissue, like the higher vertebrates, develops from coelomic epithelium of posterior end of pronephros. They are composed of small irregular, lobe-like aggregations present around the posterior cardinal veins, renal arteries and other arteries present near the mesonephros. Secretion of this gland influences the ion concentration in the adult lamprey.

In addition to this, there are found a small strips of chromaffine tissue comparable to the adrenal medulla of mammals. It extends from near the anterior part of branchial region (posterior to second gill cleft) to the tail along the course of the dorsal aorta and its branches.

5. **Pancreas :** The endocrine pancreas lies suspended in liver and intestinal wall as small masses of endocrine cells.

REFERENCES

Fange, R. (1972)— The circulatory system. In 'Biology of Lampreys (Ed. Hardisty, M.W. & Potter, I.C.) Vol. 2. Academic Press, London. pp. 241-260.

Ganguly, Sinha, Adhikari, Goswami, 2008. Biology of Animals. Vol. 2. Seventh Edition, New Central Book Agency (P) Ltd. Kolkata. India.

Hardisty, M.W. & Potter, I.C. (1972), Biology of Lampreys. Vol. 2. Academic Press, London, New York.

Hatta, S. (1923) — Uber die Entwicklung des Gefassystems des Neunauges *(Lampetra mitsukurrii)* Zool. Jb. (Ant) 44 : 1-264.

Manwell, C. (1963) — The blood proteins of cyclostomes. A study in phylogenetic and ontogenetic biochemistry. In biology of Myxine (Eds.); Brodal, A. & Fange, R. Oslo Universitets for laget. pp. 372-455.

Riggs, A. (1972) — The haemoglobins. In Biology of Lampreys. (Eds. Hardisty, M.W. & Potter, I.C.) Academic Press, London. pp. 261-286.

Young, J.Z. (1962) — The life of vertebrates. Oxford at the Clarendon Press.

Chapter 2

MYXINE (HAGFISH)

Cyclostomata belongs to superclass Agnatha. It is represented by living forms such as hagfishes and lamprays. They are not really fish. However, Ichthyologists have gladly adopted them for study because they are aquatic vertebrates that typically occur in association with true fish. Despite their primitive structural features, lampreys and hagfishes are both successful groups that are more abundant than most people realise. The origin and ancestry of this group is debatable and is not fully understood. The endoskeleton of these animals are made up of cartilage and therefore, their fossil records are not satisfactorily preserved in geological record. However, according to some workers they have descended from ostracoderms, a group of ancient extinct armoured fishes, because of the following reasons :

1. They possess a single nostril in the middle of head.
2. They do not possess lower jaw.
3. They lack paired fins.
4. There is no bony vertebral column.

The class cyclostomata is divisible into two subclasses :

1. Myxinoidea
2. Petromyzontia

The individuals belonging to Myxinoidea are commonly known as hagfishes. Linnaeus classified them as worms. They are represented by about two dozen species, the most common are *Myxine, Eptatretus (Bdellostoma),* and *Paramyxine. Myxine* is widely distributed along the coast of northern Europe and the Atlantic coast of North America. Some species occur in the coasts of Chili and Japan. *Eptatretus stoutii* is found widely distributed along the pacific coast of both North and South America, in coastal water of South Africa and New Zealand.

Biology of Hagfish

Habit and Habitat : Hagfishes are extensively ocean dwellers and generally living on the sea bottom in burrows. They not only live in muddy burrows but live into dead or dying fish and consume flesh and viscera leaving only skin and bones. They are found only in low waters of low temperature (15 °C), where the water is salty and cold. According to Jensen (1961-1966) *Eptatretus stoutii* inhabits the western continental shelf of North America from Alaska to Baja California at depth ranging from as little as 60 feets to as much as 1800 feet.

The hagfish is sluggish in nature and remains quietly in muddy burrows but when hungry it moves swiftly with snake like manner but normally the motion is sluggish or torpid or dull. The body colour is greyish brown to bluish brown above, whitish below. Hagfish has remarkable immunity against infections and disease. They do not possess thymus gland, but aggregation of lymphoid cells in velar muscles have been suggested as primitive prothymus structure (Riviere *et al.*, 1975, Good *et al.*, 1972, Ganguly, *et al.*,2008). They also do not manufacture antibody.

External Features

These are often called as slime eel because they secrete thick slippering slime in enormous quantities with wonderful rapidity. A single hagfish will fill a two gallon bucket with slime mingled with water in a few seconds and after a slight interval can repeat the operation with ease. The slime is secreted by small pores on either side of the ventral surface extending from snout to the tip of the tail. The pores secrete milky fluid which in contact with water form thread-like tenacious cocoon around hagfish. This structure is extremely helpful to avoid grasping by any predators. The hagfish comes out of this cocoon by making loops and knots of the body giving hagfish extra leverage and thus the animal pulls out of the slime and avoid chocking of gills.

Hagfishes are remarkable for their lack of striking features. Anteriorly it has a round mouth, which is provided with three pairs of barbels or fleshy tentacles (Fig. 2.1a, b).

They are tactile in nature, i.e., have the power of sense of touch. The head bears a single nostril, a peculiar feature of cyclostomata. It is having rudimentary eyes but the animal is completely blind, because of deep sea adaptation. However, they are capable of distinguishing between dark and light because of pigment present in the retina of eye region. A few sensitive areas are also present on the head and around cloaca.

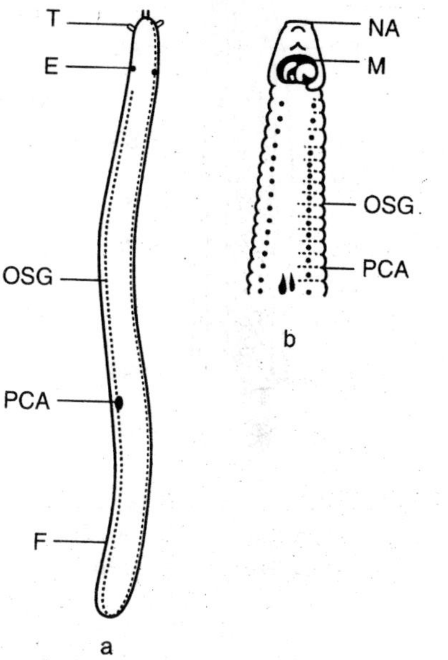

Fig. 2.1a, b : Structure of hagfish. (a) Dorsal view, (b) Ventral view. E, eye; F, fin; PCA, pharyngocutaneous aperture; M, mouth; NA, nasal aperture; OSG, openings of slime glands; T, tentacles.

There are about a dozon gill openings situated on either side on the ventral surface of the body. They look like as white spots. In *Myxine* there is only one pair. The dorsal fin is absent or ill defined narrow and undivided. The caudal fin is slightly developed. Preanal fin is present.

Alimentary Canal

The alimentary canal of hagfish is a straight tube consisting of mouth which opens into throat which leads into intestine through pharynx, which ends into anus (Fig. 2.2a, b, c).

The mouth of *Myxine* is horse-shoe shaped, bordered by two pairs of tentacles. The tongue is provided with two rows of small teeth. The teeth arise from the dental plate. The tongue and sucking apparatus are well developed. The intestine can be divided into foregut and hindgut. Near the throat there is a prominent circular band of muscle which demarcates these two regions. There is no stomach in *Eptatretus*. In some species stomach is slightly a dilated chamber. Their circular band of muscle prevents the water to enter the digestive system but while the animal consumes food the muscle of the band relaxes and allows the food to enter the digestive system. The intestine is provided with ridges which increase the surface area of the intestine for the absorption of nutrition but spiral valve is absent. The liver, gall bladder and pancreas are accessory digestive organs and open into the intestine. The pancreas of hagfish is divided into two completely separate parts, one secreting digestive juices, the other insulin. The hormone insulin presumably regulates the

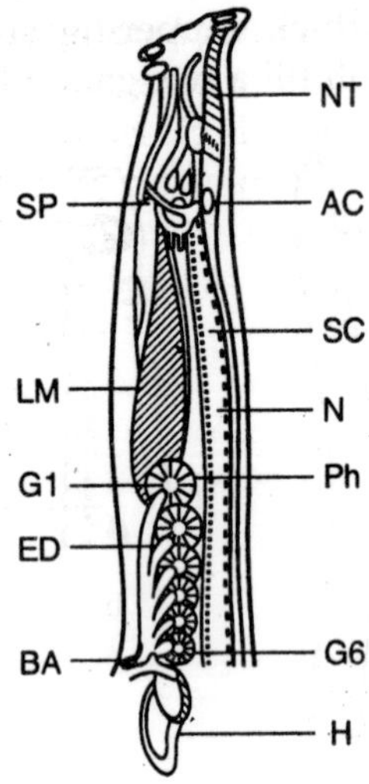

Fig. 2.2a : L.S. of hagfish. (a) Diagrammatic. AC, auditory capsule; ED, efferent duct; G1, first gill; G6, sixth gill; H, heart; LM, lingual muscle; N, notochord; NT, nasal tube; Ph, pharynx; SC, spinal cord; SP. stvloid cartilage.

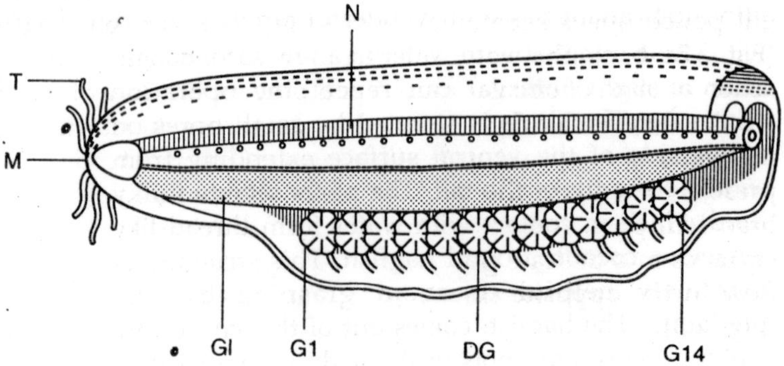

Fig. 2.2b : L.S. of hagfish. (b) Dissected, DG, duct from gill pouch; Gl, first gill; G14, fourteenth gill; GL, gullet; M, mouth; N, notochord.

blood sugar level in the hagfish as it does in other vertebrates (Jensen, 1966). The gut has dual function, on one hand it is digestive while on the other hand it is hemopoietic i.e. produce red blood cells.

Excretory System

The kidneys of Myxine are divisible into two parts, the pronephros and mesonephros which are segmentally arranged. The pronephros is retained in adult hagfish which is hardly marked off from the mesonephros. In *Eptatretus*, the pronephric duct is present, but *Myxine* lacks such ducts. There is no urine production by pronephros but has phagocytic and haemopoitic functions, (Zapapta, *et al.* 1984). In *Myxine* and *Eptatretus* the pronephros contains more than 2000 ciliated nephrostomes (peritoneal funnels). The mesonephros is functional kidney and consists of Malpighian corpuscles (Bowman's capsule), which houses glomerulus. This portion lacks tubules like those of other vertebrates. The mesonephros is the main kidney and is responsible for filtering the blood for excretion or reabsorption in the body. The hagfish cannot survive long in water that has either a very high or a very low concentration of salts. The osmotic pressure of the blood is controlled by adrenocortical hormones.

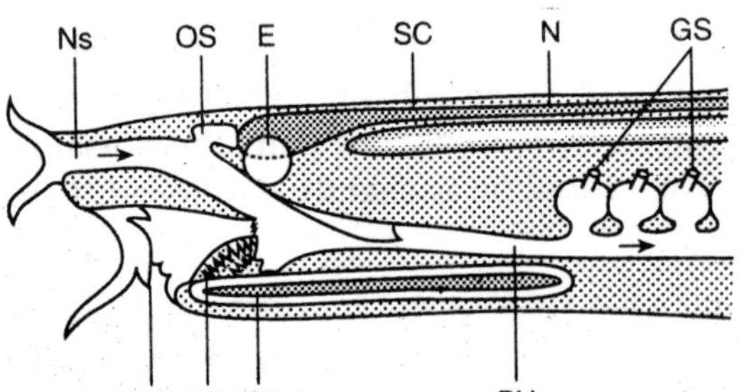

Fig. 2.2c : L.S. of hagfish. (c) dissected. E, eye; GS, gill sac; M, mouth; N, notochord; NS, nostril; OS, olfactory sac; Ph, pharynx; SC, spinal cord; TC, tongue cartilage; TT, teeth on tongue.

Respiratory System

There are a dozen or so pairs of gill pouches arranged along the body. They open into the pharynx. *Myxine* has six pairs of gill pouches while *Eptatretus* is provided with 13 to 15 pairs of gill pouches. Each gill pouch gives out exit tubes, about six exit tubes open into a common canal which opens outside by a single gill slit in *Myxine*, but in *Eptatretus*, each gill pouch opens separately, hence there are thirteen to fifteen pairs of external gill slits (Fig. 2.2a, b, c). The water enters into throat through nostril, the foregut is closed due to the action of band of circular muscle and prevent the flow of water into the intestine. The velum, a pair of muscular flaps situated in the throat direct the water into gill pouches and then the water goes out through gill by way of ducts. The duct called *pharyngocutaneous duct* is present immediately behind the last gill pouch and on the left side it opens into external branchial pore. It helps in the removal of large particles from pharynx and avoids their entrance into delicate gill pouches. The gills are provided with blood vessels and the blood flow in the opposite direction to the flow of water thus making counter-current system (Fig. 2.3).

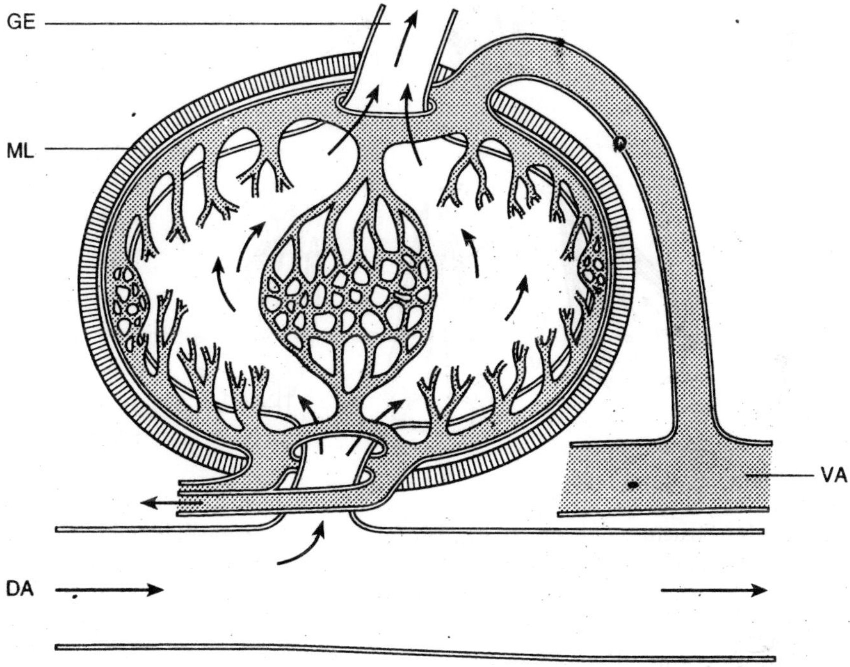

Fig. 2.3 : Counter current flow of water and blood in the hagfish gill arch. DA, dorsal aorta; GE, gill exit; ML, muscular layer; VA, ventral aorta.

Hence there is rapid exchange of oxygen gas. Such mechanism is also present in fishes.

The Cardiovascular System

The cardiovascular system consists of blood vessels, which circulate blood to and from all parts of the body, the heart which keeps the blood in motion, and the lymphatic vessels, which circulates lymph. The circulatory system is peculiar because the venous circulation is not connected to the arterial side in the ordinary way. In majority of organs, there is capillary bed and the arterial blood is discharged in the capillary bed but in some place there is no capillary bed but the arterial blood opens into sinuses through which it percolates before entering the vein. An open circulation of this kind is common among invertebrates but rare in vertebrates. The result of such circulation is that the blood pressure on the venous side is very low (Jensen 1966).

The main heart of hagfish is known as *branchial heart*. Apart of branchial heart (Fig. 2.4a, b), hagfish is peculiar because it has heart-like organs, and these are (a) caudal heart, (b) portal heart, (c) cardinal heart. The branchial heart lies closed to the bilobed liver. The branchial or systemic heart of cyclostomes lie in a cartilagenous pericardium,

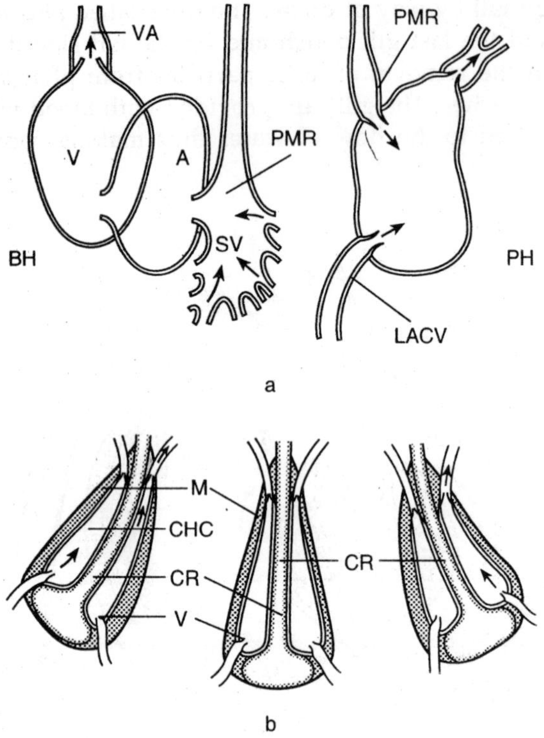

Fig. 2.4a, b : Structure of hearts of hagfish. (a) Branchial heart (BH) and portal heart (PH), (b) caudal heart. A. atrium; CR, cartilage rod; CHC, caudal heart chamber, LACV, left anterior cardinal vein; M, muscle; PMR, pace-markers region; SV, sinus venosus; V, ventricle; VA, ventral aorta.

which communicates with the peritoneal cavity in hagfish such as *Myxine* (Hardisty and Potter, 1972) via a pericardio peritoneal canal. The branchial heart consists of a sinus venosus, an atrium, a ventricle and a bulbus arteriosus similar to that of fishes (Fig. 2.5).

The coronary circulation is entirely absent in atrium and ventricle of the branchial heart of hagfish but coronary vessels do occur in the wall of the sinus venosus of *Eptatretus burgari*. The sinus venosus is a thin walled chamber, the venous blood from the various parts of the body is received through ducti Cuvieri and other blood vessels. The right ductus Cuvieri is obliterated, so only left duct of Cuvieri persists in adulthood (Goodrich, 1930). The sinus venosus consists of smooth muscle in its wall and opens into atrium. The sinuatrial opening is provided with valve which prevent the backflow. The atrium is single and lies left of the midline. It opens into ventricle through atrioventricular opening which is guarded by atrioventricular valves. The ventricle opens into bulbus arteriosus. The myocardium in the atrium and ventricle is spongy, their muscles are arranged in cords or trabeculae which are projected into lumen of the atrium and the ventricle. The trabeculae are more densely packed and larger in the ventricle than atrium. The bulbus opens into ventral aorta. It is short and non-contractile.

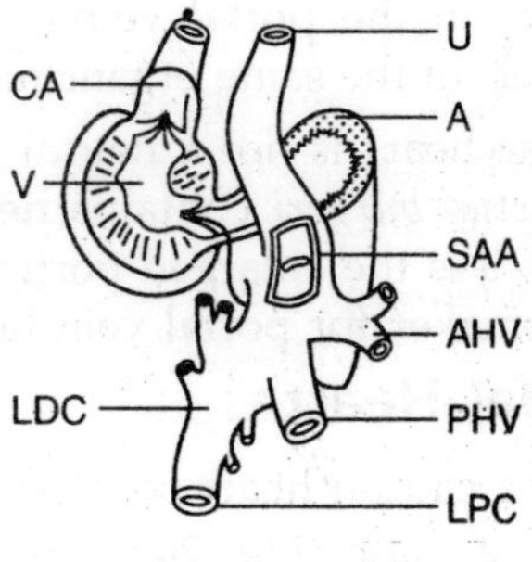

Fig. 2.5 : Dissected heart of *Myxine* to show internal structures. A, auricle; AHV, anterior hepatic vein; CA, conus arteriosus; LDC, left ductus Cuvieri; LPC, left posterior cardinal; PHV, posterior hepatic vein; SAA, Sinuauricular aperture; V, ventricle.

Electron microscopic structure of the cardiac muscle cells of the cyclostomes has been studied and reviewed by Yamauchi (1980). The fine structure of cyclostomata (lamprey and hagfish) muscle is similar to that of skeletal muscle of vertebrate. The A and I bands are clearly distinguished. I band is light and has Z-line, A-band is dark and is provided by H-band. M-line bisects the H-band and N-line are clearly visible. According to Leak (1969), the N-line is inconspicuous in the hagfish, *Myxine glutinosa* but it is quite prominent in *Lampetra* (Lignon and Le Douarin, 1978). In the *Myxine glutinosa*, Helle *et al.*, (1972) demonstrated specific myocardial granules together with the functional granules of peripheral coupling of sarcoplasmic reticulum, possess a proteinaceous matrix substance that is likely to be associated with calcium. The nucleus, Golgi complex, multicellular bodies, ribosomes and glycogen particles are present. It is to be pointed out that no criteria are yet available for discriminating, in terms of occurence or absence of any particular organelles, the cardiac muscle cells between the sinus venosus, the atrium, and the ventricle of the heart of cyclostomes (Yamauchi, 1980).

The epicardium is distinguished into a single continuous mesothelial cell and connective tissue in both *Eptatretus burgeri* and *Lampetra japonica.* The basement lamina is distinct. The nucleus and Golgi complex, endoplasmic reticulum, ribosomes and pinocytotic granules are abundant.

The endocardium lining in the hagfish shows a fairly uniform thickness, i.e., 300 nm, being entirely devoid of fenestrae or intracellular gaps and endowed with abundance of pinocytotic vesicles. The basement lamina is present in hagfish but absent in lamprey.

The myocardium cytoplasm contains ovoid or nearly spherical granules in vesicles 100-300 nm in diameter characterise the cytoplasm of secretory instestinal cells (SICs). These granules contain catecholamines (Helle *et al.* 1972) and are similar to chromaffin cells of adrenal medulla as well as chromaffin-like cells in the heart of amphibian, reptiles and mammals (Kumar, 1996; Yamauchi, 1977).

Hepatic Portal Heart

The hepatic portal vein in its course near the liver lobes becomes expanded and pulsatile and is known as hepatic portal vein heart. It is tubular and roughly T-shaped. It is located towards the lower right side of the hagfish. It is smaller and simple in structure and beats at faster rate. *Helle et al.*, (1972) was of the opinion that the SIC granules (Catecholamine) in the portal vein heart of *Myxine glutinosa* are more electron dense than the ventricle of the same organism.

The heart is not a neural because endocardial connective tissue in the portal heart of *Eptatretus burgeri* contains nerve bundles. According to Carlson (1904) and Fange *et al.*, (1963), it is the enlarged portion of the sinus supraintestinalis which is known to pulsate as a pacemaker for portal vein heart.

Caudal Heart

The caudal heart consists of paired expansion or sacs, one on each side of a median cartilage plate (Fig. 24b). Its both inlets and outlets are guarded by valves. These sacs collect the sinus fluid (Johansen *et al.*, 1962) from subcutaneous sinuses in the tail regions and pump the fluid into efferent vessels, which unite to form a median cardinal vein (Chapman *et al.* 1963, Johansen, 1963; Jensen 1965 & 1966). These sacs are not contractile in themselves but are filled and emptied alternately by the action of extrinsic somatic muscles (musculi cordis caudalis) that run on the outside of each and attach to cartilagenous plates. The caudal heart is activated by nerves. It has a reflex centre in the spinal cord. According to Jensen (1966), the stimulation of the hagfish skin will inhibit the caudal heart beat completely. A purely neurogenic contraction of the muscle on one side causes a bending of the plate toward that side which results in a filling of the sac on the same side and simultaneous emptying of the sac on the opposite side (Greene, 1900; Carlson, 1904). It has A and I bands with H,M,N, and Z lines typical to that of skeletal muscle.

Cardinal Heart

The cardinal heart are pair of sac located in the head region. It has no muscles or machinery of its own for producing rhythmic impulses; its pumping action is operated by rhythmic contraction of skeletal muscles that lie outside the sac in the head region (Jensen, 1966). The blood from these cardinal heart via the cardinal veins goes to the branchial or portal heart.

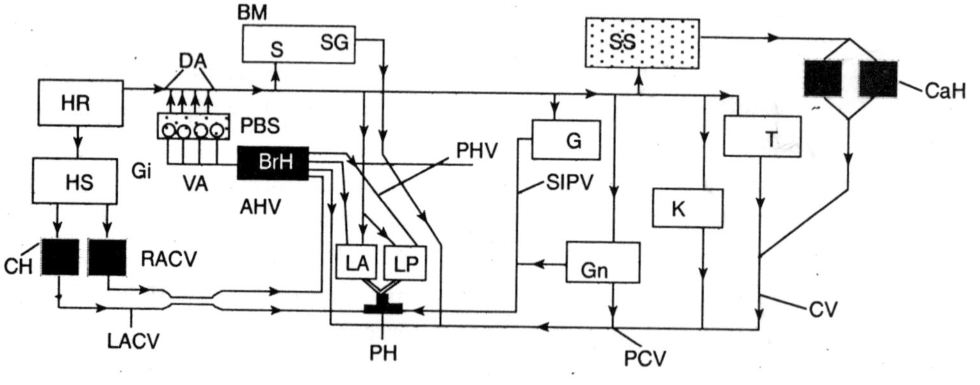

Fig. 2.6 : Diagrammatic sketch of circulatory system of the hagfish. AHV, anterior hepatic vein, BM, body muscle, BrH, branchial heart, CaH caudal heart; CH, cardinal heart; CV, caudal vein; DA, dorsal aorta; G, gut; Gi, gill; Gn, gonad; HR, head region; HS, head sinus; K, kidney; LA, liver anterior lobe; LACV, left anterior cardinal vein; LP, liver posterior lobe; PCV, posterior cardinal vein; PBS, peribranchial sinus; PH, portal heart; PHV, posterior hepatic vein; RACV, right anterior cardinal vein; S, skin; SG, slime gland; SIPV, supraintestinal portal vein; SS, subcutaneous sinus; T, tail; VA, ventral aorta.

Circulation of Blood

The blood from main branchial or systemic heart is pumped into ventral aorta through bulbus arteriosus (Fig. 2.6).

From ventral aorta the blood goes to gills through afferent branchial arteries. After purification the blood is collected by efferent branchial arteries and poured into dorsal aorta. The blood from the dorsal aorta is supplied to anterior and posterior side of the body through a series of blood vessels, the major blood vessels are carotid arteries, which supply blood to the body muscle and skin, then a branch is given off to supply blood to the anterior and posterior lobes of the liver, then a branch is given which supply blood to gonads and to kidney and then ultimately to the tail region.

The blood passes from arteries to the veins through capillaries present in the bed in gills, gut and body muscles. In the head the blood from arteries is discharged into large open spaces called *sinuses* or *sinusoids*. They are large in diameter. The sinuses in the head are known as *carotid sinus*. The venous blood from carotid sinuses reaches into the cardinal heart. The cardinal heart is devoid of muscles and its pumping action is not due to intrinsic property but is due to contraction of skeletal muscles which encloses it. The blood from the cardinal heart is received by anterior and posterior cardinal veins. The anterior cardinal vein pours its blood directly with the branchial heart while the posterior cardinal vein opens into portal heart. The venous blood from skin, body muscle and slime glands is collected and poured into posterior cardinal vein. The posterior cardinal vein also receives blood from kidney and tail region. After receiving blood from the gut, gonad, kidney and body wall and tail region, the posterior cardinal opens into the branchial heart, i.e., sinus venosus by separate opening. The blood from subcutaneous sinuses comes into caudal heart which pours its venous blood in the caudal veins, while in its turn opens into caudal vein.

Physiology and Innervation of Branchial Heart

There is a great controversy whether the heart of *Myxine* is aneural (without nerve) or provided with nerves. Greene (1902) suggested that the heart of California hagfish is free from all nervous regulation and, therefore, must be aneural. Greene (1902) concluded this, because he was unable to change the rate and the force of heart beat by electrical stimulation of the vagus nerves and of various other parts of the central nervous system. Augustinson *et al.*, (1956) reported similar results in the hagfish with vagal stimulation. Hirsch *et al.*, (1964 & 1970) reported that the heart of *Eptatretus stouti* is provided with large nerves with myelinated fibres and ganglionic cells. They also found ganglion cells with fibres in the aorta, the epicardium and endocardium. They also reported the presence of high catecholamine content of the cardiac tissues of the California hagfish. Yamauchi (1980) confirmed the presence of nerves and nerve cells in the heart of *Eptatretus burgeri* by electron microscopy. According to him, neuronal perikaryon showed the fine structural appearance quite typical of the neurons in the higher vertebrates.

It is, therefore, clear now that heart of hagfish is not aneural. However, on the basis of electron microscopy, Yamauchi (1977) held that ganglion cells along the endocardial edge of the main channel of the ventricle described by Hirsch *et al.*, (1970) are nothing but catecholamine storing, specific interstitial cells. The cardiac cells are made up of actin and myosin proteins and due to the presence of similar bands the contraction is similar to the skeletal muscles.

The sinus venosus contains muscle cells. They range from 9-15μ diameter in breadth and have well developed sarcoplasmic reticulum with peripheral coupling. The electron microscopy shows that the myofibrils possess A and I bands with H and Z lines in a sarcomere M-line is also clearly visible in the myofibril.

Reproductive System

There is a single gonad in hagfish. It is present on the right side of the intestine hanging in the abdominal cavity through a delicate membrane. It extends from the liver and reaches up to cloaca.

The gonad of hagfish is peculiar. It is protandric hermaphrodite that is in the early stages it is male and then develops into female. In very early stages of development the tissue of the gonad is neutral, but contains the elements of both sexes and it is difficult to ascertain whether the individual will develop into male or female. It is held that in the early life the anterior region has a mature ovary whereas the posterior region contains immature testicular tissue. However, when the animal matures, the male and female portion of gonad develops either into male or female and the other portion is repressed presumably under the influence of hormones from the pituitary gland. The fertilization is external as copulatory organs are absent in both sexes. The eggs are very large, keratinized with hooks. They are 2 to 3 cm in diameter. The eggs are united at both ends by tuft of filaments (Fig. 2.7a, b).

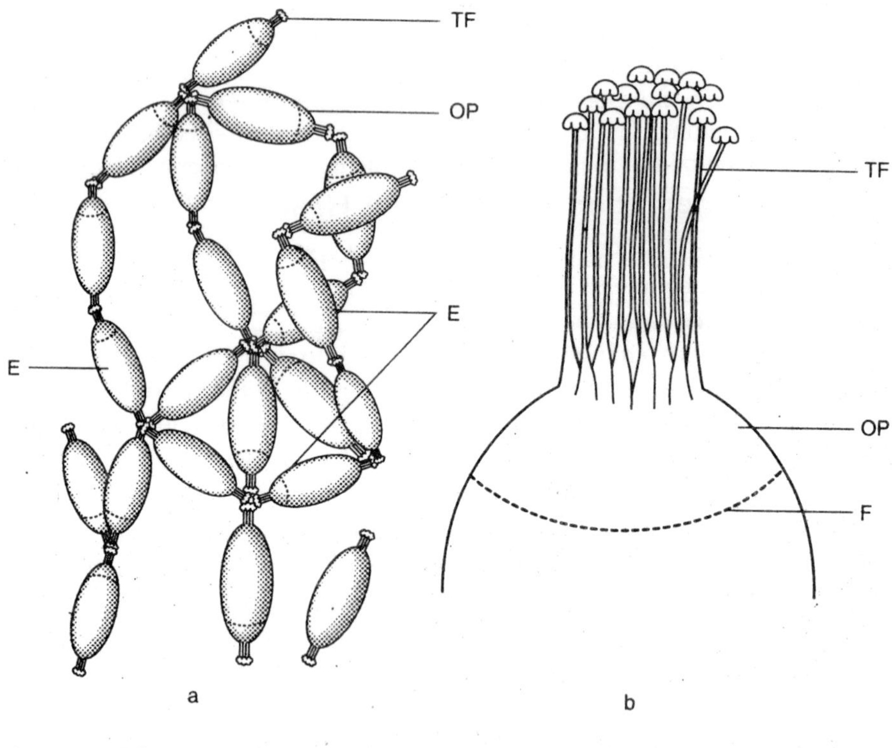

Fig. 2.7a, b : Eggs of hag fish. (a) The eggs are attached to one another by the tuft of filaments. (b) $ingle enlarged egg. E, egg; F, fissure; OP, operculum; TF, tuft of filament.

The process how sperms enter the egg for fertilization is not clearly understood. It takes many months to develop the eggs. The hagfish generally lays down about 22 eggs at a time. When the young hatches from eggs, they are essentially miniature adults, so they do not undergo the dramatic metamorphosis as that of *lamprey*.

Morphological and Anatomical Contrasts between Lampreys and Hagfishes

	Lampreys	Hagfishes
Dorsal fins	One or two; well developed in adult	None or only a trace
Pre-anal fin	Absent	Present
Eyes	Moderately developed	Highly degenerate
Oculomotor muscles & nerves	Present	Absent
Oral disc	Present	Absent
Teeth	On both tongue and disc	On tongue plus one on "palate"
Lingual laminae	One transverse; two longitudinal	Two longitudinal pairs
Pokal cell cone	Absent in teeth	Present in teeth
Barbels	Absent	Present
Intestine	Ciliated	Unciliated
Spiral fold of intestine	Present	Absent
Buccal glands	Present	Absent
Nasohypophyseal opening	On top of head	In front of head
Nasohypophyseal sac	Not opening into pharynx	Opening into pharynx
Number of gills	Always 7	5 to 14
External gill openings	7, close to head	1 or 5-14, remote from head
Internal gill opening	Into a single suboesophageal tube communicating with oral cavity	separately and directly into pharynx
Gill pouches	Ectodermal origin	Endodermal origin
Pharyngo-cutaneous duct	Absent	Present
Skull	Mostly cartilaginous with incomplete roof	Roof entirely membranous and feebly developed
Branchial skeleton	A conspicuous basketwork	Rudimentary
Neural arches	Present but rudimentary	Lacking
Dorsal and ventral roots of spinal nerves	Distinct	United
Ductus Cuvieri	Left one obliterated	Right one obliterated
Kidney	Mesonephros	Pronephros anteriorly, mesonephros posteriorly
Eggs	Small, unkeratinized and without hooks	Very large, keratinized and with hooks
Segmentation	Holoblastic	Meroblastic

* Source : Hardisty and Potter, 1972.

REFERENCES

Augustinson, K.B. Fange, R. Johnels, A.G. and Ostland, E. (1956). Histological, Physiological and biochemical studies on the heart of two cyclostomes, hagfish (*Myxine)* and *lamprey (*Lampetra*)*. J. Physiol (London). 131 : 256-276.

Carlson, J. (1904). Contributions to the physiology of heart of the California hagfish (*Bdellostoma dombeyei)* Z. Allg Physiol. 4 : 259-288.

Chapman, C.B. Jensen, D. & Wildenthal, K. (1963). On the circulatory control mechanisms in the Pacific hagfish. Circulation Res. 12 : 427-440.

Fange, R., (1972). The circulatory system. In "Biology of Lampreys" (Eds. Hardisty, M. W. & Potter, I.C.) Vol. 2, Academic Press. New York. pp. 241-259.

Fange, R., (1994). Blood cells, haemopoiesis and lymphomyeloid tissue in fish, fish a shell fish. Imm. 4 : 405-411.

Fange, R., Bloom, G., & Ostland, E. (1963). The portal vein heart of Myxinoids. In "The Biology of Myxine" (Eds. Brodal, A. & Fange, R.), Oslo University Press. Oslo. pp. 340-351.

Ganguly, Sinha, Goswami. (2008) Biology of Animals. Seventh Edition New Central Book Agency (P) Kolkata.

Goodrich, E.S. (1930). Studies on the structure and development of vertebrates. Macmillan London.

Good. R.A, Finstad, J. & Litman, J. (1972). Immunology. In : The Biology of Lampreys. (Eds. Hardisty, M.V. & Potter. I.C.) Academic Press, London. pp. 127-206.

Greene, C.W. (1900). Contributions to the physiology of the California hagfish, *Polistotrema stouti* I. The anatomy and physiology of the caudal heart. Am. J. of Physiol. 3 : 366-382.

Greene, C.W. (1902). Contribution to the physiology of the California hagfish, *Polistotrema stouti* II. The absence of regulative nerves for the systemic heart. Am. J. Physiol. 6 : 318-324.

Hardisty, M.W. & I.C. Potter (1972). The Biology of Lampreys. vol. 2, Academic press, New York.

Helle, K. B., Lonning S., & Blaschko, H. (1972). Observations on the -chromaffin granules of the entricle and the portal vein heart of *Myxine glutinosa* L. Sarsia 51 : 97-106.

Hirsch, F.F., Jellinek, M., & Cooper, T. (1964). Innervation of systemic heart of California hagfish. Circulation Res. 14 : 212-217.

Hirsch, F. F., Jellinek, and Cooper, T. (1970). The innervation of the heart in the California hagfish (*Eptatretus stouti*). In "The innervation of vertebrate heart" (Ed. Hirsch, E.F.), Thomas spring field, Illinois. pp. 28-35.

Jensen, D. (1961). Cardioregulation in an aneural heart. Comp. Biochem. Physiol. 2 : 181-201.

Jensen, D. (1965). The aneural heart of the hagfish. Ann. N.Y. Acad. Sci. 127 : 443-458.

Jensen, D. (1966). The hagfish, Sci. Am. 214 : 82-90.

Johansen, K. (1963). The cardiovascular system of *Myxine glutinosa* L. In. "The biology of Myxine" (Eds. Brodal, A. & Fange, R.) Oslo Univ. Press. Oslo. pp. 289-316.

Johansen, K., Fange, R. and Johanesen, M.W. (1962). Relation between blood, sinus fluid and lymph in *Myxine gluitinosa* L. Comp. Biochem-Physiol. 7 : 23-28.

Kumar, S (1996). The structure, function and autonomic nervous system of the fish heart. (Ed. Singh, B. R.). Advances in fish research Vol. II. Narendra Publishing House, Delhi.

Leak, L. V. (1969). Electron microscopy of cardiac tissue in primitive vertebrate *Myxine glutinosa* Morphol J, 128 : 131-158.

Lignon J., and Le Douarin, (1978). Small intensely fluorescent (SIF) cells and myocardiac cells in the ammocoete heart : a correlative histofluorescence, light and electron microscopic study with special reference to the action of rescrpine. Biol. Cell 31 : 169-176.

Rivieres, H.B. Cooper, E.L. Reddy, A.L. & Hildemann (1975). In search of the hagfish thymus. Am. J. Zool. 15 : 39-40.

Yamauchi, A. (1977). On the rece pto-endocrine property of granule-containing cells in the autonomic nervous system. Arch. Histol. Jpn. (Niigata, Jpn.) 40 : 147-161.

Yamauchi, A. (1980). Fine structure of the fish heart. In "Heart and Heart-Like organs" (Ed. Bourne, G.H.) Vol. 1, Academic Press. New York. pp. 119-148.

Zapata, A., Fange, R., Mattisson, A. & Villena, A. (1984). Plasma cells in the adult Atlantic hagfish, *(Myxine glutinosa)*. Cell Tissue Res.

Chapter 3

INTEGUMENT AND SCALES

The integument or skin is an outermost covering or wrapping of the body, hence it is the most exposed part of the body to the environment. For this reason, it plays an important role of first line of defence in a number of ways. In fishes, the skin is well-adapted for protection from injuries and diseases. It also serves for respiration, excretion and osmoregulation. In some fishes, special colouring devices and phosphorescent organs are present in the skin, which either conceal the organism or make it present or used for sexual recognition. In addition, some species have special structures like electric organs, mucous glands and poison glands.

Structure

The skin of fish is made up of two distinct layers, viz. an outermost layer, the epidermis and an inner layer dermis or corium. The epidermis originates from ectoderm and the dermis derives from mesoderm layer (Fig. 3.1).

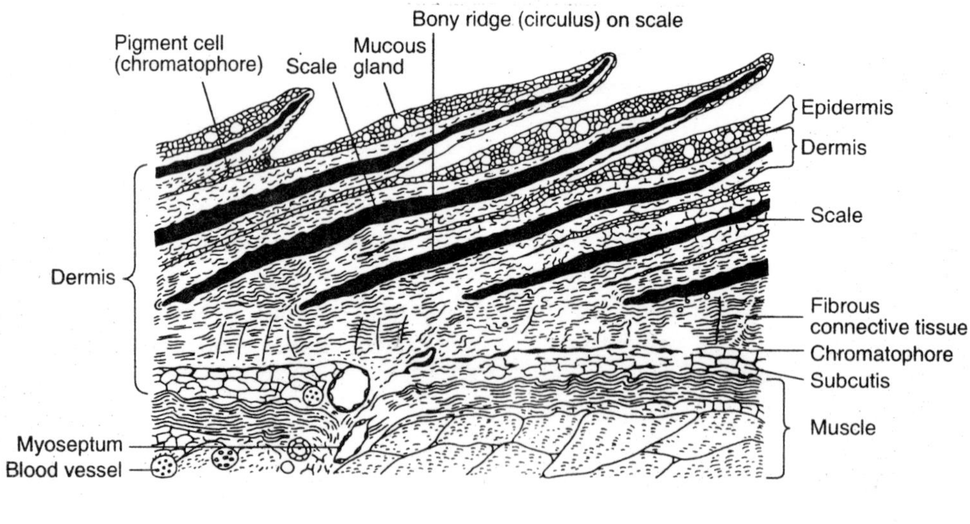

Fig. 3.1 : Section of fish skin (Source : General Biological Supply House)

The Epidermis : It is composed of many layers of flattened and moist epithelial cells. The innermost layer is called stratum germinativum. This layer is made up of active columnar cells that continuously divide by mitotic division. The newly formed cells occupy the lowermost stratum and the older cells move outside and are worn off time to time and maintain growth. These migrating epithelial cells fill the superficial wounds.

Epidermal glands : The epithelium of epidermis is modified into a variety of glands, which are :

(i) **Mucous gland** — The epidermis is provided with a number of mucous glands, which open at the surface of the skin by minute pores. These glands are flask-shaped or tubular extending to the dermis. The mucus glands secrete slippery mucus, which contain a lipoprotein, known as mucin. The slimy mucus reduces drag on fish while swimming in the water. Continuous secretion and sloughing of mucus wash away micro-organism and irritants, which may cause disease if accumulated. In some species (*Protopterus* and *Lepidosiren*), the mucus forms a cocoon-like structure around the body to avoid dry condition of weather, especially during *aestivation*. The mucus gives a characteristic fish odour. Among some fishes, mucous is used for chemical communication. Many teleosts feed their young ones on the mucus, secreted in large quantity on the surface of the body. Some species like *Macropodus* and *Gasterosteus* use their sticky mucus for preparation of nest for laying eggs. The mucus also helps in regulating to some extent, the osmotic exchange of water and ions between the body-fluids and the water. The number and size of mucus gland cells vary with species. Generally, fishes with no scales, have large numbers of mucus cells.

(ii) **Poison glands** — Venom or poison glands have evolved in different families of fishes. Glandular cells of epidermis are modified into poison glands. These glands secrete poisonous substance to protect themselves from the enemy for defence. They are also used for offence as well. The poison glands are generally present at the base of certain structures like sting, spine of dorsal fin and tooth. Poison glands open at the tip of these structures to inject poison by penetration into the prey. The most common example is the stingray, which is provided with venomous caudal sting. Similarly, Chimaeras possess venom glands in spine of the dorsal fin. The poison glands are present in the grooves of spines of dorsal, pelvic and anal fins of the Scorpion fish (Scorpionidae). In Sturgeon fish (Acanthuridae), the poison glands are found at each side of the caudal peduncle.

(iii) **Photophores** — In many marine species of fish, special multicellular glands are developed from stratum germinativum of epidermis. These glands are deeply seated into dermis and produce light. These light producing luminous organs are mostly found in deep-sea elasmobranches and in some teleosts inhabiting total darkness in sea. Each gland has an apex consisting of mucus cells that helps to magnify light, produced from the basal glandular part of the gland.

The Dermis

The dermis lies beneath the epidermis (Fig. 3.2). This layer contains blood vessels, nerves, connective tissues and sense organs. The upper layer of dermis is made of loose connective tissues and is known as stratum spongiosum, while the lower part is occupied by thick and dense connective tissues, called the stratum compactum. This layer generally has proteinaceous collagen fibres and mesenchymal cells. The dermis is well supplied by blood vessels, hence it also provides nourishment to the epidermis.

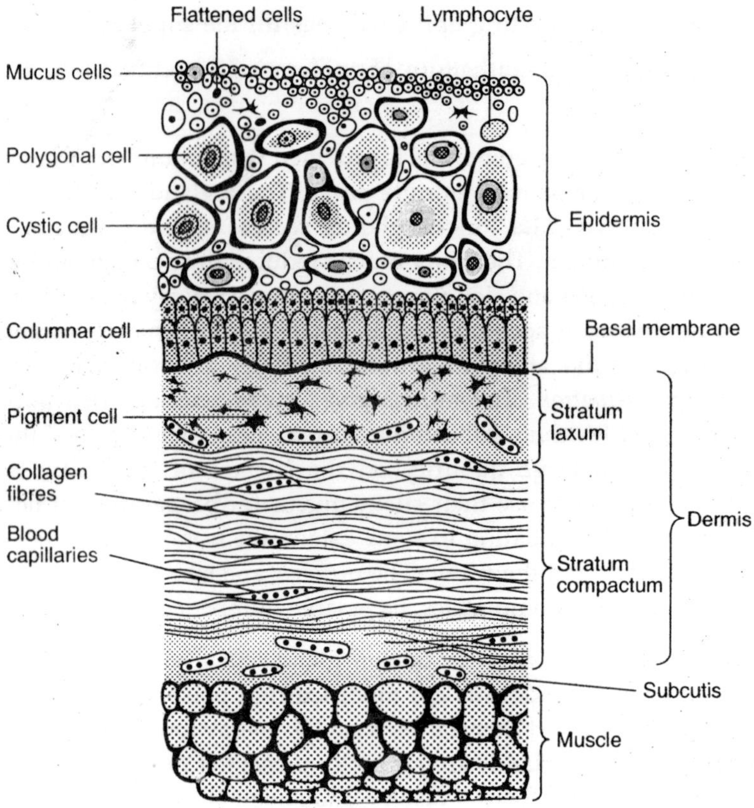

Fig. 3.2 : V. S. fish skin (generalized)

Scales — Scales are derivatives of mesenchymal cells of dermis. Some fishes are "naked" devoid of scales, e.g., freshwater catfish. Certain species exhibit an intermediate condition that are generally naked but possess scales on restricted areas. Such condition is found in paddlefish *(Polydon)*, in which scales are present in region of throat, pectoral and base of tail. (Gloria *et. al.*, 2003)

In some fishes, scales are modified into teeth, bony armour plates (Sea horse) and spiny stings (sting ray). In fresh water eel *(Anguilla)*, scales are very small and so deeply embedded that the fish appears to be naked.

Mostly scales are arranged in imbricate manner and overlap with free margin directed towards the tail that minimizes friction with water. In freshwater eel (*Anguilla*), the arrangement is mosaic, the scales unite their neighbouring ones at their margins.

Type of Scales — There are few type of scales based on their structure and shape. The different type of scales are often characteristics of the species. On the basis of shape, scales are of four types :

(i) Plate like or placoid scales commonly found in Elasmobranches.

(ii) Cycloid scales found in Burbot and soft-rayed fishes.

(iii) Rhombic or diamond shaped scales, common among gars and sturgeons.

(iv) Ctenoid scales, characteristics of spiny-rayed bony fishes (Acanthopterygii).

Scales may also be classified as placoid or non-placoid. Three basic types of non-placoid scales are — cosmoid, ganoid and bony-ridge.

Placoid Scales — Placoid scales are characteristically found among the sharks and other Elasmobranches. They are small denticles that remain embedded in the skin. Each scale has two parts, an upper part, known as ectodermal cap or spine (Fig. 3.3a). This part is

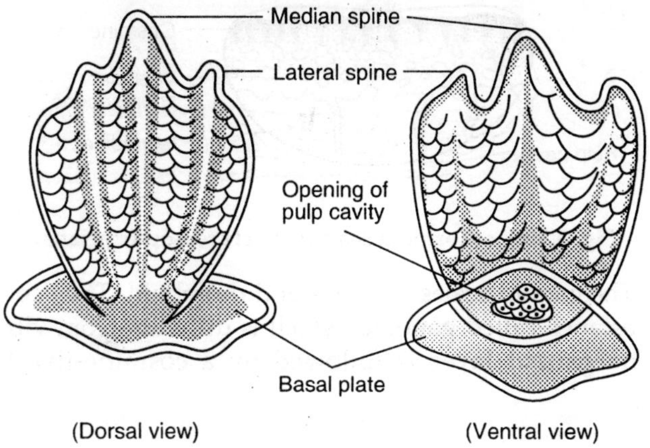

Fig. 3.3(a) : Placoid

made of enamel, like substance, known as vitreodentine similar to human tooth. Another layer of dentine that encloses a pulp cavity follows the vitreodentine. The lower part of placoid scale is a disc-like basal plate, which is embedded in dermis with cap or spine projecting out through epidermis. The basal plate has a small aperture through which blood vessels and nerves enter into pulp cavity. The placoid scales are modified in jaw teeth in sharks; in spines in dorsal fins; in *Squalus* (spiny dogfish); in sting in the stingrays and in saw teeth in the *Pristis*.

Development of Placoid Scale

The placoid scale first appears as small aggregations of dermal cells just below the stratum germinativum. These dermal cells grow upwards into an arched structure or papilla, which gradually pushes the stratum germinativum (Wallin, 1957). The cells of stratum germinativum of this region become glandular and act as enamel organ. Later on, this projecting structure differentiates into a spine and a basal plate. The outer cells of the papilla, known as odontoblasts, secrete dentine around the papilla, while the central cells do not calcify and constitute the pulp. The vitreodentine to form a cap over the spine, gradually envelopes the spine of the scale. The mesenchymal cells of the dermis secrete the basal plate. These cells secrete hard cement-like substance to cover the basal plate. Finally, the spine erupts from the cells of epidermis and projects out, while the basal plate lies embedded in the dermis.

Cosmoid scale — The cosmoid scales are found in living *(Latimaria)* and extinct lobefins (Fig. 3.3b). In dipnoi, the cosmoid scales are highly modified and appear like cycloid scale. The cosmoid scale is a plate-like structure and consists of three layers. An outermost layer is thin, hard and enamel like, called vitreodentine. The innermost layer is composed of vascularized perforated bony substance, called isopedine. The middle layer is made up of hard non-cellular and a characteristic material, called cosmine and is provided with many branching tubules and chambers. These types of scales grow at the edges from beneath by addition of new isopedine material.

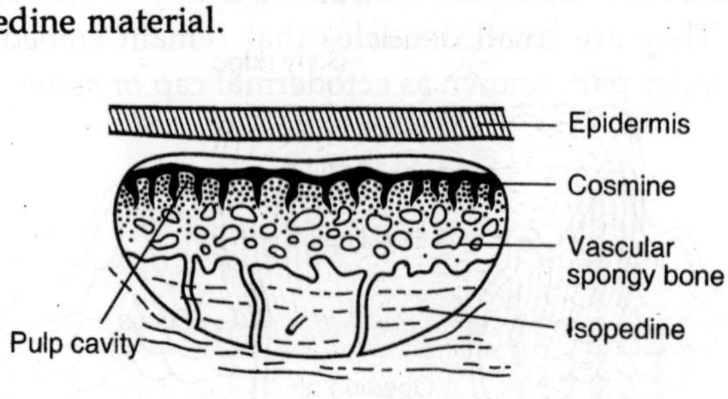

Fig. 3.3(b) : Cosmoid scale (extinct crossopterygians)

Ganoid scale — The ganoid scales are thick and rhomboid. They consist of an outer layer of hard inorganic substance called ganoine, which is different from vitreodentine of placoid scales (Fig. 3.3c). The ganoid layer is followed by a cosmine-like layer provided with many branching tubules. A bony layer of isopedine occupies the innermost layer. These scales, not only grow at the edges but also grow at the surface. The growth takes place by the addition of new layers of isopedine. The ganoid scale is best found in the *Polypterus* and *Lepidosteius*. In these fishes ganoid scales are rhombic plate-like, fitting edge to edge and invest the entire body. In *Acipencer*, the ganoid scales are modified into large bony scutes, arranged into five rows.

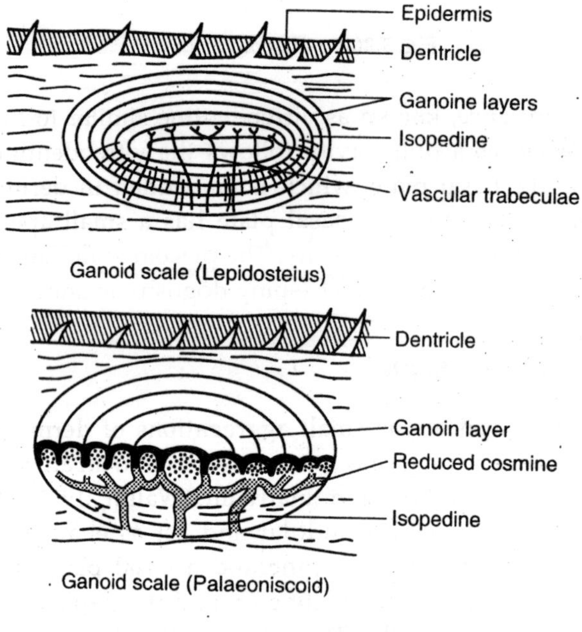

Fig. 3.3(c) : Ganoid scale *(Lepidosteius)*

Ctenoid scale — They have characteristic teeth at its posterior part (Fig. 3.3d). Ctenoid scales are found in spiny-rayed teleost. They are arranged obliquely in such a manner that the posterior end of one scale overlaps the anterior edge of the scale present behind. The chromatophores are present at the posterior part of these scales.

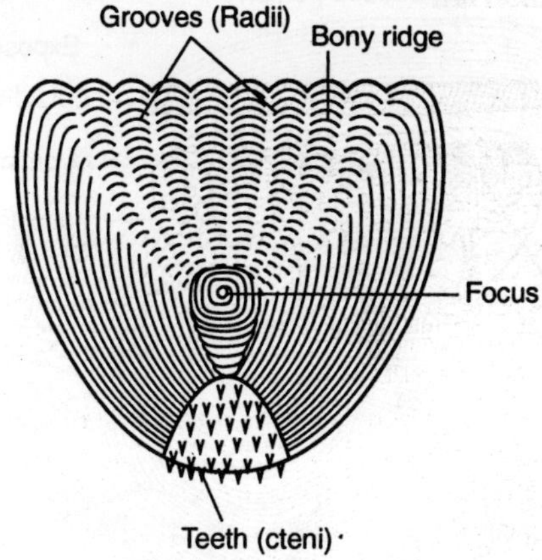

Fig. 3.3(d) : Ctenoid scale

Cycloid scale — The cycloid scales are devoid of teeth or spines, hence seem cyclic (Fig. 3.3e). They are found in soft-rayed teleost and modern lobe-finned fishes. But some spiny-rayed fishes, i.e., *Lepidosteius* show presence of cycloid scales. In *Micropterus*, both cycloid and ctenoid scales are found.

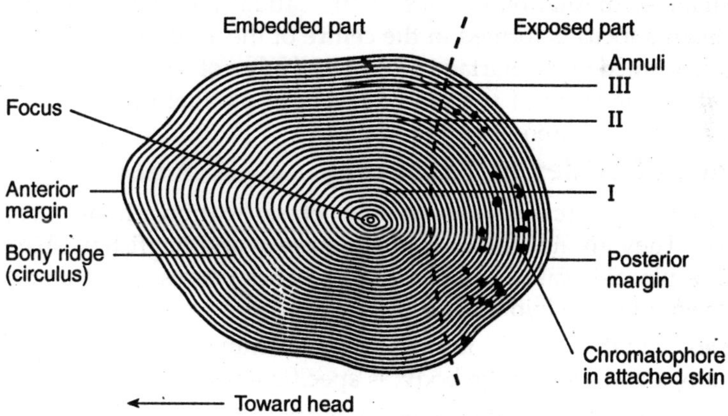

Fig. 3.3(e) : Bony-ridge scale of the cycloid type (Source: Michigan Institute for Fisheries Research).

Bony ridge scale — Bony ridges characterize the bony fishes, Osteichthyes. Bony ridge scales are thin and semitransparent because they do not possess dense enamel and dentinal layers that are found in other types of scales (Fig. 3.3f). They are of two types; cycloid and ctenoid scales. The outer surface of these scales possess bony ridges that alternate with groove-like depressions. The ridges are arranged in the form of concentric rings. The inner part of the scale is composed of fibrous connective tissue. The central zone of scale is differentiated properly and is known as focus of the scale. During the development, focus appears first and lies in the central position. When growth of scales takes place in anterior

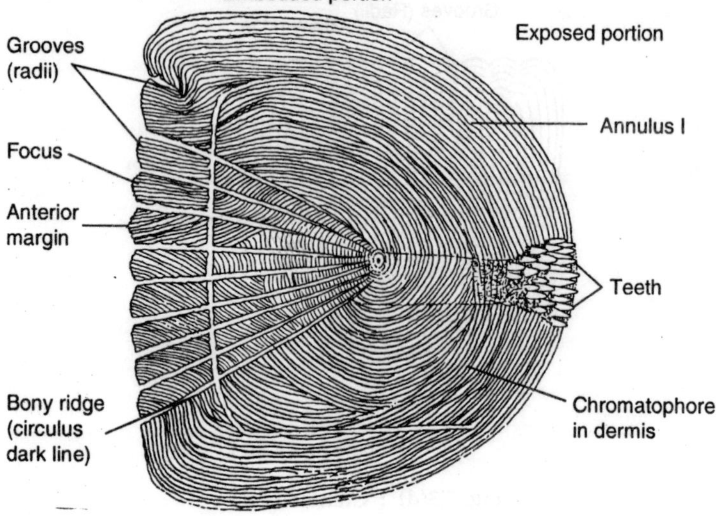

Fig. 3.3(f) : Bony-ridge scale of the ctenoid type (Source : Michigan Institute for Fisheries Research).

or posterior parts, it causes shifting of the focus anteriorly or posteriorly, respectively. Later the groves radiate from the focus towards the margin of the scales.

Development of bony ridges — The bony ridges first make their appearance in the dermis as small accumulation of cells at the caudal peduncle and then gradually spread from there. Soon a focus is formed in the centre of the accumulation of cells. Later on, ridges or circulus are formed at the surface of growing edge of scale. The deepest part of the scale, the basal plate is made up of successive layers of parallel fibres. Some calcification of this fibrillar plate occurs to strengthen the scale.

Significance of Scales in Taxonomy

Scales play significant role in the classification, hence very useful for the ichthyologists. They are not found in the lampreys and hagfishes; sharks are characterized by presence of placoid scales; primitive bony fishes possess ganoid scales; the higher bony fishes have ctenoid or cycloid scales.

The scales' count is highly important in taxonomy. The number of scales present in lateral line, along and around the body, is specific in every species.

The age of fish could be determined by measuring space in annual rings of the scales. In some species like Atlantic *Salmon*, the scales exhibit the presence of spawning marks on them. These marks indicate how many times the fish has spawned and the time of first spawning also.

REFERENCES

Gloria, A., Kapoor, B.G., Charden, M., and Diago, R. (2003). Fisheries–I, Science Publishers (USA).

Singh B.R. (1993). Development, origin and evolution of the neo-morphic air breathing organs in teleosts. In : Advances in Fish Research Vol. 1 (Ed. B.R. Singh) Narendra Publishing House Delhi. pp. 116.

Wallin., O. (1957). On the growth structure and developmental physiology of the scale of the fishes. Rep. Inst. Freshwater. Res. Drottiningholm, 38: 385-447.

Willem, V. (1947). Less manoeuvres respiratories chez les Poissons teleosteans. Bull. Mus. Hist, Nat. Belg. 23: 1. 15.

Wittenberg, J.B. and Haedrich, R.L. (1974). The choroid rate mirabile of the eye. 11, Distributor and relation to the pseudobranch and to the swim bladder rate mirabile. Biol. Bull. 149(I) : 137-156.

Woskoboinikoff, M.M. (1932). Der Apparat des kiemenatmung beiden fischen. Ein versus de syntheses in der morphologic. Zoo. Jb. (Ab. 2), 55 : 315-488.

Yadav, A.N. Prasad, M.S. and Singh, B.R. (1993). Gross structure and dimension of the gills in the Indian hill stream fish. In : 'Advances in fish research' Vol. I (Ed. B.R. Singh) Narendra Publishing House, Delhi. pp. 35-53.

Chapter 4

DIGESTIVE SYSTEM

The digestive system consists of alimentary canal and its associated glands. The digestive tube also contains numerous intramural glands which provide the tube by lubricating mucus, enzymes, water, etc.

The extramural glands are liver, pancreas and gall bladder (Fig. 4.1a, b).

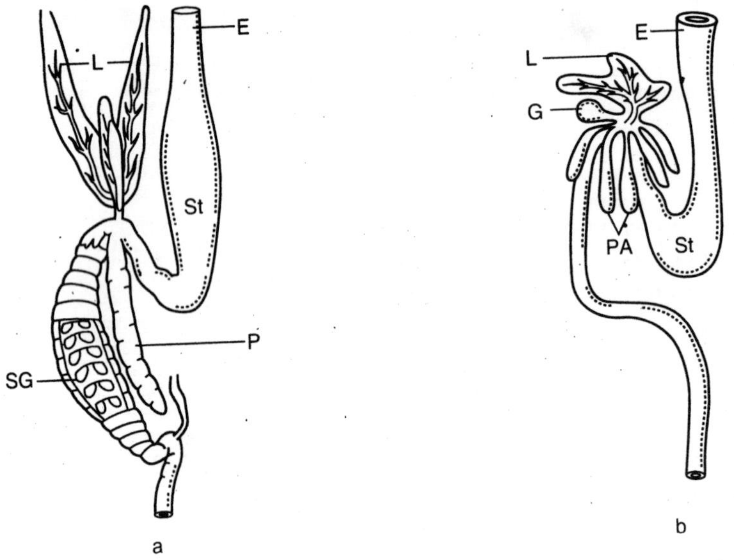

Fig. 4.1a, b : Diagram of alimentary canal of (a) Elasmobranch, (b) Teleost. E, oesophagus; G, gallbladder, L, liver; P, pancreas; PA, pyloric-appendices; SG, spiral gut; St, stomach.

The liver is present in all fishes. The pancreas which is exocrine and endocrine organ, may be a discrete organ or it may be diffused in the liver or in the alimentary canal. In sharks and rays (Elasmobranchii) pancreas is relatively compact and usually well developed as a separate organ, often two lobed, but in teleosts, the pancreas is diffused in the liver to form hepatopancreas. It is also diffused in the alimentary canal in a few fishes. It is also present in the mesenteric membranes surrounding the intestine and liver. The gall bladder is vestigeal in deep sea fishes but it is prominent in other fishes.

While passing through the alimentary canal, the food is broken down physically and chemically and ultimately solubilized so that degraded products can be absorbed. The absorption occurs chiefly through the wall of intestine. The undigested food and other substances within the alimentary canal such as mucus, bacteria, desquamated cells and bile pigments and detritus are excreted as faeces. The peristaltic movement and local contractions are important and help the food to pass through the gut. The local contraction displaces intestinal contents proximally and distally.

Parts of Alimentary Canal : The alimentary canal of fish consists of mouth, which opens into buccopharynx, which in its turn opens into the oesophagus. The oesophagus opens into the stomach/intestine. The lips, buccal cavity and pharynx are considered as non-tubular part whereas the oesophagus stomach/intestinal bulb, intestine and rectum are tubular in nature and distinguished as tubular part of the alimentary canal.

Mechanism of Feeding : In most teleosts, the food reaches the mouth by sucking it by enlarging their buccal and opercular cavities (Alexander, 1967). The pressure in the buccal and opercular cavities and pressure of water around the fish, are important for sucking. The pressure of – 50 to – 105 cm of water was recorded, as food was sucked in, and + 1 to + 9 cm of water taken in with the food was driven out through operculum. In case of black bullhead (*Ictalurus*) a negative pressure of – 80 cm was recorded at 18 °C. It is also recorded that to produce strong negative pressure, all muscles principally involved have to exert tension near maximum isometric tension (Alexander, 1970). Osse (1969) made an electromyographic study of feeding in *Perca* and confirmed the sequence of muscle action inferred from anatomical studies and by analogy with respiration (Gloria *et al.*, 2003).

Stimuli for Feeding : The mechanism of feeding behaviour in fishes is very complicated. There are generally several kinds of stimuli for feeding. The common factors affecting the internal motivation or drive for feeding include season, time of the day, light intensity, time and nature of last feeding, temperature and any internal rhythm.

The visual, chemical, taste and lateral line system also control the momentary feeding act. The interaction of these groups of factors determine when and how a fish will feed and what it will feed upon (Lagler *et al.*, 1977).

The role of visual and olfactory factors in connection with feeding behaviour has been studied by experimental conditions by Groot (1971). He found visual, chemical and mechanical sense organs in Pleuronectidae, Soleidae, and Bothidae (belongs to the family of flat fishes, Pleuronectiformes). Soleidae are polychaet mollusc feeder, feed during night, find their food mainly by olfactory clues, but still have the capability of finding their food visually, (Table 1). The barbels help the fish to locate food grubbed from soft bottom material.

Table 1: Role of visual and olfactory factors in feeding

Family	Type	Way of finding food	Form of intestine	Form of gill raker	Olfactory Lobe (After Evans, 1937)	Optic Lobes (After Evans, 1937)
Bothidae	fish-feeder	vision loop	simple toothed	heavily	small	large
Pleuro-nectidae	Crustacea-feeder	Mainly vision but also olfaction	complica-ted loop	less toothed	medium	large
Soleidae	Polychae-tamollus-cafeeder	Olfaction but vision possible	more complica-ted loop	few or no teeth	large	small

On the basis of feeding habits, the fishes are catagorised as follows:

1. Herbivorous
2. Carnivorous
3. Omnivorous
4. Detrivorous.

The fishes can be classified as 'Euryphagous' consuming mixed diet 'Stenophagous—eating limited assorted food and 'Monophagous'—consuming only one type of food. Amongst teleosts, about 61.5% are omnivorous, 12.5% are carnivorous and about 26% are herbivorous (Dutta and Hossain, 1993) (Fig. 4.2).

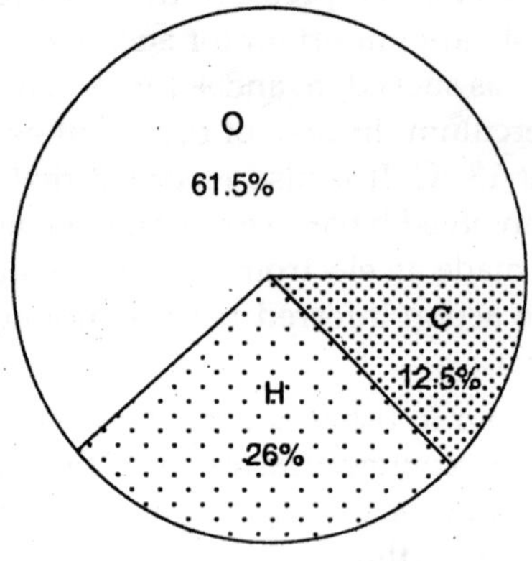

Fig. 4.2 : Diagram showing distribution of species with different feeding habits. C, carnivorous; H, herbivorous; O, omnivorous.

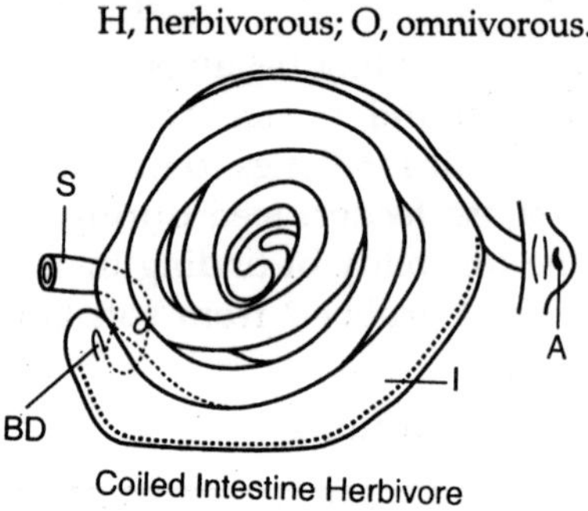

Fig 4.3 : Diagram of alimentary canal of a herbivorous catfish. A anus; BD, bile duct; I, intestine; S, Stomach.

1. **Herbivorous Fishes :** They consume about 70% unicellular algae, filamentous algae and aquatic plants. In addition to plant material these fishes also consume 1-10% of animal food and mud. The common examples are *Labeo* species *Osphronemus goramy, Sarotherodon mossambicus* etc. Herbivorous fishes have long and coiled intestine (Fig. 4.3).

2. **Carnivorous fishes :** The fishes in contrast to herbivore have shorter gut, the intestine is straight, very little coils are present. Some of the carnivores possess intestinal caecae. They prey on small organisms and consume high percentage of animals such as copepods, dafnia and insects. The examples of carnivorous fishes are *Wallago attu, Mystus seenghala, Mystus cavassius, Mystus vittatus, Channa striatus, Channa marulius, Channa punctatus, Notopterus chitala, Rita rita,* etc. (Fig. 4.4a, b, c, d).

3. **Omnivorous fishes :** Omivorous fishes like *Cyprinus carpio, Cirrhina mrigala, Puntius, Clarias,* etc. are consuming both plants and animals. The rotifers, mud and sand are also found in the alimentary canal. Their gut length intermediates between carnivorous and herbivorous fishes (Fig. 4.5).

4. **Detrivorous or Plankton feeder :** They consume detritus along with zooplanktons and phytoplanktons. The arrangement of gill rakers are such that it filters them from water. The examples are *Catla catla, Hilsa ilisa, Cirrhina reba,* and *Hypopthalmichthys molitrix*. They are both omnivorous and carnivorous.

The fishes can also be named on the basis of modification of buccopharynx :

(1) Predator,

(2) Grazers,

(3) Strainers,

(4) Suckers,

(5) Parasitic.

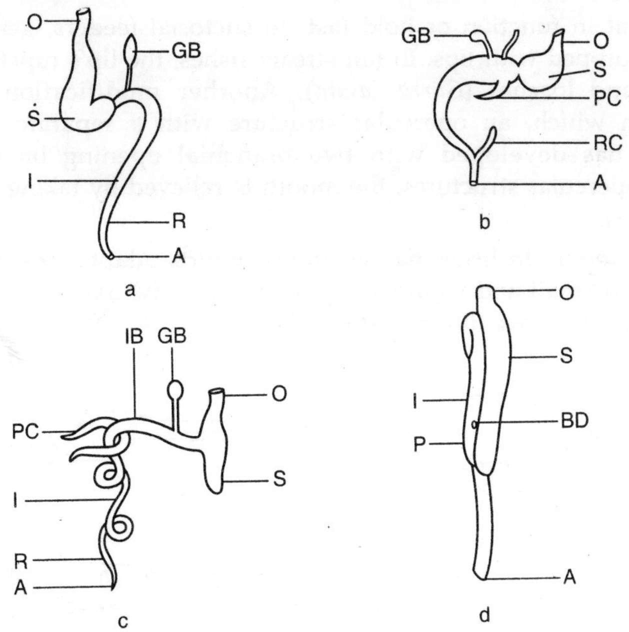

Fig. 4.4a-d : Alimentary canal of fishes. (a) *Mystus seenghala;* (b) *Notopterus chitala;* (c) *Channa striatus;* (d) A carnivore fish, *Esox lucius.*

A, anus; BD, bile duct; GB, gall bladder; I, intestine; IB, intestinal bulb; O, oesophagus; P, pylorus; PC, pyloric caecum; R, rectum; RC, rectal caecum; S, stomach.

1. **Predators :** They possess well developed grasping and holding teeth, e.g. sharks, pike and gars, etc.

2. **Grazers :** They take the food by bite. These fishes feed on plankton and on bottom organisms. e.g., bluegill (*Lepomis macrochirus*), parrot fish and butterfly fish.

3. **Strainers :** They have efficient straining or filtering adaptation due to the arrangement of gill rakers forming sieve for straining the food material. They are plankton feeders.

4. **Suckers :** The fishes have inferior mouth and sucking lips. The response depends upon stimulus of touch e.g., sturgeons, *Labeo, Osteochilus* , etc.

5. **Parasites :** Amongst fishes, the deep sea eel (*Simenchelys parasiticus*) is parasitic in nature. Lamprey and hagfish are parasitic but belongs to cyclostomata.

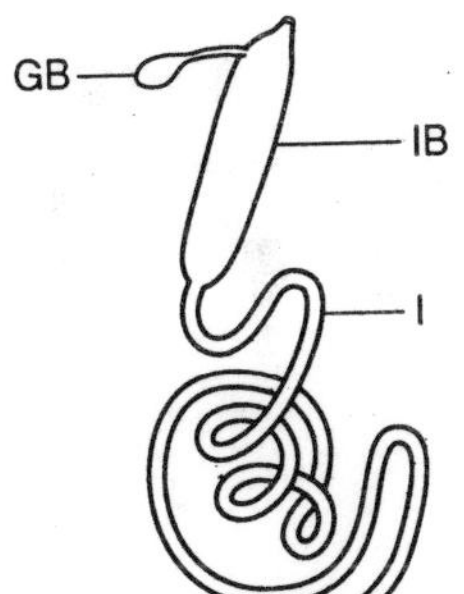

Fig. 4.5 : Alimentary canal of *Cirrhinus mrigala.* A, anus; I. intestine; IB, intestinal bulb; GB, gall bladder.

BUCCOPHARYNX AND ITS MODIFICATION

As stated earlier, the non-tubular part of the alimentary canal is buccopharynx which includes lips, buccal cavity, pharynx. There are a few modifications according to feeding habits.

Jaws : The fishes have well developed jaws— a fact, which is an important step in vertebrate evolution. The jaws hold the mouth, and the main function of the jaws in vertebrate is helping in biting of food. The jaws are provided with lips. In fishes, the lips are either suctorial in function or hold fast. In suctorial feeders, generally the mouth is inferior and is equipped with lips. In hill stream fishes, the lip's function is hold fast, e.g., *Glyptosternum* and loaches (*Garra lamta*). Another modification has developed in Gyrinocheilus, in which, an opercular structure with a separate water inhalant and exhalant device has developed with two branchial opening on each side with the development of opercular structures, the mouth is relieved by taking in respiratory water (Lagler *et al.*, 1977).

Mouth : The mouth in fishes has several structural adaptations (Fig. 4.6a-h). In some fishes, the mouth is small and its inner surface of lips have rasp-like folds to facilitate the scraping of algae from stones to which they adhere.

Fig. 4.6a-h : Some adaptation in mouth structure among fishes. (a) Elephant fish, *Gnathonemas elephas*. (b) Gar fish, *Strongylura longirostris*. (c) halfbeak, *Hyporhamphus unifaciatus*. (d) thick-lipped mojarra, *Cichasoma lobochilus*. (e) stargazer, *Zalescopus tosae*. (f) longnose butterfly fish, *Forcipiger longirostris*. (g) and (h) Paddle fish, *Polydon spathula* (source, Lagler *et al.* 1977).

In fishes with suckers, the lips are mobile and plicate having fold or papillae. The mouth in some fishes is elongated as beak. This adaptation is due to lengthening of jaws and hyomandibular bones. The examples are trumpet fishes, cornet fishes, pipe fishes (Syngnathoidae) and butterfly fishes (Chaetodontidae). In hemiramphidae, half-beaks, the lower jaw projects into beak, the beak is 1/3 of the length of fish. The mouth is situated at the top. In *Xenentodon cancila*, both the jaws are elongated.

Buccopharynx : The buccopharynx has also undergone a few modifications in relation to feeding behaviour (Fig. 4.7).

These modifications are (i) Teeth (ii) Pharyngeal pad (iii) Gill rakers. Not all fishes bear teeth. The teeth are present in some species while in some they have completely disappeared. Teeth are absent in plankton feeders and in some generalized omnivores. They are present in increasing number in predaceous fishes. Teeth are present on jaws, buccal cavity, and in pharyngeal regions. On jaws, teeth are located at maxilla, premaxilla and on dentries. At the roof of buccal cavity, the teeth are located on vomer, palatine and ectopterygoid bones on each side. In some fishes, teeth are present on tongue. Such teeth are present in *Notopterus notopterus*, *Notopterus chitala*. The major kind of teeth are classified as cardiform, villiform, canine, incisor and molariform (Fig. 4.8a-d).

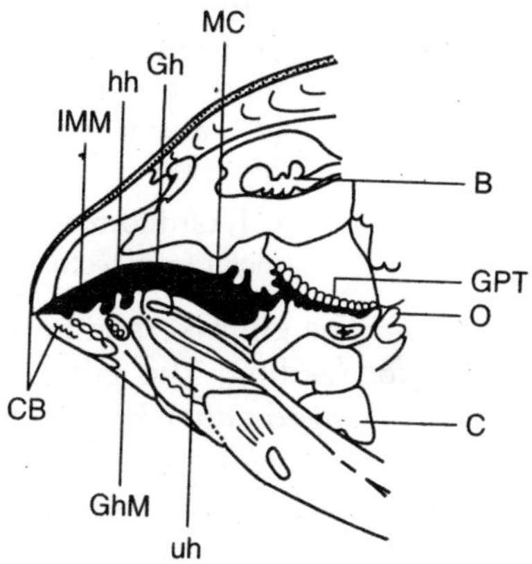

Fig. 4.7 : Sagital section of a parrot fish to show 'beak' from fusion of jaw teeth and to show pharyngeal teeth.

Source : (Grasse 1957). B, brain; C, cleithrum; CB, cutting beak; Gh, glossohyal; GPT, grinding pharyngeal teeth; GhM, Geniohyoideus muscle; hh, hypohyal; IMM, intermandibular muscle; MC, mouth cavity; O, oesophagus; Uh, urohyal.

In pharynx, the teeth are located on superior and inferior positions. The pharyngeal teeth occur as pad on various gill arch elements in many species. In *Clarias* and *Labeo*, the teeth are modified for grasping, tearing grinding (Fig. 4.9a, b) and comb-like, razor-like cutting teeth have developed in predaceous fishes such as *Serrasalmes* and *Sphyraena*. Teeth are not present on jaws and palate in *Tor tor, Puntius sarana and Catla catla*.

Pharyngeal organ : Kapoor *et al.* (1975), described pharyngeal organ in fishes which are muscular pockets situated in the roof of the pharynx as diverticulum. It has two parts, a caudal and a blind sac. The lining is provided with flagellated papillae. Another modification in a few fishes are the occurrence of pharyngeal pads or palatine organ. They are situated dorsally at the entrance of the oesophagus. The mucous membrane has generally gathered to form pads. These are found in herbivorous fishes. In *Cyprinus carpio*, pharyngeal pad is situated dorsally at the entrance of the oesophagus and is responsible for removing excess water from ingesting food (Jara, 1957). In addition to these structures, in some fishes, pharyngeal valves are present. They are hanging from the roof of the pharynx and assist in placement of pieces of corals grinded by pharyngeal teeth and lubricate the food from mucous cells in scarid fishes (Kapoor *et al.*, 1975).

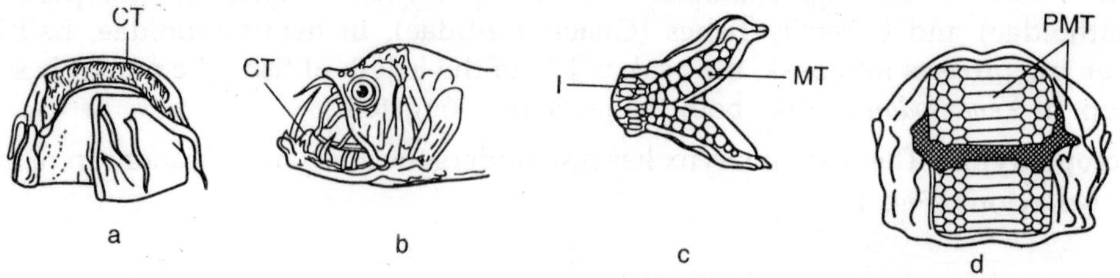

Fig. 4.8a-d : Some variations in structure and location of teeth in fishes.

(a) Upper jaw of flathead catfish, *Polydictis olivaris*. (b) Deepsea viperfish, *Chauliodus sloani*. (c) Lower jaw of a porgy, *sparidae*. (d) Pavement teeth on roof and floor of mouth of an eagle ray, *Mylisbats*.

C.T, cardiform teeth; I, incisor, PMT, pavement molariform teeth; MT, molariform teeth.

Gill rakers : The gill rakers in both the cartilaginous and bony fishes have undergone specialization for feeding. The gill rakers bear gill lamellae and project into pharyngeal cavity and are arranged either in one (row) or two rows on each arch. The main function of the gill rakers is to prevent prey from escaping through the gills but in piscivorous members (*Micropterus salmoides*), gill rakers also partially descale fish as they pass by the rakers. The gill rakers of the first arch are the largest while those of the fifth arch are the smallest in size (Khanna, 1962). The gill rakers allow the solid food to go to gullet and only water is allowed to pass through gills to outside.

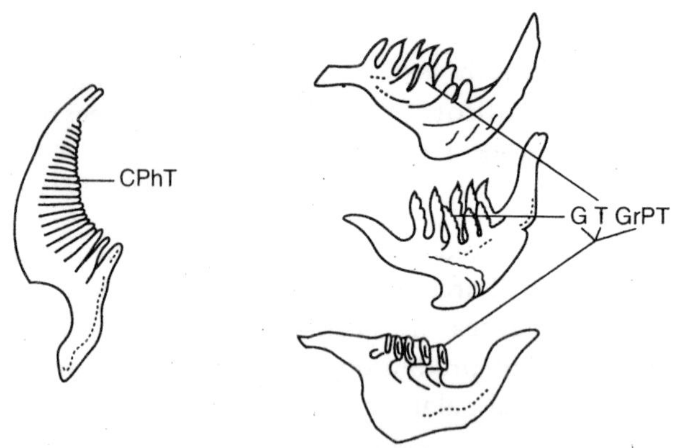

Fig. 4.9a-b : (a) Various types of pharyngeal teeth of minnows.
(b) G, grasping teeth; T, tearing teeth CPht, Cutting pharyngeal teeth.

Posteriorly, the buccopharynx opens into a circular aperture provided with sphincter-like muscle. The opening is known as gullet. The gullet in turn opens into oesophagus. The part of alimentary canal from oesophagus to anus is known as gut. The gut is distinguished into oesophagus, stomach/intestinal bulb, duodenum, intestine, rectum and anus. These parts may or may not be distinguished from each other in some fishes, while in others, there is clear morphological distinction between these chambers. Groot (1979) reported that there were great variations in the morphology of the alimentary tract in flat fish family like bothidae, pleuronectidae and solidae (Fig. 4.10a, b, c).

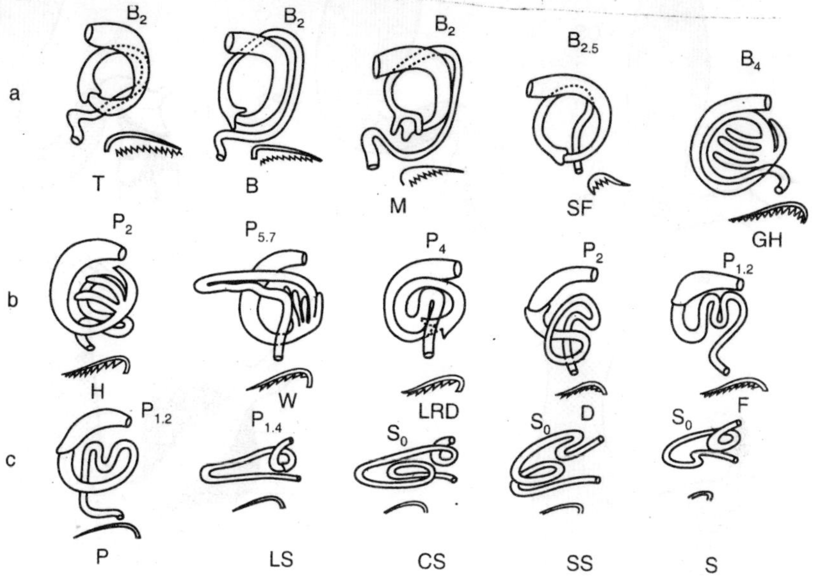

Fig. 4.10a-c : Various types of alimentary tract and structure of gill rakers in different flatfish. (a) Bothidae; (b) Plauronectidae; (c) Solidae. T, turbat; B, brill; M, megrium; SF, scaldfish; GH, Greenland halibut; H, halibut; W, witch; LRD, long rough dash; D, dash; F, flounder, P, plaice; LS, lemon sole; CS, common sole; SS, sand sole; S, solenette number of pyloric appendices are indicated as B_2........., P_2........., S_0........., etc.

Oesophagus : The oesophagus and stomach are separated by external constriction. Moreover, the size of stomach is large and may be sac-like. However, in a few fish there is no clear morphological differentiation between these two parts of the gut. They can be distinguished histochemically and physiologically, i.e., pH, digestion and enzyme secretions, etc.

Stomach : Frequently, in literature, the fishes are classified as stomach and stomachless fishes. The predatory and carnivore fishes have stomach (Sehgal, 1960; Pasha, 1964; Islam, 1951; Kapoor, 1953; Khanna, 1961; Khanna and Pant, 1964; Ray and Moitre, 1982). The cyprinidae family has no stomach but possesses intestinal bulb. (Lagler *et al.*, 1977; Grasse, 1957; Moyle and Cech, 1982). According to Dutta and Hossain (1993), 85% teleost species possess stomach. They may be I, J, U, V, Y shaped (Suyehiro, 1942). (Fig. 4.11a, b, c, d, e). In mullet and shad, gizzard is found (Fig. 4.12a & b). It is also found in *Mugil* and *Gadusia chapra*. It is stomach, which is specialized for masticating as well as for secretion purpose. It is the modification of fishes who have macrophage food habits. The true stomach is characterized by the presence of smooth muscle "muscularis mucosa". Internally, the stomach can be divided into cardiac and pyloric. In some fishes, it is distinguished into cardiac, fundus and pyloric regions.

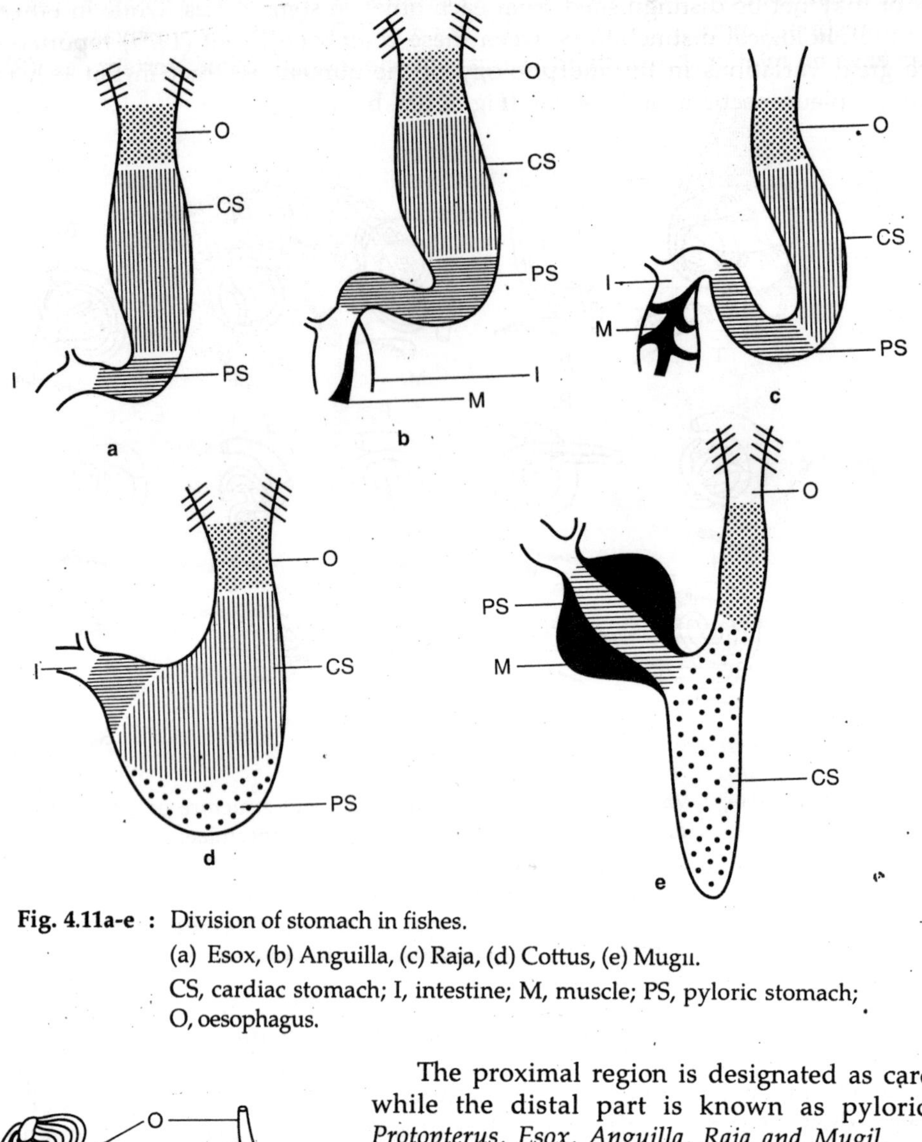

Fig. 4.11a-e : Division of stomach in fishes.
(a) Esox, (b) Anguilla, (c) Raja, (d) Cottus, (e) Mugıl.
CS, cardiac stomach; I, intestine; M, muscle; PS, pyloric stomach;
O, oesophagus.

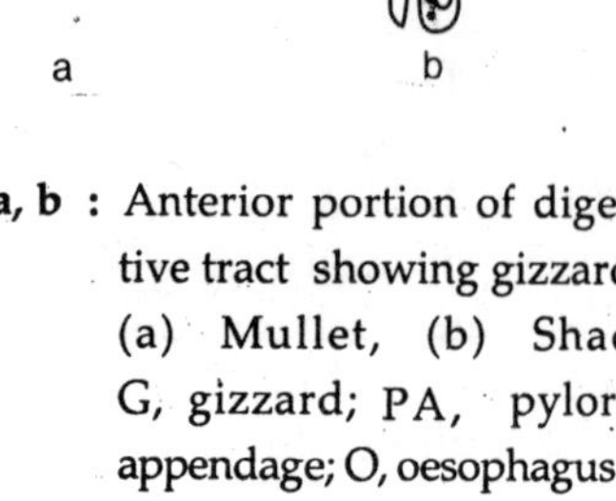

Fig. 4.12a, b : Anterior portion of digestive tract showing gizzard. (a) Mullet, (b) Shad. G, gizzard; PA, pyloric appendage; O, oesophagus.

The proximal region is designated as cardiac while the distal part is known as pyloric in *Protopterus, Esox, Anguilla, Raja and Mugil.*

Intestine : Intestine also shows many variations. It is short and straight in carnivores such as *Esox lucius* but long and thin walled and highly coiled in herbivores species. The intestine is provided in some fish with spiral valve makes four to fifty or more turns, depending upon species (Fig. 4.13a, b). Provisory the proximal part of the intestine is known as duodenum, the middle, the ileum while the distal end is rectum. However, intestine in fishes is one continuous organ and its parts are usually differentiated histologically only by gradual changes in the nature of mucosal layers.

Intestinal caecae : Fishes are only vertebrate that have appendages (caecae) at the gastro-intestinal junction (Morton, 1967; Romer; 1970; Kent, 1983; Srivastava, 2005). Fish caecae have no relationship with any caecal appendages of other vertebrates (Reifel and

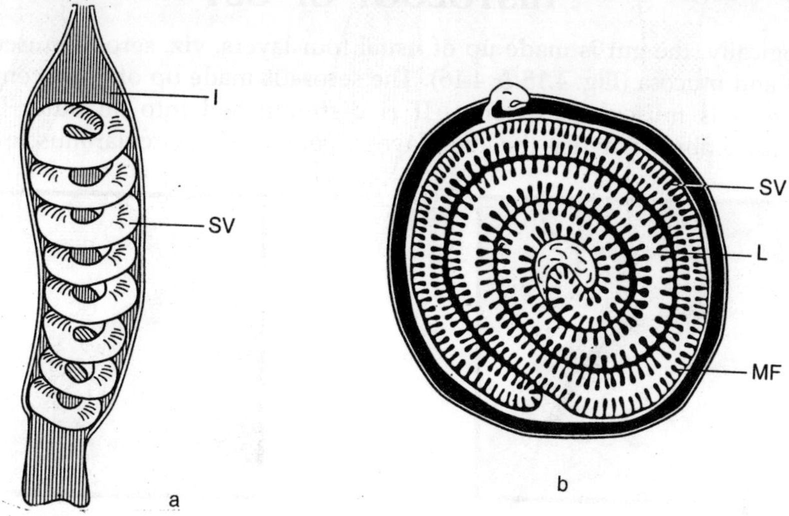

Fig. 4.13a, b : Diagram of section of intestine showing spiral valve (a) *Alopias*, (b) Carnivorous shark, *Scyllium*. I, intestine; L, lumen; MF, mucosal fold; SV, spiral valve.

Travill, 1978). The caecae are located at the junction between the stomach and intestine. Not all stomach bearing fishes have intestinal caeca. Their number of caeca varies in different species [Fig. 4.14(c)]. They are present in *Notopterus, Channa, Mastacembelus, Hilsa, & Harpodon. In Lepomis macrochirus,* they are 6-8 while in *Perca perca* they are 2-3 in number. In *Channa,* they are two in number. Some authors use the term intestinal caecae whereas some use pyloric caeca. As the caecae develop from intestinal wall, they should be designated as intestinal caeca instead of pyloric caecae (Hossain and Dutta, 1988). Kapoor *et al.*, (1975) and Stroband (1980) suggested that stomachless fishes are herbivorous, and because the size of the alimentary canal is long hence there is no necessity of the occurrence of intestinal caeca. The function of caecum is to increase the absorption area.

Rectum : Morphologically there is no distinction between the rectum and the intestine. The ilio-rectal valve is present in *Sciana, Tetradon* and *Muraenosox* (Khanna, 1961). However, there are several histological and ultrastructural specialization in this part.

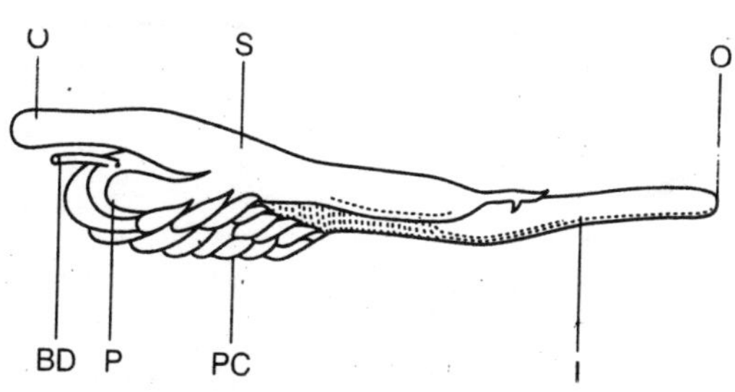

Fig. 4.14 : Alimentary canal of a carnivore, *Perca*, showing straight intestine and pyloric caeca. BD, bile duct; I, intestine; P, pylorus; PC, pyloric caeca; O, oesophagus; S, stomach.

Anus : It is the posterior opening of the alimentary canal. The internal-surface of the region near the rectum is covered with an epithelium rich in mucous cells. The muscularis is made up of an inner circular and outer longitudinal muscle layer. The circular muscle is thickly developed forming sphincter. The region of the anus facing the exterior has the epithelium continuous with the skin.

HISTOLOGY OF GUT

Histologically, the gut is made up of usual four layers, viz. serosa, muscularis externa, submucosa and mucosa (Fig. 4.15 & 4.16). The serosa is made up of loose connective tissue. Next to serosa is muscularis externa. It is distinguished into an outer longitudinally arranged muscle fibres, whereas the inner layer is composed of circular muscle fibres.

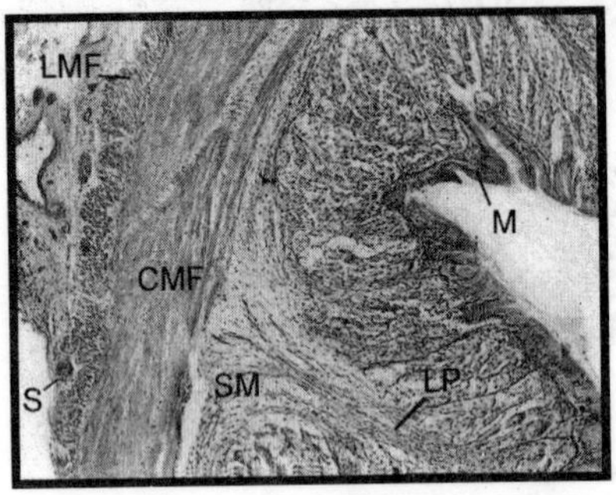

Fig. 4.15 : Microphotograph of transverse section of the oesophagus of *Clarias batrachus* showing four layers X 80. CMF. circular muscle fibre; LMF, longitudinal muscle fibre; L. P., lamina propria; M. mucosa; S, serosa.

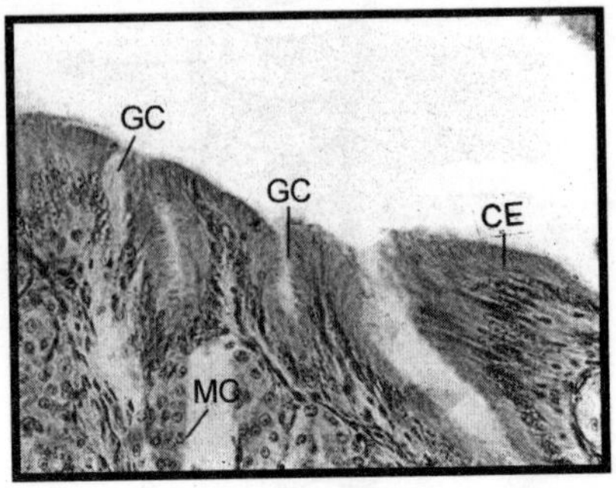

Fig. 4.16 : Microphotograph of trans-verse section of the oesophagus of *Clarias batrachus* showing mucosal folds X 320. CE, columnar epithelium; GC, goblet cell; MC, mucus cells.

The submucosa is made up of loose connective tissue, blood vessels and capillaries. The submucosa is followed by innermost mucosa, which is divisible into lamina propria and epithelial layer. The lamina propria is vascular and is made up of areolar connective tissue. The epithelial layers which lines the lumen of the gut is made up of columnar epithelium and is thrown into deep mucosal folds. Mucosa comprises of various glands.

The stomach possesses broad mucosal folds subdivided into primary and secondary folds. The mucosa contains gastric glands (Fig. 4.17a & b).

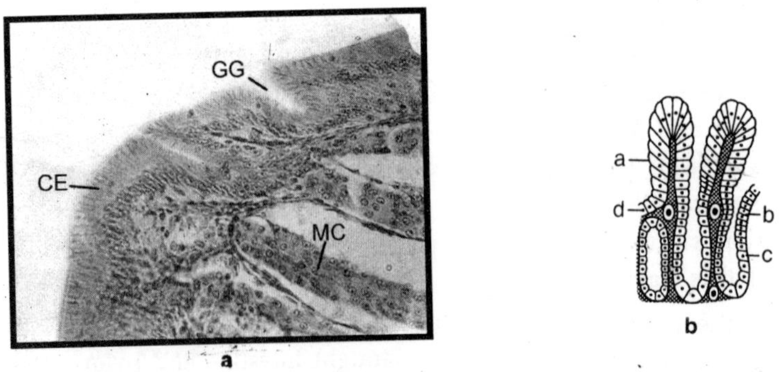

Fig. 4.17a-b : Mucosal folds of stomach. (a) Microphotograph of transverse section of stomach of *Clarias batrachus*, X 320. (b) Diagram of mucosal epithelium of the stomach. a, gastric epithelium; b, columnar epithelium, c, glandular epithelium; CE, columnar epithelium; GC, gastric gland, MC, mucus cells.

The submucosa is reduced having bundles of longitudinal muscle. Circular muscle fibre coat is well developed. Serosa is thin.

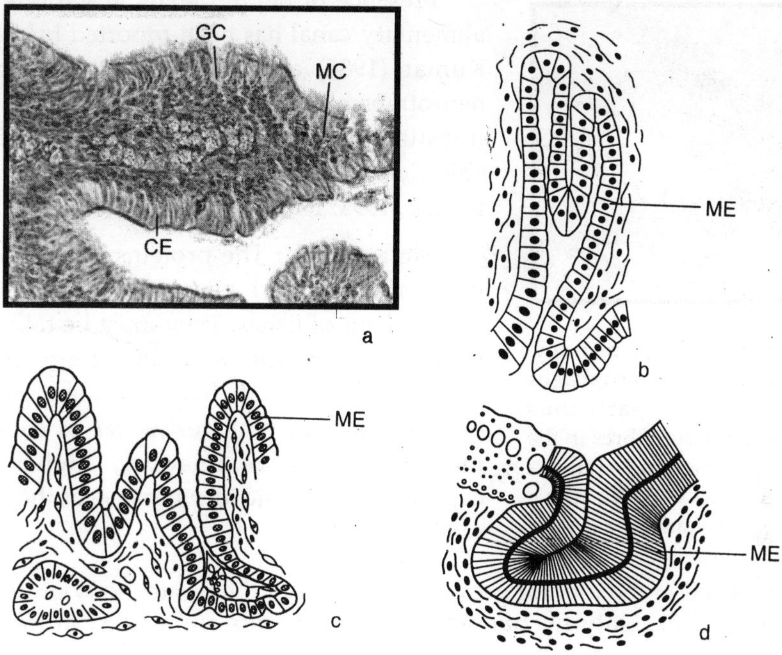

Fig. 4.18a-d : Diagrams of transverse section of intestine. (a) Microphotograph of duodenum of *Clarias batrachus*. X 320. (b), (c), (d), Diagram of mucosal folds of intestine. CE, columnar epithelium; GC. goblet cell; MC, mucus cell; ME, mucosal epithelium.

The submucosa is well developed followed by thick coat of circular muscles, which is surrounded externally by longitudinal muscle fibres. The serosa is thin comprising flattened epithelial cells.

In the intestine, the mucosal, folds are produced into prominent slender folds called villi, which have intestinal glands (Fig. 4.18a, b, c, d).

The submucosa extends into villi forming lamina propria. The circular and logitudinal muscle layers are comparatively thin than in the stomach.

The rectum has short and flat mucosal folds, provided with large number of mucous cells than the intestine. The muscular coat is thick (Fig. 4.19).

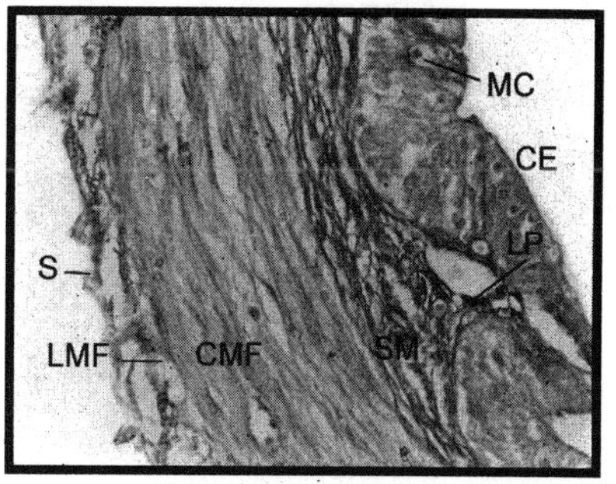

Fig. 4.19 : Microphotograph of rectum of *Clarias batrachus* showing four layers and mucus cells. X 320. CE, columnar epithelium; CMF, circular muscle fibres; LMF, longitudinal muscle fibres, MC, mucus cells; S, serosa.

Innervation of gut : The alimentary canal of fishes is innervated by sympathetic and parasympathetic components of autonomic nervous system. (Fig. 4.20).

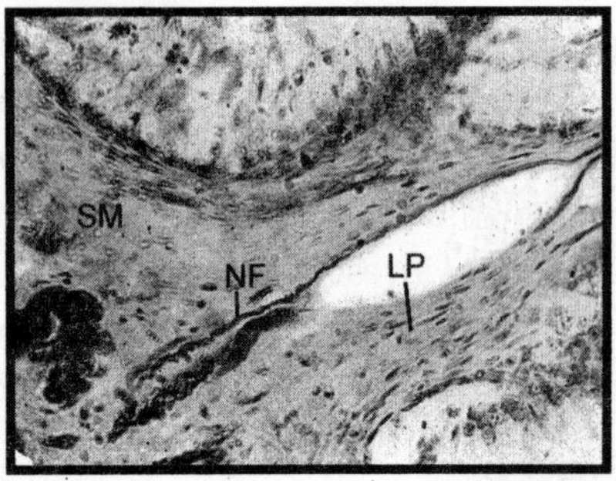

Fig. 4.20 : Microphotograph of transverse section of oesophagus of Clarias batrachus showing nerve, fibres in the submucosa and lamina propria. X 320, NE, nerve fibre; LP, Lamina propria, SM, submucosa.

Presence of nerve plexus in various part of the alimentary canal has been reported by Tembhre and Kumar (1984) and Nicol (1952). The presence of neurotransmitter acetylcholine in intestinal bulb and intestine of fish has been reported both histochemically and biochemically (Tembhre and Kumar, 1994, 1995).

Metabolism : The proteins, carbohydrates, fats, most minerals and vitamins are essential dietary requirement of fishes. They must be taken in diet for growth (anabolism) and for energy (catabolism). They take mineral from ambient water. It is generally agreed that freshwater as compared with marine fishes have relatively higher absorption capacity of inorganic ions because of the surrounding water.

Protein is needed in the diet for growth and repair of the tissue. Body protein consists of long chain amino acids. Only twenty different amino acids are needed in the body for the synthesis of protein molecule.

Out of these twenty amino acids in human beings, 8 are essential amino acids. They must be present in the diet, body cannot synthesize them. In fishes, 10 amino acids are essential. Arginine and histidine are the two amino acids which are extra and the rest 8 are similar to that of human being (Moyle and Cech Jr. 1982). The amino acids are as follows :

Non-essential Amino Acid	Essential Acids
Alanine	Arginine
Aspergine	Histidine
Aspartic acid	Isoleucine
Cystine	Leucine
Glutamic acid	Lysine
Glutamine	Methionine
Glycine	Phenylalanine
Proline	Threonine
Serine	Tryptophane
Tyrosine	Valine

Digestion of Food : For the digestion of protein, the following enzymes are required in vertebrate series.

1. Pepsin (Stomach of carnivorous fishes)
2. Trypsin (Intestine (alkaline medium), pancreas, intestinal caecae)
3. Chymotrypsin
4. Erypsin (Collection of peptidases are known as erypsin, found in intestine).

Digestion of proteins : The fishes which possess stomach are generally carnivorous and secrete pepsin enzyme from gastric mucosa. The pepsin is a protease enzyme, i.e., it can break down protein. The optimal activity is carried out at a pH 2 to 4, so HCl is required for making low pH. HCl is secreted by the gastric mucosa in carnivorous fishes creating the low pH. Both cholinergic and adrenergic nerves are present in the stomach which stimulate the secretion of gastric juices (Tembhre and Kumar, 1984; Kumar and Singh, 1990). The secretion of gastric juices (acid secretion and pepsin) depends upon temperature. At 10 °C the gastric secretion increases to three to four folds. The trypsin enzyme is present in the extract of pancreas of some Elasmobranchs such as *Mustelus cartarias, Littoralis* and *Squalus*. The trypsin is secreted by exocrine pancreatic tissue which may be concentrated in a compact organ as in mackerel (Scomber) or diffusely located in mesentric membranes surrounding the intestine and liver. It is also secreted by hepatopancreas. The inactive form of this enzyme trypsinogen is known as zymogen. It is to be converted into active enzyme, i.e., trypsin by an enzyme enterokinase. The enterokinase enzyme is exclusively secreted by intestine of fish (Barrington, 1957). In the Cyprinids, stomachless fish, protease compensation is supplemented by some intestinal enzyme known collectively as erypsin. Pepsin is absent in stomachless fishes (Kapoor *et al.*, 1975) because of the absence of true stomach. The intestine secretes aminopeptidases. These act on terminal amino acid called as exopeptidases and those act on central bonds are called as endopeptidases. Vitamins are essential constituent of the diet and a large number of vitamin deficient syndrome are noticed in fish (Halver, 1972, Table 2).

Table 2: Vitamin Deficiency Syndromes

Vitamin	Symptoms in *Salmon, Trout,* Carp, Catfish.
Thiamine	Poor appetite, muscle atrophy, convulsions, instability and loss of equilibrium, edema, poor growth.
Riboflavin	Corneal vascularization, cloudy lens, haemorrhagic eyes, photophobia, dim vision, incoordination, abnormal pigmentation of iris, striated constrictions of abdominal wall, dark coloration, poor appetite, anaemia, poor growth.
Pyridoxine acid	Nervous disorders, epileptiform fits, hyper irritability, ataxia, anemia, loss of appetite, edema of peritoneal cavity, colorless serous fluid, rapid postmortem rigormortis, rapid and gasping breathing, flexing of opercles.
Pantothenic	Clubbed gills, prostration, loss of appetite, necrosis and scarring cellular atrophy, gill exudate, sluggishness, poor growth.
Inositol	Poor growth, distended stomach, increased gastric emptying time, skin lesions.
Biotin	Loss of appetite, lesions in colon, coloration muscle atrophy, spastic convulsions, fragmentation of erythrocytes, skin lesions, poor growth.

Folic acid	Poor growth, lethargy, fragility of caudal fin, dark coloration, macrocytic anemia.
Choline	Poor growth, poor food conversion, haemorrhagic kidney and intestine.
Nicotinic acid	Loss of appetite, lesions in colon, jerky or difficult motion, weakness, edema of stomach and colon, muscle spasms while resting, poor growth.
Vitamin B_{12}	Poor appetite, low haemoglobin, fragmentation of erythrocytes, macrocytic anaemia.
Ascorbic acid	Scoliosis, lordosis, impaired collagen formation, altered cartilage, eye lesions, haemorrhagic skin, liver, kidney, intestine, and muscle.
p-Aminobenzoic acid	No abnormal indication in growth, appetite, mortality.

Digestion of Carbohydrates : The term carbohydrates was originally derived from the fact that large bulk of compounds being described fit in the empirical formula Cn $(H_2O)_n$. Though formaldehyde, acetic acid and lactic acid fulfil the formula requirement, they are not carbohydrates. The useful definition of carbohydrate might be polyhydroxyaldehyde and ketones and their derivatives. This would include deoxysugars, amino sugars and even sugar alcohols and acids.

The enzymes which break down the carbohydrates in the gut of fishes are carbohydrases. They are as follows :

1. Amylase
2. Lactase
3. Saccharsases/sucrase
4. Cellulase.

The most important enzyme is amylase which acts on starch (amylum) and which breaks down to maltose and then to glucose by the process of digestion. In human beings, the amylase is secreted from salivary glands and pancreas. The amylase is secreted from the pancreas in carnivorous fishes but in herbivorous fishes, the presence of this enzyme is reported from the whole gastrointestinal tract as well as from pancreas. Researches on the carbohydrases of fishes have largely been confined to the identification of amyloclastic activity. The pancreatic extract of *Raja*, an Elasmobranch, *Scyllium* have clearly shown the amylase activity in the pancreatic juice. Tilapia (*Sarotherodon mossambicus*), which is herbivorous, the amylase is present throughout the alimentary tract. In *Rasbora daniconius*, Saxena (1965); Kothari (1985) reported amylase in the intestinal bulb, duodenum and ileum. In reviewing the literature, it is clear that pancreas (hepatopancreas) is the chief site for the production of amylase, although the intestinal mucosa and intestinal caecae represent additional production site in various species. The enzymatic activity of these caecae is known to be lower as compared to the intestine under normal condition (Luppa, 1966).

$$\text{Starch} \xrightarrow{\text{Amylase}} \text{Maltose} \xrightarrow{\text{Maltase}} \text{Glucose}$$

In those fishes in which sucrose has been reported the effect is as follows:

$$\text{Sucrose} \xrightarrow{\text{Sucrase}} \text{Glucose} + \text{Fructose.}$$

How galactose is further hydrolyzed is not clear in fishes. Blood glucose is converted with the aid of insulin, to muscle glycogen. Although clear details are wanting, but excess of glucose enters the blood from digestive tract, the surplus is converted to glycogen in liver.

Endocommensal bacteria : Lagler *et al*, (1977) stated that in fishes such as menhaden (*Brevoortia*), silverside (*Menidia*) and silverperch (*Bairdiella*) possess endocommensal bacteria containing an enzyme, the cellulase, which breaks down the cellulose plant material. The presence of endocommensal bacteria is not fairly established in Indian fishes. Cellulose of plant material containing starch could be broken down to glucose by cellulase enzyme of these bacteria instead of passing out through faeces.

Fat digestion : The lipids are organic substances insolube in water but soluble in organic solvents like chloroform, ether and benzene. They form important dietary constituents on account of high caloric value and the fat soluble vitamins and the essential fatty acids contained in them. The main enzyme which acts on this lipid is lipase. The pancreas is also the primary site of lipase production. Vonk (1927) found lipase in the pancreas of trout but also found this enzyme in mucosa of fishes. Lipase activity has been reported in a number of Indian fishes (Saxena, 1965).

Gastrointestinal hormone : The mucosa of gastrointestinal tract of man possesses four hormones. They are secretin, cholecystokinin (CCK), gastrin and gastric inhibitory peptide. In each case the hormone is released into the blood stream by the gastrointestinal endocrine cells and as it circulates throughout the body, it is bound by receptors on the plasma membrane of the target cells.

In the teleost, the presence of gastrin and cholecystokinin are reported and are secreted by intestinal endocrine cells which are dispersed and are not grouped in clusters. The CCK affects the oxyntic cells and inhibit further gastric secretion in bony fishes.

Somatostatin is present in the stomach and pancreas of fishes. They are called as paracrine substances. It differs from hormone because it diffuses locally to the target cells instead of releasing into the blood. This inhibits other gastrointestinal and pancreatic islet endocrine cells.

The occurrence of VIP (vesoactive intestinal peptides) and PP (pancreatic peptide) have been reported in *S. aurotus* and *B. conchonius* in the gastric tract. These are classified as candidate hormone. These are gastrointestinal peptides whose definite classification as hormone or paracrine has not been established. These are designated as candidate or putative hormone. Pancreas secrete two important hormone, i.e., insulin and glucagon, insulin is secreted from β-cells while glucagon is secreted by α-cells.

In addition to acetylcholine (not peptide) known to be in nerve fibres of gastrointestinal tract, present gastric VIP (Vasoactive Intestinal Peptides) and somatostatin, met-enkephelin and substance P are reported in the teleost fishes (Romboult *et al.*, 1986).

Absorption : Inorganic ions uptake in various regions of the alimentary canal in fishes and their subsequent distribution and localization have been reported.

The iron (Fe^{++}) ions are absorbed through intestinal columnar cells and then pass into the portal blood as Fe^{++} binding protein transferitin (Khanna, 1979; Chakraborti and Sinha, 1987). The calcium is absorbed by the intestinal submucosal blood vessels. Probably. Ca^{++} after entering blood vessels in the intestinal region reaches finally in the hepatocytes where it is stored in association with vitamin D depending upon Ca^{++} binding protein (Chakraborti and Sinha, 1987).

Regarding dietary calcium and phosphorus absorption, Nakamura and Yamada (1980), Nakamura (1982) and Sinha and Chakraborti (1986) reported calcium and phosphorus in the digestive tract of *Cyprinus carpio* and *Labeo rohita*. In teleosts, ambient water also serves as external source of various dissolved minerals besides food.

REFERENCES

Alexander, R. Mc. N. (1967). Functional design in fishes. London : Hutchinson.

Alexander, R.Mc. N. (1970). Mechanism of the feeding action of various teleost fishes. J. Zool. Lond. 162 : 145-156.

Barrington, E.J.W. (1957). The alimentary canal and digestion. The physiology of fishes. 1 : 109-161. Academic Press, New York.

Chakraborti, P. P. and Sinha, G.M. (1987). Mucosal surface of the alimentary canal in *Mystus vittatus* (Bloch) : A scanning electron microscopic study. Proc. Ind. Acad, Sci. Academy B53 No. 4, 317-322.

Dutta, H.M. and Hossain A.M. (1993). Pattern of caecal distribution in teleost fish in 'Advances in fish research' Vol. 1 (Ed. B.R. Singh), Narendra Publishing House, Delhi, 267-280.

Evans, H.W. (1937). A comparative study of the brains in Pleuronectidae. Proc. R Sec. B. 122 : 308-342.

Gloria, A., Kapoor, B.G., Chardon, M., and Diogo. R. (2003). Fishesics Vol. I, Science Publishers (USA).

Grasse, P. (1957). Traite de Zoologie, T.B. Agnathes et Poissons.

Groot, S.J.D. (1971). On the interrelationships between morphology and the alimentary tract, food and feeding behaviour in flat fishes (Pisces : Pleuronectiformes). Neth. J. Sea Res. 5 : 121-196.

Groot, S. J. De. (1979). Digestive system and sensorial factors in relation to the feeding behaviour of flatfish (Pleuronectiformes) "Reading in Ichthyology" (Eds. Love S.M. and Cailliet M.G.) Prentice Hall of India Pvt. Ltd., New Delhi.

Halver, J.E. (1972). Fish nutrition. Academic Press. New York.

Hossain, A. and Dutta, H.M. (1988). Embryology of intestinal caeca in the bluegill (*Lepomis macrochirus).* Can. J. Zool. 66. 998-1003.

Islam, A.U. (1951). Comparative histology of the alimentary canal of certain fresh-water teleost fishes, Proc. Ind. Acad. Sci. 338. 297-321.

Jara, Z. (1957). On the morphology and function of the so-called palatal organ in the carp (*Cyprinus carpio*). Preegland Zoologicznv 1 : 110-112.

Kapoor, B.G. (1953). The histology of the alimentary canal of *Catla catla* a herbivorous fish. Proc. Ind. Sci. Congr. 40th Session : 209-210.

Kapoor, B.G., Smith, H. and Verighina, I.A. (1975). The alimentary canal and digestion in teleosts. In Advance in Marine biology. (F. S. Russel and M. Youngd Ed.) 13, 109-239. Academic Press, New York.

Kent, G. C. (1983). The comparative anatomy of the vertebrates, 5th edition. Mosby. Co. Publ. St. Louis.

Khanna, S. S. (1961). Alimentary canal in some teleostean fishes, J. Zool. Soc. India. 13 : 206-219.

Khanna, S. S. (1962). A study of buccopharyngeal region in some fishes. Ind. J. Zool. 3: 21-48.

Khanna, S. S. (1979). An introduction to fishes. Central Book Depot. Allahabad.

Khanna, S.S. and Pant M.C. (1964). On the digestive tract and feeding habits of some teleostean fishes. Agra Univ. Journ. Res. (Sci.) 13 : 15-30.

Kothari, S. (1985). Histochemical studies in the rectum during different stages of *Heteropneustes fossilis* (Bloch). J. Hydrobiol. 1 : 5-10.

Kumar, S and Singh, M.P. (1990). Histological and histochemical studies of the Auerbach plexus of colon of rabbit, *Oryctolagus cunniculus* Ind. J.Z. Spect. 1 : 31-35.

Lagler, K. F., Bardach, J. E., Miller, R.R. and Passino, D.R.M. (1977). 'Ichthyology' 2nd edition, John Wiley and Sons, New York.

Luppa, H. (1966). Eir Beitrag Zur Funktion der Apendices pylorica der Fische. Morphologische. histochemische and elektron enoptische undersuchungen. Morph. Jb. 109 : 315-339.

Morton, J. (1967). Guts : The forms and function of the digestive system. In : The institute of Biology's studies. No. 7 Arnold Publ. New York. pp. 1-65.

Moyle, P.B. and J.J. Cech Jr. (1982). Fishes—an introduction to Ichthyology. Prentice-Hall, Inc. New Jersey.

Nakamura, Y. (1982). Effect of dietary phosphorus and calcium contents on the absorption of phosphorus in the digstive tract of carp. Bull. Jap. Soc. Sci. Fish. 48; 409-413.

Nakamura, Y. and Yamada, J. (1980). Effect of dietary calcium levels, Ca/P ratios, and calcium components on the calcium absorption rate in carp *(Cyprinus carpio)*. Bull. Fac. Fish Hakkaido Univ. 31 : 277-282.

Nicol, J.A.C. (1952). The autonomic nervous system in lower chordates. Biol. Rev. Cambr. Philos. Soc., 27(1) : 1-49.

Osse, J.W.M. (1969). Functional morphology of the head of the perch (*Perca fluviatilis* L.) : An electromyographic study. Neth. J. Zool. 19 : 289-392.

Pasha, K. (1964). The anatomy and histology of the alimentary canal of the omnivorous fish, *Mystus gulio*. Proc. Indian Acad. Sci. 59 : 211-223.

Ray, A.K. and Moitre, S.K. (1982). On the morphohistology of the alimentary tract in the Indian climbing perch, *Anabas testudinus* (Bloch), in relation to food and feeding habits, Gegenbaurs morph. Jahrb Leipzig, 128 : 778-798.

Reifel, C.W. and Traivill , A.A. (1978). Gross morphology of the alimentary canal in ten teleostean species. Anat. Anz. 144 : 441-449.

Romer, A.S. (1970). The vertebrate body, 2nd edition. Saunders Publ., Philadelphia.

Romboult, J.S.W.M., Vander Grinten, C.P.M. (1986). Immunocytochemical identification and localization of peptide hormones in the gastro entropancreatic (GEP) endocrine system of the mouse and stomachless fish *Barbus conchonius* Histochemistry 84 : 471-483.

Saxena, A.B. (1965), The morphology and enzymology of the digestive organ in *Rasbora daniconius Proc.* of Zool. Sem. 111-118.

Sehgal, P. (1960). The anatomy and histology of the alimentary canal of *Mystus seenghala* (sykes) with a note on its feeding habits. Res. Bull. Punjab Univ. Sci. 11 : 77-86.

Sinha G.M. and Chakraborti, P. (1986). Şcanning electron microscopic studies on the mucosa of the digestive tract in *Mystus aor* (Ham.) Proc. Indian nat. Sci. Acad. 52 : 267-273.

Srivastava, C.B.L (2005). A Textbook of Fishery and Indian Fisheries. Oscar publishing, Delhi.

Suyehiro, Y. (1942). A study on the digestive system and feeding habits of fish Jap. J. Zool. 10 : 1-303.

Stroband, H.W.J. (1980). Structure and function of the digestive tract of the grass carp. Ph.D. thesis Van de Landbouwahogeschool te. Wageningen, the Netherland.

Tembhre, M. and Kumar, S. (1984). Neurohistology of alimentary canal of *Clarias batrachus* Ad. Bios. 3 (II) 83-90.

Tembhre, M. and Kumar, S. (1994). Effect of acute and chronic exposures to sublethal dose of dimethoate in the gut of *Cyprinus carpio*. J. Ecotoxicol. Environ. Monit. 4 (3) : 205-210.

Tembhre, M. and Kumar, S. (1995). Acetylcholinesterase activity and enzyme kinetics in the gut of *Cyprinus carpio* subjected to acute and chronic exposures to copper. J. of Ecobiol. 7 : 121-124.

Vonk, H.J. (1927). Die Verdauung bei fischen. Zeitschr. Vergl. Physiol. 5 : 445-546.

Chapter 5

RESPIRATORY SYSTEM

Living beings need energy to produce heat, light, electricity, etc. to permit movement, transport, transmission etc. The energy is also needed to grow and to reproduce in addition to maintain the body and its internal environment. The energy is also required to resist disease and parasitism.

The energy in the living system is achieved by the process of respiration. Respiration is an important heat and energy yielding process. For physiologists it includes the metabolic processes by which an organism uses oxygen to produce energy and this process (respiration) is mediated by multi enzyme system. The process of respiration is very much similar in all forms of life, plants, animal and man. It is important to know that in a given system, energy can neither be created nor destroyed (Law of thermodynamics—1) but it can be transferred from one system to other (i.e., from one cell to another or even from one molecule to other). There are various forms of energy (thermal, chemical, mechanical, electrical, radiant light energy) but in living organisms only some inter conversions are possible.

During respiration the oxidation of carbon and hydrogen in the food takes place in tissues to release energy for maintaining life processes. The final merger proceeds through a complicated series of steps which produce among others, a molecule called ATP (adenosinetriphosphate). In its turn ATP is required for synthesis and breakdown of food material necessary for the survival of the organism.

Table 1 : Basic plan for the production of heat and energy

Food oxidised $CO_2 + H_2O$

1. Carbon + oxygen (food)	⟶	Heat and energy + CO_2 (through water)		
2. Hydrogen + oxygen (food)	⟶	Heat and energy + H_2O (through water)		
Energy released	⟶	ATP Break down ADP	⟶	Mechanical energy, e.g., muscle contraction.
				• Osmotic energy, e.g., secretion, ions transfer.
				• Chemical energy, e.g., biosynthesis of small and large macromolecules.
				• Radiant energy, e.g., bioluminiscence.

The carefully programmed oxygen, hydrogen reaction system within the cells take place through the interventions of cytochrome, a type of conjugated protein located in arrays on mitochondria, the specialised organelles responsible for cell respiration.

For respiration the fishes like other vertebrates have ability to obtain oxygen from the external environment by means of vascularized gills, lungs or skin and then transport to tissue through haemoglobin, a conjugated protein present in RBC in the blood.

In tissues the CO_2 which is formed due to oxidation is picked up as the blood passes through the tissue capillaries, most of the CO_2 passes through plasma to red cells. In the RBC the CO_2 is rapidly converted to carbonic acid by an enzyme in RBC called *carbonic anhydrase*.

$$CO_2 + H_2O \xrightarrow[\text{anhydrase}]{\text{carbonic}} H_2CO_3 \text{ (carbonic acid)}$$

The whole process is reversed as the blood goes to lung/gill where it is converted to $CO_2 + H_2O$

$$H_2CO_3 \xrightarrow[\text{anhydrase}]{\text{carbonic}} CO_2 + H_2O.$$

The CO_2 is eliminated out.

During embryonic development, the respiration is carried out by highly vascularized foetal membranes, i.e., yolk sac while after hatching vascularized pectoral fins and operculum help in the respiration to some extent. Fish may be adapted to the life by modification of the gills, the mouth cavity, the intestine and the gas exchange. There are certain new structures (neomorphic) or air breathing organs. Fishes take oxygen dissolved in water. In the water there is only one thirtieth of volume of oxygen contained in same volume of air (Hughes, 1979), the low oxygen availability had undoubtedly contributed to the evolutionary development of gills, which are characterized by large surface area and extremely efficient gas exchange (Moyle and Cech. Jr. 1982).

STRUCTURE OF GILLS

Gill Slits : There are six or seven pairs of gills in cartilaginous fishes while four pairs in bony fishes due to the loss of spiracle (Fig. 5.1 a & b).

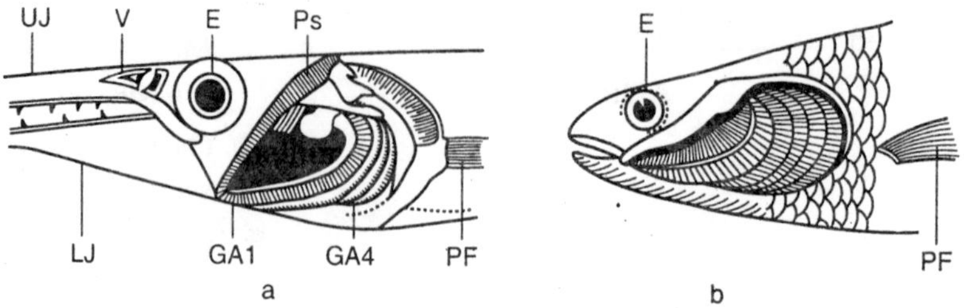

Fig. 5.1(a, b) : Position of pseudobranch and gill arch in fishes. (a) *Xenentodon*. (b) Structure of gill of teleost. E, eye; G.A1-GA4 gills; L.J. lower jaw. PF, pectoral fin; PS, pseudobranch; UJ, upper jaw; V. Value.

Gill slits of bony fishes are covered by operculum while operculum is absent in cartilaginous fishes. In sharks gill slits are laterally situated while in rays they are ventrally placed. A pair of spiracle is present in Elasmobranchii anterior to first gill which corresponds to a vestigeal primitive first gill slit. Although spiracle is absent in bony fishes, in Actinopterygii it is replaced by a pseudobranch which is free in some fishes but skin covered in others.

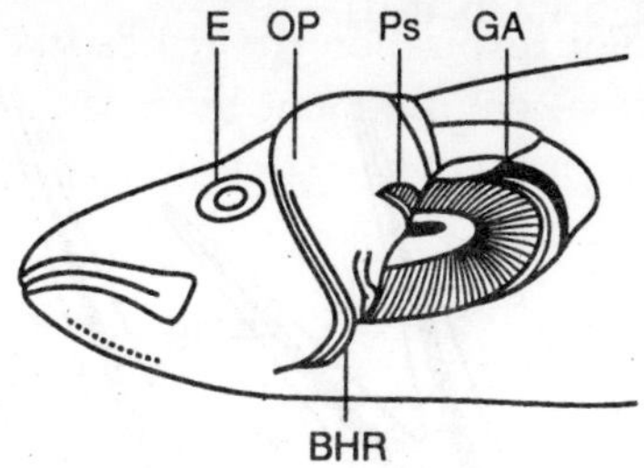

Fig. 5.2 : Location of pseudobranch in European walleye *(Stizostedion lucioperca)*. BHR, branchiostegal ray; E, eye; GA, gill arch; OP, operculum; Ps, pseudobranch.

Pseudobranch : In carp and rainbow trout the pseudobranch is embedded in submucosal connective tissue of pharyngeal wall and shows a glandular appearance due to complete conglutination of branchial filaments. (Fig. 5.1a & Fig. 5.2).

In some species, a pseudobranch with hemibranchs structure is located inside the operculum (Oguri, 1982). However, in eel the pseudobranch is not present (Whittenberg and Haedrich, 1974), it is also absent in cat fishes (Siluroidae) and featherback (Notopteridae).

In glandular pseudobranch, abundant distribution of blood capillaries is found in the parenchyma enclosed by connective tissue. It contains acidophilic cells in mitochondria and endoplasmic reticulum and is rich in enzyme carbonic anhydrase. According to Whittenberg and Haedrich (1974), the pseudobranch regulates the flow of the arterial blood to the opthalmic artery to increase the amount of blood carbon dioxide. Parry and Holliday (1960) found that in rainbow trout extirpation of pseudobranch induced melanophore expansion and body colour change, suggesting the secretion of a melanophore-aggregating hormone from tissue. It also helps in metabolic gas exchange of retina and filling of gas bladder.

Because of its direct vascular connection with the choroid gland on the eyeball, the pseudobranch has been implicated in the regulation of intracellular pressure (Lagler *et al.* 1977).

The structure of gills has been studied extensively in Indian fishes by light transmission and scanning electron microscopy. The gill comprises of gill rakers, gill arch, gill filaments (Primary gill lamellae and lamellae) (Fig. 5.3a & b).

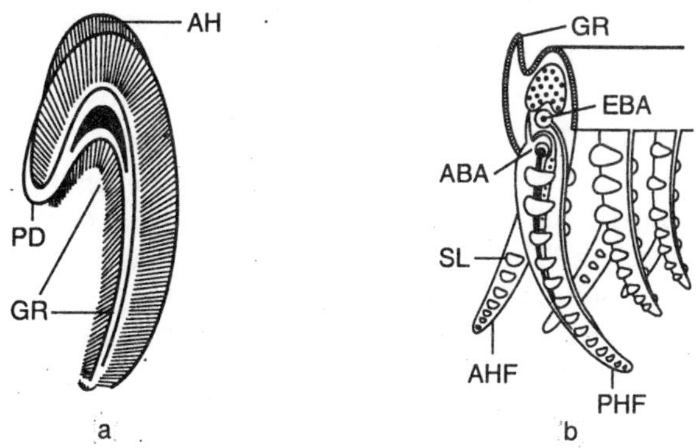

Fig. 5.3a, b : Structure of gill. ABA, afferent branchial artery, AH, hemibranch; AHF, aboral hemibranch; EBA, efferent branchial artery, GR, gill raker, PD, postero-dorsal part; PHF, posterior hemibranch filament; SL, secondary lamella.

A complete gill is known as holobranch. It consists of a bony or cartilaginous arches. The anterior and posterior part of each gill arch possesses plate-like gill filaments. Each holobranch consists of an anterior (oral) and a posterior (aboral) hemibranch (Fig. 5.4 a, b, c, d).

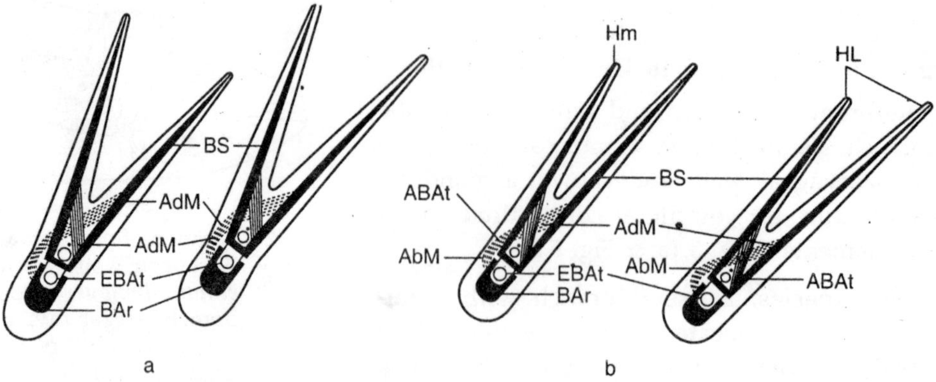

Fig. 5.4a, b : Section of adjacent holobranch and their muscles in bony fish (a) during abduction (b) during adduction. ABAt, afferent branchial artery, AbM, abductor muscles, AdM; adductor muscles; BAr, bony gill arch; BS, bony support; EBAt, efferent branchial artery, HL holobranch; Hm, hemibranch.

The architectural plan of teleostean gills shows heterogeneity in their functional unit which is due to varied osmoregulatory, feeding and respiratory behaviour and to the physicochemical status of their environment.

In teleost fishes, five pairs of branchial arches are present of which first four bear gill lamellae (Fig. 5.4a, b) but the fifth is devoid of gill lamellae and transformed into the pharyngeal bone for mastication of food. It does not play any role in respiration.

The gill arch is an important unit and bears primary (gill filament) and secondary lamellae. The branchial arch typically consists of paired pharyngobranchials, epibranchials, ceratobranchials, hypobranchials and a median unpaired basibranchial. The epibranchial and the ceratobranchial elements of each branchial arch bears two rows of gill filaments of the two hemibranchs of the holobranch, which are the seat of gaseous exchange. It encloses afferent and efferent branchial arteries and veins (Fig. 5.4a and Fig. 5.6a, b, c).

It is also provided by nerves. The branches of 9th (glossopharyngeal) cranial nerve innervate the first gill, while II, III, & IV arches are supplied by the branches of vagus (10th cranial nerve). It also contains abductor and adductor muscles. Inside it contains gill rakers, taste buds, mucous gland cells and sensory papillae.

Gill raker : It occurs in two rows on the inner margin of each gill arch. Each gill arch is short stumpy structure supported by bony elements (Fig. 5.3a & b). The gill arch projects across the pharyngeal opening. They are modified in relation to food and feeding habits. The mucous cells of the epithelium help to remove sediments from the covering epithelium in order to enable the taste buds to function effectively and to sense the chemical nature of food passing through the gill sieve.

Gill Filaments (Primary gill lamellae) : Each hemibranch consists of both primary and secondary lamellae (Fig. 5.5).

The primary gill filaments remain separated from the branchial septum at their distal end making two hemibranch in opposition which direct the water flow between the gill filaments. Amongst dual breathers the heterogeneity in the gill system is more pronounced particularly in the swamp eel, *Monopterus, Amphipnous cuchia* and climbing perch, *Anabas testudineus*. In *Monopterus*, gill filaments are stumpy and are present only in second pair of gill and lack gill lamellae. According to Munshi and Singh (1968) and Munshi *et al.*, (1990), the remaining three pairs are without functional lamellae. It is the modification for another way of exchange of gases.

The gill filaments are blade-like structures supported by gill rays. The heads of the gill rays of both the hemibranchs are connected by ligaments (Yadav *et al.*, 1993). They are provided with two types of adductor muscle units in teleosts.

The gill filaments are also lined by epithelium referred to as primary epithelium. The epithelium has glandular and non-glandular part.

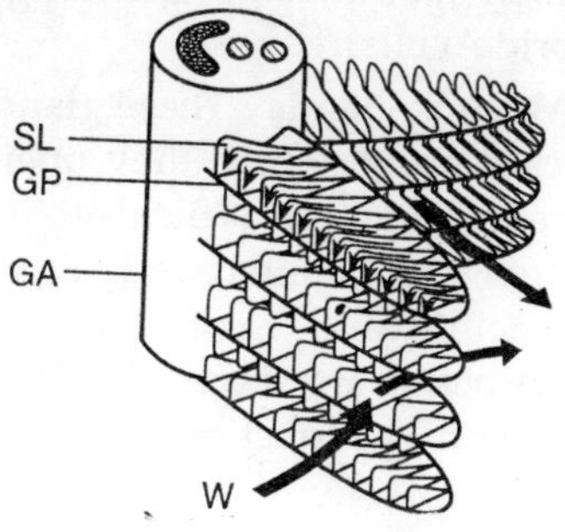

Fig. 5.5 : Diagram of gill structure of fish. B, blood flow; GA, gill arch; GP, gill filament; SL, secondary lamella; W, water.

Lamellae (secondary lamella) : The each gill filament is made up of secondary gill lamellae which are actual seat of exchange of gases. They are generally semicircular and lined up along both sides of the gill filaments. The lamellae frequency is directly proportional to the dimension and resistance of the gill sieve (Hughes, 1966 & 1979).

The secondary lamellae are having two sheets of epithelium which are separated by space and through these spaces blood circulates. The epithelial sheets are separated by a series of pillar cells. Each cell consists of central body and is provided with extensions at each end (Fig. 5.6d).

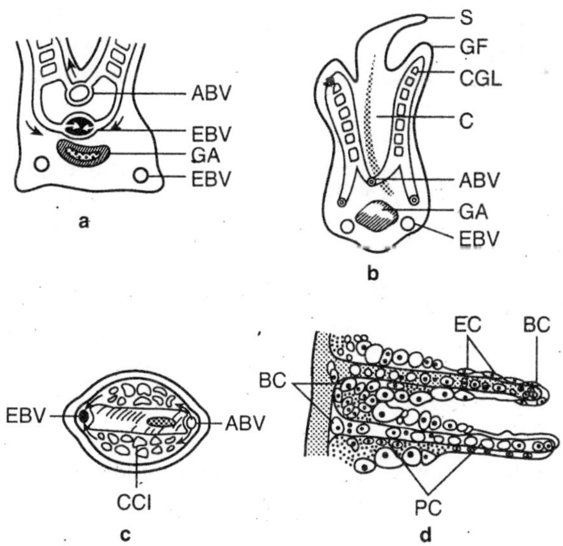

Fig. 5.6a-d : Horizontal section of gills of (a) shark, (b) bony fish, (c) cross-section, (d) epithelium of gill. ABV, afferent blood vessel; AC, acidophil cells; BC, blood cells; C, cartilaginous gill support; CLL capillaries and lacunae in lamella; CGL, capillary in gill lamellae; EC, epithelial cells; EBV, efferent branchial vessel; GA, gill arch; GF, gill filament; PC, pilaster cells; S, septum.

Branchial glands : These are specialized cells of the epithelium. They are glandular in nature and perform different functions in normal and experimental conditions. The most common specialized branchial glands are the mucous glands and acidophilic granular cells (chloride cells).

Mucous glands : These gland cells are unicellular. They may be oval or pear shaped with a neck through which they open outside the epithelium. The nucleus lies at the bottom of the cells. They are typical goblet cells. They are present throughout the epithelium, i.e., gill arch, gill filament and secondary lamellae. They secrete mucus which is glycoprotein, both acidic and neutral. According to Ojha and Mishra (1987), acidic and neutral glycoproteins are secreted by same cells suggesting that they undergo transformation from each other. The function of mucus cells are as follows :

1. Protective
2. Friction reducing
3. Antipathogen
4. Help in ionic exchange
5. Help in gas and water exchange.

Chloride cells : These cells contain granules, which take acidic stain, hence called acidophlic. They are provided with a large number of mitochondria. These cells are also known as ioncytes because they help in the ionic regulation in the euryhaline and stenohaline fishes.

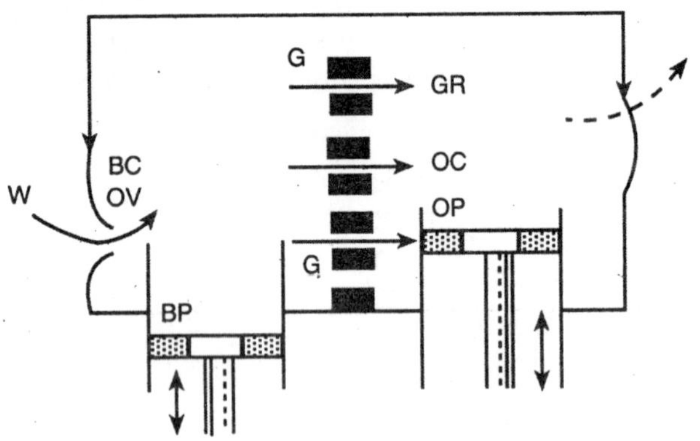

Fig. 5.7 : Diagram to show mechanism of gill ventilation by two pumps just after inspiration. BC, buccal cavity, BP, buccal pump; G, gills; GR; gill resistance; OC, opercular cavity; OP, Opening OV, oral valve; W, water.

Taste buds : They are flask shaped multicellular structure and distributed in the epithelium of the gill arch region. There are two distinct type of cells, sensory and supporting, and can be distiguished under phase contrast microscope. They help the fish to sense the nature of food particles contained in the water flowing through the gill during ventilation. They are distinguished into MGO, MREC and taste buds. The distribution and density of MGO and taste buds on the gill arch epithelium of fishes inhabiting different water bodies differ considerably. (See Advance of Fish Research, Singh, 1993).

Respiratory Mechanism : The continuous undulational flow of water over gill surface is accomplished by respiratory pump. It is now unanimously accepted that the respiratory pumps of a teleost consist of buccal cavity and two opercular cavities, caused by movements of bone of arches and operculi, resulting in the pumping action of the system (Hughes, 1960; Hughes and Shelton, 1958, 1962, Woskoboinikoff, 1932; Ballington and Hughes, 1965; Munshi, 1993).

In the beginning, the water enters into the mouth by expansion of buccal cavity. The water is then accelerated over the gill by the simultaneous contraction of buccal cavity and cavity contracts, expelling the water out through the opercular opening, the cycle begins again (Fig. 5.7).

The respiratory cycle is a complex mechanism and involves a large number of muscles, bones, ligaments and articulations. Several authors from time to time have described this system and made an attempt to understand its working (Hughes and Shelton, 1962; Kirchhoff, 1958; Willem, 1947).

Ram Ventilation : It is carried by strong abduction of the two hemibranchs of a holobranch towards each other. Interruption of this cycle causes reversal of flow or cough which fish uses to clear foreign matter or excess mucus from the gills. Drummond *et al.* (1973) stated that frequency of cough in *Salvelinus fontinalis* can be a sublethal indicator of excessive copper concentration in freshwater. There is an active and passive ventilation during the swimming movement. The transition to ram gill ventilation in fish is a graded process as swimming picks up from rest. The first indication that a critical swimming speed has been reached is singalled by the drop-out of a single cycle. The drop-out continues until only occasionally ventilatory movements and 'cough' are noticed. Return of active movements with gradual reductions in swimming velocity to below critical shows nearly the same sequence but in reverse order (Roberts, 1979).

The additional respiratory mechanism of sharks and rays can be divided into three phases, which are as follows :

(i) When coracohyoid and coraco-branchial muscles contracts to enlarge the angle enclosed by the gill arches and to increase the oropharyngeal cavity for entrance of water through mouth or spiracle during which the gill slits are kept closed.

(ii) When the contraction between the upper and lower parts of each gill occurs with the relaxation of abductors of lower jaw and gill arches, which causes mouth to serve as pressure pump. During this phase the forward flow of water through mouth is prevented by the oral valve and the water is directed backward towards the internal gill clefts. The interseptal spaces are enlarged by the contraction of the interopercular adductors to reduce the hydrostatic pressure at the inner surface of the gill and water is drawn into the gill cavities, which remain closed at the outside.

(iii) The third phase includes relaxation of interopercular muscle and contraction of their sets of muscle which cause the internal gills to narrow, and the water is forced through the gill lamellae. This follows the opening of gill clefts, and the water is forced to the outside. The mackerel sharks (Limmidae) take the sufficient respiratory water during swimming and do not show pronounced breathing movements.

Mechanism of entry of water : In all bony fishes the pressure and flow of water in the oral cavity is regulated by the muscles, which move the bases of holobranchs. "Coughing" is the process of a voilent sweeping of water over the gill lamellae to make them free from accumulated detritus. It is carried by strong abductions of the two hemibranchs of a holobranch towards each other. The diffussion of oxygen from the respiratory water is added not only by the gill filaments but also by the direction of blood and water flow.

These works a counter-current system in which the oxygenated water flows from the oral to aboral side of the gills, but the blood in lamellae flows in opposite direction, i.e., from aboral lamellar afferent to oral lamellar efferent blood vessels. During this counter-flow of oxygenated water and blood, the oxygen enters the gill and carbon dioxide leaves it. In the tench *(Tinca)* this counter-current system is so modified that the 51% of oxygen from incoming respiratory water diffuses in and 9% oxygen from outgoing water currents.

In the osteichthyes fishes, in which gills are covered by operculum, the water is propelled over the gills by suction pressure. In the beginning of inspiration, when the operculum is closed forcefully, and mouth is opened while several muscles contract including sternohyoid and elevator of palatine. The branchiostegal rays are spread and lowered simultaneously and the mouth cavity is enlarged creating negative water pressure in it. Thus water is drawn into the mouth and after a short interval the space between the gills and the operculum is enlarged as the gill covers are abducted anteriorly although the opercular skin flaps are still closed posteriorly by outside water pressure. A negative pressure develops in the gill cavity resulting flow of water over gills. This is followed by reduction of buccal and opercular cavities. At this time the oral valves presents the backflow of water out of the mouth and mouth cavity begins to function as pressure pump instead of a suction pump. Then each operculum is immediately brought towards the body, the gill flaps open and water is expelled, being prevented from flowing backwards by high pressure in the buccal cavity as compared to the epibranchial cavity.

Variations in respiration pattern : There may be variations in the basic patterns of respiration due to different life habits.

(a) Fast Swimming : In fast swimming species like mackerels (Scombridae), trouts and salmons (Salmoninae) the mouth and gill flaps remain open to bathe their gills with water currents produced during swimming. Generally, the fast swimmers have smaller gill cavities in comparison to the sedentary fishes.

(b) Bottom dwellers : The bottom dwelling fishes such as flounders (Plueuronectidae) and gossefishes (Lophiidae), possess enlarged and dilatable opercular cavities, their mouth does not open widely during inspiration causing slow and deep breathing movements. The other bottom dweller species like morays (Muraenidae) keeps its mouth open during the respiration. Some fishes such as the bonefish *(Albula vulpes)* and goalt fishes (Mullidae), when find their prey covered by sand grains, they release jet of water from the mouth with the help of strong adduction of gill covers which uncover their food or prey.

The trunkfishes (Ostraciidae) and the puffers (Tetradontidae) possess rather a compact skeleton which reduces respiratory function of the gill cover. However, these fishes have a compensatory high rate of exchange of respiratory water carried out by fast breathing (up to 180 breathing movement/min)

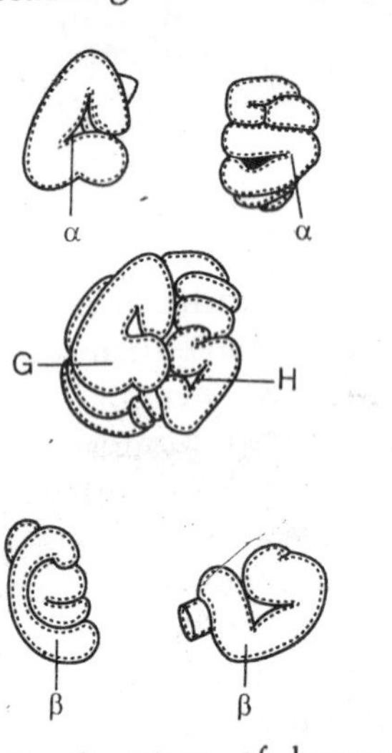

Fig. 5.8 : Structure of haemoglobin. α, alpha chain; β, beta chain; G, globin; H, heam.

(c) Hill stream : In some hill stream catfishes who attach their body temporarily to the substratum, the water intake is accomplished by developing grooves protected by barbles with slight reduction in the force of suction. Ventilation is carried out only by the opercular movement. In the andes *(Arges)*, while attaching to the substratum with their suctorial mouth, catfishes can withdraw and eliminate respiratory water via an inhalent slit in the horizontally divided gill cover. The larvae of South American lungfish *(Lepidosiren)* and the reed fishes possess "external gills" coming out through external gill slits or opercular region.

Fish blood as gas carrier : Like other vertebrates, the fishes also possess R.B.C. containing a respiratory pigment haemoglobin which has efficient oxygen carrying capacity. The haemoglobin has 15 to 25 times more capacity than water to bind with the oxygen. The 99% of total oxygen is taken by haemoglobin while only 1% by the plasma. Haemoglobin is a conjugated protein, the chromoprotein. It consists of a large protein molecule, the globin consisting of four polypeptide chains (two alpha and two beta chains), (Fig. 5.8) to each of which is attached a prosthetic haem group. Haem is based on a structure known as a porphyrin ring which include four pyrrole groups around a central ferrous iron (Fe++). The iron is joined by four of its coordination bond to N atom of the porphyrin and two bonds to imidazole N contained in histidine residues within the protein globin.

The haemoglobin contents may vary according to habit and habitate for instance, the pelagic species have higher Hb than bottom dwellers.

When haemoglobin is oxygenated, i.e., 'loading' and when transported and released, i.e., 'unloading' this process is oxygenation but not the 'oxidation'. The loading is also denoted by TI or T, sat, i.e., loading tension of blood which shows that partial pressure of oxygen at which Hb of a particular species is 75% saturated with oxygen.

The Dissociation curve of oxyhaemoglobin : The relationship between the saturation of haemoglobin and oxygen tension is studied by examination of the dissociation curve of hemoglobin in which the present satuaration is plotted against the oxygen tension. The shape of dissociation curve varies with the tension of CO_2. The curve may be hyperbolic as found in the eel (*Anguilla*). This hyperbolic curve with high O_2 affinity makes the fish enable to live in water with a low oxygen concentrations. The haemoglobin of eel becomes saturated at much lower tension than that of mammals. The amount of oxygen given to the tissues, can be determined by the difference between the **Tsat and T1/2** sat. In contrast to the hyperbolic curve, the sigmoid curve (human) shows efficiency of the blood to release more oxygen to the tissues. This can be examplified by the curve obtained in more active fish, the mackerel (Fig. 5.9).

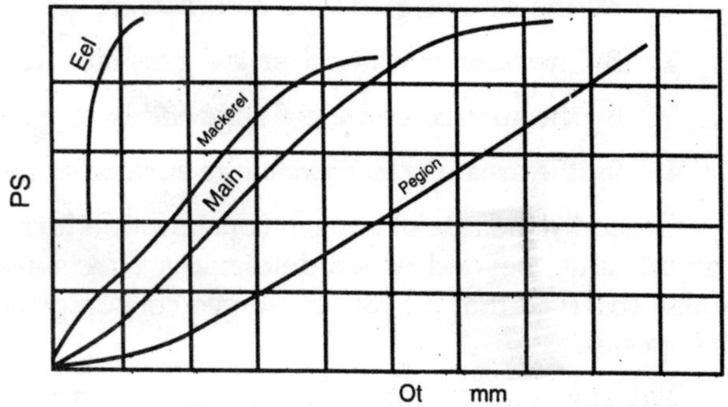

Fig. 5.9 : Graph showing comparative haemoglobin saturation curves as a function of oxygen. PS, percentage saturation; ot, oxygen tension. Source: Prosses & Brown, 1961.

Bohr's Effect : It is the phenomenon during which if the partial pressure of CO_2 (PCO_2) increases, the higher tension is needed to reach TI, and Tu is decreased proportionally. The Bohr's effect is progressive in the fishes as compared to other vertebrates, and facilitates the unloading of oxygen to the tissue cells which contain comparatively high CO_2 tension. Respiratory pigment consisting of polypeptide chain of amino acids whose carboxy and alfa amino groups are protonated accordingly to oxygenation and binding CO_2 to these group. The Bohr's effects differ according to species. The mackerel (*Scomber scombrus*) does not encounter low levels of oxygen and lives under uniform but low CO_2 tensions. They are found in high seas, whose blood is greatly affected by slight change in CO_2. However, some fishes living in stagnant water such as *Cyprinus carpio* and bullhead catfishes (*Ictalurus*) have blood insusceptible to change in CO_2 concentrations. When CO_2 tension increases it results into the formation of H_2CO_3 in blood which readily dissociates into HCO_3^- and H^+. An increase in hydrogen (H^+) causes lowering of pH and affects oxygen carrying capacity of haemoglobin. When pH falls from 9 to 6, it results the change in the shape of the oxygen dissociation curve of tuna haemoglobin, due to loss of cooperativity of the haem groups.

Effect of temperature on oxygen dissociation curve : When carbon dioxide tension is high and even with high partial pressure of oxygen as much as 100 atmosphere the blood does not saturate completely. In this so-called root effect when CO_2 tension remains high with high partial pressure of oxygen, an increase in temperature raises the partial pressure required to saturate the blood. But in some fishes the absolute oxygen carrying capacity of the blood is higher at low temperature. The warming of the water leads to increase respiration because the tissues demand more oxygen at higher temperature than at lower temperature. At the areas of human habitat having thermal and sewage effluents, mortality of fishes may occur due to asphyxiation caused by absolute lower lethal concentration of dissolved oxygen.

Oxygen Consumption : The rate of oxygen consumption is a measure of their metabolism, which can be regulated in several ways.

1. By the rate of oxidative metabolism.
2. By the flow of water over the gills causing diffusion gradient across the gill.
3. By the surface of the gill supplied by the blood.
4. By the area of gases exchange surface or the affinity of haemoglobin.

Generally the rate of oxygen consumption increases with rise in temperature up to some critical value, beyond which deleterious effects are visible and the rate falls off rapidly. It is also observed that the rate of oxygen consumption is usually lower in larger individuals of the species.

The age, activity, nutrition, disease, reproductive state and nervous and hormonal control of the animal also affect the oxygen consumption.

Air Breathing Adaptations in Fish : Fishes are modified in a variety of ways for the purpose of air breathing. Over twenty five genera of diverse families of living fishes show air breathing adaptations. It is generally accepted that there are two conditions, responsible for such modifications. Firstly, oxygen deficiency in the environment; secondly, a higher metabolic activity of fish.

However, some fishes respire air even though there is sufficient dissolved oxygen, for example *Neoceratodus* (Australian, lungfish), *Amia* and *Lepidosteus* (holostean fishes).

Of the three lungfishes, only *Neoceeratodus* is an obligatory water breather, restoring only to air breathing when the water becomes hypoxia (Johansen and Burggren, 1980), whilst *Protopterus* and *Lepidosiren* alternate between gill and lung breathing and rely entirely on lung breathing during aestivation.

Two kinds of modifications are seen in fishes for air breathing (i) those structure which does not involve gas bladder (ii) those in which gas bladder is used for respiration as a lung.

Structure different from lung : Several fresh and marine water fishes have some additional structures, which help them in respiration out of water or at the time of oxygen scarcity. Such adaptations are as follows, which do not require gas bladder as lung.

1. The gills are covered with thick film of mucus which enhances the diffusion of gases as in spiny eels (*Mastacembelus*).

2. Inner surface of gill chambers and operculum is folded and highly vascularised as in mudskippers (*Periopthalamus*).

3. A well developed mouth diverticula and pharyngeal cavities are *Channa, Clarias, Saccobranchus* and *Anabas.* The diverticula is lined by folded respiratory membrane which receives high blood supply from the afferent blood vessels. (Fig. 5.10a,b,c,d).

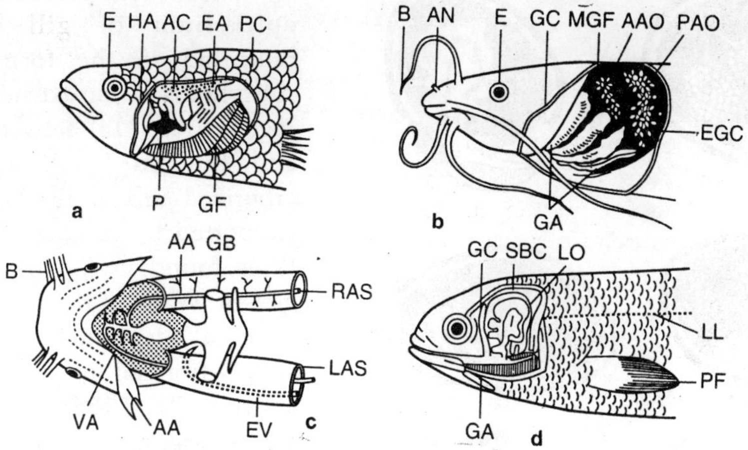

Fig. 5.10a-d : Air-breathing adaptations in fishes. (a) *Channa* (b) *Clarias* (c) *Saccobranchus* (d) *Anabas.* AA, afferent artery to left air sac; AAO, arborescent anterior organ; AN, anterior naris; B, barbles; E, eye; EA, epibranchial apophysis; EGC, extension of gill cavity; EV, efferent vessel from right air sac; GA, gill arch; GB, gas bladder in bony capsule; GC, gill cover; GF, gill filament; HA, hyomendibular apophysis; LAS; left air sac; LL, lateral line; LO, labyrinthiform organ; MGF; modified gill filament; P, connection with pharynx; PAO, posterior arborscent organ; PC, posterior chamber; PF, pectoral fin; RAS, right air sac; SBC, suprabranchial cavity, VA, ventral aorta.

4. The stomach becomes thin walled to serve for the respiration in the armoured catfishes (Loricariidae).

5. However, in certain cobitid loaches mid and hind parts of the intestine serve both for digestion and respiration, while in others both the processes alternate with each other frequently. But in some fishes (loaches) the digestive tube helps in respiration only in summer whereas it becomes non-respiratory during winters.

6. Neomorphic air breathing organs are developed to utilize atmospheric oxygen (Fig. 5.11a, b, Fig. 5.12a to d).

Indian fishes which are provided with neomorphic breathing organs are *Anabas testudineus, Colisa fasciatus, Osphronemus nobilis, Heteropneustes fossilis, Clarias batrachus, Channa punctatus, Channa striatus, Channa marulius, Channa gachua, Monop-*

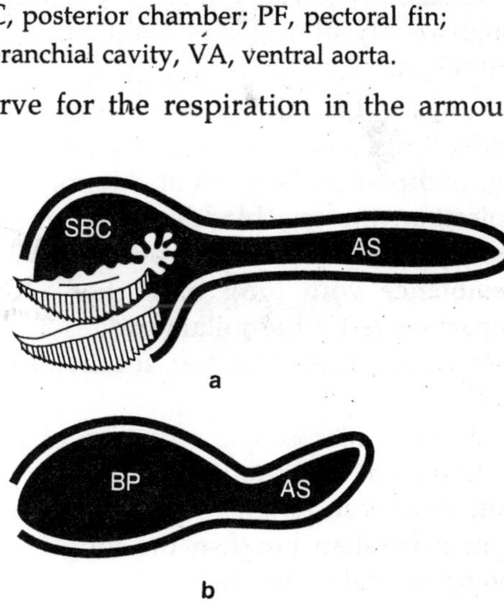

Fig. 5.11a-b : Diagram to show neomorphic air breathing organ of monopterus (*Amphipnous cuchia*). AS, air sac; BP, buccopharynx; SBC, suprabranchial chamber.

terus cuchia and Monopterus albus. Pseudapocryptes lanceolatus, Periophthalmodon schloisteri and Boleophthalmus boddari are estuarine genera (Singh, 1993).

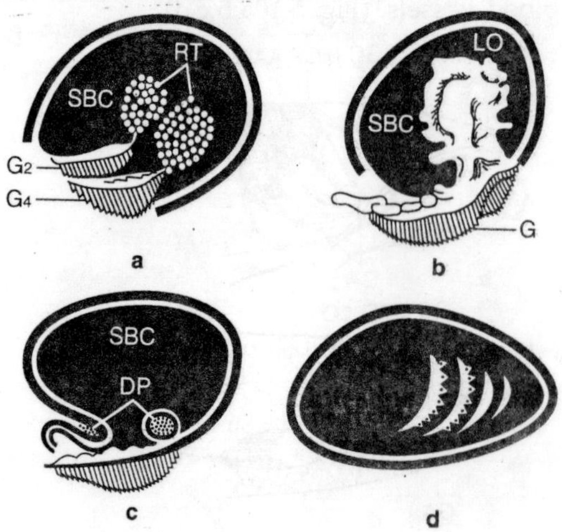

Fig. 5.12a-d : Diagram to show neomorphic air breathing organ of Clarias batrachus. BP, buccopharynx; DP, dendritic plates; G, gill; GP, gill plates; LO, labyrinthine organ; RT, respiratory tree; SBC, supra-branchial chamber.

7. In Saccobranchidae, the epithelium of gill chamber is modified in the form of a long subcylindrical sac extending to caudal region, (Fig. 11a) between the second and the third gill arches. Small arborized and shrub-like enlargement supported by cartilage of gills II and IV is found in *Clarias* and related genera. Walls of these structures help in aerial respiration receiving blood supply from all four efferent branchial arteries (Fig. 5.12d).

In *Anabas* a highly branched structure called labyrinth is developed from first gill arch and lies in moist pocket of dorsolateral head region prior to the operculum and is covered with respiratory epithelium (Fig. 5.12b).

The development of such air breathing structures makes the fish capable to live out of water for more than 24 hours and enable them to restore to normal life even after the skin has become dry *(Channa)* (Fig. 5.12c). Some species depend more on the atmospheric air even though plenty of water is available. It happens probably because the aquatic gills are insufficient.

Lungs : Like the lungs of higher vertebrates, the lungs of fishes (dipnoi) have gas bladder lungs having no clear homology with true lungs. It is, however, considered that the lungs of dipnoi and the gas bladder of higher fishes have common stem of evolution. Besides respiration, the gas bladder also functions for hydrostatic equilibrium, gravity maintenance, sound production and reception. The gas bladder of dipnoi shows structural resemblance with lung of higher vertebrates. Air chambers are divided into small compartments by the pillars, ridges and internal septa communicating to median cavity. These compartment further subdivides into smaller alveoli-like sac yield by the blood vessels.

Electron microscopic structure of lungs resembles the lung of higher vertebrates. It develops from embryonic fore gut as a median ventral pouch. In teleost and in *Lepidosiren* (South American lungfish) paired and bilobed lungs originate from the ventral of the gut, but in Australian lungfish the lung is single and dorsally placed with a pnumatic duct opening ventrally into the gut.

Lungs receive blood from last efferent branchial artery and after aeration the pulmonary vein returns the blood to left atrium of heart.

Conus arteriosus is subdivided by a spiral valve keeping oxygenated and unoxygenated blood partially separated while distributing to the branchial vessels.

The African lungfishes *(Protopterus)* aestivate generally in a cocoon of brown parchment likehardened slime. At the time of aestivation the body is kept coiled with mouth upwards communicated to the air passage from the surface. They may spents 3 to 4 months to one year inside the cocoon by reducing their metabolism and oxygen consumption. With onset of favourable conditions of rains the animal emerges out of the cocoons and respire with gills. The South American lungfish aestivate in the same way as that of African lungfish. But Australian lungfish neither burrows nor forms a cocoon.

GAS BLADDER

Gas bladder is one of the characteristic feature of the true fishes. It is quite often regarded as swim bladder or air bladder and found to be highly developed in Acanthopterygii (spiny rayed teleosts).It is an accessory respiratory organ, which also helps in sound production and sound perception, fat storage (e.g., in the gonostomatid species). It is an important hydrostatic organ, which contains a gas secreting complex, is composed of a gas gland covered with blood vessels.

The gas bladder as a respiratory device : Respiration is supplemented by the gas bladder in many physostomous fishes with an open duct. The gas bladder has undergone several modifications in various species of bony fishes (Fig. 5.13a to f).

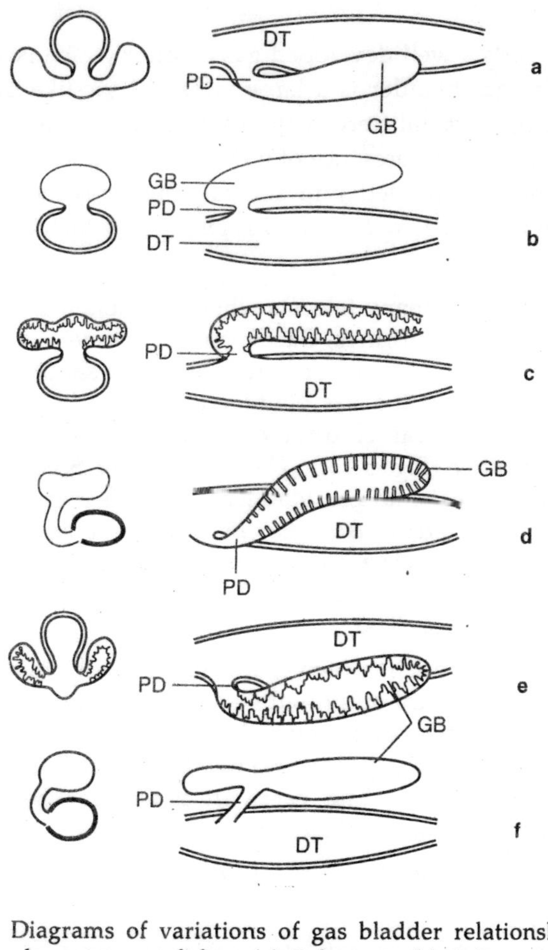

Fig. 5.13a-f : Diagrams of variations of gas bladder relationships to gut in physostomous fishes. (a) *Polypterus* (b) *Acipencer* (c) *Lepidosteus* (d) *Neoceratodus* (e) *Protopterus* (f) Climbing perch. DT. digestive tract; GB, gas bladder; PD, pneumatic duct.

In chondrostei fishes such as *Polypterus,* the gas bladder is in the form of an unequal bilobed structure with a small left lobe and large right lobe communicating with ventral part of the pharynx (Fig. 5.13a). Both the lobes join together to a small opening called 'glottis' provided with a muscular sphincter. However, the *Acipenser* comprises oval shaped bladder of wide opening into the oesophagus (Fig. 5.13b).

The holostean fishes like *Lepidosteus* has an unpaired sac which opens into the oesophagus by a glottis. (Fig. 5.13c). The wall of bladder is composed of the fibrous bands produced in the alveoli arranged into two rows. Each alveoli is further subdivided into smaller sacculi. In *Amia,* the gas bladder is very large and its wall is highly sacculated. These fishes can survive in water depleted with oxygen, if they are capable to swallow air, which then passes into the gas bladder through a pneumatic duct. In *Amia* the gas bladder is relatively important as it lives in the temperate regions of North America. This fish frequently ascends for air when the temperature of well aerated water increases to 25 °C. Since the gas bladder of physostomous fishes contains more carbon dioxide than the atmospheric air, it has been considered that removal of this waste gas is performed there also.

The dipnoi fishes possess a well developed gas bladder which is structurally similar to the amphibian lungs. The gas bladder is a large unpaired sac-like in *Neoceratodus* which contains one dorsal and one ventral fibrous ridges projecting in to this cavity (Fig. 5.13b). Many alveoli are formed due to the presence of transverse septa in between these ridges. The alveoli in turn are further subdivided into several smaller sacculi. The complexities in the gas bladder increase in the *Protopterus* and *Lepidosiren* which have lung-like bladder. (Fig. 5.13e).

The gas bladder is present in many teleosts while in others it is completely absent such as in Echeneiformes, Symbranchiformes, Saccopharyngiformes and Gobeisøciformes. If present, the gas bladder may be oval, fusiform, tubular, heart shaped, horse-shoe shaped or dumb-bell shaped. In Cyprinidae the gas bladder lies freely in the abdominal cavity or may be attached to the vertebral column by fibrous tissue. It has two chambers, interconnected with each other (Fig. 5.13f). The members of Sparidae, Notopteridae and Scombridae possess paired caeca-like gas bladder extended into the tail. In some fishes, like *Clarias batrachus* and *Heteropneustes fossilis,* the gas bladder is reduced and lies enclosed in bone. The fishes living in torrential waters of the hills have contain rudimentary gas bladder having only small anterior lobe enclosed in bone and no posterior lobe *(Psillorhynchus* and *Nemacheilus).*

In most of the sound producing fishes, the gas bladder is provided with caecal outgrowths. In *Gadus* a pair of caecal outgrowths arises from the gas bladder and projects in the head region while in *Otolithus* each antero-lateral side of the gas baldder gives off a caecal outgrowth which immediately divide into two branches. One branch runs anteriorly while another goes to the posterior. The caeca are much branched in *Corviva lobata* and arises from whole periphery of the gas bladder.

The gas bladder is rarely divided completely by the septum. Most often it is partially subdivided by incomplete septum.

All teleosts in the beginning generally have an open duct of gas bladder, i.e., they are physostomous but in later stages it closes in many teleosts and they become physoclistic (Fig. 5.14).

Blood supply of gas bladder : The gas bladder is supplied with blood from the posterior branches of dorsal aorta or from coeliaco-mesenteric artery. In some fishes the venous blood is collected by a vessels of hepatic portal system while in others the gas bladder vein collects the venous blood and discharges it into posterior cardinal vein. The vascularization of gas bladder differs from species to species. In physostomous carps, the inner surface of the bladder is covered at frequent places by blood vessels arranged in a fan-like manner. These vessels form red patches of various shapes and sizes, are known as 'red bodies' which is a countercurrent arrangement of small arterioles and venules that constitute a 'rete mirabile' (Fig. 5.15a, b).

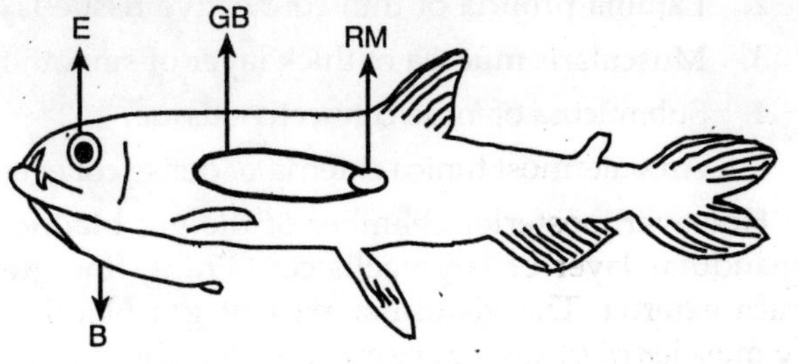

Fig. 5.14 : Position of gas bladder and rete mirabilis in deep sea fish *Astronesthes.* B, barbels; E, eye, GB, gas; bladder; RM, rete mirabile.

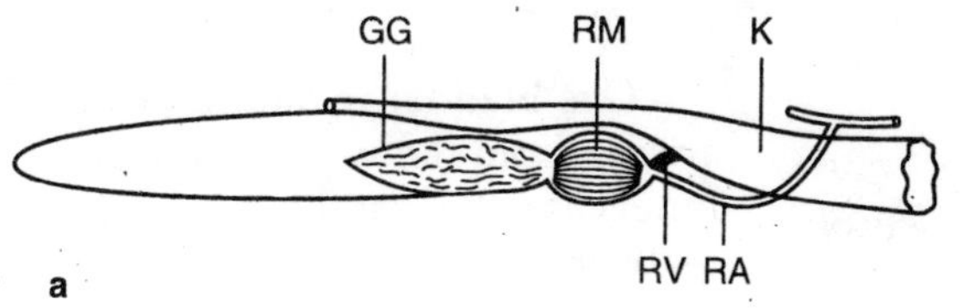

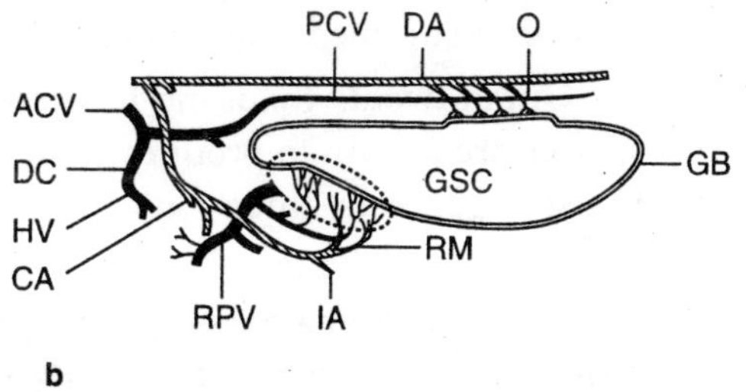

Fig. 5.15a, b : Gas bladder and rete mirabile in fishes. (a) In *Astronesthes* (b) blood supply of gas bladder in physostomous fish. ACV, anterior cardinal vein; CA, coeliacomesenteric artery, DA, dorsal aorta; DC duct of Cuvieri; GB, gas bladder; GG, gas gland; GSC, gas secreting complex; HV, hepatic vein; IA, intestinal vein; K, kidney; PCV, posterior cardinal vein; RA, retinal artery; RM, rete mirabile O, oesophagus RPV, renal portal vein; RV, retial vein.

Before entering the tissue, artery divides into a large number of small capillaries, they are parallel to a series of venous capillaries leaving the tissue. The 'arterial' capillaries are surrounded by 'venous' capillaries and vice versa, forming an extensive exchange surface between inflowing and outflowing blood. Retail capillaries serve to transfer heat or gases between arterial blood entering the tissue and venous blood leaving it.

In physostomous fishes, the rete mirabile is rather primitive and is covered with flattened epithelium, known as 'red bodies', while in physoclitus fishes the capillaries are covered by thick glandular folded epithelium and is called as 'red gland'. In some fishes like Clupidae and Salmonidae the blood vessels are uniformly distributed over the bladder and do not form a rete mirabile.

Histology of gas bladder : In cyprinids the anterior chamber of gas bladder comprises.

1. An innermost epithelial layer.
2. Lamina propria of thin connective tissue layer.
3. Muscularis mucosa of thick layer of smooth muscle fibres.
4. Submucosa of loose connective tissue.
5. An outermost tunica externa of dense collegenous muscle fibres.

However, posterior chamber of the gas bladder differs histologically, and comprises of a glandular layer of large cells containing fine granulated cytoplasm which lie inside the tunica externa. The glandular part of gas bladder is richly supplied by blood capillaries. The muscles of posterior chamber is also known to have regulatory function of gas gland and to control the volume of gas bladder.

In some fishes the anterior chamber of gas bladder contains a gas gland, which secretes gas whereas the posterior chamber is thin-walled and helps in gas diffusion as in Synganthidae species. In these fishes the gas bladder is closed and partially subdivided into two chambers. However, in the Cyprinids it has pneumatic duct and gas gland is present in the posterior chamber, which performs hydrostatic function, while the anterior chamber plays auditory function (Fig. 5.16).

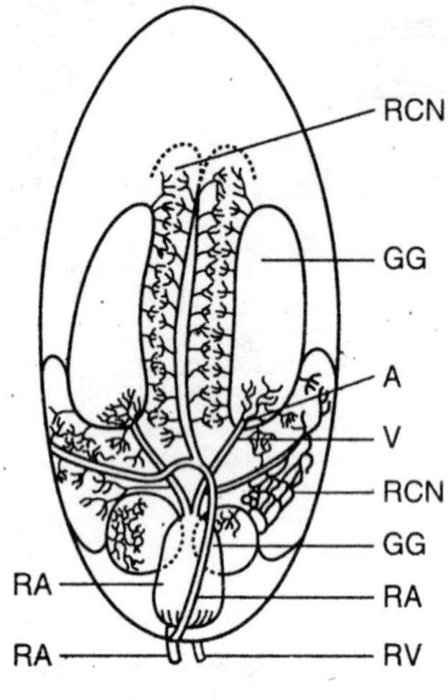

Fig. 5.16 : Diagram of ventral view of gas bladder in deepsea mouthfish with detail blood supply. A, artery; GG, gas gland; RA, retail artery; RCN, resorbent capillary network, RV, retail vein; V, vein.

Gas bladder in sound production : Various branches arising from the vagus nerve and from the coeliac ganglia innervate the gas bladder. These nerves terminate in reabsorbent area, the oval, the rete and in the secondary epithelium. The muscular wall of the bladder is also very well supplied with nerves.

Out of twenty thousand fish species only few hundred species are known to produce sound of various intensities. In fishes generally three sonic mechanisms work for sound production.

1. Hydrodynamic — Sound produced as a result of swimming movements particularly when rapid changes in direction or velocity occurs.

2. Stridulatory — Sound produced by rubbing of teeth, fin spines and bones. Ex. grunts, pomadasyidae.

3. By gas bladder — Sound is produced by vibrations of striated muscle, which originates from dorsal body wall and inserts on the gas bladder. Ex. grenadieres (melanonidae), drums (Sciaenidae).

Toad fishes are able to produce sound by rapid change in the volume of the gas bladder. Sound produced by the gas bladder usually has low pitches, however, sound made by teeth or bones has higher frequencies. Sound plays important roles in breeding behaviour and in the defence as well.

Gas Bladder in Sound Reception : The sound waves easily pass from the sea water to fish body because of similar densities. But these sound waves are discountinued by the gas bladder and therefore, the gas bladder acts as a sound conductors or resonator. In fishes like cods (Gadidae) and orgies (Sparidae) the gas bladder is extended in such a way that it touches the bones near the sacculus of the internal ear, the variation in pressure due to sound waves may be transmitted directly to the perilymph. Extension of the gas bladder grows in the form of cartilaginous capsule, i.e., prootic and pterotic bullae, lies closely to the perilymph spaces of superior and inferior part of the internal ear. In the order Cypriniformes the gas bladder transmits the sound waves to the internal ear by as special apparatus which consists of series of paired bones or ossicles and is known a Weberian apparatus, which connects the gas bladder to the internal ear. These ossicles originate from the apophysis of anterior vertebrae.

Weberian apparatus consists of five ossicles, i.e., claustrum, scaphium, intercalarium and the tripus, which do not show homology with the mammalian ear, hence called 'Weberian ossicles'. The posterior most ossicle is tripus, which is largest and triangular piece. Posteriorly it touches anterior wall of the gas bladder while anteriorly it articulates to the ligaments of next bone, i.e., intercalar. But when the latter is absent it is attached to scaphium which in turn is attached to the minute anterior most claustrum. The claustrum touches a membranus atrium sinus impar, that lies in the basioccipetal bone of the head and is an extension of the perilymph system of the internal ear. In Gymnotids (Gymnotidae), the scaphium touches the atrium sinus impar due to absence of claustrum. The intercalarium also shows variations in the structure and mode of its development. It may be a small nodule like bone in the ligament, separated from the vertebral column as found in the siluroids (Siluridae). Sometimes it may develop as rod-like extension touching the centrum of second vertebra as in the carp *(Abeo, Cirrhina and Tor)*. The Weberian ossicles provide a connection between the gas bladder and the internal ear by a series, i.e., gas bladder → Weberian ossicle → sinus impar → sinus endolymphaticus → transverse canal → sacculus.

At the time of functioning of Weberian ossicles, volume of gas bladder changes due to which the gas bladder moves in such a manner that pressure changes are transmitted to the perilymph and hence to the sensory cells of inferior part of the labyrinth which is the seat of sound reception. In some species the gas bladder is enclosed in a bony capsule or connective tissue and projects through a small aperture to attach the tripus. A change in the volume of gas bladder due to its rhythmic compression causes its wall to bulge out and push the ossicles forwards.

Among the Cypriniformes fishes wide range of sound perception and better sound discrimination are seen than those fishes which do not possess Weberian apparatus. Removal of gas bladder in fish-like minnows, greatly reduces the auditory range.

The Gas Bladder as a Hydrostatic Organs : The density of fish flesh is greater than that of water. To make the body weightless and to minimize the energy consumption in maintaining the body position, the fish stores fats and oils in muscle and liver, fills the oxygen in the gas bladder. In this way fish can reduce its body weight. In bony fishes the gas bladder brings the density of fish closely to that of surrounding water. In sharks and rays the air bladder is absent and they maintain their body buoyancy by regulating the 'water ballast' present in the body cavity and operated through their abdominal pores. In marine fishes the gas bladder can make up 4 to 11 percent of the body volume while in freshwater fishes 7 to 11 percent of the body volume is maintained by the gas bladder.

The fishes may be divided into physostomous (bladder with opening into gut) an physoclitous (bladder closed) on the basis of its functional and morphological differences. The change of one condition to another is a gradual process and is concerned with gas secreting and resorbing structures. In many physostomous species the gas bladder loses the pneumatic duct which was open outside in the young. The condition is known as paraphysoclistious as found in lanternfishes (Myctophidae).

The soft-rayed fishes (Malacoptergii) are physostomous and the spiny rayed ones (Acanthopterygii) are physoclitous. In true physoclistous teleosts the pressure in the gas bladder is adjusted through secretion or resorption of gasses from or to the blood.

The position of the gas bladder in relation to the centre of gravity of fish plays an important role in swimming and maintaining its position. Normal swimming position of fish is maintained effortlessly with the help of gas bladder. Some fishes can displace their gas bladder to achieve their normal position from unusual upside down swimming position of the body.

The filling and emptying of the gas bladder : Gas bladder has unique character that it stores 500 times oxygen and 30 times nitrogen. Physostomous fishes such as trouts and salmon fill their gas bladder by gulping air at the time of removal of yolksac. Although adults of these fishes are able to secrete and absorb gas through the blood supply but at initial stage they have to depend on atmosphere to fill their gas bladder. Many physoclistous fishes like sticklebacks (*Gastrosteus*), guppy (*Lebistes*) and seahorse (*Hippocampus*) possess pneumatic duct in larval stage, thus first filling of the gas bladder takes place from atmospheric air.

Some deep sea fishes such as grenadiers (Melanonidae) have functional gas bladder with different mechanism for initial filling of gas bladder unless they are pelagic in early life stages. Fishes are able to change gas content in such a way that the gas volume is almost constant regardless of hydrostatic pressure. Boyle's law which states that the volume of gas changes inversely with pressure, is also applicable to the gas bladder.

Secretion of gas from blood to lumen of bladder : The gases contained in the blood is released into the cavity of the gas bladder through highly vascular regions, called 'gas secreting complex' present in the wall of the bladder. The 'gas secreting complex' consists of (i) gas gland and (ii) rete mirabile. The gas gland is the region of the bladder epithelium and may be one layered, folded or made up to multilayered stratified epithelium. Rete mirabile is small blood vessels underlying the epithelium. Arteries and veins of the bladder make intimate diffusional contact with each other and forms countercurrent multiplier system which ensures concentration difference of many substances from one end to another end of the organ (Fig. 15). Deep sea fishes such as searobins (Trigla) usually fill their gas bladder with oxygen.

Resorption of gas from bladder : It is accomplished as follows.

1. The gas from bladder may be diffused into the blood vessels present throughout the gas bladder wall, other than the gas secreting complex as found in killifishes (Cyprinodontidae) and sauries (Scombresocidae).

2. Generally the gas is drained from single chambered or posterior sac of gas bladder through a thin area of the bladder wall which comprises a network of capillaries separated from the lumen of bladder through a very thin area contained in capillaries in the bladder wall is known as oval organ. Sphincter surrounds the oval organ and regulates that rate of gas resorption by dilating and contracting the oval orifice, e.g, cods and spiny rayed fishes, i.e., Acanthopterygii.

REFERENCES

Ballington, C.M. and Hughes, G.M. (1965). The muscular basis of the respiratory pumps in the trout, J. Exp. Biol. 43 : 349-362.

Drummond, R.A., Spoor, W.A., and Olson, G.F. (1973). Some short term indicators of sublethal effects of copper on brook trout, *Salvelinus fontinallis* . J. Fish. Res. Bd. Canada. 30 : 689-701.

Hughes, G.M. (1960). A comparative study of gill ventilation in marine teleosts. J. Ex. Biol. 37 : 28-48.

Hughes, G.M. (1966). The dimensions of fish gills in relation to their function J. Exp. Biol. 45 : 177-195.

Hughes, G.M. (1979). How a fish extracts oxygen from water. In : Readings in Ichthyology. (Eds. Love, M.S. and Cailliet, G.M.) Prentice Hall of India Pvt. Ltd., New Delhi pp. 228-239.

Hughes, G.M., & Shelton (1958). The mechanism of gill ventilation in three freshwater teleosts. J. Exp. Biol. 35 : 807-23.

Hughes, G.M. and Shelton, G. (1962). Respiratory mechanism and their nervous control in fish. Adv. Com. Physiol. Biochem. 1 : 275-364.

Johansen, K. and Burggren, W. (1980). Cardiovascular function in the lower vertebrates. In Heart and heart like organs. Vol. 1 Comparative anatomy and development (Ed. Bourme, G.H.), Academic Press, New York, pp. 161-177.

Kirchhoff, J. (1958). Functional anatomische untersuchung des Visceralapparates von *Clupea haregus* L. Zool. Jb. (Abst). 76 : 461-540.

Lagler, K.F., Bardach, John E. Miller, R.R., and Passino, D.R. May (1977) – 'Ichthyology' John Villey and Sons, New York.

Munshi, J.S.D. (1993). Structure and function of the air breathing organs of *Heteropneustes fossilis*. In : 'Advances in fish research' Vol. I. (Ed. B.R. Singh) Naredra Publishing House, Delhi. pp. 99-137.

Munishi, J.S.D. and Singh, B.R. (1968). On the respiratory organs of *Amphipnous cuchia*. J. Morph. 124 : 423-444.

Munshi, J.S.D., Olson, K.R., Ghosh, T.K. and Ojha, J. (1990). Vasculature of the head and respiratory organs in an obligate air breathing fish, the swamp eel *Monopterus cuchia*. J. Morph. 203 : 181-201 p.

Moyle, P.B. and Cech, J. Jr. (1982). 'Fishes - An introducton to Ichthyology. Printed Hall International, Inc, London.

Ojha, J. and Mishra, A.K. (1987). Micromorphology and cytochemistry of the branchial gland of the freshwater mullets, *Rhinomugil corsula* and *Sciamugil* cascasia. Proc. Indian Acad. Sci. (Anim Sci) 96 : 417-442.

Oguri, M. (1982). 'An atlas of fish histology normal and pathological features' (Ed. T. Hibiya) Kodanasha Ltd. Gustav, Fischer Verlag, New York, P. 54.

Parry, C. and Holliday, F.G.T., Holliday, (1960). An experimental analysis of function of the pseudobranch in teleosts. J. Expt. Biol. 37 : 344-354.

Roberts, R.J. (1979). Active branchial and ram gill ventilation in fishes. In : readings in Ichthyology, (Eds. Love, M.S. and Caillie G.M.). Prentice Hall of India. Pvt. Ltd. New Delhi. pp. 248-266.

Chapter 6

CARDIOVASCULAR SYSTEM

The cardiovascular system consists of heart and blood vessels, the heart keeps the blood on motion and blood vessels circulate the blood to and from all parts of the body. The lymphatic vessel, an ancillary set of vessels, which circulate lymph and lymphocytes through the circulatory system is closed in fishes.

HEART

The heart of fishes is known as branchial heart, because its main function is to pump venous blood to ventral aorta into gills (branchial) and then to somatic vasculature. Thus branchial and systemic vascular beds are arranged in series with heart. Apart from heart, heart-like organs are present only in Agnatha (*Myxine* and *Petromyzon*).

The heart of fishes consists of four chambers, a sinus venosus, an atrium, a ventricle and a conus or a bulbus arteriosus (Fig. 6.1a-b).

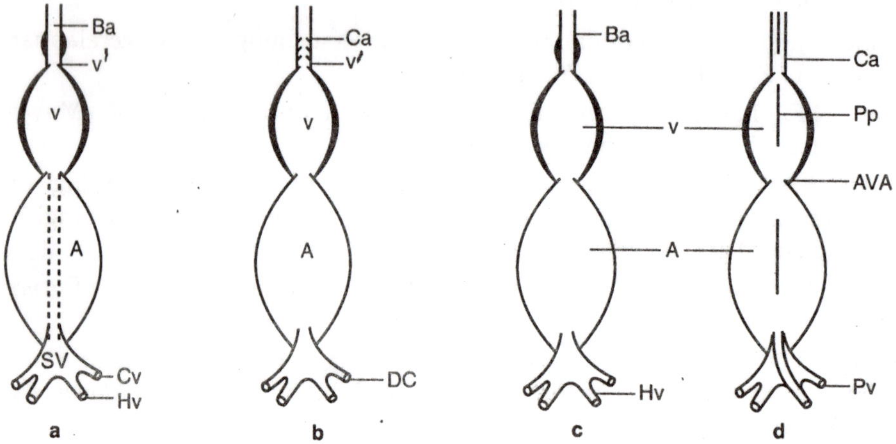

Fig. 6.1a-d : Diagrammatic sketch of structure of the heart of fishes. (a) Lamprey heart showing bulbus arteriosus. (b) Cartilaginous fish heart showing conus arteriosus having many valves. (c) Heart of bony fish showing both conus and bulbus arteriosus. (d) Dipnoan heart showing partial septum in the atria and ventricle. AVA, atrioventricular aperture; A, atrium; Ba, bulbus arteriosus; Ca, conus arteriosus; Cv, Cuvieri vein; DC, ductus Cuvieri; Hv, hepatic vein; Pv, post caval; SV, sinus venosus, v, valve. v', Ventricle.

Some authors considered atrium and ventricles as the chambers of heart while some considered sinus venosus and conus arteriosus also as the chambers of the heart. There is some confusion in bulbus and conus arteriosus in fishes. In elasmobranchs the fourth chamber is designated as conus arteriosus whereas it is known as bulbus arteriosus in teleost, a specialized ventral aorta in teleosts. The distinction between the two is that the conus consists of cardiac musculature similar to the ventricle and is generally provided by a large number of valves arranged in successive rows (Fig. 6.1b) while bulbus arteriosus comprises only smooth muscle fibres and elastic tissues (Boas, 1880; Smith, 1918; Danforth, 1912; Parson, 1929; Karandikar & Thakur, 1954). According to Torrey (1971), the heart of *Cyprinus carpio* a teleostean fish, contains both conus and bulbus arteriosus. However, later workers held that in teleosts only bulbus arteriosus is present (Kumar, 1974; Santer, 1977). Elasmobranch and aganthan have conus arteriosus instead of bulbus arteriosus.

Heart rate and Stroke volume : The heart performance basically depends on two factors; the heart rate and the stroke volume. At each heart beat, the ventricle pumps out blood. The volume is termed stroke volume and the time of the heart beat is known as heart rate. These are controlled both by aneural factors such as extent of cardiac filling (Starlings law of the heart) or circulatory substances (hormones) and by the innervation of the cardiac pacemaker and muscle (Laurent *et al.*, 1983). The fish atrium is filled by suction created by the rigidity of the pericardium and surrounding tissue. Venous blood return to the atrium is aided by ventricular contraction in systole which causes a fall in intrapericardial pressure that is transmitted through the thin wall of the atrium to create an aspiratory or *via a fonte* effect (Johansen and Burggren, 1980). It is contrary to the situation in mammals where the central venous pressure determines the atrial filling during diastole (*vis a tergo*, driving force from behind.)

Sinus venosus : The sinus venosus is not an active part of the heart although pacemaker properly starts in this chamber (Fig. 6.2a, b).

It is actually a continuation of the venosus vessels and its main function is to receive blood and to pass it on to the atrium. Sinus venosus receives blood through two duct Cuvieri, hepatic veins pour the blood from liver. The ventral ductus Cuvieri receives blood from anterior and posterior cardinal veins (Laurent *et al.*, 1983; Farrel & Jones, 1992).

The sinus venosus is distinguished histologically into tunica intima, tunica media and tunica adventitia. Normally the sinus venosus is purely amuscular in some fishes. The matrix of this chamber is made up of elastic and collagen fibres. The muscles are restricted around the sinuatrial opening in circular fashion forming sinuatrial ring. The sinus venosus opens into atrium by a sinuatrial ostium, which is provided by two sinuatrial valves. Farrel and Jones (1992) reported single atrioventricular valve in teleost fishes.

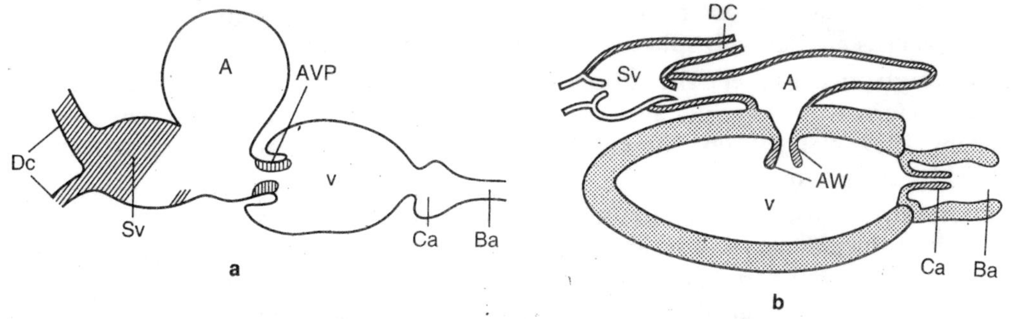

Fig. 6.2a, b : Distribution of pacemaker in the sinus venosus of eel. The pacemaker area is shown by stripes. (a) Suggested by Bielig. (b) As suggested by Grodzinski. A, atrium; AVp atrioventricular pad; Ba, bulbus arteriosus; Ca, conus arteriosus; Dc., ductus Cuvieri; Sv, sinus venosus; v, ventricle.

Atrium : The atrium is a large muscular contractile chamber. It is situated dorsal to the ventricle in almost all fishes (Fig. 6.3). In fishes, the atrium is also known as auricle

(Yamauchi, 1980), but actually the appendages of the atria are called as auricles. The atrium is undivided single chamber in elasmobranch and teleosts but in dipnoi, the atrium is partially divided by an incomplete interatrial septum (Fig. 6.1d).

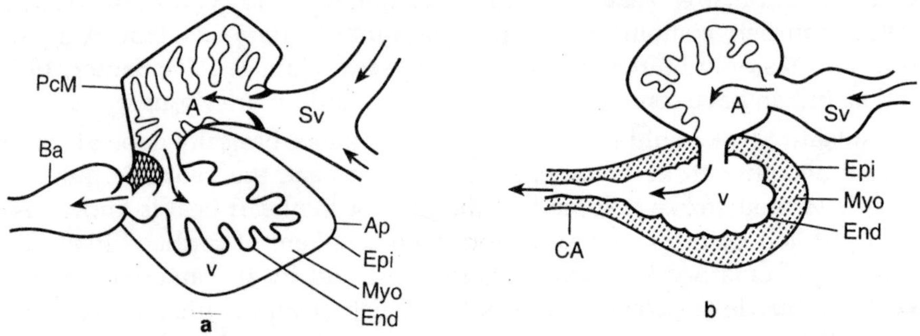

Fig. 6.3a-b : Diagram of heart showing various histological layers of the heart chambers. (a) Trout (teleost). (b) Shark (elasmobranch). A, atrium; Ap, apex; CA, conus arteriosus; Epi, epicardium; End, endocardium; Myo, myocardium; PcM, pacemarker region; Sv, sinus venosus; v, ventricle.

Pulmonary blood drains directly into the left side of the atrium, whereas the systemic venous blood is collected in the sinus venosus through ducti Cuvieri. The blood from the sinus venosus goes to the right side of the atrium.

Internally, the atrium is divisible into two parts, a sinuatrial canal and atrium proper. The former is rather thick-walled semi-cylindrical rigid tube and the latter is a thin-walled distensible spongy cavity (Kumar, 1974; Laurent *et al.*, 1983). The significance and functional importance of this funnel is due to the pressure of blood in the sinus venosus and atrial filling.

The spongy portion of the atrium contains pectinate muscles (Fig. 6.3a-b). The trabeculae at the atrioventricular ostium form mesh-like network. When they contract, they pull the roof and sides of the atrium towards atrioventricular ostium (Satchell, 1971). The atrial mass constitutes 0.25% of ventricular mass and 0.01-0.03% of body weight (Farrel & Jones, 1992).

The atrium histologically is distinguished into epicardium, endocardium and myocardium. The endocardium is the innermost layer, lining the lumen of the atrium. The endothelial cells are flat with spheroid or more often elongated nuclei.

Atrioventricular funnel : The atrium communicates with the ventricle through a tubular structure referred to as canalis auricularis or atrioventricular funnel (Laurent *et al.*, 1962; Abraham, 1969; Nair, 1970b; Kumar, 1976). The atrioventricular opening is round and guarded by atrioventricular valves.

Regarding the disposition and number of AV valves in the heart of fishes in general and teleosts in particular are still much disputed. Generally, in teleosts two atrioventricular valves are present but Farrel and Jones (1992) described a single atrioventricular valve. The atrioventricular valves in all the three genera of dipnoans, lungfish, *i.e., Protopterus* (Africa), *Lepidosiren* (South America) and *Neoceratodus* (Australia) are replaced by another structure known as atrioventricular plug (Fig. 6.2a). The atrioventricular plug which guards the horseshoe-shaped atrioventricular opening, the functions is similar to atrioventricular valve. It is in the form of inverted cone with its apex pointing into the atrial lumen. It is projected dorsally with the atrial lumen and reaches upto pulmonalis fold and due to this, there is partial septation of the atrium. It is made up of hyaline cartilage encircled by fibrous connective tissue. In *Neoceratodus*, the hyaline cartilage is absent and the plug is made up of fibrous connective tissue.

Ventricle : The teleost ventricle is either tubular, pyramidal or sac-like in appearance (Fig. 6.4).

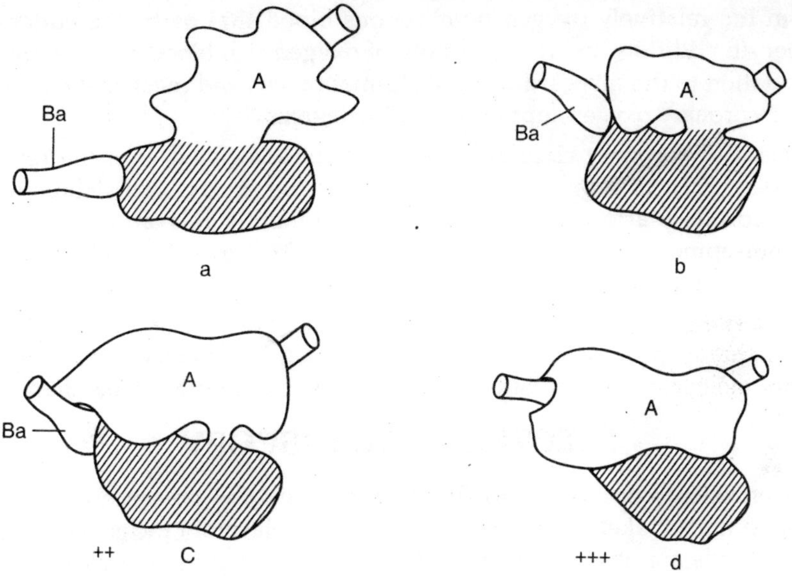

Fig. 6.4 : The degree of position of atrium to the bulbus arteriosus varies amongst teleost from a condition (-) where the two chambers are not in contact to a condition (+++) where the most anterior part of the atrium almost completely covers the bulbus arteriosus. A, atrium; Ba, bulbus arteriosus

It is relatively large muscular chamber. It is undivided in elasmobranch and teleost, but it is partially divided into left and right chambers by a muscular septum in Dipnoi. The muscular septum is posterior to the atrioventricular plug in all the three genera but extends anteriorly along the ventral surface in *Lepidosiren*. Its anterior and dorsal margins are free. In majority of Indian teleost fishes the ventricle is sac-like.

Histology : The layers constituting the ventricle wall are fairly well differentiated in epicardium, myocardium and endocardium (Fig. 6.3a & b). These layers are essentially similar to the atrium except that the myocardium is substantially thicker than that of the atrium.

The ventricular myocardial architecture is different in different fishes (Tota, 1978, 1983, 1989). The arrangement may be compact, mixed, i.e., compact and trabeculated or very weakly compact but well developed trabeculate (Spongiosa). In compact myocardium, layers of muscle bundles are arranged orderly within ventricular wall. In elasmobranchs, the compact myocardium, at the level of the atrioventricular orifice is continuous with the trabeculated myocardium. In teleosts, the compact myocardium is independent of the trabeculated myocardium and a large number of fibres insert into the bulboventricular fibre ring (Fernandez Teran & Hurle, 1982; Sanchez Quintana & Hurle, 1987). No such detailed description had been given regarding myocardial arrangement in the ventricle of any Indian fish, but in majority of Indian teleosts both compact and trabeculated situation occur. The ventricular myocardium is entirely trabecular in lungfishes. The arrangement of myocardial layer helps in developing high blood pressure in compensating atropic effect of low temperature and accomodation of a large cardiac stroke volume (Davie and Farrel, 1991 a, b).

Coronary Circulation : The working myocardium of the fish's heart, like other tissues require a blood supply to provide oxygen. There are two routes for oxygen supply and they are utilized to different degrees among fishes. Since the heart pumps venous blood, oxygen is available from the relatively oxygen poor venous blood that bathe the endocardial lining of the chamber. In addition, an arterial supply of oxygen rich blood may be provided by the coronary circulation to the myocardium. All elasmobranchs and most active teleosts use both the venous and coronary oxygen supply to varying degrees.

Development of coronary circulation is generally associated with the relatively larger ventricle (Hesse, 1921). In rainbow trout, *Onchorhynchus mykiss*, acetylcholine helps in the contraction of coronary arteries and there is predominantly relaxation with isoproterenol, epinephrine, nor-epinephrine and serotonin (Farrel, 1987; Davie & Farrel 1991, b).

Coronary vascular resistance increases exponentially as coronary flow rate decreases. Coronary resistance was also influenced by cardiac metabolism and acclimatization. Farrel (1987) experimentally produced vaso-constriction of the coronary vessels by injection of adrenaline into the coronary circulation. He held it as temperature dependent.

ELECTRON MICROSCOPY

The electron microscopy of the cardiac tissue shows the presence of Z, I, A, H, M, N bands similar to that of skeletal muscle in teleost and elasmobranchs. T-tubular system in fish myocardial cells are absent (Santer, 1985). Epicardium is made up of mesothelial cells supported by basal lamina. The cytoplasm of mesothelial cells contains rough endoplasmic reticulum, Golgi bodies and electron dense vesicles.

CONTRACTILE PROTEINS

The available evidence suggests that properties of the contractile proteins from lower vertebrate are broadly similar to those found in the skeletal and cardiac muscle of mammalian species (Johnston, 1983). However, adult cardiac muscles contain isotypes of myosin, tropomyosin and troponin which have distinct chemical structures and somewhat different properties from those found in skeletal muscle (Perry, 1979; Stull *et al.*, 1980). The complex orientation of fibres and the presence of a large proportion of non-muscle cells in cardiac tissues make it difficult to obtain multicellular preparations for the study of their contractile properties. The myosin isolated from fish and amphibians skeletal muscles are of unstable type which readily lose their ATPase activity of storage (Connell, 1960, 1961; Ferenczi et al., 1978 a & b). The fish actomyosin preparations are orders of magnitude more stable than corresponding myosin preparations (Johnston and Walesby, 1979). It is now of common belief that in common with myosin there have been selective modifications in the sequence of tropomyosin and troponin to permit efficient regulation of contraction at different body temperatures (Malencik *et al.*, 1973, Lehaman & Szent-Gyorgyi, 1975; Altringham & Johnston, 1983; Davie & Ferrel, 1990).

PATHOLOGY

The heart muscles are infected with bacteria and viruses. The bacterial infection is due to aeromonas and vibrios. They form colonies in the myocardium resulting that the endocardium becomes swollen and their nuclei become pycnotic. The viral infection commonly affecting the heart muscle are rhabdovirus. The infection causes myocardial necrosis resulting in inflammation in all the three layers, i.e., epicardium, endocardium and myocardium. The inflammation of cardiac muscle is known as myocarditis (Takashima, 1982). A few reports deal with atrioventricular valve diseases (Roberts, 1978). Like the higher vertebrates, the regeneration ability of cardiac muscle is nil and any injury or myocardial infarction develops into fibrous connective tissue.

The cardiac conducting system (Specialized tissues) : The cardiac conducting system of homoiothermal vertebrates is responsible for the initiation and conduction of electrical impulse at right place and at right time. This system is also often called "Purkinje system" or "Specialized tissues". In higher vertebrates this system is well developed and consists of a sinuatrial node (pacemaker muscle) situated in the right atrium, an atrioventricular node placed at the caudal end of the interatrial septum near the coronary sinus and the atrioventricular bundle placed above the interventricular septum (bundle of His) and its two branches along with Purkinje fibres situated subendocardially both in the atria and the ventricle.

It has been unanimously accepted that the Purkinje fibres similar to those of higher vertebrates are absent in the heart of fishes (Choidi and Bortolami, 1967; Laurent *et al.*, 1983). Whether the heart beat in fishes is generated by means of muscles or by nerves has not been clearly understood so far. The physiological investigations are few and also as controversial as morphological ones. The cardiac beat is originated at the ostial part of the sinus (Mc William, 1885) and there are three groups of pacemaker (Skramlik, 1935) in eel, while four groups were reported by Grodzinski (1954). A few investigators found histologically specialized structures such as sinuatrial and atrioventricular plugs in the heart of fishes (Prakash, 1953; Sahai & Chawla, 1967). Presence of histologically specialized muscles which take lesser stain than the working cardiac muscles in fishes, have been reported in a few species (Gyevai, 1960; Choudhry & Das, 1963; Saxena & Bakshi, 1966). On the other hand, majority of workers denied the presence of histologically specialized tissues in any part of the heart of fishes (Davies & Francis, 1941; Nair, 1970a, Thomas, 1976 and Kumar, 1976).

Nodal tissue : Keith and Flack (1907) and Keith and Mackenzie (1910) found nodal tissue at the base of venous valve. The criterion allowing a distinction of the nodal cells from other cardiac muscle cells in higher vertebrates is the relative poorness in myofibrils within cytoplasm as revealed by electron microscope. This characteristic feature is reported in some part of the sinuatrial myocardium of loaches (Yamauchi, 1969) in catfish (Laurent, 1962) and trout (Yamauchi & Burnstock, 1968). These authors reconfirmed the existence of nodal tissue as reported by Keith and Flack (1907) and Keith and Mackenzie (1910).

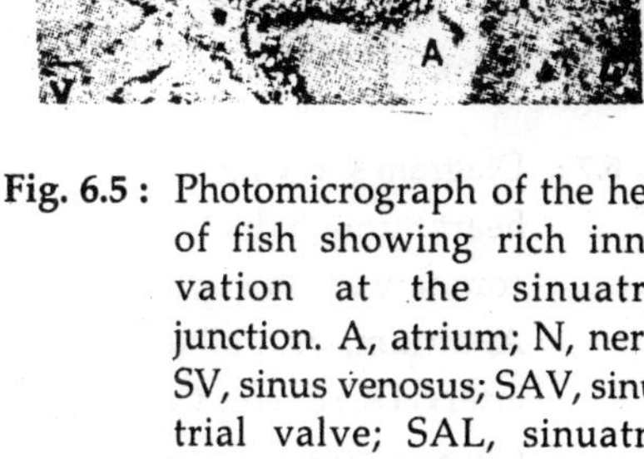

Fig. 6.5 : Photomicrograph of the heart of fish showing rich innervation at the sinuatrial junction. A, atrium; N, nerve; SV, sinus venosus; SAV, sinuatrial valve; SAL, sinuatrial Ligament; V, ventricle.

There is no unanimity regarding the occurence of nodal tissue in true histological sense, but almost all investigators in this field found heavy nerves and intimate nerve connection at the sinuatrial junction where pacemaker potential has been described (Saito, 1969, 1971, 1973). There is a muscular continuity in various chambers of the heart and the chamber are not interrupted by nodes, bundles and Punkinje fibres (Kumar, 1978 a,b). Nair (1970) described ganglion cells and nerve plexus in the sinus venosus of *Protopterus aethiopicus* (Fig. 6.5).

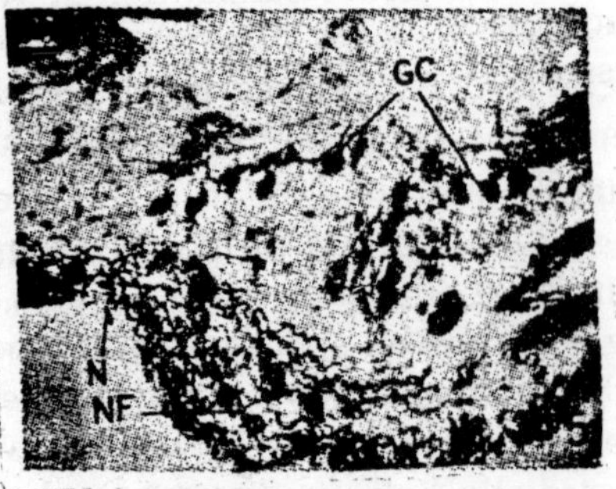

Fig. 6.6 : Microphotograph showing the presence of ganglion cells and ganglionated nerves in the heart of fish. GC, ganglion cell; N, nerve; NF, nerve fibre.

The distribution of nervous (Fig. 6.6) connection corresponds fairly accurately with electrophysiologically defined pacemaker region and it is, therefore, likely that there is a cholinergic vagal influence on pacemaker activity of dipnoans sp.

Like other fishes dipnoans heart is also not provided with sympathetic innervation. From the sinuatrial region the contraction wave successively invades the atrium, the atrioventricular funnel and then to the ventricular myocardium (Saito and Temma, 1976). It is generally believed that the cardiac conducting system of the heart of fishes is neither purely myogenic nor entirely neurogenic, but is the complex combination of the two (Kumar, 1976, 1978a, b, 1979).

INNERVATION

The heart of fishes is innervated by a pair of cardiac branch of vagosympathetic trunk (Nicol, 1952), (Fig. 6.7) except in myxinoid heart, which receives no extrinsic innervation.

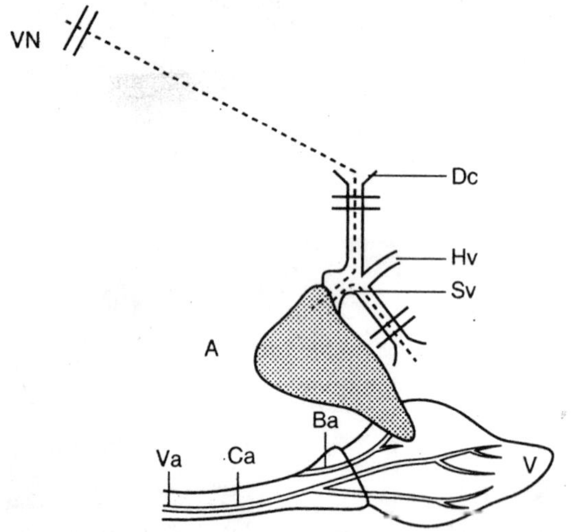

Fig. 6.7 : Diagram showing vagus nerve supplying to the heart through ductus Cuvieri and also showing coronary artery supplying to the trout ventricle. A, atrium; Ba, bulbus artereosus; Ca, coronary artery, DC, ductus Cuvieri; Hv, hepatic vein; Sv, sinus venosus; V; ventricle; VA, ventral aorta; VN. vagus nerve.

Like other vertebrates, the heart is under autonomic control. The autonomic nervous system in teleost is sympathetic and parasympathetic. There is no direct sympathetic nerves going to the heart. The vagus at its origin is parasympathetic (cranial outflow) but it receives post-ganglionic autonomic fibres from sympathetic chain into the head region (Young, 1931; Nilsson *et al.*, 1976; Nilsson, 1983).

The various chambers of the heart are richly innervated by both cholinergic and adrenergic nerve fibres (Abraham, 1969; Ganon & Burnstock Kumar, 1974 a,b,c 1976, 1979).

The different nerve endings (intracardiac mechano receptors) present in the heart (Laurent, 1962; Kumar, 1979) when receive adequate stimulus transmit impulses to CNS. This information then processed in CNS and then subsequently transmit impulses through autonomic (efferent) fibres to the heart which helps in atrial filling a cardioventilatory coupling. (Smatresk, 1986; Burleson *et al.*, 1992).

The heart of the fish, as of higher vertebrates, is under inhibitory control by cholinergic vagal fibres (Laurent, 1962).

Cholinergic nerves present in the heart secrete ACh, a neurotransmitter at their termination is essential for impulse transmission and action potential. It is now accepted that the hydrolysis of acetylcholine to choline and acetic acid is catalysed by an enzyme, cholinesterase in animal system. The enzyme prevents the excessive accumulation of acetylcholine at cholinergic synapse and at the neuromuscular junction.

The cholinesterase at a neuromuscular junction is capable of hydrolysing some 10^{-9} molecules (2.4×10^{-7}) of acetylcholine in one milli second. The enzyme kinetics of cholinesterase in cardiac tissue studied by Nemcsok (1990) and its inhibitory kinetics by using pesticides has been studied in the cardiac as well as in other tissues of fish by several investigators (Main, 1969; Mohan & Brink, 1970; Ratha & Ramanujam, 1986; Seto & Shinohara, 1986; Hande & Pradhan (1990 a, b) Gaur; 1992; Kumar *et al.*, 1995). The Km in normal heart of *Cyprinus carpio* is 1.37×10^{-3} M and 1.87×10^{-3} M in *Channa punctatus* (Nemcsok, 1990 & Gaur, 1992).

It is reported that Km changed to 1.83×10^{-4} M and 2.86×10^{-4} M when the fish was subjected to 4.6×10^{-6} and 2×10^{-4} concentrations of methidathion (Nemcsok *et al.*, 1990). Similar increasing trends were reported by Gaur (1992) and Gaur & Kumar (1993) in the heart of *Channa punctatus*. It goes to 2.78×10^{-3} M when artificial infarction was produced in the *Channa* heart. When normal heart is treated with 2 ppm dimethoate, the Km is increased to 3.30×10^{-3} M and Km further increased to 4.07×10^{-3} M when incised heart is subjected to 2 ppm dimethoate. The constant V_{max} in all experiments indicate that inhibition is competitive in nature (Fig. 6.8).

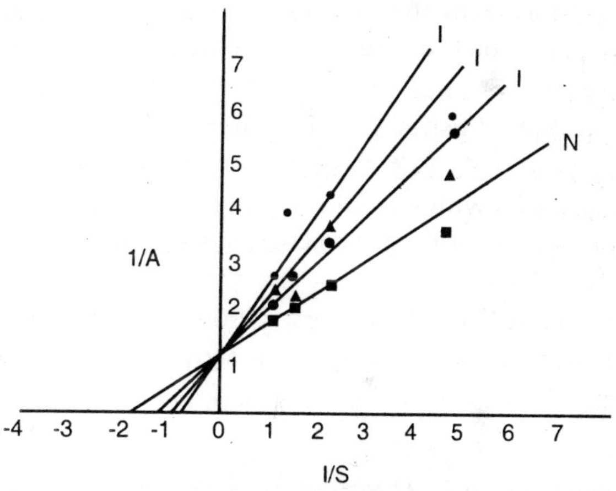

Fig. 6.8 : Lineweavers-Burk plot of acetylcholinesterase of heart of *Channa punctatus* showing inhibition with a pesticide dimethoate. 1/A, absorbance; 1/S, substrate concentration; I, inhibitor; N, normal.

These experiments support that infarction and treatment with pesticide show that in these cases there is inhibition of acetylcholinesterase enzyme in the heart tissue.

REFERENCES

Abraham, A. (1969). Microscopic innervation of the heart and blood vessels in vertebrates including man. Pergamon Press Ltd., London.

Altringham, T.D. & Johnston, I.A. (1983). The p Ca-tension and force velocity characteristics of skinned fast and slow fibres isolated from fish myotomes. J. Physiol.

Boas, J.E.V. (1880). Uber den Conus arterious bei Butirinus aund bei andern Knochenfischen. Morph. Jb. 6 : 527-533.

Burleson, M.L. Smatresk, N. J. and William, K. Milson (1992). Afferent inputs associated with cardioventilatory control in fish. In : Fish Physiology. Vol. XII, Part B. (Eds. Hoar, W.S. Randall, D.J. and Farrell, A.P.), Academic Press Inc., New York.

Choudhury, H.S. and Das, A.B. (1963). The heart of Indian cuchia eel *Amphipnous cuchia* (Ham.). Proceedings National Academy Sciences India, 33 : 647-650.

Chiodi, V. and Bortolami, R. (1967). The conducting system of the vertebrate heart Edizioni Calderini, Bolgna, Italy, PP. 22-29.

Connell, J.J. (1960). Studies on the proteins of fish skeletal muscle. Denaturation and aggregation of cod myosin. Biochem, J. 75 : 530-538.

Connell, J.J. (1961). The relative stabilities of the skeletal muscle myosins of some animals. Biochem. J. 80 : 503-509.

Danforth, C.H. (1912). The heart and arteries of *Polyodon*. J. Morph., 29 : 409-454.

Davie, P.S. and Farrel, A.P. (1990). The coronary and luminal circulations of the myocardium of fishes. Can. J. Zool. 69 : 1993-2001.

Davie, P.S. and Farrel, A.P. (1991a). Cardiac performance of an isolated heart preparation from the dog fish *(Squalus acanthias)*. The effects of hypoxia and coronary artery perfusion. Can J. Zool. 69 : 1822-1828.

Davie, P.S. and Farrel, A.P. (1991b). The coronary and luminal circulations of the myocardium of fishes. Can. J. Zool. 69 : 1993-2001.

Davies, P.S. and Francis, E.T.B. (1941). The heart of a salamander (*Salamandra salamondra L.)* with special reference to its conducting (connecting) system and its bearing on the phylogeny of the conducting system of the mammalian and avian hearts. Philos. Trans. 231(B) : 99-130.

Farrel, A.P. (1987). Coronary flow in a perfused rainbow trout. J. Exp. Biol., 129 : 107-123.

Farrel, A. & Jones, D. R. (1992). The heart. In *Fish Physiology*. The Cariovascular system. (Eds. Hoar, W. and Randall, D.J. and Farrell. A.P.) Academic press. London. XII : 1-88.

Ferenczi, M.A. Homsher, E. Trentham, D.R. & Weeds, A.G. (1978 a). Preparation and characterization of frog myosin subfragment I and actin. Biochem. J., 171 : 155-163.

Ferenczi, M.A., Homsher, E., Simmons, R.M. & Trentham, D.R. (1978b). Reaction mechanism of the magnesium ion-dependent adenosine triphosphate of frog muscle myosin and subfragment I, Biochem. J., 171 : 165-175.

Fernandez-Teran, M.A. & Hurle, J.M. (1982). Myocardial fibre architecture of the human heart ventricles. Ant. Rec., 204 : 137-147.

Ganon, B.J. & Burnstock, G. (1969). Excitatory adrenergic innervation of the fish heart. Comp. Biochem. Physiol., 29 : 765-773.

Gaur, M. (1992). A comparative study of AChE enzyme kinetics of the normal and infarcted heart of *Ophiocephalus punctatus*. M. Phil. Thesis, Barkatullah University, Bhopal, India.

Gaur, M. & Kumar, S. (1993). Effect of an organophosphorus pesticide on AChE enzyme kinetics in normal injured myocardium of the heart of *Channa punctatus*. Biomed. Res., 4 : 171-179.

Grodzinski, A. (1954). Initiations of contractions in the heart of seatrout, *Salmo trutta* L. embryo. Bulletin Acad. Polon des Sci., C1 II : 127-130.

Gyevai, A.T. (1960). Herzstudien III. Histologische and histochemische Untersuchungen am Herzen. Von *Abramis brama* ; Acta Biol. Hungaricae 11.

Hande, R.S. and Pradhan, P.V. (1990a). In vivo effect of dimethoate of AChE from a freshwater teleost, *N. notopterus*. Proc. Indian Acad. Sci. (Animal Sci.) 99 : 53-56.

Hande, R.S. and Pradhan, P.V. (1990b). Effect of sumicidin of AChE from a freshwater teleost, *N. notopterus*. J. Comp. Anim. Physiol., 8 : 55-59.

Hesse, R. (1921). Das Herzgewicht der Wirbeltiere. Zool. Jb. Zool. Physiol. 38 : 243-364.

Johansen K. & Burggren, W. (1980). Cardiovascular function in lower vertebrates. In : Hearts and Heart-like organs. (Ed. Bourne, G.h.) Vol. 1. Academic press, N.Y.

Johnston, I.A. (1983). Comparative studies of contractile proteins from the skeleton and cardiac muscles of lower vertebrates. Comp. Biochem. Physiol. 76A (3) : 439-445.

Johnston, I.A. and Walesby, N.J. (1979). Evolution temperature adaptation and calcium regulation of the actomyosin ATPases. J. Comp. Physiol., 129 : 169-173.

Karandikar, K.R. and Thakur, S.S. (1954). Anatomy with notes on distribution and bionomics of *Sciaenoides brunneus*. Day Zoological Memoir, University of Bombay (First edition) 3 : 54-63.

Keith, A. and Flack, M. (1907). The form and nature of muscular connections between primary division of the vertebrate heart J. Anat. 41 : 172-189.

Keith, A. and Mackenzie, I. (1910). Recent researches on the anatomy of the heart. Lancet 1 : 101-103.

Kumar, S. (1974d). Neurohistological studies on the heart of *Labeo rohita* (Ham.) with special reference to the conducting system. Geobios, 1 : 127-130.

Kumar, S. (1976). Neuroanatomy of the heart of *Garra lamta* Ham. Buch. J. Anat. Soc., India, 25; 1-5.

Kumar, S. (1978a). Cardiac conducting system in fishes. Proc. All India Sem. Ichthyol., 36-40.

Kumar, S. (1978b). Innervation of the heart of *Notopeterus notopterus*. Proc. All India Sem. lchthyol. 126-128.

Kumar, S. (1979). Nerve endings in the heart of teleosts. Z. mikrosk anat. Forsch Leipzig, 93 : 1051-1056.

Kumar, S. Tembhre, M. & Satyadevan, S. (1995). Acetylcholinesterase inhibitory kinetics for monitoring the pesticide toxicity to fish. In : Pollution and Biomonitoring (Ed. B.C. Rana). Tata McGraw Hill Publications, New Delhi. pp. 426-448.

Laurent, P. (1962). Contribution a Fetud emorphotogique et physiologique de innervation du coeur des Teleosteens. Arch. Anat. Muir. Morph. Exp., 51 : 339-548.

Laurent, P., Halmgren, S. and Nilsson, S. (1983). Nervous and humoral control of the fish heart structure and function. Comp. Biochem. Physiol., 76 : 525-542.

Lehman, W. & Szent-Gyorgyi, A.G. (1975). Regulation of muscular contraction. Distribution of actin control and myosin control in the animal kingdom. J. gen. Physiol., 1-30.

Main, A.R. (1969). Kinetic evidence of multiple reversible cholinesterase based on inhibition of organophosphates. J. Biol. Chem. 244 : 829-840.

Malencik, P.A., Heizmann, C.W. and Fishcher, E.H. (1973). Structural proteins of dogfish skeletal muscle. Biochemistry 11 : 715-721.

McWilliam, J.A. (1885). On the structure and rhythm of the heart in fishes with special reference to the heart of eel. J. Physiol. Lond., 6 : 192-245.

Mohan, P.J. and Brink, J.J. (1970). Inhibition of AChE in vitro by Penthylenetetrazol. J. Neurochem., 17 : 949-953.

Nicol J. A. (1952). Autonomic Nervous System in lower Chordates, Biological Review, 27 : 1-49.

Nemcsok, J. Asztalos, B. and Szabo, A. (1990). The effect of methidathion, paraquat and CuSO4 singly or in combination of AChE activity of carp. Biol. (Personal communication).

Nair, M.G.K. (1970a). The structure and development of heart of the Chinese carp. *Cyprinus carpio* (Linnaeus). Proc. Zool. Soc., Bengal, 23 : 169-179.

Nair, M.G.K. (1970b). The development of the nervous system in the heart of Chinese carp, *Cyprinus carpio* (Linnaeus), with special reference to its conducting system. Microscopie, 29 : 1-7.

Nilsson, S., Abrahamsson, T. & Grove, D.J. (1976). Sympathetic nervous control of adrenaline release from the head kidney of the cod, *Gadus morhua*, Comp. Biochem. Physiol., 55C : 123-127.

Nilsson, S. (1983). Autonomic nerve function in the vertebrate. Springer-Verlag, Berlin.

Parson, C.W. (1929). The conus arteriosus in fishes, Q.J. Micros. Sci., 73 : 145-176.

Perry, S.V. (1979). The regulation of contractile activity in muscle. Biochem. Soc. Trans. 7 : 593-617.

Ratha, B.K. & Ramanujam, S.N. (1986). Variation in AChE from brain and muscle of a freshwater air breathing teleosts, *Heteropneustes fossilis*. Acta Biochem. Biophys. Hung., 21 : 381-390.

Prakash, R. (1953). The heart of common Indian catfish, *Heteropneustes fossilis* (Bloch) with special reference to the conducting system. Proc. Zool. Soc. Bengal, 6 : 113-118.

Roberts, R.J. (1978). Fish Pathology. Bailliere, Tindall, London.

Sahai, R. and Chawla, D.S. (1967). The heart of *Labeo rohita* (Ham.) with special reference to the conducting system. Ichthyologica, 6 : 31-36.

Saito, T. (1969). Electrophysiological studies on the pacemaker of several fish hearts. Zool. Mag., 78 : 291-296.

Saito, T. (1971). Effect of cardiac nerve stimulation of the spontaneous activity of carp hearts. Zool. Mag., 274-278.

Saito, T. (1973). Effects of vagal stimulation on the pacemaker action potentials of carp heart. Comp. Bochem. Physiol., 44A : 191-199.

Saito, T. and Tenma, K. (1976). Effects of left and right vagal stimulation on excitation and conduction of the carp heart *(Cyprinus carpio)*. J. Comp. Physiol. 111 : 39-53.

Sanchez-Quintana and Hurle, J.M. (1987). Ventricular myocardial architecture in marine fishes. Anat. Rec., 217 : 263-273.

Santer, R.M. (1977). Monoaminergic nerves in the central and peripheral nervous system of fishes Gen. Pharmac., 8 : 157-172.

Santer, R. M. (1985). Morphology and innervation of the fish heart. Adv. Anat. Embryol. Cell Biol., 89 : 1-102.

Satchell, G.H. (1971). Circulation in fishes, Comb. Monogr. Exp. Biol. No. 18.

Saxena, D.B. and Bakshi, P.L. (1966). Cardiovascular system of some fishes of the torrential streams in India. Part III. Heart of *Garra mullya* (Sykes) and *Schizothorax curvifrons* Niger Heckel. Acta Societatis Zoologica Bohem. 30 : 323-326.

Seto, Y. and Shinohara, T. (1987). Inhibitory effects of paraquat and its related compounds in the acetylcholinesterase activities of human erythrocytes and electric eel *(Electrophorus electricus)*. Agric. Biol. Chem., 61 : 2131-2138.

Skramlik, E.V. (1935). Ueber den Kreislauf bei den. Fischen. Ergebnisse der Biologie, 2 : 1-130.

Smatresk, N.J. (1986). Ventilatory and cardiac reflex responses to hypoxia and NaOH in *Lepisosteus osseus* and air breathing fish. Physiol. Zool., 59 : 385-397.

Smith, W.C. (1918). On the process of disappearance of the conus arteriosus in teleost, Anat. Rec., 15 : 65-71.

Stull, J.T., Manning, D.R. High, C.W. & Blumenthal, D.K. (1980). Phosphorylation of contractile proteins heart and skeletal muscle. Fedn. Proc. 39 : 1552-1557.

Takshima, F. (1982). Vascular system. In : Atlas of Fish Histology, normal and pathological features. (Eds. Hibiya, T.) Kodansha Ltd., Tokyo, Japan.

Taylor, E.W. (1992). Nervous control of the heart and cardiorespiratory interaction. In : Fish Physiology. (Eds. Hoar, W.S. Randall, D.J. & Farrell A.P.). Vol.12. A. Academic Press, London.

Torrey, T.W. (1971). Morphogenesis of the vertebrates. 3rd Ed., Wiley, New York.

Tota, B. (1978). Functional cardiac morphology and biochemistry in Atlantic bluefin tuna. In : "The Physiological Ecology of Tunas" (Eds. G.D. Sharp and Bizon A.C.) Academic Press, New York. pp. 89-112.

Tota, B. (1983). Vascular and metabolic zonation in the ventricular myocardium of mammals and fishes. Comp. Biochem. Physiol., 76A : 423-437.

Tota, B. (1989). Myoarchitecture and vascularization of the elasmobranch heart ventricle. J. exp. Zool. (Suppl.), 2 : 122-135.

Thomas, E.T. (1976). The minute anatomy of the heart of *Anabas testudineus*. Zool. Anz., 196 : 397-404.

Yamauchi, A. (1969). Innervation of the vertebrate heart as studied with electron microscope. Arch. Histol. Jpn. (Niigata, Jpn), 31 : 83-117.

Yamauchi A. (1980). Fine structure of the fish heart. In : Heart and heart-like organs. (Eds. Geoffrey H. Bourne,) Academic Press, London.

Yamauchi, A. and Burnstock, G. (1968). An electron microscopic study on the innervation of the trout hearts. J. Comp. Neurol., 132 : 567-588.

Young. J.Z. (1931). On the autonomic nervous system of Selachians. Q. Microsc. Sci., 75 : 571-624.

Chapter 7

BLOOD

The blood of fishes is similar to that of any other vertebrate. It consists of plasma and cellular (blood cells) components. The cellular components are red blood cells (RBC), white blood cells (WBC) and thrombocytes designated as formed elements (Ross and Reith, 1985). The plasma is liquid portion and consists of water. It acts as the solvent for a variety of solutes including proteins, dissolved gases, electrolytes, nutrients, waste material and regulatory substances. Lymph is the part of plasma that perfuses out of the capillaries to bathe the tissue.

Composition of Blood Plasma : The plasma composition is as follows :

Water

Proteins (fibrinogen, globulin, albumin)

Other solutes

Small electrolytes (Na^+, K^+, Ca^{++}, Mg^{++}, Cl^-, HCO_3^-, PO_4^{---}. SO_4^{--})

Non-protein nitrogen (NPN) substance (urea, uric acid, creatine, creatinine, ammonium salts).

Nutrients (glucose, lipid, amino acid)

Blood gases (oxygen, carbon dioxide, nitrogen)

Regulatory substances (hormones, enzymes).

Plasma and Serum

If the blood is prevented from clotting, it separates out into cells and plasma, if it is allowed to clot it separates into clot and serum. Serum and plasma are very similar, the only difference being that serum has lost the clotting factors prothrombin and fibrinogen which are present in plasma.

If the blood is collected in a vial containing an anticoagulant, the blood will not coagulate and if centrifuged, the blood cells will be separated and settle down, the liquid portion is known as 'plasma'. If the blood is collected in the vial without any anticoagulant, then the blood will coagulate and if this is centrifuged, then the liquid portion is known as 'serum'. Actually the serum has lost the clotting factor prothrombin and fibrinogen but the plasma contains clotting factor proteins also. The marine teleosts living in polar and sub-polar regions contain antifreeze proteins (AFP) or antifreeze glycoprotein (AFGP). They lower the freezing point of plasma without affecting the melting point (McDonald and Milligan, 1992).

The fishes which live at low temperature as low as – 1.9 °C do not freeze due to glycoprotein containing amino acids alanine and threonine in the ratio of 2 : 1 with molecular weight between 2600 to 33000.

Fish plasma contains albumin, the protein which control osmotic pressure. It also contains lipoprotein whose main function is to transport lipid. Ceruloplasmin, fibrinogen and iodurophorine are some important proteins of fish blood. Ceruloplasmin is a copper binding protein.

The total plasma protein in fish ranges from 2 to 8 g dl^{-1}. The thyroid binding proteins such as T_3 and T_4 are present in the blood circulation in free form. Thyroxine binds to vitellogenin in several Cyprinid species (MacKenzie et al., 1987).

Enzyme such as CPK, alkaline phosphatase (Alk PTase), SGOT, SGPT, LDH and their isoenzymes are reported in fish plasma.

Formed Elements of the Blood

There are three varieties of cells or corpuscles present in blood :

(a) Red blood corpuscles or Erythrocytes

(b) White blood corpuscles or Leucocytes.

1. Agranulocytes

   (a) Lymphocytes (b) Monocytes

2. Granulocytes

   (a) Neutrophils (b) Eosinophils (c) Basophils

(c) Platelets or Thrombocytes. (Fig. 7.1a-j)

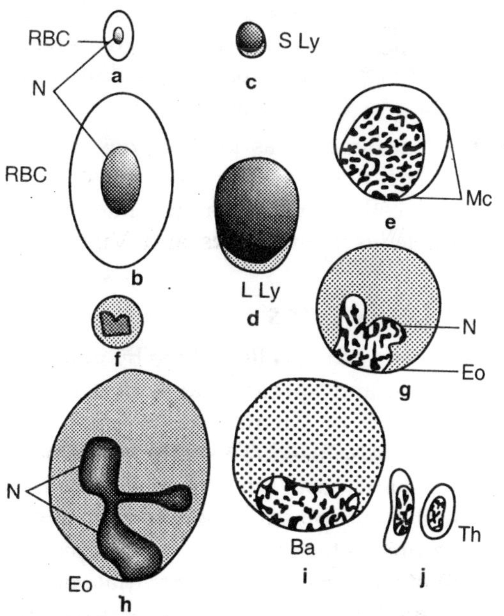

Fig. 7.1a-j : Diagrammatic sketch of blood cells of fishes. (a) Erythrocyte (RBC) of elasmobranch (Small in fig.). (b) Erythrocyte (RBC) teleost (smaller). (c) Lymphocyte (small) (SLY). (d) Lymphocyte (large) (LLy). (e) Monocyte (Mc). (f) Neutrophil (g) Eosinophil (Eo). (h) Eosinophil (Eo). (i) Basophil (Ba). (j) Thrombocyte (Th). N, nucleus; RBC, red blood cells.

Erythrocytes

Dawson (1933) classified immature erythrocytes into five catagories according to structure, distribution and quantity of basophilic substances within the cell. The cytoplasm of RBC is coppery purplish, coppery pinkish or light bluish in Indian freshwater teleosts (Joshi, 1987). Mature erythrocytes contain abundant haemoglobin and are pink or yellowish on preparation stained with Giemsa.

During the growth of RBC firstly the cytoplasm shows strong basophilia because of the presence of polyribosomes. Due to the accumulation of ultra cellular protein, the changes in the staining reaction of cytoplasm occurs because of haemoglobin which stain with eosin. The cytoplasm takes eosin stain due to haemoglobin and basophilia due to staining of the ribosomes. Because of double staining the cell is called a polychromatophilic erythroblast. These polychromatophilic erythroblast are often called reticulocytes.

In adult teleosts, the blood normally contains a certain percentage of immature red cells or senescent cells (growing cells) called proerythrocytes or reticulocytes (Boomker, 1980; Keen *et al.*, 1989). The number of erythrocyte in blood varies according to the species as well as the age of the individual, season and environmental conditions. However, under similar conditions, a fairly constant number of reticulocytes are present in species (Hibiya, 1982).

The nucleus is centrally placed and round or oblong in shape (Fig. 7.1a, b).

The RBC size is larger in elasmobranchs in comparison to teleosts. According to Shrivastava and Griffith (1974), the brackish water species of *Fundulus* have smaller blood cell than freshwater species. Glazova (1977) reported that erythrocytes are slightly smaller in active species than in non-active. In deep sea teleosts, the size of the RBC is larger than normal teleosts. The mature erythrocytes of fishes vary greatly in their shape and outline, and in peripheral blood they are mostly mature. The shape is generally circular in *Clarias batrachus, Notopterus notopterus, Colisa fasciatus, Tor tor* but ellipsoid, oval or oblong in *Labeo rohita* and *Labeo calbasu*.

Abnormal forms of RBC designated as poikilocytes, microcytes, macrocytes, karyorrhexis, basophilic punctation and nucleated forms were also reported (Saunders, 1966 a,b). However, non-nucleated red cells (erythroplastids haemoglobin packets) are formed in *Maurolicus milleri, Valencienellus tripunctatus* and Vincignerria species as reported by Fange (1992) in fish blood.

White Blood Cells or Leucocytes

The study of different types of blood cells is done by making a smear on the slide. A drop of blood is placed on the slide and spread thin with another slide. The slide is stained generally with Leishman, Wright or Giemsa stains, often supravital stains like brilliant cresyle blue and neutral red are also used. The stain contains methylene blue (a basic dye), related azures (also basic dyes) and eosin (an acid dye). The basic dye stain nuclei, granules of basophils and RNA of the cytoplasm whereas the acid dye stains granules of the eosinophils. The basic dyes are metachromatic, they impart a violet to red colour to the material they stain. It was originally thought that neutral dye is formed by the combination of methylene blue and its related azures with eosin, which stains neutrophil granules is not clear.

Although the fish white blood corpuscles have been well investigated, there is no unanimity regarding their classification. The fish leucocytes in peripheral blood are generally (i) Agranulocytes (ii) Granulocytes. The nomenclature is based on affinity of acid and basic dyes and depends upon human haematology. Plasma cells, basket and nuclear shades are also present.

Agranulocytes : They have no granules in the cytoplasm. The most important distinguishing character is unlobed nuclei. Thus they are distinguished from granulocytes, which posses specific segmented nucleus. Agranulocytes have two varieties.

(a) Lymphocytes, large and small

(b) Monocytes.

(a) Lymphocytes : They are most numerous types of leucocytes. The nucleus is round or oval in shape. They constitute 70 to 90% of the total leucocytes. They are rich in chromatin, although its structure is obscure, and is deep reddish violet in colour in preparation with Giemsa (Rampure, 1984). The teleost lymphocytes measure 4.5 and 8.2 as reported by Ellise (1977), Joshi (1987) and Saunders (1966, a & b) noticed large and small lymphocytes in peripheral blood smears of teleosts, fresh and marine fishes similar to that of mammals. The cytoplasm is devoid of granules but cytoplasmic granules are occasionally present.

In large lymphocytes there is large amount of cytoplasm but in small only little quantity of cytoplasm is evident and nucleus constitutes most of the cellular volume (Fig. 7.1c & d).

In fresh samples wandering activity of lymphocytes is rather rare, but leaf like pseudopods may occasionally be seem protruding from the cell.

Functions of Lymphocyte : The main function of fish lymphocytes is to produce immune mechanism by the production of antibody. The T and B lymphocytes are present similar to T and B lymphocytes of mammals. Teleostean lymphocytes respond to nitrogen, such as PHA, concanvalin A (Con. A) and LPS which are considered as specific to mammalian subclass of lymphocytes (Fange, 1992).

Klontz (1972) demonstrated antibody forming cell in the kidney of rainbow trout while Chiller *et al.*, (1969 a,b) found antibody forming cells in the head kidney (Pronephros) and spleen of *Salmo gairdneri*.

Plasma cells which synthesize and secrete antibody and immunoglobins are reported in fishes both in light and electronmicroscopy (Zapata *et al.*, 1984). In mammals, when B lymphocytes are activated by an antigen then they transform into immunoblasts (plasmoblast) which proliferate and then differentiate into plasma and memory cells.

Monocytes : It consists of much less proportion of WBC population often absent in few fishes. It is suggested that they originate in the kidney and become apparent in the blood when foreign substances are present into the tissue or blood stream. The cytoplasm usually stain smoky bluish or pinkish purple. The nucleus of monocyte is fairly large and varied in shape (Fig. 7e). The function of monocyte is phagocytic.

Macrophage : They are of large size, the cytoplasm is occasionally finely or coarsely granulated. They belong to mononuclear phagocyte septum. According to Bielek (1980), they are abundant in renal lymphomyeloid tissue and spleen in *Oncorhynchus mykiss*. The tissue bound macrophage are termed as reticuloendothelial septum (RES), which is a system of primitive cells, from which monocytes originate. Macrophages are present in various other tissues of fishes such as pronephros and olfactory mucosa, etc. Macrophages system of the spleen, bone marrow and liver play a role in phagocytosis of RBC which undergo degradation. Iron separated from haemoglobin molecule are remove by the liver.

Granulocytes : These cells possess specific granules in large numbers and they retain their nucleus. They are of three types.

(a) Neutrophils

(b) Eosinophils

(c) Basophils.

(a) Neutrophils : The neutrophils in fishes are most numerous of the white blood cells and constitute 5-9% of total leucocyte in *Solvelinus fontinalis*. They are 25% of total leucocytes in brown trout (Blaxhall and Daisley, 1973). They are named for their characteristic cytoplasmic staining. They can be easily identified by the multilobed shape of their nucleus and therefore they are segmented or multilobed but in some fishes neutrophils are bilobed (Rowley *et al.* 1988; Ellsaesser *et al.* 1985) (Fig. 7.2) & (Fig. 7.1f).

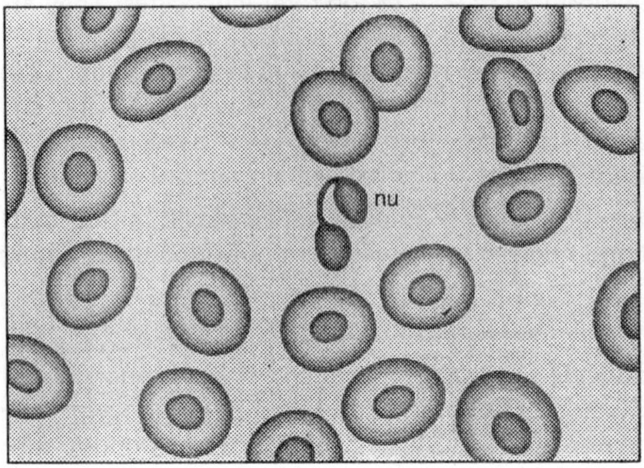

Fig. 7.2 : Microphotograph showing neutrophils in the blood of *Notopterus notopterus* X 1000.

Their cytoplasmic granules are pink, red or violet in peripheral blood smear (Ellis, 1977; Hightower *et al.*, 1984; Roubal, 1986) or azurophilic (Haider, 1967; Joshi, 1987). Peroxide is present in the azurophil granules of neutrophils and is reponsible for bacterial killing. Neutrophils possess Golgi apparatus, mitochondria, ribosomes, endoplasmic reticulum, vacuoles and glycogen. In *Carassius auratus*, three types of granules are reported in the neutrophils (Fujimaki and Isoda. 1990).

The nucleus often looks like the human kidney. In Giemsa stained smear the nucleus is reddish violet in colour and usually exhibits a reticular structure with heavy violet colour. Neutrophils show peroxidase and sudan black positive reaction. The neutrophil is an active phagocyte. It reaches to inflammation site and inflammation refers to local tissue responses to injury.

(b) Eosinophils : According to Ellis (1977), there is much controversy regarding the presence or absence of eosinophils, but both eosinophils and basophils function as antigen sensitivity stress phenomenon and phagocytosis (Ellis, 1977). They occur in low percentage (Jakowska, 1956; Shrivastava, 1968 a, b). These cells are generally round and cytoplasm contains granules which has affinity to acidic dye and they take deep pinkish orange or orange red with purple orange background. The nucleus is lobed, takes deep orange purple or reddish purple stain (Fig. 7.1g & h).

(c) Basophils : The basophils are round or oval in outline. The cytoplasmic granules take deep bluish black stain. They are absent in anguilled and plaice (Ellis, 1977) (Fig. 7.1i).

Thrombocytes or spindle cells : These are round, oval or spindle shaped cells, hence called thrombocytes but in mammals they are disc-like and are called platelets (Fig. 7.3 & Fig. 7.1j).

They occupy as much as half of the total leucocytes in fish. It constitutes 82.2% of WBC in herring (Sherburne, 1973) but only 0.7% in other teleosts (Boyer, 1962). The cytoplasm is granular and deeply basophilic in centre and pale and homogenous on the periphery. The cytoplasm takes pinkish or purplish colour. The thrombocytes help in clotting of blood.

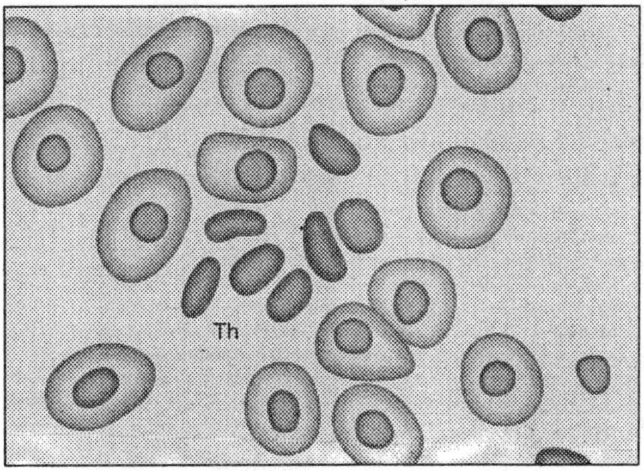

Fig. 7.3 : Microphotograph showing thrombocytes in the blood of *Notopterus* X 1000.

Formation of Blood Cells (Haemopoiesis) : The formation of cells and fluid of the blood is known as haemopoiesis, but usually the term haemopoiesis is restricted to cells.

In mammals the first phase of haemopoiesis or haematopoiesis in developing individual occurs in "blood island" in the wall of yolk sac. It is followed as hepatic phase, i.e., the haemopoietic centres are located in the liver and in lymph tissues.

The third phase of foetal haemopoiesis involves bone marrow and other lymphatic tissues. After birth haemopoiesis occurs in red bone marrow of the bones and other lymphatic tissue.

It is generally agreed that fishes lack haemopoietic bone marrow. Hence erythrocytes and white blood corpuscles are produced in different tissues.

There are two theories, the monophylatic theory and dualistic or polyphylatic theory regarding the origin of vertebrate blood. According to polyphylatic theory, the blood cells arise from a common stem cells whereas according to the monophylatic theory each blood cell arises from its own blood cells. Recent experiments indicate that monophylatic theory is correct or widely accepted. The vertebrate blood arises from pleuropotent stem cells (Cline and Golde, 1970 : Hoffbrand and Pettit, 1980) in haemopoietic tissue. Such experiments are lacking in fishes. Fange (1992) stated that piscine stem cells founded on analogy with mammalian haemopoiesis is an indirect morphological evidence.

Both RBCs and WBCs are originated from lymphoid haemoblast or haemocytoblast usually mature after they enter the blood stream. (Jordan 1938; Moyle and Cech Jr. 1982).

In fishes apart from spleen and lymph nodes many more organs take part in the manufacture of the blood cells.

In elasmobranch fishes, the erythrocyte and granulopoietic tissues are produced in the organ of Leydigs, the epigonal organs and occasionally in the kidney. The Leydig organ is wheatish tissue and analogous to bone marrow-like tissue found in oesophagus but the main site is spleen. If the spleen is removed, then the organ of Leydig takes over erythrocyte production. In teleost, both erythrocytes and granulocytes are produced in kidney (pronephros) and spleen. The teleost spleen is distinguished into a red outer cortex and white inner pulp, the medulla. The erythrocytes and thrombocytes are made from cortical zone and lymphocytes and some granulocytes originate from medullary region. In higher bony fishes (actinopterygii) red blood cells are also destroyed in the spleen. It is not known whether other organs also function into blood decomposition or how blood destruction comes about in the jawless fishes (Agnatha) or in the basking shark and rays (Elasmobranchii) (Lager *et al.*, 1977).

In Chondrichthyes and lungfishes (Dipnoi), the spiral valve of the intestine produces several white blood cell types.

RBC and WBC are formed from haemocytoblast precursor cells which originate from a variety of organs but usually mature after they enter the blood stream (or immature blood corpuscles are usually of two types, large and small).

Function of blood cells : The blood of fishes like other vertebrates consists of cellular components suspended in plasma. It is a connective tissue and is complex non-Newtonian fluid. The blood is circulated throughout the body by cardiovascular system. It is circulated chiefly due to the contraction of heart muscles. The blood performs several function. A few important functions are mentioned as follows :

1. **Respiration :** An essential function is the transport of dissolved oxygen from water and from the gills (respiratory modifications) to the tissue and carbon dioxide from the tissue to the gills.

2. **Nutritive :** It carries nutrient material, glucose, amino acids and fatty acids, vitamin, electrolytes and trace elements from alimentary canal to the tissue.

3. **Excretory :** It carries waste materials, the products of metabolism such as urea, uric acid, creatine, etc. away from cell. Trimethylamine oxide (TMAO) is present in all fishes. It is in high concentration in marine elasmobranchs (Griffith, 1981).

Creatine is an amino acid which is the end product of metabolism of glycine, arginine, methionine whereas creatine is formed by spontaneous cyclisation of creatine. Its level in plasma is 10-80 μm (Wells *et al.*, 1986) and is excreted by kidney.

4. **Haemostasis** of water and electrolyte concentration : The exchange of electrolyte and other molecules and their turn over is the function of blood. Blood glucose levels often cited is being a sensitive physiological indicator of stress in fish and there is no unanimity about the blood glucose levels amongst fishes.

5. **Hormones** and humoral agent : It contains regulatory agent such as hormones and also contains cellular or humoral agent (antibodies). The concentration of various substances in the blood is regulated through feedback loops that sense changes in concentration and triggers the synthesis of hormones and enzymes which initiate the synthesis of substances needed in various organ.

REFERENCES

Boyar, H.C. (1962). Copeia. II, 463-465

Bielek, K, (1980). Zool. Jb. Anat. 103, 498-509.

Blaxhall, P.C. and Daisley, K. W. (1973). J. Fish Biol. 5, 771-782.

Boomker, J. (1980). Onderstepoort J. Vet. Res. 47, 95-100.

Cline, M.J. and Golde, D.M. (1979). Nature, 277-181.

Chiller, J.M., Hodgkins, H.O., Chambers, V.C. and Weister, R.S. (1969,a). J. Immunol. 102. 1193-1201.

Chiller, J.M., Hodgkins, H.O., Chambers, V.C. and Weiser, R.S. (1969,b) J. Immunol. 102, 1202-207.

Dawson, A.B. (1933). Biol. Bull. 64, 33-43.

Ellis, A.E.(1976). J. Fish Biol. 8, 143-156.

Ellis, A.E. (1977). J. Fish Biol. 11, 453-491.

Ellsaesser, C.F., Miller, N.W., Cuchens, M.A. Lobb., C.J., and Clem, L.W. (1985). Trans. Am. fish Soe. 114, 279-285.

Fange, R. (1992). Fish blood cells. In : "Fish Physiology", vol. XIIB (Eds. Hoar, W.S., Randall, D.J., and Farrell. A.P., Academic Press, INC. New York. pp. 1-54.

Fujimaki, Y., and Isoda, M. (1990). J. Fish Biol. 36. 821-831.

Glazova, T.N. (1977). Ichthyologia (Beograd) 9. 65-73.

Griffith, R.W. (1981). Biol. Bull. 160. 250-264.

Haider, G. (1967) - Zool. Anz. 179, 384-409.

Hibiya, T. (1982). An altas of fish histology normal and pathological features. Kodansha Ltd. Gustav Fischer Verlag Stuttgart.

Hightower, J.A., McCumber, L.J., Welsh, M.G., Whatley, D.S., Hartivigsen, R.E. and Sigel. M.M., (1984). J. Fish Biol. 24, 587-598.

Hoffbrand, A.V., and Pettit, J.E. (1980). Essential Haematology. Blackwell, Oxford.

Jakowska, P. (1956). Revue Hemat. 11, 519-539.

Jordan, H.E. (1938). In : Handbook of Haematology (Ed. H. Downey) Vol. 2. Hamilton Medical Books, London, pp. 715-744.

Joshi, B.D. (1987). J. Env. Zool. I. 98-113.

Keen, J.E., Steele, A.M.C., and Houston, A.H. (1989). Comp. Biochem. Physiol. A. 94, 699-711.

Klontz, G.W. (1972). Symp. Zool. Soc. London. 3. 89-99.

Lagler, K.F., Bardach, J.E. Miller, R.R., Passino, D.R.M. (1977). Ichthyology. John Willey and Sons. New York.

McDonald D.G. and Milligan, C.L., (1992). Chemical properties of the blood, In Fish Physiology Vol. XII Part B (Eds. Hoar, W.S. Randall. D.J. and Farell A.K.). Academic Press. New York. pp. 55-133.

MacKenzie, D.S., Warner, J. and Thomas, T. (1987). Thyroid reproductive relationship in the channel catfish, *Ictalurus punctatus* evidence for estradiol-induced Changes in plasma thyroid hormone binding. In : Reproductive physiology of Fish (Eds. Idler, D.R., Crim, L. W., and Walsh, J.M.). Memorial Univ. Press, St. John's Newfoundland. pp. 205-211.

Moyle, P.B., Cech, J.J.Jr. (1982). 'Fishes' an introduction to Ichthyology. Prentice-Hall Inc. New Jersey, New York.

Rampure, V. (1984) M. Phil. Dissertation. Bhopal Univ. Bhopal.

Ross, M.H. and Reith, E.J. (1985). 'Histology' A test and atlas. Harper and Row, Publishers : J.B. Lippincott Company. New York.

Rowely, A.F., Hunt, T.C., Page, M. and Mainwaring : (1988). In "Vertebrate blood cells" (Eds. A.F. Rowley and N.A. Aatclife). Cambridge University Press, Cambridge. pp. 19-127.

Saunders, D.C. (1966, a). Trans. Amer. Miscrose. Soc. 87 : 427-449.

Saunders, D.C. (1966, b). Copeia 2, 348-351.

Sherburne, S.W. (1973). Fishery Bull. 71, 1011-1017.

Shrivastava, A.K. (1968, a). Anat. Anz. 123-233.

Shrivastava, A.K. (1968, b). Anat. Anz. 123-520.

Shrivastava, A.K. and Griffith, R.W. (1974). Copeia 1, 136-141.

Wells, R.M.G., McIntyre, R.H., Morgan, A.K. and Davie, T.S. (1986). Comp. Biochem. Physiol 84. (A), 565-571.

Zapata, A., Fange, R. Mattisson, A., Villena, A. (1984). Cell Tissue Res. 235, (691-693).

Chapter 8

CIRCULATORY SYSTEM

In fishes, the blood circulates through a well defined tubular system of heart and vessels. The heart pumps blood towards gills for oxygenation. Through aortic arches arterial blood is distributed to various parts of the body. The venous blood from the body after passing through the kidneys (renal portal system) and the liver (hepatic portal system) returns to the heart arterial system carries blood away from the heart, however, the venous system, towards it. (Randall, 1970, Satchell, 1971).

ARTERIAL SYSTEM

The heart pumps blood through bulbus arteriosus to the ventral aorta. The ventral aorta is a straight tube having thick walls. It lies on the mid-ventral line of the floor of pharynx. In Elasmobranchii and Dipnoi, the ventral aorta seems to be tubular extension of the conus arteriosus. Histologically, both structures are different. The conus arteriosus comprises of cardiac muscle, however, ventral aorta is composed of smooth muscle. In bony fishes (Teleostomi), the ventral aorta has a muscular swelling of smooth muscle at its point of origin from conus arteriosus. In *Xenentodon*, the ventral aorta is thin walled tube (Gupta, 1969).

Arteries to head and gill : The gills of lungfishes (Dipnoi) has four or five pairs of afferent branchial arteries, whereas sharks (Elasmobranchii) (Fig. 8.1) and bony fishes (Teleostomi) have four pairs (Fig. 8.2). The ventral aorta supplies venous blood to the gills through these afferent branchial arteries. These arteries form loops around the gills which divide into arterioles than into capillaries and lacunae into the gill lamellae, which are the main seat of solute, and gases exchange between blood and water. Thus in the capillary bed of the gills reoxygenation of blood and removal of carbon dioxide occur gradually. Oxygenated blood from the gills is collected by the efferent branchial arteries. Only one efferent branchial artery is present in each gill arch of bony fishes (teleostomi). In cartilaginous fishes (Elasmobranchii), they may be paired, i.e., one for each hemibranch of the gill arch, except the anterior hyoid arch, which possesses only one branchial artery. Lungfishes (Dipnoi) show intermediate conditions. The first two efferent branchials meet dorsally and form the anterior epibranchial artery on each side. The third and fourth efferent branchial vessels also arise from their respective holobranchs and join together to form a short post second epibranchial. Thus both the epibranchials run posteriorly and merge together to form the median dorsal aorta. Thus, the dorsal aorta is paired in the gill region and becomes median and unpaired just behind the gill region (Lagler, *et al.*, 1977 & Moyle & Cech. Jr. 1982). In *Xenentodon*, the first and second epibranchial arteries are communicated by a longitudinal connection.

The right posterior epibranchial artery differs from the left, as the former gives out coeliaco-mesenteric artery and another artery to interrenal body and nearby muscles and vertebrae. Coeliaco-mesenteric artery shows variation in its origin in teleosts. In *Ophiocephalus striatus*, it originates from the fourth right branchial artery (Marathe and

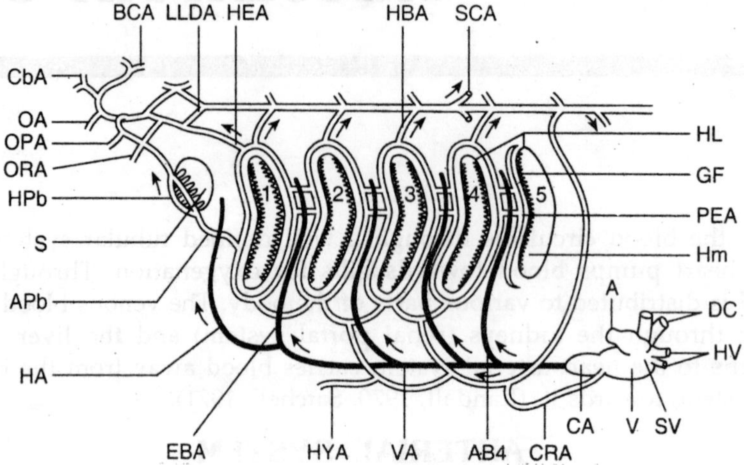

Fig. 8.1 : Schematic diagram of branchial circulation in shark (Elasmobranch). A, atrium; AB4, afferent branchial; APb, afferent pseudobranch; BCA, basal cerebral artery, CA, conus arteriosus; CbA, cerebral artery; CRA, coronary artery, DC, ductus Cuvieri; EBA, efferent branchial artery; GF, gill filament; HA, hyoidean branch; HEA, hyoidean efferent artery; HL, holobranch; Hm., hemi branch; HPb, hyoidean pseudobranch; HV, hepatic vein; HYA, hypobranchial artery; LLDA, left lateral dorsal aorta; OA, optic artery; OPA, ophthalmic artery; ORA, orbital artery; PEA, posterior efferent artery; S, spiracle; SCA, subclavian artery, SV, sinus venosus; V, ventricle; VA, ventral aorta.

Kulkarni, 1957), and in *Caranx* (Dater, 1954) from the right side of the dorsal aorta. In *Sciaenoids* (Karandikar and Thakur, 1953) it arises from the place between the posterior branchial artery and the dorsal aorta. In some fishes there are separate coeliac and mesenteric arteries. In Polynemus (Karekar, 1954) coeliaco-mesenteric artery comes out from the right epibranchial artery formed by the third and the fourth branchial arteries.

The anterior epibranchial artery runs forward to form common carotid artery on both the sides. The common carotids immediately divides into an external and an internal carotids. From the external carotid artery gives out two arteries, the first is afferent pseudobranchial artery to pseudobranch and the second is mandibular artery to the lower jaw. The spiracles (Elasmobranchii) and the hyoidean pseudobranch (Teleostomi) are supplied with oxygenated blood only from a mandibular artery and pseudobranchial capillary respectively. The latter present a special arrangement as the efferent pseudobranchial artery carries the blood that has passed the pseudobranchial capillaries through ophthalmic artery to the choroid gland of the eye. The pseudobranch may possess receptors sensitive to oxygen and carbon dioxide tension.

Both the right and left internal carotids join together and form circular cephalicus just below the pituitary body or hypophysis in *Xenentodon*.

Circular cephalicus gives off many arteries on its anterior and lateral sides.

From each side of hypophysis, two branches come out anteriorly which run forward and unite behind the optic chiasma forming another sinus. This anterior sinus gives a pair of maxillonasal arteries, which supply first the olfactory organs and than run as maxillary artery up to the tip of the snout.

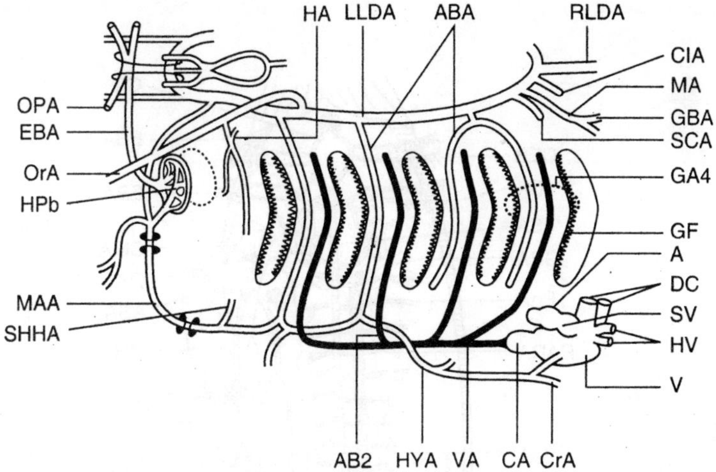

Fig. 8.2 : Generalized diagram of branchial circulation in bony fish (Osteichthyes). A, atrium; ABA, afferent branchial arteries, CA, conus arteriosus; CIA, coeliaco-mesentric artery, CrA, coronary artery. DC ductus Cuvieri; EBA, efferent branchial artery, GA4, gill 4 GF, gill filament; GBA, gas bladder artery; HA, hyoidean branch; HPb, hyoidean pseudobranch; HV, hepatic vein; HyA, hypobranchial artery; LLDA, left lateral dorsal aorta; MA, mesenteric artery; MAA, mandibular arterial arch; OPA, ophthalmic artery; OrA, orbital artery; RLDA, right lateral dorsal aorta; SCA, subclavian artery, SHAA, secondary hyoidean afferent artery, SV, sinus venosus; V, ventricle; VA, ventral aorta.

A large single cephalic artery arises from each lateral side of circular cephalicus. It divides into orbital and cerebral arteries. The orbital artery divides into optic and an opthalmic artery. The former extends along the optic nerve and latter further divides into various branches supplying the surface of eye wall.

The cerebral artery also divides immediately into the anterior and posterior cerebral arteries. The posterior cerebral artery of each side comes down after piercing the brain and unite on the mid-ventral side of the brain and form the basillary artery, which runs posteriorly below the spinal cord in the form of the spinal artery.

Abdominal Arteries

The trunk receives blood through two main arteries and their branches, which are dorsal aorta and coeliacomesenteric artery (Fig. 8.3).

Dorsal aorta : The dorsal aorta gives out four branches soon on entering the abdominal region. These are as follows.

1. **Subclavian arteries :** It arises from dorsal aorta, near the second epibranchial. They run anteriorly and bifurcates into two branches to supply the pectoral girdle, pectoral fin and its musculature.

2. **Parietal arteries :** The dorsal aorta gives out numerous parietal arteries on each lateral side, which supply to lateral muscle, vertebrae and integument.

3. **Renal arteries :** The dorsal aorta passes through the substance of kidneys and gives off a number of renal arteries to the kidney, some of which continues into pelvic fins, urinary bladder and corpuscles of Stannius

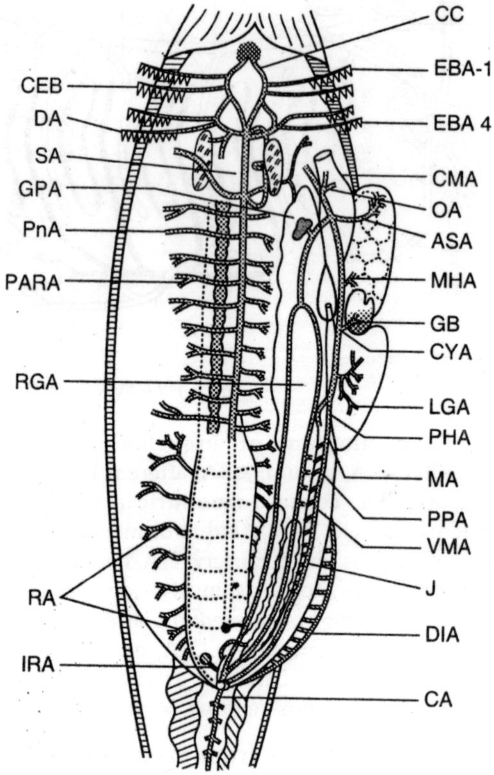

Fig. 8.3 : Diagram of arterial system of *Xenentodon*. ASA, anterior splenic artery; EBA-1, caudal artery; CMA, coeliaco-mesenteric artery, CC, common carotid; CYA, cystic artery; DA, dorsal aorta; DIA, dorsal intestinal artery; EBA-4, efferent branchial arteries; GB, gall bladder; GPA, gonado-pneumatic artery; LGA, left gonadal artery, MA, mesenteric artery; MHA, median hepatic artery, OA, oesophrangeal artery; PARA, parietal artery; IRA, inter renal artery; PnA, post hepatic artery; PHA, pneumatic artery; PPA, posterior pneumatic artery; RA, renal artery; RGA, right gonadal artery, SA, subclavian arteries.

4. **Caudal arteries :** The dorsal aorta then continues behind as caudal artery and runs along with the caudal vein inside the haemal canal. The caudal artery supplies blood to the dorsal, anal and caudal fins and the muscles of this region.

Coeliaco-mesenteric artery : This artery takes its origin from the right posterior epibranchial artery and supplies the whole digestive tract, liver, air bladder and gonads through thin branches. All thin branches are present on the dorsal side except the gonads and pneumatic arteries, which are ventral to them. The branches of coeliaco-mesenteric artery from the beginning are as follows.

1. **Oesophageal artery :** As soon as the coeliaco-mesenteric artery arises from the right posterior epibranchial artery, it divides into two branches, one extends forwardly to the pharynx whereas the other goes to oesophagus.

2. **Anterior hepatic artery :** It arises from just behind the oesophageal artery, then enters the anterior lobe of liver, where it divides into capillaries.

3. **Gonado-pneumatic artery :** This artery originates from just opposite to the origin of the anterior hepatic artery. Soon it divides into following branches.

(a) Gastric artery to stomach.

(b) Anterior pneumatic artery to air bladder, particularly to its oval gland.

(c) Right and left gonadial arteries run to the gonads and thin ducts.

(d) Posterior pneumatic arteries : Number of these arteries originate from the left gonadial artery and supply to the air bladder.

4. **Splenic arteries :** The anterior and posterior splenic arteries supply blood to the spleen.

5. **Median hepatic and pancreatic arteries :** They go to hinder part of the anterior lobe of liver and also the pancreatic follicles buried in the hepatic tissues.

6. **Cystic artery :** It takes its origin from just opposite to posterior splenic artery. It enters the gall baldder and breaks up into capillaries in its wall.

7. **Posterior hepatic artery :** After a short distance from cystic artery, the coeliaco-mesenteric artery gives out a large posterior hepatic artery, which divides into capillaries in the posterior lobe of the liver.

8. **Mesenteric artery :** After giving out posterior hepatic artery, the coeliaco-mesenteric artery is considered as mesenteric artery, which divides to form dorsal and ventral intestinal artery extending to the intestine and rectum.

Coronary Arteries

Median hypobranchial artery and coronary network is present over the ventricle. The median hypobranchial artery sends fine capillaries to the bulbus arteriosus. The origin of hypobranchial artery is still unknown.

VENOUS SYSTEM

The sinus venosus receives venous blood from the different parts of the body through paired Cuvierian ducts and single hepatic vein. The right Cuvierian duct receives venous blood through right subclavian, right superior jugular, inferior jugular and right posterior cardinal veins. However, the left Cuvierian duct collects venous blood through superior jugular vein and left subclavian vein only. Blood from various parts of the body is collected by the following veins (Fig. 8.4).

Hepatic Vein

It collects blood from the liver. The hepatic vein receives blood also from pharyngeal muscles and skin through a pair of right and left anterior lateral cutaneous veins.

The inferior jugular vein is single and present above the heart and ventral aorta. It brings blood into the right Cuvierian duct near the sinus venosus. It has following branches:

1. **External mandibular vein :** It collects blood from the intermandibularis muscles, labial folds and the lateral line.

2. **Internal or submandibular vein :** It runs on the each lateral side of median line of lower jaw.

3. **Common median mandibular vein :** External and internal mandibular veins unite mid-ventrally to form common median mandibular vein.

4. **Branchiostegal vein :** It collects blood from branchiostegal rays (Gupta, 1969).

5. **Pharyngeal vein :** Blood from the infrapharyngeal part and buccal cavity is collected by these veins.

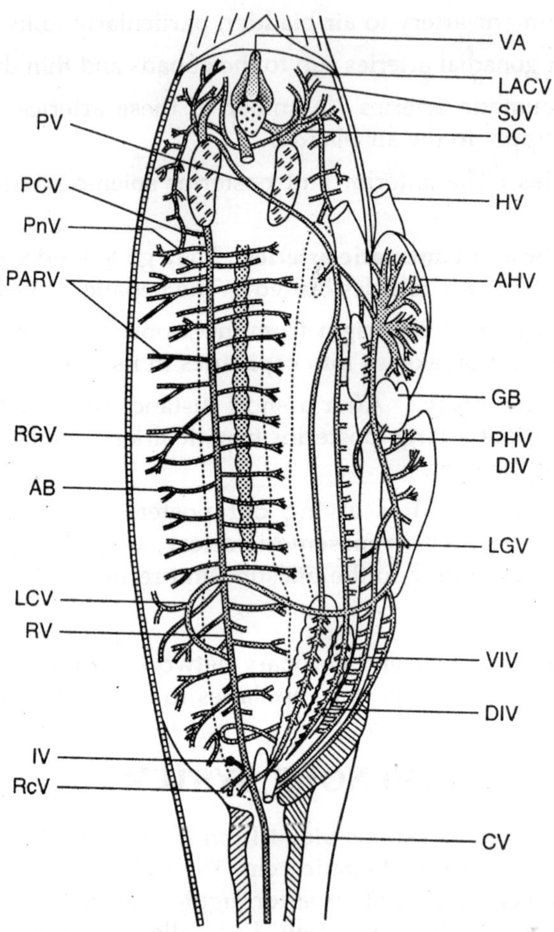

Fig. 8.4 : Diagram of venous system of *Xenentodon*. AB, air bladder, AHV, anterior hepatic vein; CV, caudal vein; DC, ductus Cuvieri; DIV, dorsal intestinal vein; GB, gall blader; HV, hepatic vein; LACV, lateral anterior cutaneous vein; LCV, lateral posterior cutaneous vein; LGV, left gonadal vein; PARV, parietal veins; IV, interrenal vein; PHV, post hepatic vein; PV, pancreatic vein; PnV, pneumatic vein; PCV, post cardinal vein; RV, renal vein; RcV, rectal vein; RGV, right gonadal vein SJV, superior jugular vein; VA, ventral aorta; VIV, ventral intestinal vein.

Subclavian Vein

It collects blood from the pectoral girdle and thin muscles.

Posterior Cardinal Vein

It is a large vein, runs posteriorly on the right side of the body and receives the following branches :

1. **Caudal vein :** It lies extended between the kidney and the base of caudal fin. It brings blood from caudal, anal and dorsal fins and adjacent muscles.

2. **Rectal vein :** When the caudal vein passes through the kidney it receives a rectal vein.

3. **Post cardinal vein :** It receives blood from the paired corpuscles of Stannius. It lies just anterior to the rectal vein.

4. **Parietal veins :** They collect blood from throughout the abdominal integument and the lateral muscles and discharge into the postcardinal veins.

5. **Renal veins :** A number of renal veins collect blood from the kidney tissues and open into post cardinal vein.

6. **Interrenal vein :** It collects blood from the interrenal body and pour into postcardinal vein.

HEPATIC PORTAL SYSTEM

The veins from stomach and intestine discharge venous blood into the hepatic portal vein. In the liver, the portal vein gives rise to innumerable sinusoids that facilitate transfer of dissolved food substances to the liver tissue. The blood is then collected by hepatic veins which are as follows (Fig. 8.4, and Fig. 8.5) :

1. **Gastric vein :** It is a small vein. It arises from behind the oesophagus and soon opens into the main hepatic portal vein.

2. **Splenic vein :** They arise laterally from the spleen and join the hepatic portal vein.

3. **Cystic vein :** It leaves the hinder part of the gall bladder and opens into the hepatic portal vein.

4. **Dorsal intestinal vein :** It runs dorsally on the posterior part of the alimentary canal. It has several fine branches. It joins the hepatic portal vein at the level of the posterior lobe of the liver.

5. **Ventral intestinal vein :** It is anastomosingly branched into capillaries in the rectal region and also has small blood vessels on the ventral side of the intestine. It collects blood from the rectal and the intestinal region. It then turns around the alimentary canal to meet the hepatic portal vein opposite to the cystic vein.

6. **Pancreatic vein :** It is a small vein leaving pancreatic follicles and opens into the hepatic portal vein.

7. **Posterior cutaneous vein :** It takes blood from the skin and lateral muscles of hinder part of the body and discharge into the hepatic portal vein.

Branches of Postcaval Vein

The post caval vein runs parallel to coeliacomesenteric artery and opens into the posterior cardinal vein. It collects blood through the following (Fig. 8.4, and Fig. 8.5) :

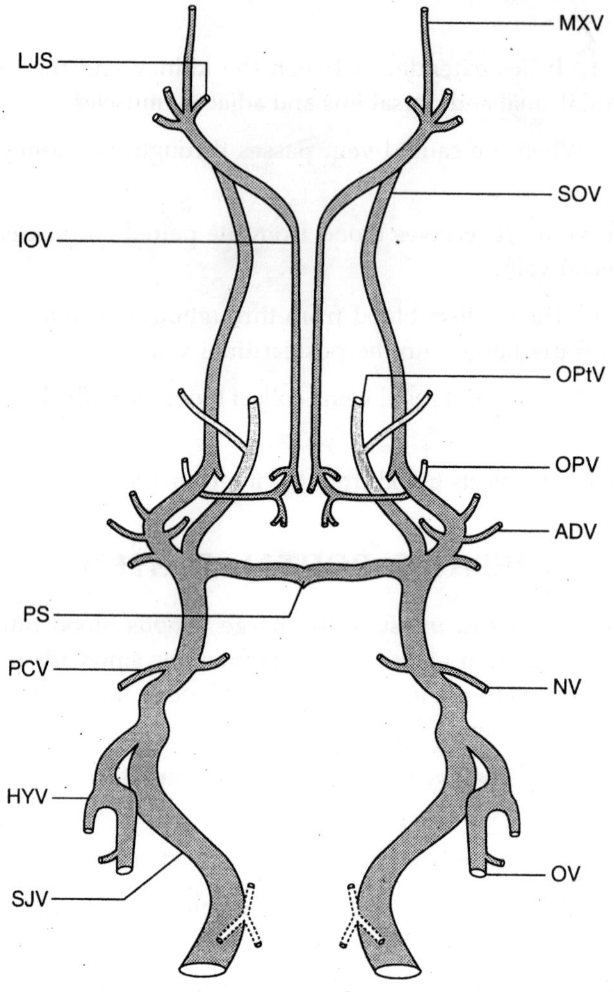

Fig. 8.5 : Diagram of cephalic veins of *Xenentodon*. ADV, vein from adductor mandibular muscle; HYV, hyoidean vein; IOV, infraorbital vein; LJS, lateral jugular sinus; MV, mandibular vein; MXV, Maxillary vein; NV, nasal vein; OV, veins from opercular muscles; OPV, Optic vein; OPtV, ophthalmic vein; PCV, postcerebral vein; SJV, superior jugular vein; SOV, supraorbital vein

1. **Common gonadial vein :** The right and left gonadial veins unite to form common gonadial vein.

2. **Anterior pneumatic vein :** It arises from the air bladder and opens into post caval vein just behind the opening of common gonadial vein.

3. **Anterior splenic vein :** It collects some blood from the spleen into post caval vein. The major part of the blood from the spleen is collected by four lateral splenic veins opening into the portal vein.

Cephalic Veins

The blood from the head region is collected through a pair of large paired superior jugular vein. Several veins open into each superior jugular vein, which opens into Cuvierian duct. Its branches are as follows (Fig. 8.5).

Maxillary vein : The capillaries of snout region merge to form maxillary vein. These are :

1. **Nasal vein :** Two or three nasal veins join the maxillary vein near olfactory organ. It also receives a small vein from lateral line organ.

2. **Infraorbital vein :** It brings blood from the eye muscles and optic vein and opens into maxillary vein.

3. **Optic vein :** It opens into infraorbital vein and runs along the optic nerve.

Supraorbital vein : The main stem of maxillary vein after uniting with the infraorbital vein makes the supraorbital vein. It brings blood from dorsal part of the snout and the adductor mandibular muscles. It brings blood from dorsal part of the snout and the adductor mandibular muscles. It also collects blood from the brain through a small cerebral vein. It opens into lateral jugular sinus. It receives the following veins :

1. **Ophthalmic vein :** It brings blood from the surface of eye-wall and discharge into suprabranchial vein.

Lateral jugular sinus : The supraorbital vein, the ophthalmic vein and veins from the adductor mandibular muscles continues as lateral jugular sinus. It receives blood from pituitary sinus by a transverse vein.

1. **Pituitary sinus :** It lies behind the hypophysis and the arterial circulus cepalicus. Blood from this sinus is taken to lateral jugular sinus by transverse veins.

Pituitary sinus : It lies behind the hypophysis and the arterial circulus inside the skull. It also drain blood from the brain by the posterior cerebral vein and lower jaw by mandibular vein.

Hyoidean vein : It brings blood from the opercular muscles and part near the hyomandibular bone and pour into the superior jugular vein.

THE LYMPHATIC SYSTEM

A system of paired and unpaired lymph ducts and sinuses collect lymph from all parts of the body which ultimately taken to the main blood stream.

The lymphatic system of cartilaginous fishes (Elasmobranchii) comprises lymph vessels. The sinuses and contractile 'lymph hearts' are absent in them. But they are present in bony fishes. The lymphatic fluids are collected from the tail region by subvertebral lymph trunks passing through the haemal canal of tail vertebrae. It then opens into abdominal lymph ducts which form ducts network of vessels with renal and gonadial lymphatic systems. Subvertebral lymph trunk also receives lymph vessels from the intestine and segmental musculature and then open into the cardinal sinus near the subclavian artery. The subvertebral lymph trunk also receive lymph from cranial and branchial regions.

The lymphatic system of fishes is originated from the venous system. It shows increasing complexity through elasmobranchii fishes to teleostomi fishes. The bony fishes (Teleostei) have well formed lymphatic system like the terrestrial vertebrate. They possess highly branched subcutaneous lymph vessels; lymph from the head is collected by branching sinuses and pours into a subscapular sinus in the pectoral region where it is joined by fluid from the three main lymph ducts of the body—the lateral, dorsal, and ventral subcutaneous lymph trunks. Tissue fluid from the body musculature is collected by neural, ventral and haemal submuscular trunk. The lymph ducts of the visceral organs divide into a superficial and a deep system. The latter collects the fat absorbed from the intestinal mucosa (chyle) into a coeliaco-mesenteric lymph trunk. The pararenal lymph vessels that end in the pericardial sinus collect the lymph from the gas bladder, the abdominal part of the kidney and other parts of the body.

REFERENCES

Dater, G.G. (1954). A study of *Caranx rottleri* (Day) Univ. M.Sc. Thesis Bombay.

Gupta, O.P. (1969). The blood vascular system of *Xenentodon* (Ham.) Sonderbdruck aus, Morphologisches Jahrbuch. Band 114. Heft 2. Seiten 177-203.

Karandikar, K.R., and S.S. Thakur (1953). *Sciaenoids brunneaous* (day), Zoological Memoirs 3, The University of Bombay.

Karekar, P.S. (1954). A study of polynemids of Bombay. Ph.D. Thesis Univ. Bombay.

Marathe, V.B. and R.S. Kulkarni (1957). The circulatory of *Ophiocephalus striatus* (Bloch), J. Univ. Bombay N.S. 25B.

Legler, K.F. Bardach., J.E. Miller., R. May Passion (1977). Ichthyology, John Wiley & Sons, New York.

Moyle, P.B. & Cech, J.J.Jr. (1982). Fishes— An introduction to Ichthyology. Prentice-Hall International Inc. London.

Randall, D.J. (1970). The circulatory system. In Fish Physiology. Academic Press, New York.

Satchell, G.H. (1971). Circulation in fishes. Cambridge Univ. Press, London.

Chapter 9

LYMPHOMYELOID TISSUES

The lymphatic system is also called immune system. It consists of small cells with round nuclei having a non-granulated basophilic cytoplasm. These are called lymphocytes. Lymphocytes constitute one of the most abundant cell types in vertebrates. The lymphatic tissues are present in bone-marrow, the thymus, the spleen, the lymph nodes and nodules of lymphatic tissues in the wall of the alimentary canal, respiratory passage and genitourinary tract. Lymph ducts are not distinct in case of teleosts. Spaces containing lymphocytes can be recognized in the intestinal mucosa and mesenteries, the neural and haemal arches of the vertebrae (subvertebral lymphatics) and the sub-cutaneous layer of the skin (subcutaneous lymphatic). A continuos migration of lymphocyte units with the different structures forming "lymphomyeloid complex" (Yoffey, 1985). There are three subcutaneous lymphatics, lateral—dorsal, and ventral.

The lymph finally flows into blood stream through lymph sinus located at the pectoral fin base. A lymphatic heart with valve and cardiac muscle like contractile fibres is recognized in the tail of eel and Atlantic salmon (*Salmo trutta*). The following lympho-myeloid tissues are present in various fishes :

Thymus : The thymus is present in all fishes except cyclostomes (Fig. 9.1). However, the lymphoid cells are reported in velar muscles of Myxinoids. The diffused accumulation of lymphocytes are found in the ammocoetes larvae of *Petromyzon*. Riviere *et al.* (1975) and Good *et al.* (1972) called it as primitive prothymus.

In holocephalans and chondrosteans the thymus is well developed lobulated and is distinguished histologically into cortex and medulla on the pattern as in mammals. It differs from mammals as Hassall's corpuscles are absent. The medulla of mammals contains unusual structure, the thymic Hasall's corpuscles as isolated masses of closely packed epithelial reticular cells.

In teleosts the thymus as a rule regresses at sexual maturity. It is usually present in close association with the epithelium of the branchial cavity, but in carp fish (Hammer, 1908), the thymus during its development moves away from the branchial mucus (Burne, 1927). In a few teleosts a close association is observed between the thymus and another lymphoidal tissue, called the head kidney. There is no clear demarcation of cortical and medullary segments and is of mixed nature.

The thymus contains lymphocytes, reticular, myeloid and mucous cells. Fange (1984 and 1992) stated that the function is unknown. In mammals it is connected with autoimmune disease 'mysthenia gravis'. The function of the thymus is in maturation of lymphocytes and probably in the production of antibody. The plasma cell are found in the thymus of the teleost (Zapata, 1981, Pulsford and Fange, 1984). The plasma cells of mammals are transformed from B-lymphocytes which synthesize and secrete antibody.

The spleen : The spleen is not reported in the cyclostomes. In cyclostomes particularly *Myxine*, the haemopoietic tissue in the intestinal mucosa acts as spleen. In ammocoetes larva, the lymphohaemopoietic tissue is found in the spiral valve or typhlosole of the intestine. The spleen of dipnoans is situated in the stomach wall and in the spiral valve of the intestine. In selachians the spleen is an elongated, lobated mass situated in the duodenosplenic omentum. The spleen of *Alopias vulpes* is large and about one metre in length. The size and shape in teleosts vary in different species. The surface of spleen in teleost is covered with a serous membrane which is made up of connective tissues. It contains connective tissues and capillaries. In the space between them is filled with erythroblast, mature and immature erythrocytes, lymphocytes and monocytes.

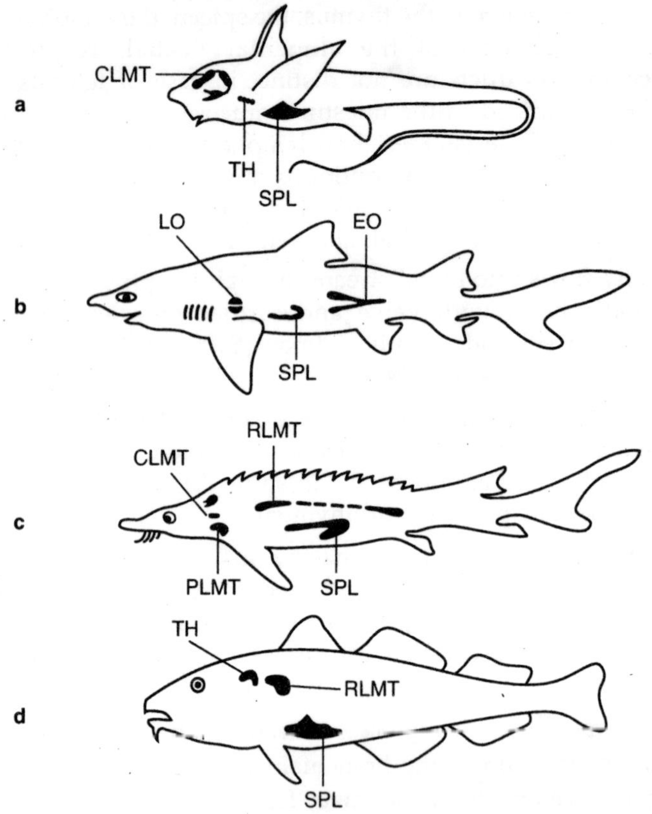

Fig. 9.1a-d : Lymphomyeloid tissues in gnathostomatous fishes. (a) Chimaera (holocephalan). (b) Elasmobranch. (c) Acipencer (chondrostean). (d) Teleost. CLMT, cranial lymphomyeloid tissue; EO, epigonal organ; LO, Leydig organ; PLMT, pericardial lymphomyeloid tissue; RMLT, renal lymphomyeloid tissue; SPL. spleen; TH, thymus.

Histologically, the spleen of fishes resembles the mammalian spleen but white pulp (Malpighian bodies) are less developed. There is no clear distinction between red pulp and white pulp in Osteichthyes.

Kidney : Lymphomyeloid tissue is present in kidney of fishes. In cyclostomes, the phagocytic system of cells are present in mesonephros. The lymphoid masses are present in the kidney of *Chimaera* and *Hydrolagus*. In *Amia calva* the lymphoid tissue is present in the renal tubules (Robenson, 1932). In dipnoans granulocytopoietic tissue is associated with layers of the kidney. In teleosts the lymphoid tissue is present in the head kidney (Sharma and Kumar, 1982). The detail may be seen in the kidney chapter.

Digestive tract : The lymphoid follicles in the form of variable Peyer's plaques are present in the intestine of *Myxine* and *Polydon* and many elasmobranchs (Jacobshagen, 1915; Bucke, 1971). The gut associated lymphoid tissue (GALT) is considered as an important immune system of vertebrates (Fange, 1984, Mattisson, *et al.*, 1990).

In teleosts, intestinal wall in some fishes is known as 'Stratum granulosum' characterized by a large number of granulocytes. They resemble mast cells (Arvy, 1955).

Leydig organ is a large pink or yellowish-white lymphoid mass present in the oesophagus of elasmobranchs. It has a resemblance to haemopoietic bone marrow.

Epigonal organ present in the gonads of elasmobranch is also a lymphoid organ. The detail is given in the reproductive system.

Pericardium of ganoid fishes such as *Acipencer* is rich in lymphoid tissue.

REFERENCES

Arvy, L. (1955). Characters histologiques du stratum granuosum de oppel la truite-C.R. Seanc. Sec. Biol. Paris 149 : 2131-2134.

Bucke, D. (1971). The anatomy and histology of the alimentary tract of the carnivorous fish the Pike, *Esox lucius*. J. Fish, Biol. 3 : 421-431.

Burne, R.H. (1927). A contribution to the anatomy of the ductless glands and lymphatic system of the angler fish (*Lophius piscatorius*). Phil. Trans. R. Sec. B 215 : 1-54.

Fange, R. (1984). Lymphomyeloid tissue in fishes. Vidensk. Medder dansk natur. Foreign 145 : 143-162.

Fange, R. (1992). Fish blood cells. In fish Physiology (Eds. Hoar W.S., Randall, D.J. & Farrel, A.P.) Academic press. INC NY. pp. 1-54.

Good, R.A., Finstad J. & Litman, J., (1972). Immunology (Eds. M.V. Hardisty and Potter, I.C.). In : The biology of Lampreys, Vol. 1 Acad. press, London 127-206.

Hammer, J.A. (1908). Zur kenntnis der Teleostieithymus. Arch. mikrosk. Anat. Entwmech. 73 : 1-68.

Jacobshangen, E. (1915). : Zur morphologic des spiraldams-Anat. Anz. 48. 188-201, 220-235, 245-254.

Mattison, A., Fange, R., & Zapata, A. (1990). Histology and ultra structure of the cranial lymphohaemopoietic tissue in *Chimaera monstrosa* (Pisces, Holocephali). Acta Zoologia (Stockholin), 71 : (2) 97-106.

Pulsford, A. and Fange, R. (1984). Light and electron microscopic study of the thymus of the Anglerfish, *Lophius piscatorius* (Pisces), Teleostei-Pruc. Fish Immunology Symposium July. 11-13. 1983 *in Pay mouth*.

Rivieri, H.B., Cooper, E.L., Reddy, A.L. and Hildemann, W.H. (1975). In search of hagfish thymus. Amer. Zool. 15 ; 39-49.

Robenson, J.M. (1932). Haemopoiesis in *Amia calva* - Z. Zelf. mikr. Anat.16.

Sharma, R. and Kumar, S. (1982). Distribution of thyroid follicles and nerves in the kidney of a teleost. *Clarias batrachus* in (Limn.) Z. Mikrosk. Anat. Foruch. Leipzig 96 : 1969-1977.

Yoffey, J.M. (1985). Cellular migration streams. The integration of the lymphomyeloid complex lymphology : 18, 5-21.

Zapata, A. (1981). Ultra structure of elasmobranch lymphoid tissue. 2. Leydig's and epigonal organs— Devl. comp. Immum. 5 : 43-52. Press. London 127-206.

Chapter 10

OSMOREGULATION

Osmoregulaion in teleost fishes, whether they live in freshwater or sea, its physiological activity is very closely related to their survival, yet inspite of the importance of osmoregulation surprisingly little is known about how fish deals with physiological problems inherent in living in hypoosmotic and hyperosmotic environments. The ability of some fish (e.g., salmon) to regulate in both environments during migration is of great' interest. The classical review of osmoregulation in aquatic animals has been done by Krogh (1939), and Pyefinch (1955).

In fishes the kidneys play an important role in osmoregulation, but major portion of the osmoregulatory functions are carried out by other organs such as the gills, the integument and even the intestine (Lagler *et al.*, 1977).

Osmoregulation may be defined as "the ability to maintain a suitable internal environment in the face of osmotic stress". As a consequence there is always difference between the optimal intracellular and extracellular concentrations of ions. In the fish body, number of mechanisms take place to solve osmotic problems and regulate the difference. Of which most common are (i) between intracellular and extracellular compartment (ii) between extracellular compartment and the external environment. Both are collectively called 'osmoregulatory mechanisms', a term coined by Rudolf Hober.

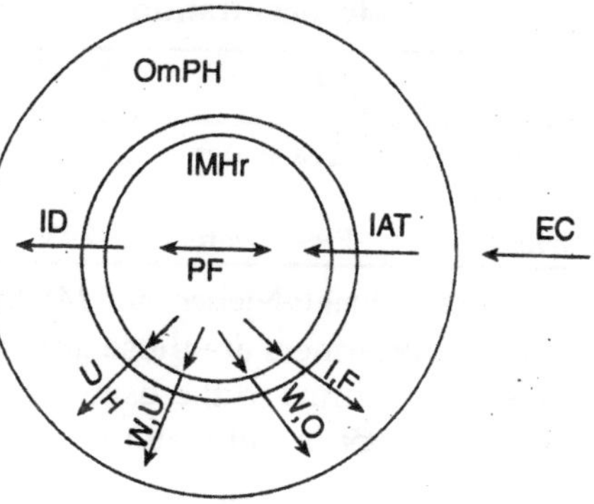

Fig. 10.1 : Principal process of osmoregulation in freshwater fishes. EC, ecological condition; I, ion; IAT, ion active transfer; ID, ion diffusion; IMHr, inner medium hypertonic; O, osmosis; OmPH, outer medium hypotonic; PF, physiological factor; U, urine; W, water.

Problems of Osmoregulation : Generally fish lives in an osmotic steady state in spite of frequent variations in osmotic balance. That is, on the average, the input and output being equal over a long period sum up to zero (Fig. 10.1).

The osmotic exchanges that take place between the fish and its environment may be of two types (i) obligatory exchange. It occurs usually in response to physical factors over which animal has little or no physiological control and (ii) regulatory exchange—these are the exchanges which are physiologically well controlled and help in the maintenance of internal homeostasis.

Factors Affecting Obligatory Exchanges

1. **Gradient between the extracellular compartment and the environment :** The greater the ionic difference between the body fluid and external medium, the greater the tendency for net diffusion to low concentrations. Thus, a bony fish in a sea water is affected by the problem of losing water into the hypertonic sea water.

2. **Surface/Volume ratio** : Generally the animal with small body size desicates (or hydrates) more rapidly than a larger animal of the same shape.

3. **Permeability of the gills** : Fish gills are necessarily permeable to water and solutes as they are the main site of exchange of oxygen and carbon dioxide between the blood and the water. Active transport of salts also takes place in the gills. Euryhaline fishes (who have tolerance of wide range of osmolarity) are well adapted to saline water by reduced permeability to water.

4. **Feeding** : Fishes take water and solute along with the feeding. A gill takes high quantity of salt than water at the time of feeding on seashore invertebrates, these fishes, therefore, must have some special device to excrete excess of salt. However, a freshwater fish ingests large amount of water than salt and thus needs special means of salt conservation.

Osmoregulators and Osmoconfirmers : Osmoregulators are those animals who can maintain the internal osmolarity different from the medium in which they live. The fishes, except the hagfish which migrates between fresh and saline waters, the changing osmotic stress due to environmental changes is overcome with the help of endocrine mechanism (Table 1).

Table 1: Approximate composition of extracellular fluids of teleostean fishes (concentration in millimoles per litre of water)

Species	Habitat	Milli	Na^+ mole	K^+	Ca^{++}	Mg^{++}	Cl^-	SO_4^{--}
Paralichthys (flounder)	Sea-water	337	180	4	3	1	160	0..2
Carrassius	Fresh-water	293	142	2	6	3	107	–

Source : Schmidt-Nielsen, and Mackay 1972; Prossor, 1973.

Osmoconfirmers are those animals who are unable to control osmotic state of their body fluids but confirms to the osmolarity of the ambient medium. Majority of fishes either live in freshwater or in salt water (a few live in brackish water). Due to various physiological processes, metabolic wastes are removed from the body in vertebrates by gut, skin and kidneys. But in fishes and aquatic animals their gills and oral membranes are permeable both to water and salts in marine environment, salt is more in water against the salt inside the body fluid, hence water moves out due to the process of 'osmosis'. The 'osmosis' may be defined as "if two solutions of different concentrations are separated by a semipermeable membrane, the solvent from the less concentrated part will move through the membrane into more concentrated solution." Hence to compensate the loss of water marine fishes drink water. The salt will enter the body due to concentration gradient and so salt will be more inside the body. On the other hand, in freshwater fishes, the salt will go out to the environment as the salt concentration will be more inside the body fluid. The water

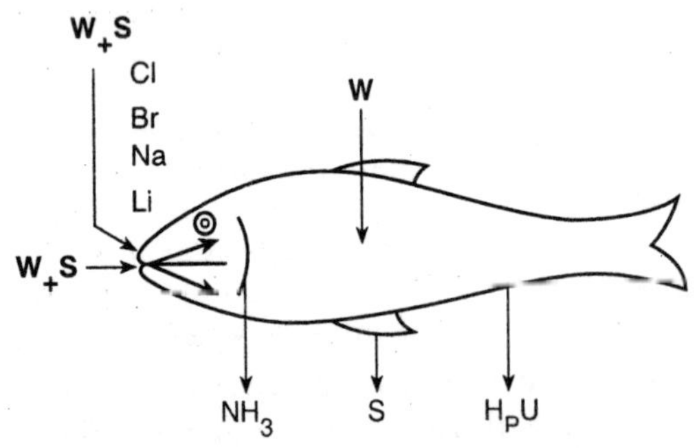

Fig. 10.2 : Osmoregulatory inflow and outflow of salts and water in a fresh water fish. HpU, hypotonic urine; S, salt; W, water, W + S, water and salt.

will move inside the body due to osmosis through partially permeable membrane. This means solvent will pass into more concentrated solution, but solute will also pass in the opposite direction. There will be, however, a difference in the rate dependent upon the relative permeability for two types of molecules usually solvent pass rapidly.

Osmoregulation in freshwater fishes : The body fluid of freshwater fishes is generally hyperosmotic to their aqueous medium. Thus they are posed with two types of osmoregulatory problems.

1. Because of hyperosmotic body fluid they are subjected to swelling by movement of water into their body owing to osmotic gradient.

2. Since the surrounding medium has low salt concentration, they are faced with disappearence of their body salts by continual loss to the environment. Thus freshwater fishes must prevent net gain of water and net loss of salts. Net intake of water is prevented by kidney as it produces a dilute, more copious (i.e., plantiful hence dilute) urine (Fig. 10.2).

The useful salts are largely retained by reabsorption into the blood in the tubules of kidney, and a dilute urine is excreted. Although some salts are also removed along with urine which creates torrential loss of some biologically; important salts such as KCl, NaCl, $CaCl_2$ and $MgCl_2$ which are replaced in various parts. Freshwater fishes have remarkable capacity to extract Na^+ and Cl^- through their gills from surrounding water having less than 1 m M/L NaCl, even though the plasma concentration of the salt exceeds 100 m M/L NaCl. Thus NaCl actively transported in the gills against a concentration gradient in excess of 100 times. In these fishes the salt loss and water uptake are reduced by the integument considerable with low permeability or impermeability to both water and salt also by not drinking the water (Fig. 10.3).

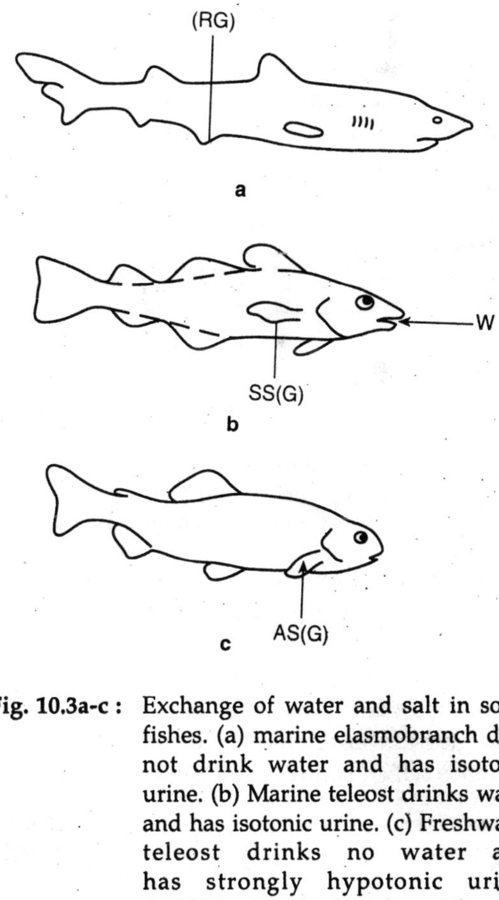

Fig. 10.3a-c : Exchange of water and salt in some fishes. (a) marine elasmobranch does not drink water and has isotonic urine. (b) Marine teleost drinks water and has isotonic urine. (c) Freshwater teleost drinks no water and has strongly hypotonic urine; AS(G), absorbed salt with gill; HrNaCl (RG), hypertonic NaCl from rectal gland; SS(G), secretes salts from gill; W, water.

Osmoregulation in marine water fishes : In marine fishes, the concentration of body fluid and marine water is almost similar. Therefore, they do not require much energy for maintenance of osmolarity of their body fluid. The classic example is hagfish, *Myxine* whose plasma is isoosmotic to the environment. Hagfish maintains the concentraion of Ca^{++}, Mg^{++} and

SO_4 significantly lower and Na^+ and Cl higher in comparison to sea water. Other marine water fishes such as sharks, rays, skates and primitive coelacanth, *Latimaria,* have plasma which is isoosmotic to sea water. They differ from the hagfish in having capacity to maintain very lower electrolyte (*i.e.,* inorganic ions) concentrations. They also have difference with organic osmolytes like urea and trimethylamine oxide. Kidneys of coelacanth and elasmobranchs excrete excess of inorganic salts such as NaCl. Also rectal gland located at the end of alimentary canal takes part in the excretion of NaCl. Modern bony fishes (marine teleosts) have the body fluid hypotonic to sea water, so they have tendency to lose water to the surroundings particularly from gill via epithelium. The lost volume of water is replaced by drinking salt water (Fig. 10.3). About 70—80% sea water containing NaCl and KCl enters the blood stream by absorption across the intestinal epithelium. However, most of the divalent cations like Ca^{++}, Mg^{++} and SO_4 which are left in the gut are finally excreted out. Excess salts absorbed along with sea water is ultimately received from the blood with the help of gills by the active transport of Na^+Cl^- sometimes K^+ and eliminated into the sea water. However, divalent ions are secreted into the kidney (Fig. 10.4).

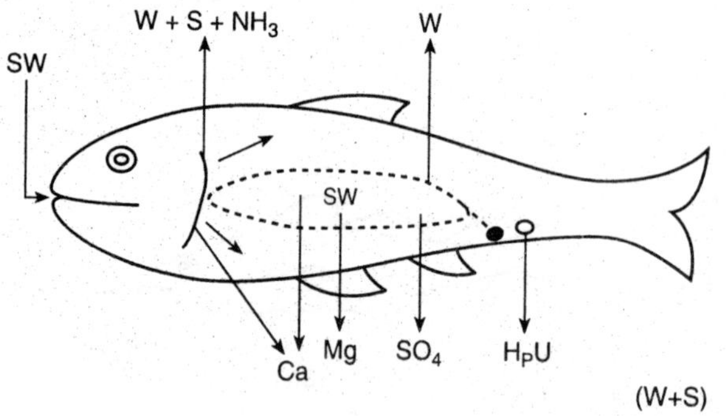

Fig. 10.4 : Osmotic regulation in marine bony fishes. HpU, hypotonic urine; SW, sea water; W + S + NH_3, water, salt and ammonia; W, water.

Thus, urine is isosmotic to the blood but rich in those salts, particularly Mg^{++}, Ca^{++} and SO_4^{--} which are not secreted by the gills. Combined osmotic action of gills and kidney in marine teleosts resulted in the net retention of water that is hypotonic both to the ingested water and urine. By using similar mechanism some teleost species such as the salmon of pacific northwest maintain more or less constant plasma osmolarity inspite of being migratory between marine and freshwater environment.

According to Moyle and Cech. Jr. (1982) the fishes can be divided into four groups on strategies of regulation of internal water and total solute concentrations.

Control of Osmoregulation : The concentration and dilution of urine is controlled by hormones, which affects the rate of renal filtration by changing the blood pressure and thus control the quantity of urine. Hormones also influence the rate of diffusion and absorption across the gill epithelium. Thyroid gland and suprarenal bodies secrete adrenocortical hormones which control osmoregulation in fishes.

REFERENCES

Krogh, A. (1939). Osmotic regulation in aquatic animals. Cambridge University Press. pp. 242.

Lagler, K.F., Bardach, J.E, Miller R.R. and Passion D.R.M. (1977). Ichthyology. John Wiley and Sons. New York.

Moyle, P.B. and Cech J.Jr. (1982). Fishes—An introduction to Ichthyology. Prentice Hall International Inc. London.

Prossor, C.L. (1973). Comparative animal physiology. Vol. 1., Philadelphia, Saunders.

Pyefinch, K.A. (1955). A review of the literature on the biology of Atlantic salmon (Salmo salar). Fresh water and salmon fisheries research, Scottish Home Department No. pp.24.

Schmidt-Nielsen, B.M. and Mackay W.C. (1972). Comparative physiology of electrolyte and water regulation with emphasis on sodium, potassium, chloride, urea and osmotic pressure. In clinical disorders of fluid and electrolyte metabolism. M.H. Haxwell and C.R. Kleeman. eds. New York. McGraw Hill.

Chapter 11

EXCRETORY SYSTEM

In vertebrates, the excretory and reproductive organs are morphologically interrelated because certain excretory ducts are used for the discharging of gametes also. So it has been convenient to treat them together as urinogenital system. In fishes, the association is restricted to the pseudocopulatory papilla through which both the excretory and generative products leave by a common vent. The association is more intimate in male than in female. Here, these systems are dealt separately as excretory and reproductive organs.

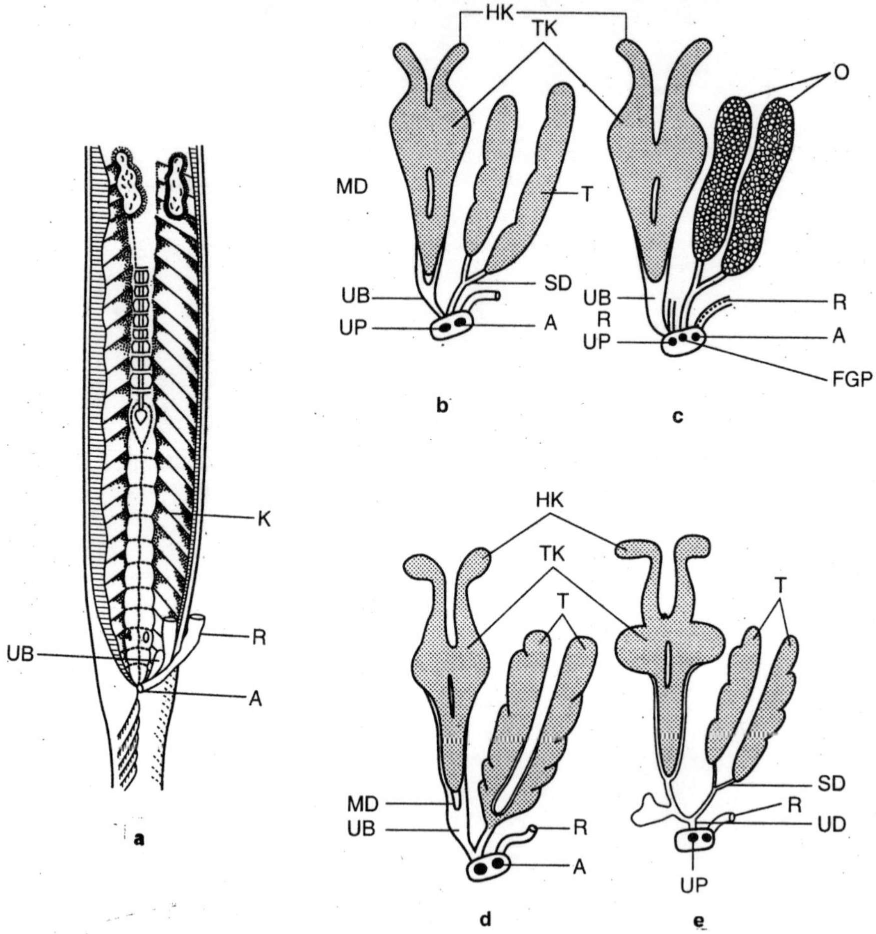

Fig. 11.1a-e : Excretory system of (a) *Xenentodon* (b) *Cirrhina* (male) (c) *Cirrhina* (female) (d) *Labeo* (male) (e) *Barbus* (male). A, anus; FGP, female genital pore; HK, head kidney; MD, mesonephric duct; O, ovary, R, rectum; SD, sperm duct; T, testis; TK, trunk kidney; UB, urinary bladder, UD, urinary duct; UP urinogenital pore.

EXCRETORY ORGANS

The excretory organs consist of kidneys, ureters and urinary bladder. The urinary bladder is not homologous to that of higher vertebrates (Fig. 11.1a-e).

Kidney : Kidneys of vertebrates are made up of nephron or kidney tubules. In ancestral vertebrates, kidney possesses one nephron for each of those body segments that lay between the anterior and posterior end of the coelom. The nephron drained into a duct called *Wolffian* or *archinephric* duct located posterior to the cloaca. This sort of kidney is known as holonephros because it extends to the entire length of the body. The holonephros is found today in the larvae of certain cyclostomes but not in any adult. In fish and amphibians the most anterior tubules have been lost, some of the middle tubules are associated with tests and there is a concentration and multiplication of tubules posteriorly. Such a kidney is known as a posterior kidney or opisthonephros. Generally in fishes, the tubules of the anterior region become functional in early life and are designated as pronephors and the

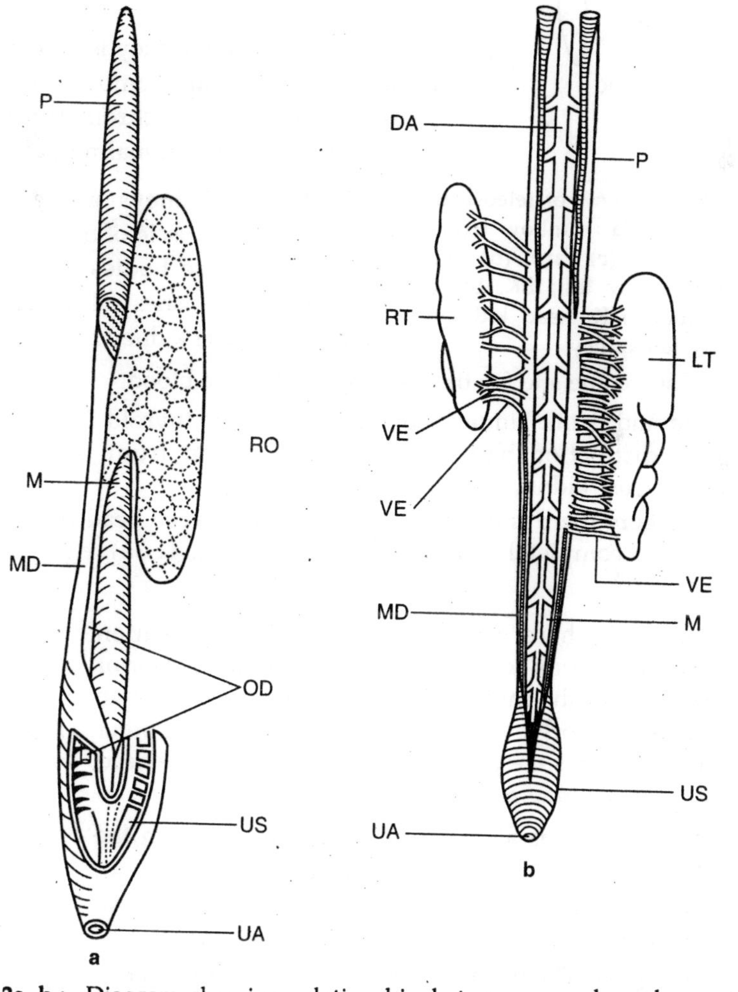

Fig. 11.2a, b : Diagram showing relationship between gonads and excretory organs in (a) Female *Lepisosteus*. (b) Male *Lepisosteus*. DA, dorsal aorta; LT, left testis; M, mesonephros; MD, mesonephric duct; OD, oviduct; P, pronephros; RO, right ovary, RT, right testis; UA, urogenital aperture; US, urogenital sinus; VE, vasa efferentia.

tubules present in the posterior regions take up excretory function throughout life. This region of the functional tubules is known as mesonephros (Fig. 11.2a, b).

There are great variations in the external structure of the kidney in fishes. The shape varies according to species. The kidneys occupy dorsal position in the body cavity and is placed just ventral to the vertebral column. In teleosts, the kidney is distinguished into head and trunk regions. Such distinction is clearly discernible in carps, but in other fishes the macroscopic differentiation into head and trunk kidneys is not prominent. Head kidney is nonexcretory and endocrine in function whereas trunk kidney (posterior kidney) is excretory in nature. So the kidneys of fishes are peculiar in comparison to other vertebrates. The peculiarities are mentioned as follows :

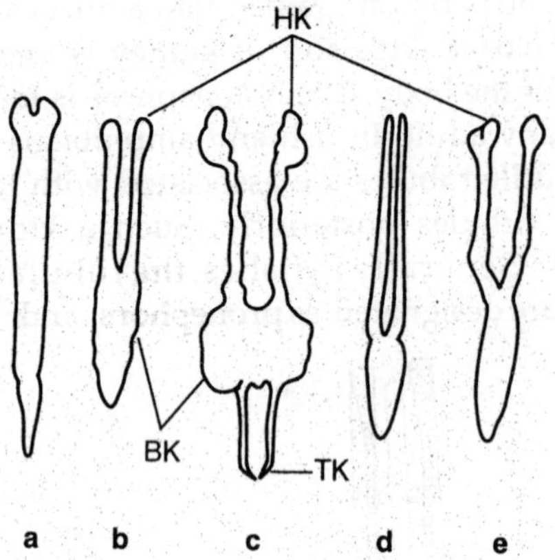

Fig. 11.3a-e : External view of teleostean kidney. (a) rainbow trout. (b) *Ayu*. (c) carp. (d) Eel. (e) yellow tail; BK, body kidney, HK, head kidney; TK, tail kidney.

1. The head kidney is endocrine in nature. It has interrenal gland homologous to adrenal cortex of mammals. It has chromaffin cells also which are similar to the adrenal medulla of mammals.

2. In the kidneys are embedded yellow bodies called corpuscle of Stannius. It is endocrine in function. These bodies are visible macroscopically in some fishes while in others they are intermingle in the kidney tissues.

3. Head kidney is the site for the development of the blood.

4. Both head and trunk kidneys contain heterotopic thyroid follicles. On the basis of morphology and distinction between head kidney and trunk kidney, Ogawa (1961) classified marine teleost kidney into five categories.

In the first category there is no clear demarcation between head and trunk kidneys and the two kidneys are completely fused throughout (Fig. 11.3a). *e.g.*, rainbow trout and salmon.

In the second type there is clear macroscopic demarcation between the head and trunk kidneys. The middle and posterior portions are fused. From the middle fused part are given out two tube like structures anteriorly, which are separate from each other and at the tip of these tubes are present sac-like structure, the head kidneys. *e.g.*, *Ayo*, Cyprinidae and carps.

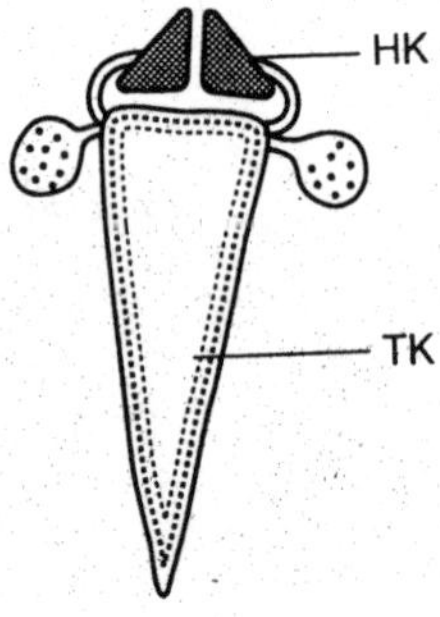

Fig. 11.4 : Structure of kidney of *Clarias batrachus*. HK, head kidney; TK, trunk kidney.

In the third type, there is also a clear distinction between the head and trunk kidneys. The kidney is distinguished into head, trunk and tail portions. The tail kidneys are fused, while the trunk and head kidneys are separated and are located at the tip of the anterior most region. The head kidney, is generally globular in shape, e.g., *Notopterus notopterus*.

In the fourth type there is no morphological demarcation between head and trunk kidneys. The two kidneys are separate except at the posterior most region where the kidneys are fused.

In the fifth type, the two kidneys are completely separate from each other. The tail kidneys are thin tube-like while the anterior trunk kidneys are thick. There is no morphological distinction between head and trunk kidneys. Regarding freshwater teleosts, Ogawa (1961) suggested that kidneys can be grouped into first three of the five types described above.

Kidneys of some Indian fishes : The kidneys of *Clarias batrachus* are situated in the abdominal cavity. It is placed in retroperitoneal position against the ventral aspects of the vertebral column and dorsal to the alimentary canal and the gonads. They are dark reddish brown in colour and is covered by a thin transparent membrane. The trunk kidneys are fused throughout their length. They are broadened anteriorly and become gradually narrow posteriorly. The head kidney are in the form of two triangular lobes which are separated from each other by a very narrow gap. They are not connected with the trunk kidney in *Clarias batrachus*. The triangular head kidneys is present in *Channa marulius*, *Channa punctatus* and *Channa gachua*. The apex of each lobe is pointed and placed anteriorly, while the base is flat and posteriorly directed. The anterolateral side of the trunk kidneys on either side is provided with rounded structures and are known as mesonephric lobes (Fig. 11.4), (Sharma and Kumar, 1982).

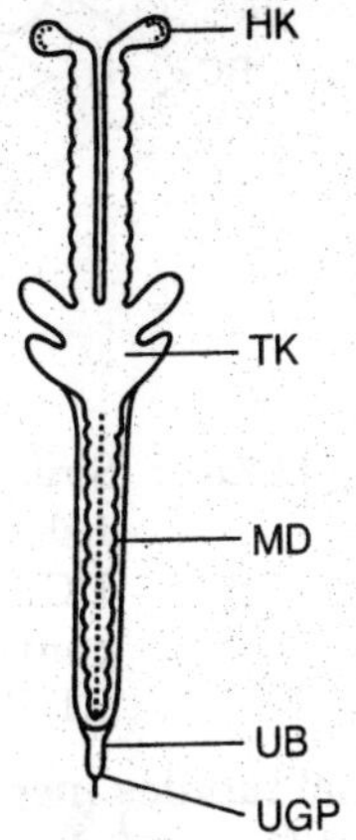

Fig. 11.5 : Morphology of kidney of *Labeo rohita*. HK, head kidney, MD, mesonephric duct; TK, trunk kidney, UB, urinary bladder, UGP, urinogenital pore.

Mesonephric lobes are also present in *Heteropneustes fossilis*. In *Labeo rohita*, the kidneys are also paired structures. They are situated ventrally to the vertebral column. It is very clearly attached by the connective tissue. The peritoneal layer separates them. They are elongated structures which run from the vent and reach very close to the gills. They are reddish brown in colour. The two kidneys are fused in the middle, forming a flattened wing-like middle portion of the trunk kidneys. From this wing-like structure, two tube-like structures are given out anteriorly which are separate from each other. At the anterior end of these tubes are present sac-like structures known as head kidneys. The two head kidneys are separate. Posteriorly, from the middle part of the trunk kidney, the kidneys become narrow. This part is called as tail kidneys. Externally, a depression is visible which indicates that they are two structures. The tail kidneys are also fused. The trunk kidneys are very big in comparison to the length of a fish. Dorsally, the trunk kidneys show several annulations whereas ventrally it is smooth in outline (Fig. 11.5).

Fig. 11.6 : Microphotograph of transeverse section of kidney of *Clarias batrachus*. X 320. BC. Bowman, capsule; BV, blood vessel; G, glomerulus; UT, uriniferous tubule.

In *Xenentodon cancila*, the trunk kidney in its anterior region shows ladder-like appearance.

Ureter : The mesonephric ducts or ureters lie closed together in the median line. Anteriorly, they are separate, posteriorly the two mesonephric ducts open separately into urinary bladder. In some species a sac-like enlargement is clearly visible in the posterior region of the ureter. This is known as urinary bladder, but it is not homologous to that of higher vertebrates. The urinary bladder usually opens to the exterior by a common urinogenital aperture in the male fish but a separate urinary aperture is present in female fish as found in *Mystus*.

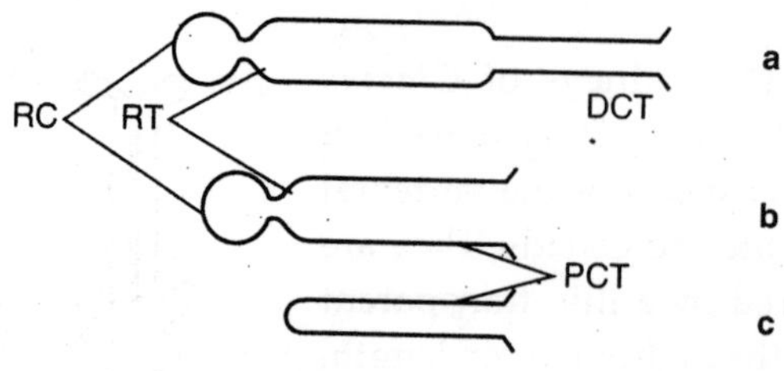

Fig. 11.7a-c : Structure of kidney of teleosts. (a) Freshwater fish. (b) Marine fish. (c) Aglomerular fish. DCT, Distal convoluted tubule; PCT, Proximal convoluted tubule; RC, Renal corpuscles; RT, Renal tubule.

Histology of Trunk Kidney : The trunk kidney or body kidney, like other vertebrates, contains renal tubules (nephrons) and interstitial lymphoid tissue. The number of renal tubules varies in different fishes. In teleosts, the trunk kidney consists of a large number of nephrons. The functional unit of kidney is nephron. Each nephron consists of two parts, the renal corpuscles (Malpighian body) and the renal tubule (urinary tubule). the renal corpuscle or Bowman's capsule is double layer cup-like structure of uriniferous tubule which contains tuft of capillaries known as glomerulus (Fig. 11.6).

The remaining segment of urinary tubule (renal tubule) is divided into proximal convoluted segment (which further divides into segment I and segment II), the intermediate and distal segments (Fig. 11.7a, b, c).

The distal segment is absent in marine fishes (Oguri, 1982). The segment of Henle found in higher vertebrates is also absent in the fishes (Oguri, 1982).

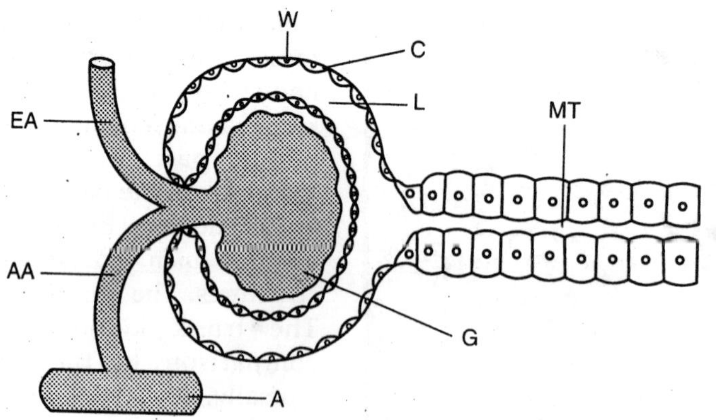

Fig. 11.8 : Diagram of renal corpuscles of fish. A, aorta; AA, afferent arteriole; C, capsule; EA, efferent arteriole; G, glomerulus; L, lumen; MT, mesonephric tubule; W, wall.

The glomerulus and Bowman's capsule together constitutes the renal or Malpighian capsule. It is a filtration apparatus of kidney. The glomerular capillaries, which are the vascular part of corpuscle, is the afferent arteriole which divides and forms capillary loops. The loops reunite and leave the capsule as efferent arteriole (Lagler, *et al.*, 1977) (Fig. 11.8).

The renal corpuscle contains an additional group of cells known as mesangial cells. Mesangial cells are also present in the space between the loops of glomerular capillaries. They are most obviously present at the vascular stalk.

The function of mesangial cells are not known, although experimental data show that they can remove large proteins from the glomerular basal lamina (Ross and Reith, 1985). Oguri (1982) reported the presence of juxtaglomerular cells in the wall of afferent arterioles. These cells contain secretory granules and are specialized muscular cells. They are the source of the hormone renin. Renin is the hormone active in increasing blood pressure. In mammals, the juxtaglomerular apparatus is thought to provide feedback information for the control of glomerular filtration.

The glomeruli of freshwater teleosts are numerous and large in size. In marine teleosts, glomeruli are reduced in size and number. In extreme cases, the glomeruli disappear completely from the kidney of some marine fishes. The examples are Sea horse (*Hippocampus coronatus)*, Pipefish (*Syngnaths schelegeli*) and frogfish (*Anternnarius tridens*), and these fishes are called *aglomerular fishes*.

The renal tubules are thin and short in the neck segment and consist of single layer of low epithelial cells with long cilia. The proximal convoluted segment I and segment II are provided with cuboidal epithelial cells. The nuclei are large, round or oval. The cytoplasm contains secretary granules. The intermediate segment is well-developed in carp kidney but absent in several species of fishes. The distal convoluted segment could be distinguished because of coarse granules in the cytoplasm. The distal convoluted segments are absent in the kidney of marine fishes.

Ureter : The function of the ureter is to conduct urine up to urinary bladder. Histologically, it is made up externally by tunica adventitia, the middle layer contains lamina propria and smooth muscles and outermost layer is columnar epithelial cells. Urinary bladder is a thin walled sac-like structure. It is also made up of three layers similar to the ureter.

Head kidney : Embryologically the head kidney originates from pronephros. It is made up of lymphoid tissue containing reticular cells (supporting framework of lymphatic tissue) (Fig. 11.9) and numerous capillaries.

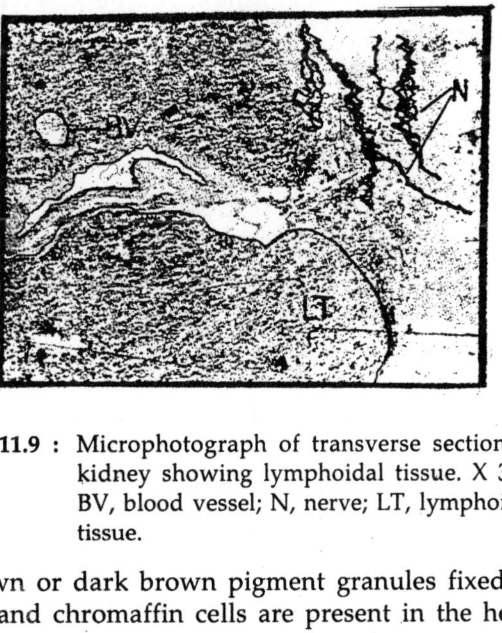

Fig. 11.9 : Microphotograph of transverse section of kidney showing lymphoidal tissue. X 320. BV, blood vessel; N, nerve; LT, lymphoidal tissue.

The chromaffin tissue often shows brown or dark brown pigment granules fixed in bichromate solution. The inter-renal gland and chromaffin cells are present in the head kidney. The name inter-renal body was adopted by Balfour in 1878 and since then commonly accepted. In *Xenentodon cancila*, the inter-renal body lies as a compact mass surrounded by a capsule of connective tissue. It lies at a short distance behind the septum transversum and is an elongated mass about twice the size of a wheat grain (Gupta and Shrivastava 1971). The cells are basophilic with large nuclei. The inter-renal gland of fish is homolgous to the adrenal cortex of mammals. In eel *(Anguilla japonica)* the gland is located in the wall of the postcardinal vein running closely along the head kidneys. Chromaffin tissues are also present, this tissue is homologous to adrenal medulla of mammals (Sharma and Kumar, 1982). Both inter-renal and chromaffin tissues are present in many fishes either discrete structure or intermingled in the tissue.

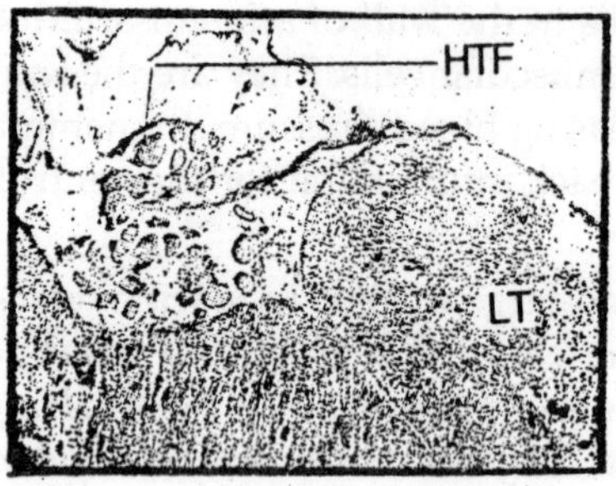

Fig. 11.10 : Microphotograph of transverse section of head kidney of *Clarias* showing heterotopic thyroid follicle. HTF, heterotopic thyroid follicle; LT, lymphoid tissue.

The histochemical examination of teleostean inter-renal cells shows 3-B hydroxysteroid dehydrogenase and glucose 6 phosphate dehydrogenase. These are important in biosynthesis of steriod hormones. The function of chromaffin cells is to secrete adrenalin and noradrenalin whereas inter-renal secretes cortico-steriod. Cortisole is the main corticoid in teleosts but cortisone and cortico-steroid are also produced by inter-renal glands. Whether aldosterone is secreted by teleostean inter-renal gland is not clear (Oguri, 1982).

Urine : The freshwater teleosts have to excrete large amount of water which is taken through mouth. The urine of freshwater fishes contains creatine, unidentified nitrogenous compounds some of which are amino-acids, little amount of urea and ammonia. The urine is copious and have very low in concentration of electrolyte. Urine contains nitrogen amounting to 2 to 25% of total nitrogen excreted by freshwater fishes. The bulk is removed out through gills as ammonia (Lagler *et at.* 1977). Marine fishes produce scanty urine, which contain Ca^{++}, Mg^{++}, $SO4^{--}$, SO_4^{--} and PO_4^{--}. In addition to the creatine, creatinine and TMAO (*i.e.*, trimethylamine oxide) are also excreted out. Ammonia, urea and monovalent electrolytes (Na^+, Cl^-), however, are excreted mainly through gills.

CORPUSCLES OF STANNIUS

The corpuscles of Stannius is a small ductless gland (endocrine gland) situated partly or completely embedded in the kidney on its dorsal, dorsolateral and ventrolateral side. It was discovered by Stannius (1939), but it is recently considered that they are the subject of physiological interest specially for their role in calcium metabolism. It is now established that corpuscles of Stannius work in conjuncture with pituitary gland, which produces a distinct hypercalcemic effect, to maintain a relatively constant level of calcium in *Fundulus heteroclitus* (Pang *et al.* 1973, Scheribman and Pang, 1975). Idler and Freeman (1966) suggested that it is associated with steroidegenic activity, while Oguri (1966) reported that they are responsible in producing some hormone like polypeptides. Chaster Jones *et al.* (1969) were of the opinion that they had pressor activities and helped in electrolyte metabolism. They also suggested that corpuscle of Stannius might belong to renin angiotensin system. The corpuscles of Stannius is situated near the middle portion of mesonephros in salmnoid fishes and *Altherninopsis calciforniensis* (Krishnamurthy & Burn, 1969) but in majority of fishes they are located in the posterior region of the kidney. There is a great variation regarding the number of corpuscles of Stannius in the kidney, they may be single

(Heteropneustes setani, Notoptarus notopterus, Lepidocephalichthyes quntea) or may be two in number *(Lepadocephalichthyes)* as described by Bose and Ahmad (1975). Histologically, the corpuscles of Stannius contain two types of cells (Ford, 1959; Nadkarni and Gorbman, 1966; Bose and Ahmad, 1975) while other authors held that there is only one population of cells (Olivereau, 1961).

HETEROTOPIC THYROID FOLLICLE

The thyroid in fishes is not a discrete organ but fused in the kidney also, hence it is known as heterotopic. The thyroid follicles are scattered in the haemopoietic tissue of the head, mesonephric lobe (air breathing fishes) and in trunk kidneys (Fig. 11.10).

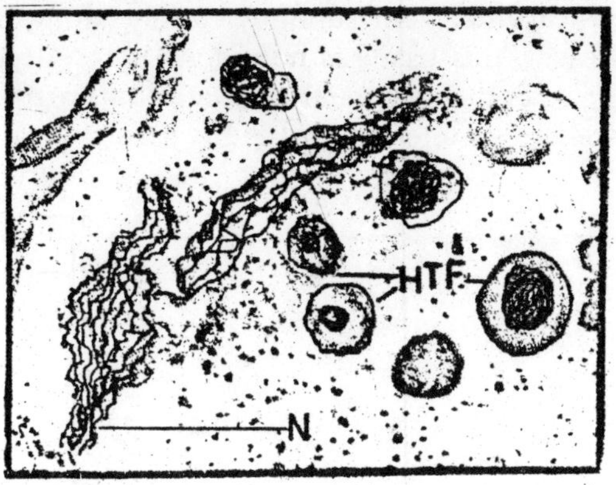

Fig. 11.11 : Microphotograph of transverse section of kidney of *Clarias* showing heterotopic thyroid follicle. X 320. HTF, heterotopic thyroid follicle; N, nerve.

The follicular epithelium is visible and contains colloid which is strongly repeated acidophilic, dense, homogenous and non-vacuolated (Fig. 11.11). An aggregation of a large number of follicles has been present at the junction where the post-cardinal vein opens. It appears that the thyroid follicles are migrating from the pharyngeal region to the kidney (Baker, 1958 a & b). They appear only after the fish becomes two months old. They are present in all parts of the kidney, but the highest concentration is near the cardinal vein. Thick and thin nerves both myelinated as well as unmyelinated are present close to the heterotopic thyroid follicles (Sharma and Kumar, 1982).

INNERVATION

The kidneys are richly innervated by autonomic nervous system. The nerves generally penetrate in the kidney through blood vessels. They divide and redivide to form nerve plexuses (Fig. 11.12).

Fig. 11.12 : Microphotograph of section of kidney of *Clarias* showing innervation. X-320. N. Nerve.

Scattered ganglion cells are also present in the trunk kidneys. Both cholinergic and adrenergic nerves are present.

HORMONES AND ENZYMES

Renin is the hormone secreted from juxtaglomerular cells. The hormone is active in increasing blood pressure and controls the glomerular filtrate. The cholinergic nerve endings secrete an enzyme acetylcholinesterase (AChE). The enzyme kinetics of AChE of the head kidney of *Labeo* is 1.11×10^{-3} M and V_{max} is 0.222 A/mg proteins/30 minutes. While the Km of the middle portion of trurk and tail kidney is

3.33×10^{-3}M and V_{max} is 5.0 A/mg proteins/30 minutes. The lower Km in head kidneys indicates higher enzyme activity (Fig. 11.13).

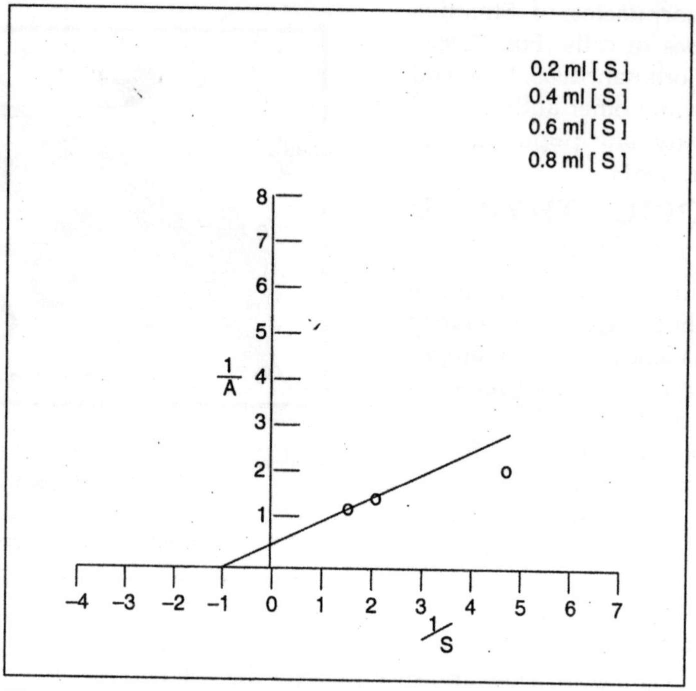

Fig. 11.13 : Lineweaver-Burk plot of AChE activity in the kidney of *Labeo rohita*.

BLOOD SUPPLY OF KIDNEY

The kidney of fish receives blood supply by the renal artery and renal portal vein. The renal artery supplies blood to glomeruli, where high blood pressure helps to separate glomerular filterate. The renal portal veins are connected to capillary network around the kidney.

REFERENCES

Baker, K.F. (1958, a) . Heterotopic thyroid tissue in fishes. I. Origin and development of heterotopic thyroid tissue in platyfish. J. Morph. 103 : 91-133.

Baker, K.F. (1958, b). Heterotopic thyroid tissue in fishes. II. The effect of iodine and thiourea upon the development of heterotopic thyroid tissue in platyfish. J. Exp. Zool. 138 : 329-353.

Bose, K.C. and Ahmad, M.F. (1975). Studies on corpuscles of a hillstream Loach, *Lepidocephalichthyes quntea* (Ham). Zoologische Beitrage, 26 : 101-105.

Chaster Jones, I., Chan, D.K.O. Henderson I.W, and Ball, J.N. (1969). The adrenocorticol steroids, adrenocorticotropin and corpuscles of Stannius. In fish physiology. (Eds. W.S. Hoar and D. J. Randall). Vol. 2. New York Academic Press pp. 321-376.

Ford, P. (1959). Some observations on the corpuscles of Stannius. In : Comparative endocrinology. Ed. A. Gorbman. New York. J. Wiley. pp. 728-734.

Gupta, O.P., and Shrivastava, R.K. (1971). Morphological studies of inter-renal bodies, mesonephric kidneys and urinary bladders of the Indian catfish *Xenentodon cancila* (Ham.) Okaj. Fal. Anat. Jap. 48(4) : 2 : 205-211.

Idler, D.R. and Freeman, H.C. (1966). Steroid transformation by corpuscles of Stannius of the Atlantic cod (*Gadus morhua*) J. Fish Res Bd. Canada 23 : pp. 1249-1255.

Krishnamurthy, V.G. and Bern, H.A. (1969). Correlative histological study of the corpuscles of Stannius and the juxtaglomerular cells of teleost fishes. Gen. Comp. Endocrinol. 13, pp. 313-325.

Lagler, K.F., and Bardach, J.E. Miller, R.R. and Passino, D.R.M. (1977). Ichthyology. John Willey and Sons. New York.

Nadkarni, V.B. and Gorbman, A. (1966). Structure of the corpuscles of Stannius normal and radiothyroidectomized chinook fingerlings and spawning salmon. Acta. Zool. (Stockholm) 47 : 61-66.

Ogawa, M. (1961). Comparative study of the external shape of the teleostean kidney with relation to phylogeny. Sci. Rept. Tokyo Kyoiku Daigaku B. 10 : pp. 61-68.

Oguri, M. (1966). Electron microscopic observation on the corpuscles of Stannius in gold-fish. Bull. Jap. Soc. Fish (32) pp. 903-908.

Oguri, M. (1982). Kidney. In : An atlas of fish histology—normal and pathological features. (Ed. T. Hibiya), pp. 95-101, Kodansha Ltd. Gustav Fischer Verlag Stuttgart.

Olivereau, M. (1961). Corpuscles of Stannius et reporduction chez languille male, Compt. Rend. 253 : pp. 541-543.

Pang. P.K.T., Schreibman, M.P. & Griffith, R.W. (1973). Pituitary regulation of serum calcium level in the killifish, *Fundulus heteroclitus* L. Gen. Comp. Endocrinol. 21, pp.536-542.

Ross, M.H. and Reith, E.J. (1985). Histology. A text and atlas. Harper and Row Publisher, J.B. Lippincott company.

Schreibman, M.P. and Pang, P.K.T. (1975). The histophysiology of transplanted corpuscles of Stannius in the killifish, *Fundulus heteroclitus* Gen, Comp. Endocrinol. 26. pp. 186-191.

Sharma, R. and Kumar, S (1982). Distribution of thyroid follicles and nerves in the kidney of a teleost. *Clarias batrachus* (Linn). Z. Mikrosk. Anat. Forsch Leipzig 96-1969-1977.

Stannius, H. (1939). Uber Nebennieren bei Knochemfischen. Arch. Anat. Physiol. Wiss. Med. 6, pp. 97-101.

Chapter 12

NERVOUS SYSTEM

With the advancement of cephalization and aggregation of ganglionic units in the head, the fishes have achieved a well developed nervous system. There is a great interaction between nervous system and endocrine system and in the area of interchange between the two systems. Some neurons display the combination of both nervous and hormonal systems. It is termed as neurosecretory or neuroendocrine.

On the basis of anatomy, the nervous system can be divided into a central nervous system (CNS) and peripheral nervous system (PNS). The central nervous system consists of brain and spinal cord. All nervous tissues other than brain and spinal cord are known as peripheral nervous system. It consists of nerves, ganglia and receptors.

The peripheral nervous system is divisible into somatic and visceral. The term visceral is sometimes used in referring to the autonomic nervous system.

Nervous System

- CNS
  - Brain
  - Spinal cord
- PNS
  - Somatic
  - Visceral autonomic
    - Sympathetic
    - Parasympathetic

The nervous system is made up of two chief types of cells, the nerve cells and supporting cells. The nerve cells are known as neurons and they are functional units. They have characteristic feature that they can conduct impulses. The receiving of information and conduction to other part is known as impulse. The message having travelled through the body of nerve cell is adapted for transmission of the impulse to the next functional unit, be it another nerve cell, a muscle, a gland and so forth.

The transmission of information from one neuron to the next in the chain is called synapse. The information is transmitted due to difference of potential between the inside and the outside of nerves. In the resting stage it is about 70 mV.

Neuroglia or simple glia are the supporting cells in the CNS while Schwann cells and satellite cells of ganglia in peripheral nervous system. In addition to neurons and glia there are numerous blood vessels both in the CNS and PNS.

The neuroglia cells are classified as astrocytes, oligodendrocyts and microglia.

The Neuron : A typical neuron has cell body or perikaryon, the dendrites and the axon. The cell body contains inclusions and **organelles** such as mitochondria, rER, free ribosomes, Golgi bodies, neurofilaments, lysosomes, etc. The nucleus is prominent with significant nucleolus (Fig. 12.1).

The cytoplasm contains Nissl's bodies which are basophilic in nature. The Nissl's bodies are characteristic feature of the nerve cell. Another important feature is that neuron never divides. It contains numerous enzymes and complex molecules. During the course of nerve degeneration, changes also occur in the cell body, the most conspicuous being the disorganization of Nissl's bodies, a process called *chromatolysis*.

The renewal of these enzymes and complex molecules and their transport to distant location is done by the process known as axonal transport.

The dendrites are generally branched and are the sites of synaptic contact.

The neuron gives out one.axon, it may be long. The motor neurons are generally very long. The junction of the axon with soma is raised as cone-shaped structure called as *axon hillock*. This cone of the cytoplasm is the discharge region.

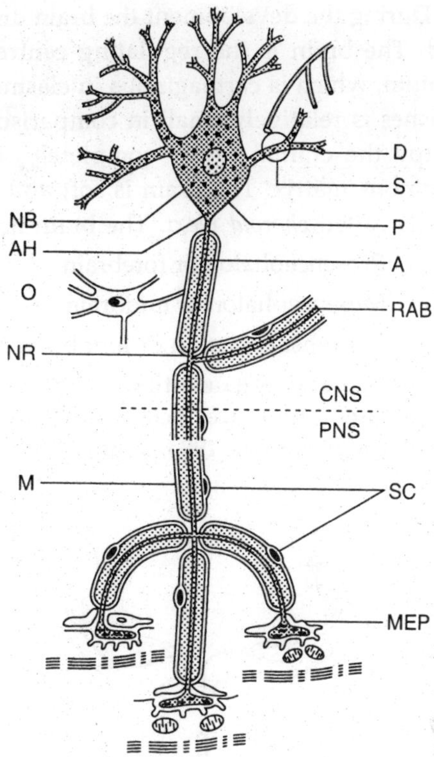

Fig. 12.1 : Diagram of motor neuron. A, axon; AH, axon hillock; CNS, central nervous system; D, dendrites; M, myelin; MEP, motor end plate; NB, Nissl bodies; NR, node of Ranvier; O, oligodendrite; P, perikaryon; PNS, peripheral nervous system; RAB, recurrent axonal branch; S, synapse; SC, Schwann cell.

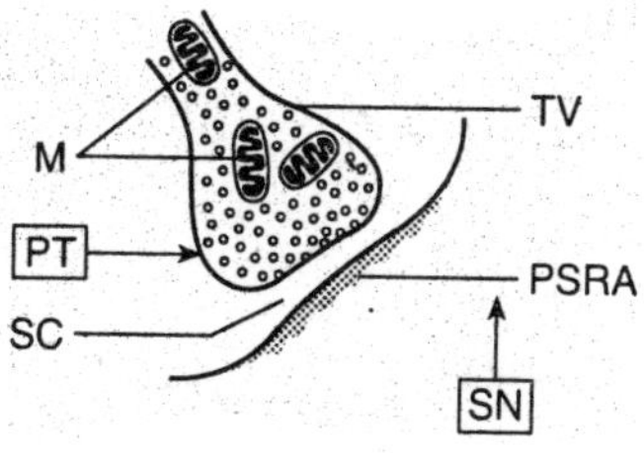

Fig. 12.2 : Diagram of synapse. M, mitochondria; PSRA, postsynaptic receptor area; PT, presynaptic terminal; SC, synaptic cleft; SN, synaps of neuron; TV, transmitter vesicles.

The axon finally terminates and this synaptic contact is called end bulb or terminal bouton. The synapse has three regions, presynaptic knob (terminal bouton), the synaptic cleft, a narrow space between the presynaptic knob and post synaptic membrane (Fig. 12.2). The third is the postsynaptic membrane or the receptor membrane of the next neuron.

The bouton contains synaptic vesicles and vesicles contain transmitter substances such as ACh (Acetylcholine), norepinephrine (NE), dopamine (DA), gamma aminobutyric acid (GABA), serotonin, glutamic acid and glycine. Two enzymes, the AChE (Acetylcholinesterase) and catechol-O-methyltransferase (COMT) act on substrate ACh and NE respectively.

BRAIN

During the development the brain distinguishes as an anterior enlargement of the spinal cord. The brain is the regulating centre for all the receptors. The brain is enclosed in a cranium, which is cartilaginous in elasmobranchs and bony in teleosts. Generally, the brain of fishes is relatively small in comparison to their body size and hence the brain does not occupy the cranial cavity completely, leaving small gap, which is filled with a sort of gelatinous matrix. The brain is soft and white and covered by extensive network of blood vessels called *choroid plexi*. The brain has three main divisions i.e.

1. Prosencephalon or forebrain
2. Mesencephalon or midbrain
3. Rhombencephalon or hindbrain (Fig. 12.3a, b)

The general organization in the major groups of fishes is not dissimilar, although considerable differences occur in the form and enlargement of different parts of the brain (Herrick, 1991, Singh, 1993, Evan, 1990).

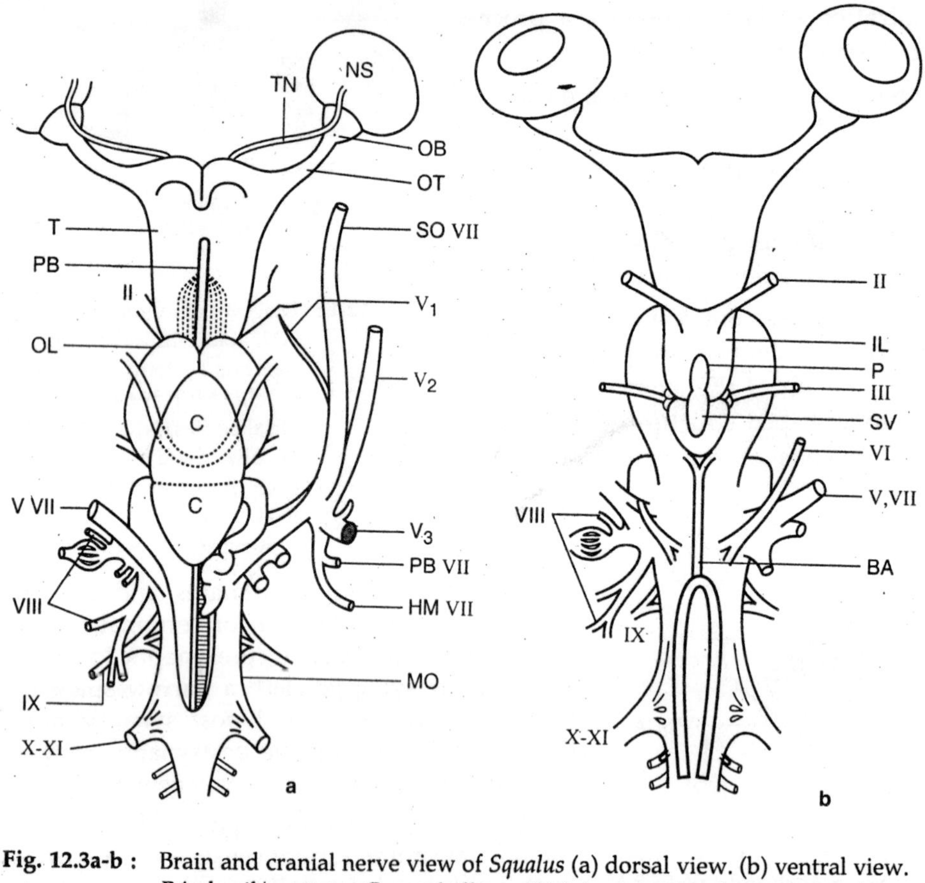

Fig. 12.3a-b : Brain and cranial nerve view of *Squalus* (a) dorsal view. (b) ventral view. BA, basilar artery; C, cerebellum; HM, hyomendibular; NS, nasal sac; IL, inferior lobe; MO, medulla oblongata; OB, olfactory bulb. OL, optic lobe; OT, olfactory tract; P, pituitary; PB, palatine branch; SO, superficial ophthalmic; SV, Saccus vasculosus; T, telencephalon; TN, terminal nerve; II-XI-cranial nerves.

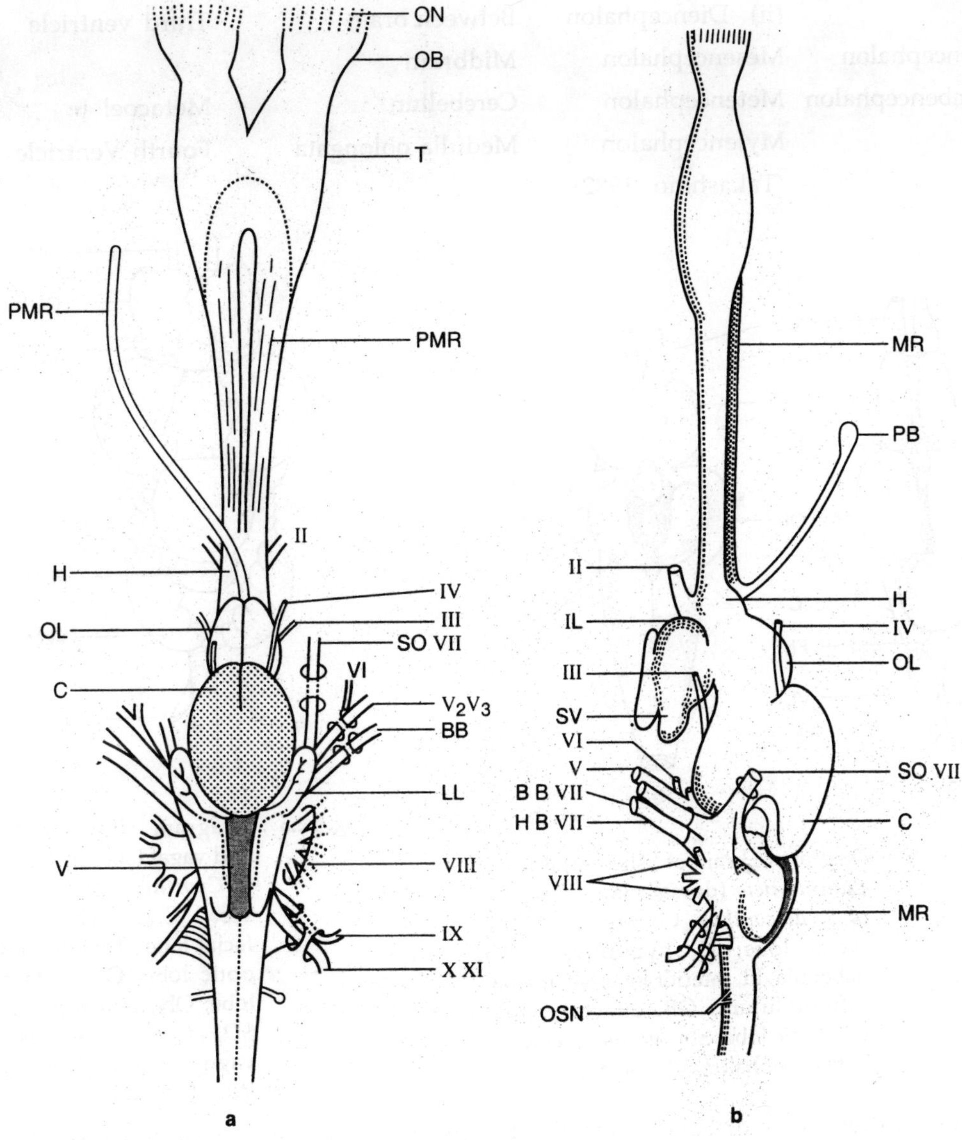

Fig. 12.4a-b : Brain and cranial roots of *Hydronagus*. (a) Dorsal view. (b) lateral view. BB, buccal branch; C, cerebellum; FV, fourth ventricle; H. hebenula; LL, lateral-line lobe; MR, membranous roof; OB, olfactory bulb; ON, olfactory nerve; OL, optic lobe; OSN, Occipetostinal nerve; PB, pineal body; PMR, paraphysis in membranous roof; SO, superficial opthalmic VII. SV, saccus vasculosus; T, telencephalon. II-XI are cranial nerve.

The homologies of brain parts are summarized by Lagler *et al.*, 1962

Embryonic Parts		Definite Brain Parts	Brain Cavities
Prosencephalon	(i) Telencephalon	Forebrain (Cerebral Hemisphere)	Paired lateral ventricles

Embryonic Parts		Definite Brain Parts	Brain Cavities
	(ii) Diencephalon	Between brain	Third ventricle
Mesencephalon	Mesencephalon	Midbrain	
Rhombencephalon	Metencephalon	Cerebellun	Metacoel
	Mylencephalon	Medulla oblongata	Fourth Ventricle
	(Takashino, 1982)		

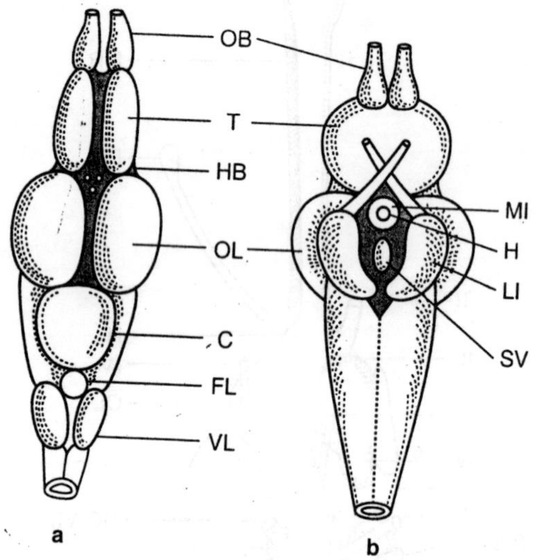

Fig. 12.5a, b : Diagram of brain of fishes. (a) Brain of *Puntius ticto* (dorsal view). (b) brain of *S. diaconthus.* C, cerebellum; FL, facial lobe; H, hypophysis; Hb, habenula; LI, inferior lobe; MI, median infundibulum; OB, olfactory bulb; OL, optic lobe; SV, saccus vasculosus; T. telencephalon; VL, vagal lobe.

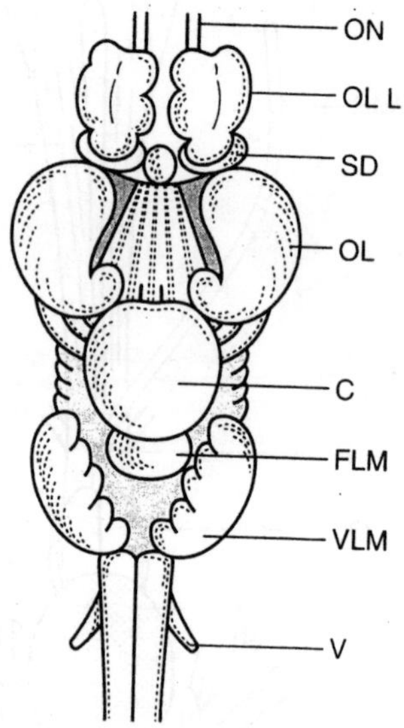

Fig. 12.6 : Diagram showing facial and vagal lobes in the brain of *Cyprinus carpio* (dorsal view). C, cerebellum; FLM, facial lobe of medulla; OL, optic lobe; OLL olfactory lobe; ON, olfactory nerve; SD, saccus dorsalis; V, vagus nerve, VLM, vagal lobe of medulla.

Some authors have divided brain directly into five parts, viz., telencephalon, diencephalon, mesencephalon, metencephalon and mylencephalon.

I. Prosencephalon or Forebrain

The forebrain consists of two parts, viz.,

(a) Telencephalon,

(b) Diencephalon.

(a) **Telencephalon :** It is the most anterior part of the forebrain, mainly responsible for reception, conduction of smell impulses (DeBruin, 1980). Telencephalon consists of two parts. The anterior paired olfactory bulb and posterior two large cerebral hemispheres. The olfactory lobes are anteriorly projected and slender and are known as olfactory bulbs which possess internal cavities called brain having lateral ventricles I and II.

The olfactory lobes are composed of mass of nerve cells, which transmit smell impulses to other centres. There is considerable variations in the structure of olfactory part. Elasmobranch and some bony fishes who depends mainly on smell for searching the food or social activities have remarkably enlarged olfactory lobes. Some fishes possess both parts i.e., olfactory lobes and bulbs (*Puntius ticto*). In others, only olfactory bulbs are present (*Tor tor, Mystus seenghala*) whereas some species contain only olfactory lobes (*Channa striatus*).

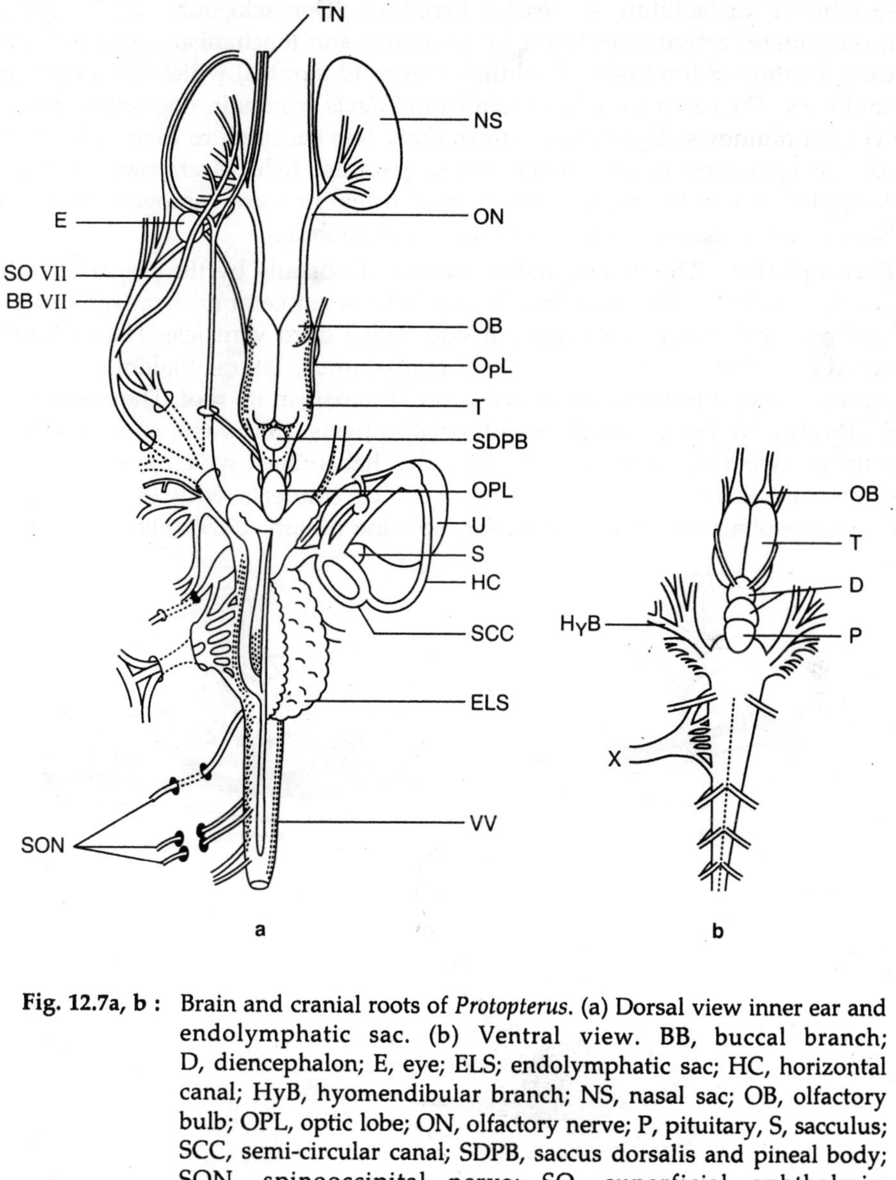

Fig. 12.7a, b : Brain and cranial roots of *Protopterus*. (a) Dorsal view inner ear and endolymphatic sac. (b) Ventral view. BB, buccal branch; D, diencephalon; E, eye; ELS; endolymphatic sac; HC, horizontal canal; HyB, hyomendibular branch; NS, nasal sac; OB, olfactory bulb; OPL, optic lobe; ON, olfactory nerve; P, pituitary, S, sacculus; SCC, semi-circular canal; SDPB, saccus dorsalis and pineal body; SON, spinooccipital nerve; SO, superficial ophthalmic; T, telencephalon; TN, terminal nerve; U, utriculus; VV, vertebral vein.

The second part of the telencephalon is the cerebral hemisphere. They are two in numbers and solid mass-like, jointed on mid-line. The anteriormost large bundle of nerve fibres connects the two hemispheres and is known as anterior commissure. They are also connected with the diencephalon by other tracts. The cerebral hemisphere are covered by a non-nervous membranous sheath called 'pallium', which is dorsolaterally situated and subpallium is basomedially situated (Nieuwenhuys, 1966).

The forebrain of actinopterygains has a single, median ventral ventricle in contrast to the paired ventral (basal) ventricles in lower fishes. (Droogleever, 1961; Morgan, 1975).

In addition to the olfaction, the telencephalon is thought to serve other generalized function such as to regulate aggression, sexual activity and parental behaviour, maintain body equilibrium to facilitate successful breeding. (Demaski et.al., 1975). The telencephalon also facilitates activities of lower brain centres and mechanism (Aronson, 1970). It is evident from ablation of forebrain of goldfish (*Carrasius auratus*) which shows less response than normal ones. The removal of forebrain from *Tilapia* results in neglecting their youngs, while in certain minnows (Cyprinidae) make them less cautious to their new situation. In stickleback, the spawning is suppressed and in paradise fish (*Macropodus opercularis*) the nest building behaviour is affected, when their forebrains are removed. In goldfish the colour vision is suppressed as result of extirpation of forebrain.

(b) **Diencephalon** : The diencephalon is covered dorsally by the posterior bulgings of the cerebral hemispheres. Its small area is visible between cerebral hemispheres and optic lobes. The diencephalon possesses cavity inside, called third ventricle. The diencephalon is distinguished into three parts, i.e., a dorsal epithalamus, lateral thalamus and a ventral hypothalamus. The epithalamus contains a choroid plexus in its roof. The posterior part of the roof elevates to form pineal gland or epiphysis (Shrivastava *et al.*, 1980). The epithalamus has two ganglionic masses, called the 'habenulae', which are of equal size and also the nerve fibres from the telencephalon which connects with the thalamus, hypothalamus and the olfactory areas of the telencephalon (Singh, 1969) (Figs. 12.9a, b).

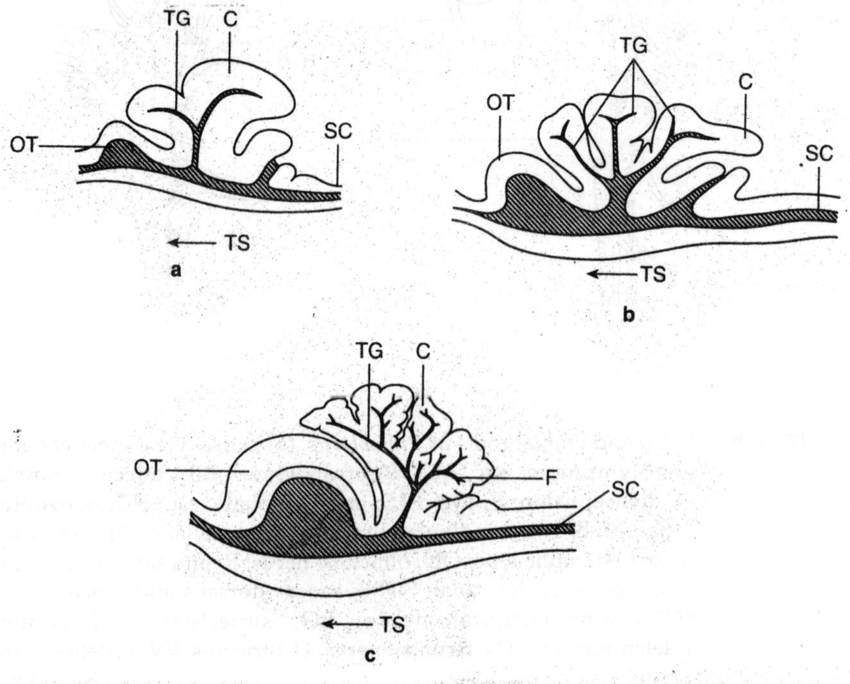

Fig. 12.8a-c : Diagram of cerebellum of sharks. (a) Cat shark. (b) Smooth dogfish. (c) Sand shark. C, cerebellum, F, folds; OT, optic tectum; SC. spinal cord; TG, transverse groove; TS towards snout.

Regarding the function of pineal body, two theories have been put forward. According to the first hypothesis, the pineal body has sensory roles while the second hypothesis states its secretory function. In sensory roles it acts as a baro or chemoreceptor for cerebrospinal fluid or helps in facilitating the olfactory response to sex hormones. The secretory functions involves the external secretion related to the chemical composition of the cerebrospinal fluid or the metabolism of brain tissue. It is also expected to be an endocrine gland with internal secretions.

The thalamus consists of lateral walls of the diencephalon. The thalamus acts as a relay centre for transmitting olfactory and strait body impulses to thalamomedullar and thalamospinal tract. The ventral part of thalamus bears geniculate nuclei or ganglia (Singh, 1971). These ganglia are well developed in sharks and is known as geniculate lobes.

The hypothalamus constitutes the floor of the diencephalon. It is the largest and the most important part of diencephalon. Anterior part of the hypothalamus contains preoptic area. However, from the ventral part of hypothalamus projects a pouch-like downgrowth, known as the infundibulum. The tip of infundibulum bears hypophysis or pituitary gland. There are two important nuclei centres present in the hypothalamus, one is nucleus preopticus and the other is nucleus lateralis tuberis. Both are made up of nerve cells of secretory nature. In Osteichthyes, the lower part of the hypothalamus extends laterally which are known as inferior lobe. The saccus vasculosus is highly vascular protrusion of the ventral wall of the diencephalon of fishes. It is regarded as benthic organ detecting changes in water pressure (Dammerman, 1910). It plays a secretory role also (Sundara Rajan and Prasad, 1964).

The diencephalon acts as an important correlation centre for incoming and outgoing message concerned with internal homeostasis. The hypothalamus plays an important role in regulating the endocrine system through pituitary gland (Savage, 1968).

II. Mesencephalon or Midbrain

The mesencephalon or midbrain is relatively large. It is made up of the dorsal optic tectum and the ventral tegumentum. Fishes that feed by sight possess large optic lobes. (Khanna and Singh, 1966).

The optic tectum appears as two optic lobes from the dorsal side. The ventricle communicates the mesencephalon as narrow cavity, called aquaductus mesencephali or aqueduct of Sylvius. There is considerable variation in the histological structure of the optic tectum. According to Ariens Kriens Kapper *et al.* (1960) the optic tectum is composed of six layers which are as follows :

1. The superficial layer, the stratum opticum which is innervated by optic nerve fibres.

2. Stratum fibrosum, griseum superficial, which is an important seat of visual sense and possess nerve fibres and nerve cells.

3. Stratum griseum centrale consists of nerve cells of efferent nerves.

4. Stratum album centrale, it consists of nerve fibres of efferent nerves and is associated with oculomotor and thalamus.

5. Stratum griseum periventricular comprising nerve cells which is connected with the fibres of the periventricular system.

6. Stratum fibrosum periventricular, lies in front of aquaductus mesencephalic and possess numerous nerve fibres.

Because of these multi-layered arrangement of the nerve cells in optic tectum, it is often considered to be homologous with the cerebral cortex of mammals.

The midventral part of the optic tectum contains a protuberance, called torus longitudinalis, which helps in integration between the sense of equilibrium and the sense of vision (Singh, 1970). There is a strong evidence for optic tectum as eye body co-ordinating centre and compensate the absence of a true optic chiasma of bony fishes (Dhillon and Tandon, 1987). Electric stimulation of tegumentum results in uncoordinated locomotory response. Optic tectum is an important correlation centre for various types of impulses of the brain (Singh and Khanna, 1970; Schnitzlein, 1964).

III. Rhombencephalon or Hindbrain

The hindbrain consists of cerebellum or metencephalon and medulla oblongata or mylencephalon.

(a) **Cerebellum or Metencephalon :** The cerebellum develops as prominent dorsal outgrowths from the medulla. Its anterior part projects forwardly into the cavity of optic lobes and is known as valvula cerebelli which is composed of cortex and medulla. The cortex is made up of three distinct layers comprising differently shaped nerve cells.

(i) The molecular or chief receptive layer comprising Purkinje cell dendrites and glia cells.

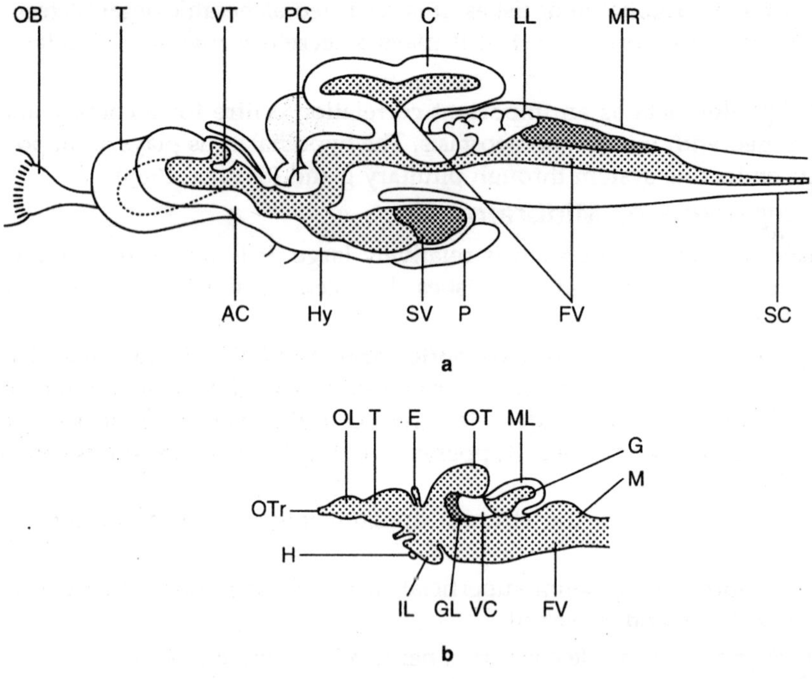

Fig. 12.9a-b : Lateral view of brain of fishes. (a) *Squalus*. (b) *Puntius ticto*. AC, anterior commissure; C, cerebellum; E, epiphysis; FV, fourth ventricle; GL, granular layer; Hy, hypophysis; IL, inferior lobe; LL, lateral lobe; M, medulla; MR, membranous roof; OB, olfactory bulb; OL, olfactory lobe; OT, optic tectum; OTr, olfactory tract; P, pituitary; PC, posterior commissure; SC, spinal cord; SV, saccus vasculosis; T, telencephalon; VC, vulvula cerebelli; VT; velum transversum.

(ii) The granular cell layer consists of small granular cells and also receives the Purkinje cell axons. Fishes having well developed gustatory (carp) and lateral line (catfish) system have a more developed vulvula cerebelli. They are the largest component of the brain and project forwardly beyond the forebrain in the mormyrids. Since mormyrids generate and respond to weak electric currents, it is considered that their enlarged cerebellum is involved in the reception of electrical impulses (Khanna and Singh, 1967).

Elasmobranchs shows an increase in size of cerebellum as compared to the Cyclostomes. The dogfish (small shark) possess simple bilobed cerebellum. However, the mackerel sharks which are large and swift, have the larger cerebellum occupying much part of the brain. In these fishes the cerebellum becomes convoluted to increase surface area (Figs. 12.8a, b, c).

The electric current discharging fishes such as mormyrids and the electrical catfish (*Malapterurus*) contains very well developed cerebellum dealing with electrical impulses.

The cavity of the cerebellum is known as metacoel, which is prominent in sharks and rays (Elasmobranchs) and is completely disappeared in cerebellum of higher bony fishes. The main functions of the cerebellum is to control swimming equilibration, maintenance and co-ordination of muscular tonus, and orientation in space (Singh, 1993).

(b) **Medulla oblongata or mylencephalon or brain stem :** The medulla oblongata or mylencephalon is the posterior most part of the brain, which can be distinguished from the spinal cord. The medulla is divisible into columns of nerve fibres based on the types of information transmitted. Thus there are visceral and somatic sensory and visceral and somatic motor columns.

The medulla contains nuclei of cranial nerves from III to X, arranged anteroposteriorly. (Figs. 12.3 & 12.4).

The different parts of medulla are enlarged with the development of various senses. The medulla contains a cavity inside, known as the fourth ventricle. Some fishes such as *Culupea* and Mugil possess prominent paried swellings called 'cristae cerebelli' present on the anterolateral boundary of the fourth ventricle. These cristae are associated with schooling behaviour of these fishes. In goldfish prominent vagal lobes are present behind the cristae cerebelli, from which IX and X cranial nerves arise. This fish also possesses a palatal organ found in the roof of mouth, helping in testing the food by taste and touch. Another prominent structure, the facial lobe or tuberculum impar (Fig. 12.6). found behind the cerebellum in *Cyprinus carpio*. In the facial lobes the gustatory and tactile impulses are corrected with visceral sensory ones as it arises from the fusion of root elements of the facial (VII) and (X).

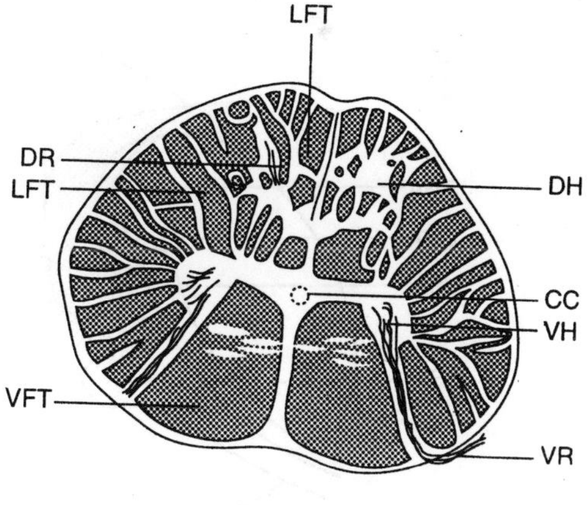

Fig. 12.10 : Section of spinal cord of shark. CC, central canal; DFT, dorsal fibre tract; DH, dorsal horn; DR, dorsal root of spinal nerve; LFT, lateral fibre tract; VFT, ventral fibre tract; VH, ventral horn; VR, ventral root fibres of spinal nerve.

The medulla of higher bony fishes (Actinopterygii) have a pair of large neurons called the giant cells of Mauthner. These cells are present at the level of cranial nerve VIII. In *Anguilla* and *Mola* the Mauthner cells are absent whereas less prominent in bottom dwellers such as gobies and scorpion fishes. These cells are motor co-ordinator for transmitting multiple sensory impulses mainly from lateral-line centres to the caudal and body swimming muscles.

The medulla is sensory and sensory co-ordinating centres and a relaying area between the remaining brain and spinal cord. Medulla also has certain centres that control some somatic and visceral functions (Lagler *et al*, 1962). Among bony fishes these functions includes respiratory, paling of body colour and osmoregulatory.

SPINAL CORD

The basic structure of spinal cord in fishes resembles to that of higher vertebrates. It runs dorsal and lengthwise of the fish body in the neural canal of the vertebral column. In transverse section of spinal cord two regions are clearly distinct, i.e. central and peripheral. The central region is composed of numerous nerve cells, hence looks grey and called grey matter, which roughly resembles a letter 'Y' or 'X'. The peripheral region, which surrounds the grey matter, contains large number of myelinated nerve fibres and is known as white matter (Fig. 12.10). The centre of grey matter is occupied by a central canal, which is filled with cerebrospinal fluid (CSF). The cerebrospinal fluid is secreted by the brain. The central canal is lined with ependymal cells. The central canal is formed by extension of brain ventricles. The grey matter has an anterior, posterior and lateral columns or horns from where the spinal nerve fibres enter or leave the spinal cord. The anterior or ventral roots arise from the anterior columns to innervate somatic muscles or visceral organs (Fig. 12.11).

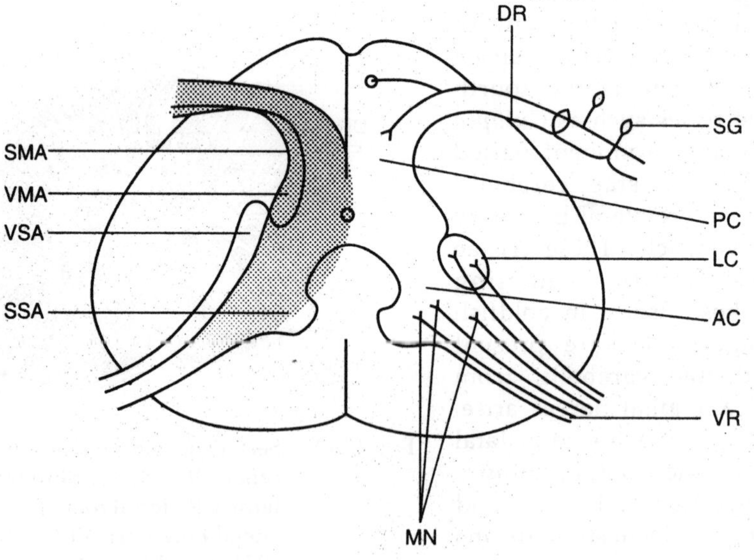

Fig. 12.11 : Diagram of spinal nerve showing sensory and motor areas. AC, anterior column; DR, dorsal root; LC, lateral column; MN, motor neuron; PC, posterior column; SG, spinal ganglion; SMA, somatic motor area; SSA, somatic sensory area; VR, ventral root; VSA, visceral sensory area.

Those nerves which innervate visceral effectors comprise a pair of sympathetic ganglion present along their pathways. However, the peripheral stimuli transmit through another ganglion found in the posterior root of the posterior column of spinal cord. Then these stimuli are conveyed to nerve cells in the central nervous system. The fish spinal cord possess two types of cells in the grey matter. One kind of cell generally found in the dorsal region and plays an important role in controlling the movements of trunk musculature. The another type of cells are present in the ventral region of the grey matter. The fibres originating from these cells innervate certain effectors such as the fins. The spinal nerve fibres are arranged into two roots, one is the posterior sensory root and another is anterior motor root.

PERIPHERAL NERVOUS SYSTEM (P.N.S.)

The P.N.S. consists of cranial nerves, spinal nerves and their ganglion. The bundle of nerve fibres originating from the brain and spinal cord is called cranial and spinal nerves respectively.

Cranial Nerves

There are ten pairs of cranial nerves found in the fishes (Figs. 12.12, 12.13 and 12.14). These are as follows :

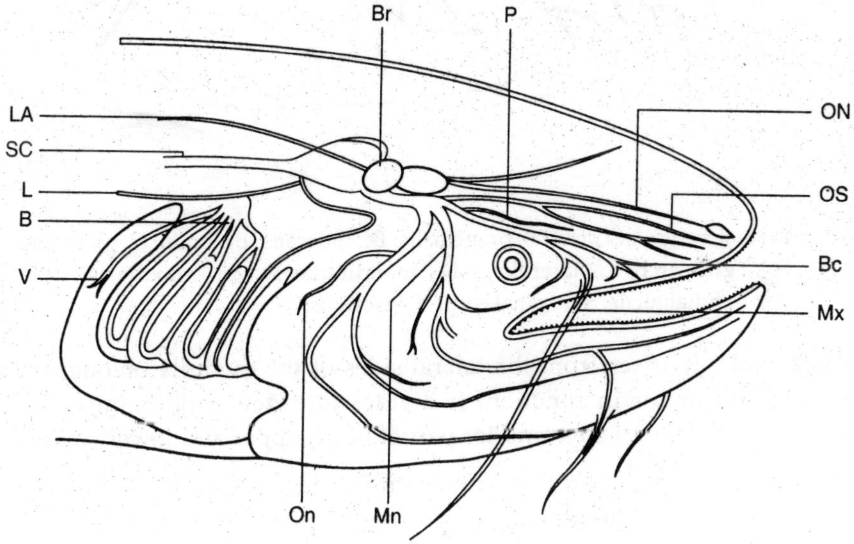

Fig. 12.12 : Cranial nerves of *Wallago attu*. B. branchialis; Bc. buccalis; Br, brain; L, lateralis; LA, lateralis accessories; Mn, mandibularis; Mx, maxillaris; ON, olfactory nerve; OS, ophthalmicus superficialis; P, palatinus; SC, spinal cord; V, visceralis.

I. **Olfactory nerve :** It originates from the olfactory bulb particularly from lamina terminalis. It innervates snout and is a special sensory which carries smell impulses to the brain.

II. **Optic nerve :** It takes origin from ventral side of optic tectum of midbrain. It innervates retina of eye. It is also a special sensory nerve and carries visual impulses.

III. **Oculomotor :** It arises from optic tectum of midbrain (anterior brain stem under optic tectum), passes through the optic foramen and enters the orbit to supply superior, inferior, anterior, rectus and inferior oblique muscles of eyeball. It is somatic motor, innervates four of the six striated eye muscles and muscles inside eye.

IV. **Trochlear** : It originates from dorsolateral side of brain stem under cerebellum. It innervates one of the six striated muscle of the eye, i.e., superior oblique muscle of eyeball.

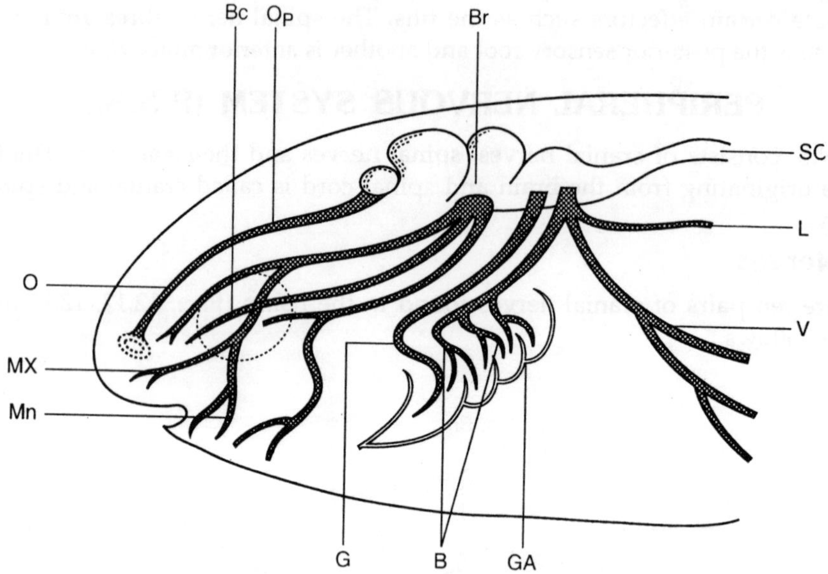

Fig. 12.13 : Cranial nerves of *Labeo*. B, branchialis; BC, buccalis; Br. brain; G, glossopharyngeal; GA, gilll arch; L, lateralis; MX, mandibularis; Mn. maxillaris; O, olfactory, Op, ophthalmicus; SC, spinal cord; V, visceralis.

V. **Trigeminal** : It arises from the lateral sides of the medulla oblongata. It is mixed somatic sensory and motor in function. It divides into deep ophthalmic, maxillary and mandibular branches. It innervates anterior part of head, upper and lower jaw regions.

VI. **Abduscens** : It originates from the ventral sides of the medulla oblongata and runs just behind the trigiminal. It innervates sixth striated muscle, i.e., posterior rectus muscle, which moves the eyeball.

VII. **Facial** : It arises from the sides of medulla oblongata. It is truely mixed nerve. It divides into three branches, supra-orbital, infra-orbital and the hyomandibular. The supra-orbital branch further divides into :

(i) Ophthalmicus superficialis trigeminalis and

(ii) Ophthalmicus superficialis facialis. They innervate lateral line system.

The infra-orbital branch runs ventral to supra-orbital and divides into four branches—maxillaries, buccalis, mandibularis and palatine.

The hyomandibular branch runs just behind the mandibular nerve and innervates lower jaw.

VIII. **Auditory** : It originates from sides of medulla and runs behind the facial. It gives two branches :

(i) Vestibular branch : goes to utricular and ampullae of internal ear.

(ii) Saccular branch : goes to sacculus and lagena.

The facial VII and acoustic VIII arise from the medulla so intimately that they are often considered as acoustic facialis nerve (Lagler *et al.*, 1962).

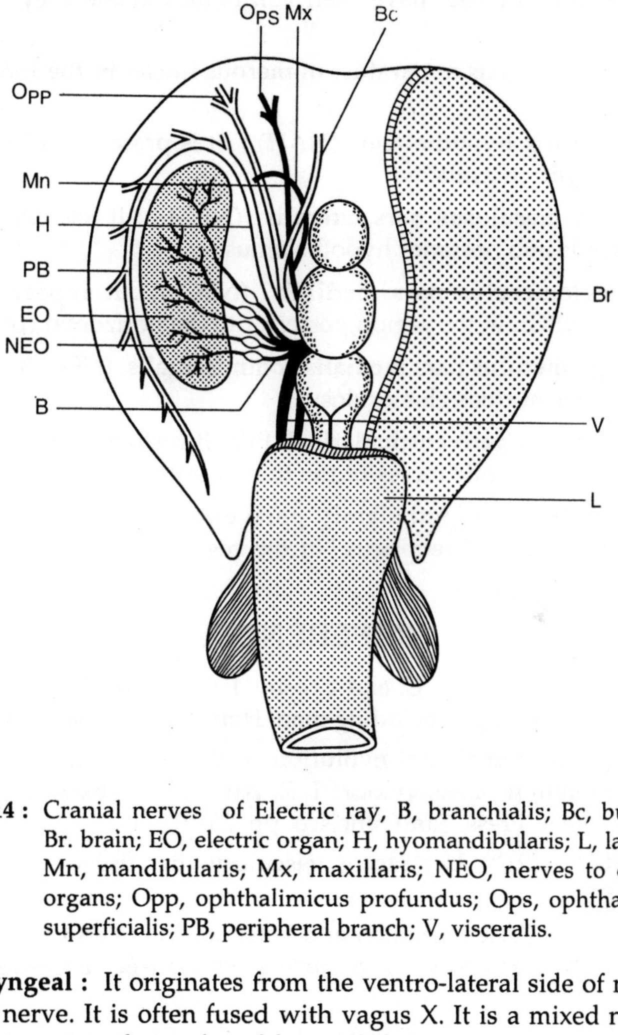

Fig. 12.14 : Cranial nerves of Electric ray, B, branchialis; Bc, buccalis; Br. brain; EO, electric organ; H, hyomandibularis; L, lateralis; Mn, mandibularis; Mx, maxillaris; NEO, nerves to electric organs; Opp, ophthalimicus profundus; Ops, ophthalmicus superficialis; PB, peripheral branch; V, visceralis.

IX. **Glossopharyngeal :** It originates from the ventro-lateral side of medulla oblongata, just behind the VIII nerve. It is often fused with vagus X. It is a mixed nerve. It innervates ventral pharyngeal mucosa and muscles of first gill slit.

X. **Vagus :** It takes origin from medulla just after the glossopharyngeal. It is mixed nerve. It divides into five branches :

(i) Supratemporal branch—It innervates skin and lateral line organs of supratemporal region

(ii) Dorsal recurrent branch which—It with body test bud innervation of VII.

(iii) Body lateral line branc —It goes to lateral line.

(iv) Visceral branch—It innervates visceral organ

(v) Branchial branch—It supplies and receives nerves from four posterior gill slits.

Spinal Nerves

Each spinal nerve contains two roots : The posterior sensory root with its ganglion and anterior motor root. The posterior root is composed mostly of sensory fibres from nerve cells in the spinal nerve ganglion and connects it with the C.N.S. However, the motor fibres passing through the anterior root have their cell bodies in the grey matter of the spinal cord.

With the help of stereotaxis apparatus, numerous nuclei in the forebrain are reported. These are as follows :

(i) **Nucleus anterioris hypothalami** (NAH) : It is present on the ventrolateral to the posterior part of the optic chiasma.

(ii) **Nucleus preglomerulosus pars lateralis** (NPGL) : It lies lateral to the horizontal commissure and extends into posterior hypothalamus.

(iii) **Nucleus preglomerulus pars medialis** (NPGm) : It appears medially and then ventromedially to the NPGL in the region posterior to the horizontal commissure.

(iv) **Nucleus preglomerulus pars medialis commissuralis** (NPGC) : It is placed medially at the posterior end of the nucleus glomerulosus.

(v) **Nucleus preopticus periventricularis** (NPP) : It is situated around the borders of the anterior extension of the preoptic recess of the third ventricle.

(vi) **Nucleus anterioris** (hypothalami) **periventricularis** (NAPv) : This nucleus is present ventral to the nucleus preopticus in the region of the posterior end of the optic chaisma.

REFERENCES

Arien Kappers, C.V., Huber, G. C. and Crosby, E.C. (1960). The comparative anatomy of the nervous system of vertebrates, including man, Hafner, Pub. Company, N.Y.

Aronson, L.R. (1970). Functional evolution of the forebrain in lower vertebrates. In development and evolution of behaviour (L.R. Arson, E. Tobach, D.S. Lehrman and J.S. Rosenblatt, Eds.). Freeman Press. San Francisco. pp. 75-107.

Dammerman, K.W. (1910). Der saccus vasculosus der fische, ein tiefeorgan. A. Wiss. Zool 96 : pp. 654-726.

DeBruin, J.P.C. (1980). Telencephalon and behaviour in teleost fish—a neuroethological approach. In comparative neurology of the telencephalon (Ed. Ebbesson Co. E) pp.175-201 Plenum Publ. Corp.

Demaski. L.S., Bauer, D.H. and Gerald, J.W. (1975). Sperm release evoked by electrical stimulation of the fish brain, a functional anatomical study. J. Exp. Zool. 191 : 215-221.

Dhillon, A.K. And Tandon, K.K. (1987). Tori longitudinales, a compensation of true optic chiasma. Curr. Sci. 56 : 280-281.

Droogleever Fortuyn, J. (1961). Topographical relations in the telencephalon of the sunfish, *Eupomotis gibbosus*. J. Comp. Neur, 116 : 249-263.

Evan. H.M. (1990). Brain and body of fish. The Blakiston Co. (McGraw Hill Book Co. N.Y.).

Herrick, C.L. (1991). Contribution to the morphology or the brains of bony fishes. II Studies on the brains of some American freshwater fishes. (Cont.) J. comp. Neurol 1 : 333-357.

Khanna, S.S, and Singh, H.R. (1967). On the morphological pecularities of the valvula cerebelli in some teleosts. Zool. Anz. 178 : 413-419.

Khanna, S.S., and Singh, H.R. (1966). Morphology of the teleostean brain in relation to feeding habits. Proc. Nat. Sci. India B. 36 (3) : 306-316.

Lagler, K.F., Bardach, J.E. and Miller, R.R. (1962). Ichthyology John Wiley and Sons. Inc. N.Y.

Morgan Jr. G.C. (1975). The telencephalon of the sea catfish *Galeichthys felis*. J. Hirenforsch. 16 : 131-150.

Nieuwenhuys, R. (1966). The interpretation of the cell masses in the teleostean forebrain. In evolution of forebrain, phylogenesis and ontogenesis of the forebrain, (Eds. R. Hasler and Stephan), Georg Thieme Verlag, Stuttgart. pp. 32-39.

Savage, G.E. (1968). Function of the forebrain in the memory system of the fish. In the central nervous system and fish behaviour. (Ed. Ingl, D.) Univ. of Chicago Press. Chicago. pp. 127-138.

Schnitzlein, H.N. (1964). Correlation of habit and structure of the fish brain Am. Zoologist. 4 : 21-32.

Singh, H.R. (1969). On the structure and relation of the habenula in some teleosts. Zool. Beitrage. 15 : 457-464.

Singh. H.R. (1970). The structure and relation of the torus longitudinalis in some teleosts. Zool. Beitrage. 16 : 83-91.

Singh, H.R. (1971). On the thalamic and hypothalamic nuclear areas of some teleosts. Acta. Anat 78 : 574-590.

Singh, H.P. (1993). Fish brain, a fascinating structure. In 'Advances in Fish Research' Vol. 1 (Ed. Singh, B.R.) Narendra Publishing House. Delhi. pp. 281-294.

Singh, H.R. and Khanna, S.S. (1970). The cytoarchitecture and the fibre connections of the optic tectum of some teleosts. Zool Beitrage. 16 : 129-140.

Shrivastava, V.M.S., Shrivastava, D.K. and Tripathi R.S. (1980). Morphological and histochemical studies of the pineal organ in *Wallago attu*. Intl. J. Acad. Ichthyol. 1 : 31-34.

Sundararajan, B.I. and Prasad, M.R.N. (1964). The histochemistry of the saccus vasculosus of *Notopterus chitala* (Teleostei) Quart. J. Micr. Sci. 105 : 91-98.

Tokashino, F. (1982). Nervous system. In : An atlas of fish histology, normal and physiological features (Ed. Hibiya, T.) Kodansha Ltd. Goustav Fischer Verlag, Stuttgart ; pp. 32-41.

Chapter 13

AUTONOMIC NERVOUS SYSTEM

The autonomic nerves in fishes, control the aperture of iris, blood pressure, blood flow through gills for oxygenation and blood supply to various parts of the body autonomically. It controls heart performance, gastric motility and control the function of swim bladder. It also controls the colour changes and release of catecholamines from the chromaffin tissue (Nilsson *et al.*, 1983).

The autonomic nervous system is sub-divided into sympathetic and parasympathetic nervous system. The autonomic nerves of teleost are much similar to that of land vertebrates (Nicol, 1952). The reviews on the autonomic nervous system of fishes is made by Burnstock (1969), Campbell (1970), Santer (1977), Holmgren and Nilsson (1981, 1982), and Nilsson *et al.*, (1983) Nilsson (1983).

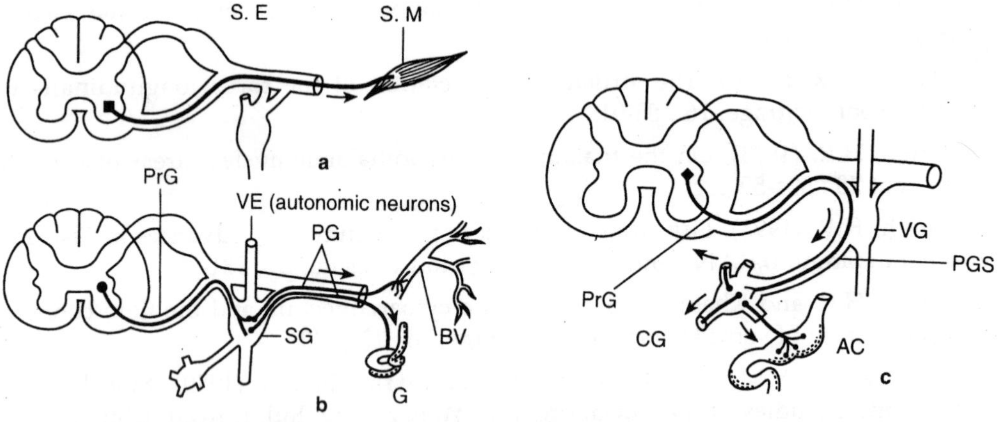

Figs. 13.1a, b, c : Scheme of somatic and autonomic efferent fibres. (a) Efferent somatic neurons supplying to skeletal muscle (no interuption ganglion). (b) Efferent (motor) neurons of autonomic (sympathetic) system (two neurons involve) supplying gland cells and blood vessels. (c) Efferent neurons of autonomic (sympathetic) nervous system supplying through splanchnic nerve to abdominal organs. AC, alimentary canal; CG, celiac ganglion; BV, blood vessel; G, gland cell; SE, somatic efferent; SG; sympathetic ganglion; SM, striated muscle PrG, Preganglionic; PG, postganglionic; VE, visceral efferent; VG, vertebral ganglion.

The autonomic nervous system in strict sense can be defined as that portion of the peripheral nervous system which conduct impulse to visceral organs, glands and blood vessels or in other words smooth muscles, cardiac muscles and glandular epithelium. The effectors are present in the organs. There are two types of neuron, the sensory and the motor. The sensory neurons (afferent) leave the organs to conduct impulses from the organ to the central nervous system. They have the same arrangement in visceral and somatic systems.

The difference in the organization pattern is observed in the motor or efferent fibres. The impulses flow in the autonomic efferent fibres from the CNS passes to the affector organs through two neuron systems (Fig. 13.1a, b, c).

- The impulses from the first neuron which is located in the CNS is carried by preganglionic fibre to the second neuron, which is situated outside the central nervous system. The impulse is then carried to the target organ by the postganglionic fibres. These postganglionic fibres by secreting the chemical transmitter/modulator at its postganglionic nerve fibres on the receptors in the effector organs. The transmitters are acetylcholine, adrenaline/ nor-adrenaline. In mammals a number of substances suggested as non-adrenergic, non-cholinergic such as ATP, 5HT, VIP (vasoactive intestinal peptides), substance P, nerotensin, somatostatin and gastrin have been reported in autonomic nerves (Furness & Costa, 1980). However, in fish autonomic nerves only, VIP, substance P, enkephalin and 5HT have been reported (Holmgren &Nilsson 1981).

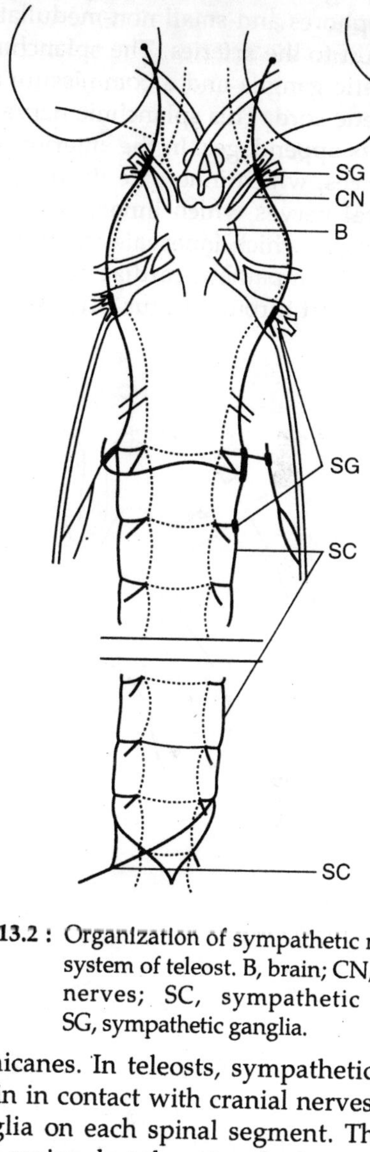

Fig. 13.2 : Organization of sympathetic nervous system of teleost. B, brain; CN, cranial nerves; SC, sympathetic chain; SG, sympathetic ganglia.

SYMPATHETIC NERVOUS SYSTEM

There are two sympathetic chains extending from the first spinal nerves to the end of the tail. On each sympathetic chain there is ganglionic swellings, known as sympathetic chain ganglia or paravertebral ganglia (Fig. 13.2). In dipnoans the ganglionic swellings are not distinct on the sympathetic trunk. Actually the sympathetic cells are scattered along the cord or grouped near the junctions with the remi communicanes. In teleosts, sympathetic chains continue into the head region and the ganglia remain in contact with cranial nerves V, VII, IX, & X. Generally there are two sympathetic ganglia on each spinal segment. There are two separate sympathetic cords in the anterior trunk region, but these trunks fuse to form a single cord between the kidneys. When the sympathetic cord enters haemal canal, it again bifurcates. Several authors have described the variation in the extent of fusion of the two cords in different species of fishes (Stannius, 1949; Young, 1931 & 1936).

In each spinal segment, the sympathetic trunk joins the spinal nerve by rami communicanes (Fig. 13.3). According to young (1931), they are two, white and grey rami. The white ramus contains medullated preganglionic fibres, while the grey ramus comprises non-medullated postganglionic fibres. The first two sympathetic ganglia give out rami communicanes, which are made up of postganglionic fibres, most of the preganglionic fibres to these ganglia run out at more posterior levels, and travel forwards in sympathetic chain. Preganglionic fibers do not terminate in the nearest ganglion but may cross over to the opposite sympathetic cord to run to higher or lower levels in the sympathetic chain.

Postganglionic fibres in the grey rami run through spinal nerves to blood vessels and chromatophores and small non-medullated nerves may innervate directly from the sympathetic chain to the arteries. The splanchnic nerve arises on the right side from the first two sympathetic ganglia and a commissure at the same level to contribute fibres from the left sympathetic cord. This splanchnic nerve provides the entire sympathetic innervation of the gut and its appendages. In the anterior trunk region the sympathetic ganglia give out many small nerves, which innervate. In the posterior trunk region the sympathetic ganglia give out genital nerves which innervate the gonads. Vesicular nerve also arises from sympathetic ganglia which innervate the urinary bladder and mesonephric ducts. In teleosts the presence of sympathetic cardiac nerves is not evident (Young, 1931). The sympathetic fibres enter the heart through vagus (Kumar, 1979).

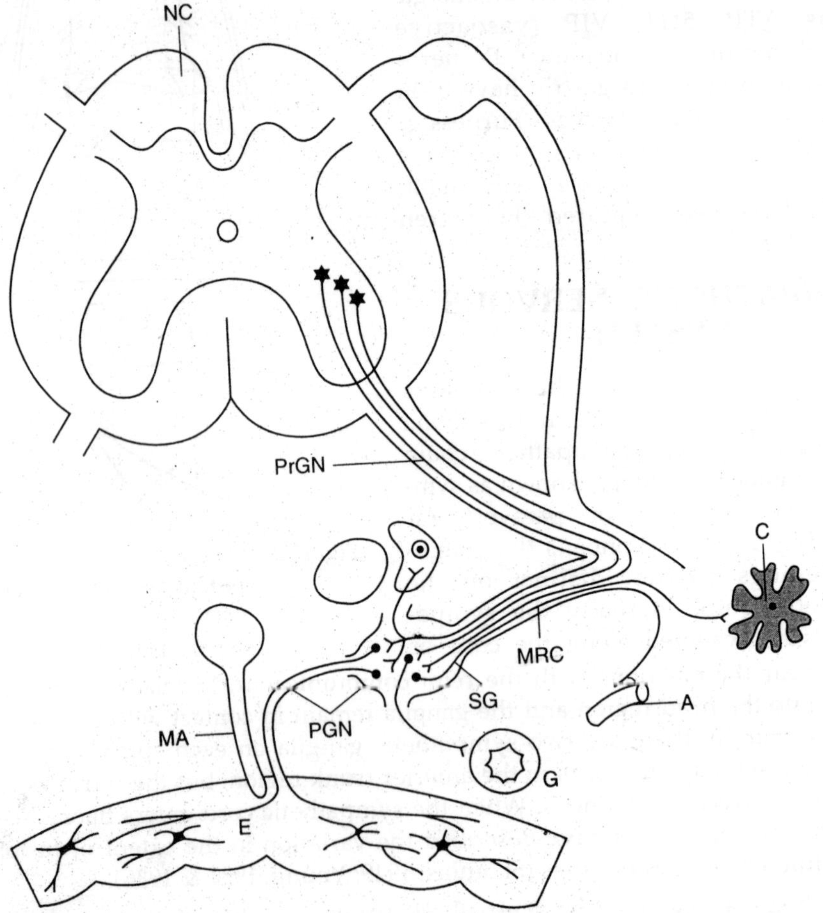

Fig. 13.3 : Sympathetic pathway in teleost fish. A, artery; C, chromatophore; MA, mesenteric artery; MRC, mixed ramus communicans; NC, nerve cord; PrGN, pre-ganglionic neuron; PGN, post-ganglionic neuron; G, gut; SG; sympathetic ganglion.

PARASYMPATHETIC NERVOUS SYSTEM

The parasympathetic components of autonomic nervous system comprise the cranial outflow. The preganglionic parasympathetic fibres leave the central nervous system from the brain as components of cranial nerves III, VII, IX and X in elasmobranchs similar to that of higher vertebrates (Fig. 13.4). There is no evidence of spinal parasympathetic outflow in fish (Campbell, 1970). III or oculomotor cranial nerve gives out preganglonic nerve fibres to the ciliary ganglia and from the ciliary ganglia postganglionic nerve fibres innervate the eyeball. Oculomotor nerves are absent in forms with reduced eyes. In *Polydon* an ciliary ganglion has been observed, but in *Scohirhynchus* there are ciliary ganglia on the ventral division of the oculomotor nerves (Nicol, 1952).

In teleosts, parasympathetic autonomic fibres are present only in oculomotor III and X (vagus) (Fig. 13.5). In dipnoans, the cranial autonomic fibres are present only in vagus (X) in *Protopterus* and *Lepidosiren* while the *Neoceratodus* oculomotor and vagus are given out from the cranial outflow. The ocular outflow has preganglionic fibres, which form synaptic relations with postganglionic ciliary neurons, proceed to the eyeball and neighbouring arteries. A large number of sympathetic fibres join the vagi forming vagosympathetic trunk, which carries both cranial and spinal autonomic fibres to gills, heart, stomach and swim bladder.

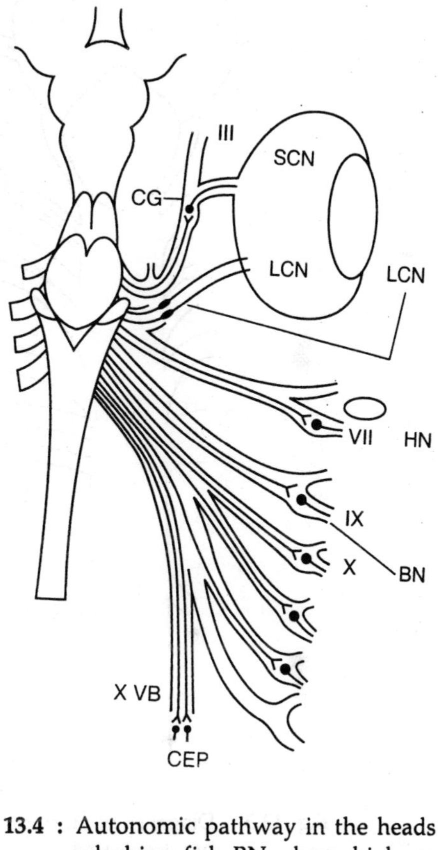

Fig. 13.4 : Autonomic pathway in the heads of selachian fish BN, branchial nerve; CEP, cardiac and enteric plexuses; CG, ciliary ganglion; HN, hyomendibular nerve; LCN, long ciliary nerve; SCN, short ciliary nerve; VB, visceral branch.

The vagus nerve is motor to stomach, and its electrical stimulation brings strong contractions in the stomach. Electrical stimulation of splanchnic nerves results in the movements of pyloric caeca and peristalsis of intestine and rectum (Young, 1936).

Adrenaline causes inhibition of gastric and intestinal contractions and falls in tonus; acetylcholine causes an increase in the tonus and inhibition of contractions.

Adrenergic : The adrenergic neurons in teleost fish contain both adrenaline and noradrenaline with predominance of adrenaline. The adrenergic transmitters act as adrenoreceptor (adrenergic receptors) either the alpha or beta type in effector organs (Ahlquist, 1948).

Cholinergic : Generally the pre- and postganglionic nerves secrete ACh. The receptors of the postganglionic neurons are of the nicotinic type. ACh released from postganglionic cholinergic nerve endings acts as muscarinic receptor in effector organs in teleosts, elasmobranch and dipnoans in all higher vertebrates (Holmgren and Nilsson, 1981).

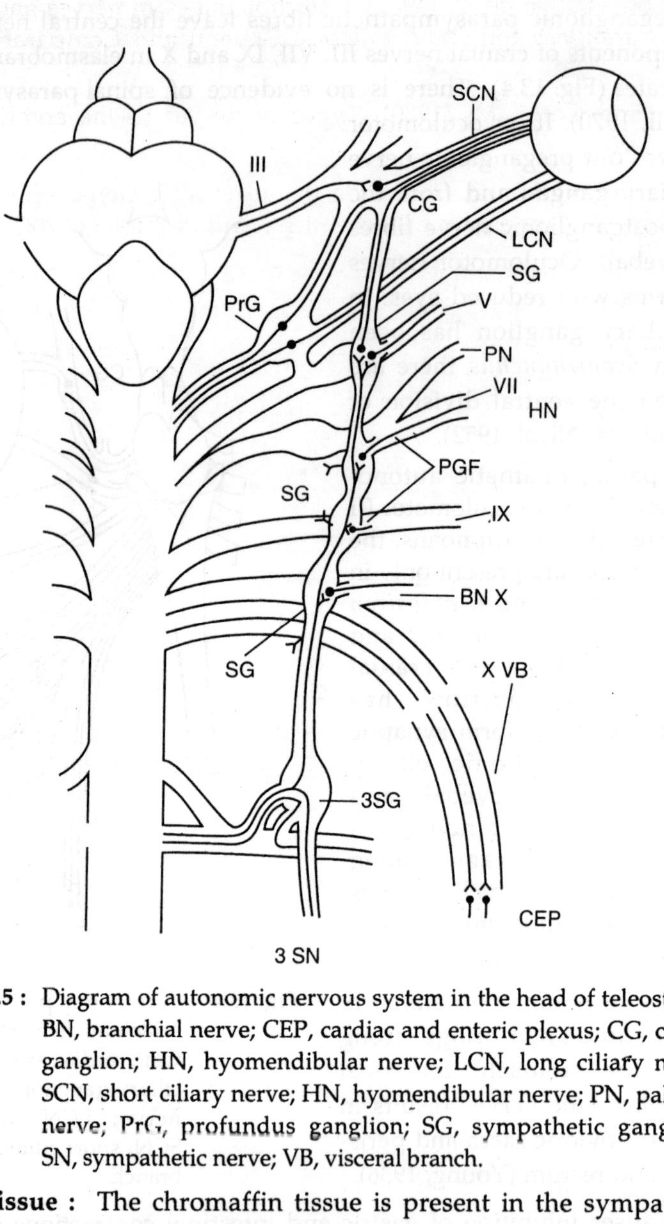

Fig. 13.5 : Diagram of autonomic nervous system in the head of teleost fish. BN, branchial nerve; CEP, cardiac and enteric plexus; CG, ciliary ganglion; HN, hyomendibular nerve; LCN, long ciliary nerve; SCN, short ciliary nerve; HN, hyomendibular nerve; PN, palatine nerve; PrG, profundus ganglion; SG, sympathetic ganglion; SN, sympathetic nerve; VB, visceral branch.

Cholinergic tissue : The chromaffin tissue is present in the sympathetic ganglia of elasmobranch and contains more of noradrenaline than adrenaline (Euler and Fange, 1961).

REFERENCES

Ahlquist, R.P. (1948). A study of the adrenotropic receptors. Am. J. Physiol. 153 : 585-600.

Burnstock, G. (1969). Evolution of the autonomic innervation of visceral and cardio vascular system in vertebrates, Pharamacological Reviews, 21 : 247-324.

Campbell, G. (1970). Autonomic Nervous System. pp. 109-132. In : Fish Physiology (Eds. Hoar, W.S. and Randall, D. J.) Academic Press, New York.

Euler, U.S. Von and Fange, R. (1961). Catecholamines in nerves and organs of *Myxine glutinosa*, *Squalus acanthias* and *Gadus callarias*. General and Comparative Endocrinology; I : 191-194.

Furness, J.B. and Costa, M. (1980). Types of nerves in the enteric nervous system. Neuroscience, 5 : 1-20.

Holmgren, S. and Nilsson, S. (1981). On the adrenergic, non-cholinergic innervation of rainbow trout stomach. Comparative Biochemistry and Physiology, 70C : 65-69.

Holmgren, S. and Nilsson, S. (1982). Mini review Neuropharmacology of adrenergic neurons in teleost fish. Comparative Biochemistry and Physiology.

Kumar, (1979). Nerve endings in the heart of teleosts. Z. Mikrosk. Anat. Forsh. 93 : 1051-1056.

Nicol, J.A.C. (1952). Autonomic nervous system in lower chordates. Biological reviews, 27 : 1-49.

Nilsson, S. (1983). Autonomic nerve function in vertebrates. Springer Verlag, Berlin.

Nilsson, S., Holmgren, S. and Fange, R. (1983). Autonomic nerve function in fish. pp. 1-22. In : Control Processes in Fish Physiology (Eds. Rankin, J.C., Pitcher, T. J. and Duggan, R.T.) Croom Helm, London.

Santer, R.M. (1977). Monoaminergic nerves in the central and peripheral nervous system of fishes. General and Comparative Pharmacology, 8 : 157-172.

Stannius, H. (1949). Das peripherische Nervensystem der Fische Rostock Stiller.

Young. J.Z. (1931). Autonomic nervous system of the teleostean fish *Uramoscopus scaber*. Quart. J. micro Sci., 74 : 491–535.

Young, J.Z. (1936). The innervation and reaction to drugs of the viscera of the teleostean fish. Proc. Roy. Soc. B. 120 : 303-318.

Chapter 14

AGE AND GROWTH OF FISHES

AGE AND GROWTH OF FISHES

Determination of age and growth of fishes are one of the important aspects in the development of fisheries. Both age and growth are closely related with one other. As the fish ages, it grows, but after attaining a particular size, growth stops. Age gives an idea about sexual maturity, spawning time, catchable size, growth rate and longevity. Knowledge of all these parameters is essential in fisheries production (Bengal 1974). Growth of fish or any organism is the change in length and weight with increase of age as a result of metabolism of nutrition (Wallin, 1957). Hence growth is an index of healthy food and oxygen supply in the water-body. Proper growth of fish also indicates that the waterbody is devoid of any pollution. Knowledge of age and growth of fish has many applications :

1. We can calculate the time of sexual maturity of different species.
2. Further, we can know their spawning time.
3. Growth rate of fish also indicates the suitability of particular species for a particular type of the waterbody.
4. Growth rate and age of fish also indicate the size of fish at different stages, e.g., fry, fingerling, and adult of different species.
5. The study of age and growth is helpful in catching fishes by using nets of desirable mesh size.

AGE DETERMINATION OF FISH

Various methods employed for determination of age of fish are :

1. Scale Method

This method is most commonly used for determination of age of osteichthyes (bony fish), which are provided with cycloid and ctenoid scales. The structure of scale and its development is useful in the interpretation of growth zones. The structure of scale can be seen very easily under the microscope after washing with dilute solution of caustic soda followed by staining with borax carmine. A well-developed scale has the following structures :

1. **Focus** — It is a clear area in the centre, but may be shifted from the center due to irregular growth of anterior or posterior parts of scale caused by unusual overlapping of scales.

2. **Circuli** — These are concentric rings present around the focus, they run parallel at regular intervals or distances. They appear as ridges.

3. **Grooves** — The grooves are found between the ridges of circuli and they are responsible for maintaining the regular space between them.

4. **Radii** — These are grooves found radially, viz., they run from focus to margin of scale. Radii cut the circuli present in their path.

5. **Annuli** — These are wide circular troughs found in aged fish over one year. Each trough contains a few incomplete and narrow circuli different from the circuli outside it, which are complete and more widely spaced. The number of annuli represents age of a fish in years (Fig. 14.1).

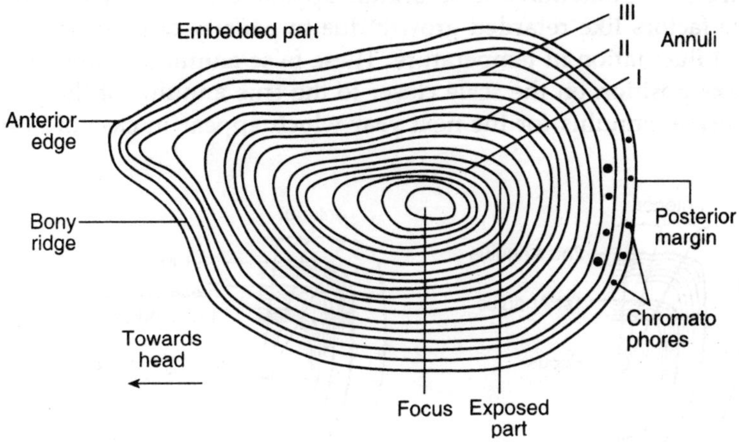

Fig. 14.1 : Scale method fish scale showing growth rings.

At the time of development of scale, focus is established first and represents the original size of the scale. As scale grows older, other structures are added and perform their functions. The grooves and circuli represent growth activity. They also indicate osteoblastic activity as a result of which secreted material is deposited around the focus. In this way every year many such circuli and grooves are formed. A characteristic bone material, *ichthylepidin,* is deposited in circuli and thus their height increases which depends on the calcification. Annuli show slow growth in a year but in many fishes, during winters, annuli grow remarkably and are added yearly as fish grows. Thus annuli are very useful in counting the age of fish and serves as year-marks on the scale for age determination. The annuli are best seen at anterior part of the scale.

Types of Annuli

1. **True Annuli** — The true annuli has following characteristic features :

(i) In cycloid scales true annuli is represented by a closely situated circuli, which is covered by widely spaced circuli.

(ii) Two complete circuli surround the trough, which is wider on the antero-lateral and posterior side.

(iii) The wide part of the trough contains incomplete circuli that do not grow completely around the scale.

(iv) The trough remains narrow at the antero-lateral side. In the ctenoid scales specially, the outer circulus cuts across or crosses over the incomplete circuli lying in the anterior part of the trough.

Annuli are considered as year-marks in the age determination of fish in the case of the following facts.

1. When there is a correlation between the calculated age from the scale and the size of the fish.
2. The length frequency distribution should coincide with the calculated age from the scales.
3. The calculated age should be in agreement with the age determined by other methods.

2. **False Annuli** — Sometimes false annuli appear on the scales of fishes due to undesirable factors like retarded growth due to paucity of food, starvation, injury, disease, and fluctuation in temperature. These false annuli resemble the true annuli but they take position on the scale closer to the true annulus of the preceding year than the normal annulus for the next year which appear in case of normal growth (Fig. 14.2).

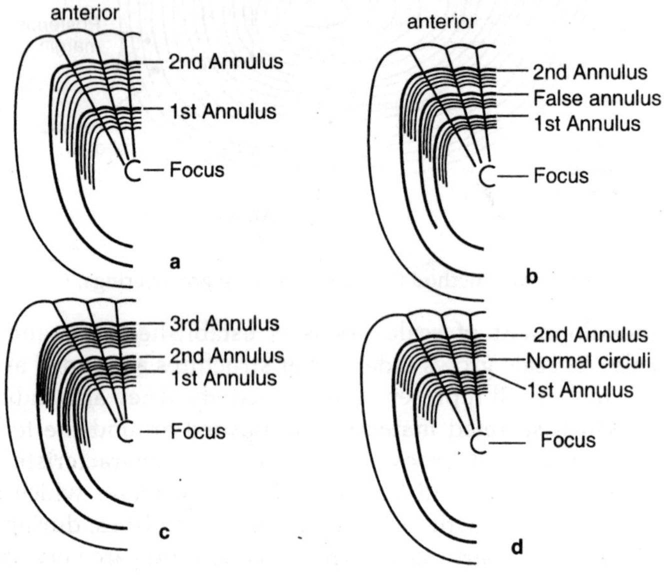

Fig. 14.2a-d : Annuli on scale in four cases of a certain species of fish. a. normal growth for a fish of a certain weight and of 2 years of age, showing annuli; b. abnormal growth for a fish of same weight and of same age, showing false annulus; c. abnormal growth in a 3-years old fish, showing skipped annulus (at 2nd annulus); d. abnormal growth for a 3-year old fish, showing overlapping annulus (at 2nd annulus)

3. **Overlapping Annuli** — The position of these annuli in posterior field coincide with the annuli of the preceding year while, in anterior part, it is separated from the preceding year's annuli by 4 or 5 circuli. The overlapping of scales can occur due to a slow growth during the growing period, which is represented by an increase in the length but not in the weight of body.
4. **Skipped Annulus** — This type of annulus by position coincides with the annulus of the preceding year, with no normal circuli forming in between. This abnormal function is due to the fact that the fish has not grown during one growing season (one summer) either in length or in weight.

Applications of Scale Method

1. Fish of temperate regions shows clear rings, which are true marks. This is because there is a sharp difference between the temperatures of two seasons—summer—the period of faster growth, and winter—the period of slow growth or no growth. Therefore, the calculation of the age of fish by annuli is most reliable in temperate fish.
2. This method is more reliably applicable in case of salmons, carps, cod and herrings. established a method of estimating age of fish based on scales, which is as given on next page:

Number of Annuli on the Scale	Estimated Age	Designation of Age Group
None	Less than one year old	0+
One	One year old	1
One + a few circuli	More than one year old but less than two years old	1+
Two	Two years old	2
Two + a few circuli	More than two years old, but less than three years old	2+

Limitations of Scale Method

1. More than one annulus are added in the extreme conditions of life, e.g., extreme cold (causes cessation of feeding), change in food quality or starvation at the time of spawning. These additional rings are called supplementary rings, which cause problem in age determination by the scale method.
2. This method cannot be applied to those fishes which live in the water with more or less uniform temperatures (tropics). This may be because in these places fishes spawn more than once and, due to fluctuations in food and chemical compositions of water due to rains and floods, formation of annulus may not be an annual feature. Thus, in the fishes of tropical regions, the growth rings do not actually represent year-marks.
3. **Bone Method** — Annuli are also present on some bones. The important bones such as operculum, vertebra, supra occipital and scapula are provided with annulations. These annuli are increased in number with the age of fish. The growth rate is different in different seasons. Number of annual rings are helpful in calculating the age of fish. Similarly, the centrum of fish is also helpful in calculating the age of fish. The centrum of fish vertebrae possess rings, which are used in age determination. For counting the rings on the centrum, it is exposed by removing the tissues attached to it using solution of 0.7% pepsin in 0.2% of hydrochloric acid. The rings on the centrum are counted under the microscope.
4. **Otolith Method** — The otolith or ear-stone is present in the internal ear of the fishes and helps in balancing the body. The otolith has annulations formed by the regular deposition of calcium salts. The formation of annulations varies with the growth of fish, and the growth of fish varies with the season. The growth rate of fish is faster in the summer as compared to the winters, thus annulations formation is faster in the summer. To get the otolith, fish is killed and dissected and otolith is taken out and annulations are counted. For this, first of all the otolith is broken and cut in transverse plane followed by polishing with liquid of high refractive index,

such as immersion oils, creosote, etc. The otolith thus prepared is examined under the microscope to count the annulations for age determination and the growth rate analysis Graeme *et al.*, (2007).

5. **Known Age Method** — The determination of growth rate through this method requires the knowledge of age, which can be known from time of breeding. The hatchlings are kept in a tank in appropriate conditions for two or three seasons. The growth is measured periodically. The fishes are marked with tags and reintroduced in the water. The fishes are recaptured at regular intervals and idea of growth rate in relation of the time is taken.
6. **Length Frequency Distribution Method** — Peterson introduced this method in the nineteenth century. In this method, in the sample of fishes of a particular fish, frequency analysis shows length of individuals of one age varying around the mean length according to normal distribution. Data of the sample is plotted and peaks are counted which represent grouping of fishes of successive lengths. Age groups are thus separated and the age of a given species is determined. This method is suitable for determining the age of younger fishes of 2 to 4 years. This method is not reliable for calculating the age of older fishes because of overlapping of length frequencies in individuals of different ages.
7. **Pectoral Spine or Fin Ray Method**— Spines of the pectoral fin is also useful in age determination. For this 3μ to 4 μ thick sections of spines are cut and mounted in glycerin and then observed in the microscope. 1% to 2% of indistinct annuli or fin ray consistency are compared to 15% to 20% of annuli on the scales and, thus age is calculated.

GROWTH OF FISHES

Growth is a characteristic feature of living beings. Every organism grows and attains its normal size. Growth is actually an increase in any dimension of any organism in relation to time. Growth is the process of addition of flesh as a result of protein synthesis (Bron 1957). Knowledge of fish growth is of vital importance for obtaining high yield of fish. The rate of growth varies from species to species and sometimes it varies even among species also. Rate of fish growth is influence by many factors:

1. **Different Localities** —Individuals of same species grow at different rate in different localities.
2. **Seasonal Effects** — The different temperatures in different seasons affect growth of fish. In summer the fish grow fast, however, in winter growth is slow.
3. **Availability of Food and Oxygen** —This is the most important factor influencing growth of fish. In ecologically balanced waterbody, growth rate is high because of ample supply of oxygen and food.
4. **Population Density** —Higher population density shows slow growth while low density shows high growth rate.
5. **Age** — Fish continue to grow throughout their life but with increasing age, growth rate is slow and in extreme old age fish growth is extreme slow. Growth slows down at the onset of sexual maturity when large amount of nutrient material periodically go into egg or sperm formation.
6. **Adaptive Character** — Sharks and Sturgeon grow very fast into large size because of constant food supply, which provides protection against their predators. However, in Goby growth is slow, resulting in small size of the body. This is an adaptation to

overcome problem of limited food supply and is related to increased reproductive capacities as they have more danger of destruction by predators (Hile ,1941).

Growth Curve

Growth curve is obtained when either length or weight of a fish is plotted against time periods. The curve obtained is sigmoid or S-shaped. This curve illustrates the rate of growth. If rate of growth is plotted against time, such curve is known as growth rate curve (Fig. 14.3a, b).

Linear Growth

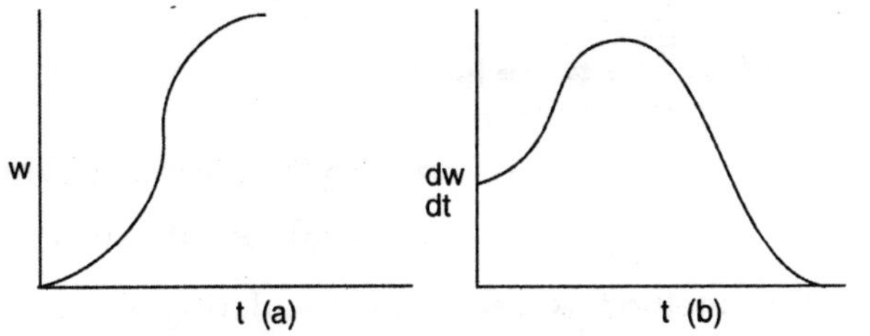

Fig. 14.3a, b : (a) Growth Curve, (b) Growth rate Curve

Fish growth varies according to various periods. Maximum growth rate is found just before attaining the sexual maturity of fish. After maturation of gonads the growth slows and decreases. A linear growth pattern is seen during the postnatal development of fish, specially during the period of hatching to the sexual maturity. This may differ from species to species, and depends on various factors, viz., appropriate and balanced food, optimum temperature for vital activities like metabolism, vitamins and trace elements for proper growth. Fluctuation in these factors affects growth rate of the fish. If there is scarcity of food the size of fish will be small, and, if the availability of food is restored to normalcy, the fish grows to normal.

Determination of Length (Linear Growth)

The scales of many fishes have permanent marks, which is considered as confirmation of both age and growth. This evidence provides the base of estimating yearly changes in

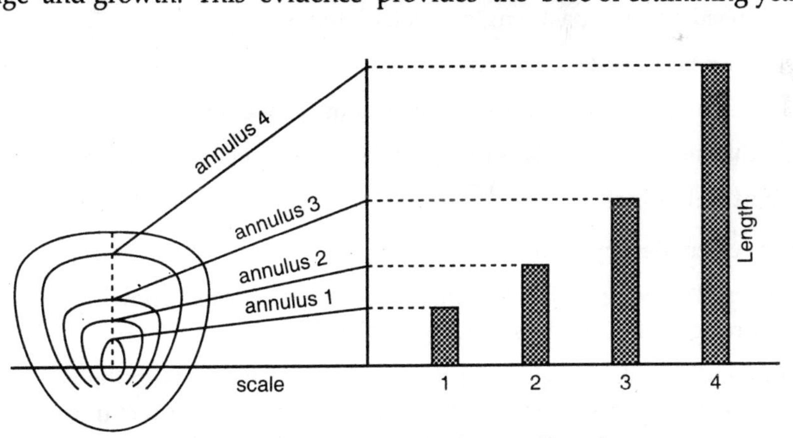

Fig. 14.4 : Back calculation of length

growth and also of calculating the length of body (Ozadin *et al.*, 2007). Annuli formed during winter shows zone of retarded growth. There exists a relationship between the annuli and the growth of the fish. Thus body length can be estimated by studying the relationship between the scale size and the body size. Some important methods are as follows :

1. Direct Proportion — Body/Scale Relationship Method

This method is based on the facts of isogonic growth rate, which shows that the growth of the scale and the body occurs at the same proportional rate. However, the calculated body length is found to be less than its empirical length. This is because of the fact that fish attains some length before scale formation. Following formula is applied to calculate the length of a fish.

$\frac{S^+}{S} = \frac{L^+}{L}$ where S^+ = length of scale radius to annulus X.

S = length of total radius of scale

or,

$L^+ = \frac{S^+}{S} \times L$ where

L^+ = length of fish when annulus X was formed

L = length of fish when scale sample was obtained.

This method is useful in calculating the length of fish at early stages. The above formula works well when fish shows a linear and directly proportional body. For example if a fish has L = 600 mm, S=10 mm, S^+ = 4 mm, then the length of the fish at the time of formation of first annulus can be calculated as follows :

$$L^+ = \frac{S^+}{S} \times L = \frac{4 \times 600}{10} = 240 \text{ mm.}$$

Here the fish shows a non-linear but not directly proportional growth of body and scale.

2. Dahl-Lea Method

This method states that there is a curvilinear (sigmoid curve) relationship between body length and scale length and not a straight line (linear) relationship. Therefore, a correction factor is used and the past length is calculated as:

L = C S, where L = body length, S = scale length and C = a constant.

or, log L = log C + n log S, where C and n = constants from data.

It is observed that body/scale relationship varies from species to species.

Length-Weight Relationship

Weight can be calculated by a formula:

$W = KL^3$, where W = weight, L = length, K = constant ; for fish showing symmetrical or isometric growth throughout (Ludwig and Lange, 1975).

The body form and specific gravity of a fish change constantly with the advancement of age. Thus simple cube W expression therefore does not seem to be correct throughout the life history of fish, as the value of K is not constant but subject of great variation. A more satisfactory formula is:

$W = L^n$, in which W = weight, L = length, n = constants for log W = log C + n log L.

Linear Growth and Condition of Fish: Coefficient of Condition Factor

The ratio of length and growth represents the condition at any given time. When this ratio is calculated and if the obtained value is large, it indicates better condition of fish. The Fulton's formula is used to express condition of the fish.

$Q = \frac{W100}{I^3}$, where Q is coefficient of condition: W is the weight of fish and I is the length

Coefficient factor or coefficient condition is that state in which a fish remains for a certain period. This value is useful in elucidating the following :

1. Difference in conditions of the individuals of same species in different water.
2. Differences among individuals of same length.
3. Differences occur due to seasonal changes in relation to the age and sex of the fish.

There is another way to know the condition of the fish by estimating the fat contents, which is related to linear growth of fish. During the period of winters, migration and spawning, the major fat contents are saturated fatty acids. This type of fat is used for long-term use. On the other hand, unsaturated fatty acids are present in fat when the fat is deposited for use in daily maintenance (Demirhan *et al.*, 2007).

Food Consumption and Growth of Fish

Food plays an important role in the growth of fish. Food consumption rate is estimated by the following methods:

1. Samples of fish are collected from the nature. Then quantity of food present in the stomach is determined followed by estimation of rate of gastric digestion.
2. By determining the intake of nitrogen, which helps in the growth, and the loss of nitrogen in excretion through kidney and gills.
3. By comparison of food consumption and growth rate between the fish from natural habitat and the fish kept in captivity. It has been found that the total amount of food consumed is not used for growth. Only a small amount of food is utilized for growth. Hence it is necessary to know about the fate of consumed food, which is as follows:
   - (i) Some part of food is neither digested nor absorbed and is passed out along with faeces.
   - (ii) A part of ingested food is digested and absorbed by the gut.
   - (iii) Some part of absorbed food is excreted through the kidney, gall bladder and skin.
   - (iv) Some part of assimilated food is used for maintenance of metabolism.
   - (v) Some part of assimilated food is used for growth.

Life Span of Fishes

It has been observed that generally species with large individuals seems to exhibit older age than species consists of small fish (Bagenal, 1974). Many fishes that attain size of 12 inches show a life span of at least 4 to 5 years. However, age of fishes in captivity, i.e., in the aquarium may be expected to be more but do not exceed to 5 years (Griffiths, 2001).

Factors Affecting Fish Growth

1. *Abiotic Factors:* Fishes show a considerable variation in their growth rate and age of sexual maturity in northern and southern waters. Temperature plays an important role as it affects metabolism and food consumption in fishes. The seasonal increase in the

temperature of freshwater bodies cause increase in natural food for fishes. Fishes grow at faster rate at the optimum temperature, at which their appetite is high and requirement of maintenance is low. The optimum temperature varies from 7°C to 19°C in case of brown trout (*Salmo trutta*).

Light also affects the rate and pattern of fish growth. Intensity of light depends on the length of the day. Light causes increased thyroid activity resulting the increased swimming activity and reduction in growth rate in experiments with the brown trout (*Salmo trutta*) in mid summer. Longer days are advantageous where feeding is accomplished by sight whereas shorter days are beneficial when the food is abundant.

2. *Biotic Factors* : The population density is also of a vital importance. Among the fish population there is competition for food, nest site and so occur, which may affect the growth of fish. High density of fish in a place can also affect the fish growth. It has been described that the fish in a crowded condition may release a chemical substance known as "crowding substance", which prevents growth.

3. *Genetic Factor*: Growth and size of the fish are genetically determined. Size difference between the sexes may be due to genetic factors.

4. *Endocrine factor*: Several metabolic hormones like insulin, thyroxin, gonadal and adrenal steroids produce synergy which affects the growth. In carefully controlled experiments, Higgs *et al.* (1977) treated groups of yearling coho salmon with bovine growth hormone thyroxin and methyltestosteron, given separately and with the combination of these three substances maximum growth was obtained.

REFERENCES

Bagenal, T. B. (1974). Proc. Intern. Symp. Aging of Fish. Unwin Bros., England.

Bron, M.E. (1957). Experimental studies on growth. The physiology of fishes : 361-400. Academic Press, New York.

Demirhan, S. A. Can M. F. and Seyhan. K. (2007). Age and growth of stargazer (*Uranoscopus scaber* L., 1758) in the southeastern Black Sea. Journal of Applied Ichthyology, Volume 23 (6), 692-694.

Griffiths, S. P. (2001). Recruitment and growth of juvenile yellowfin bream, *Acanthopagrus australis* Günther (Sparidae), in an Australian intermittently open estuary J. of Applied Ichthyology. Vol. 17(5) 240

Graeme P. Ewing, Jeremy M., Lyle, a, Raymond J. Murphy, John M. Kalish,. and Philippe E. Zieglar (2007). Validation of age and growth in a long-lived temperate reef fish using otolith structure, oxytetracycline and bomb radiocarbon methods. Marine and Freshwater Research 58(10)944-955.

Higgs, D.A. Fagerlund, U.H.M., Macbride, J. R., and Donaldson, E.M. (1977). Influences of bovine growth hormone, 17 α-methyltestosteron and L-thyroxin on growth of yearling coho salmon *(Onnchorhynchus kisutch)*. Can. J. Zool.55:1048-1056.

Hile , R. (1941). Age and growth of the rock bass, *Ambloplites rupestris* , in Nebish lake, Wisconsin Acad. Sci., Arts and Lett., 33: 198-337.

Ludwig, G.M. and Lange, E.L. (1975). The relationship of length, age and age length interaction to the fecundity of northern mottled *Sculpin cottus.*

Özaydin, O. Uckun, D. Akalin, S. Leblebici, S. Tosunoglu, A. (2007). Length-weight relationships of fishes captured from Izmir Bay, Central Aegean Sea. Journal of Applied Ichthyology. Vol. 23(6). 695.

Wallin, O. (1957). On the growth structure and developmental physiology of the scale of fishes. Rept. Inst. Freshwater Res. Drotiningholm, 38: 385-447.

Chapter 15

SENSE ORGANS

The number of highly specialized organs, which receive physical or chemical stimuli from the environment are called as sense or sensory organs. They are associated with the nervous system. Most obvious sense organs are eye, nose, ear and skin which receive stimuli for vision, olfaction, audition, and touch respectively. Less obvious are lateral line, sensory crypts and papillae, ampullae of Lorenzini and vesicles of Savi.

THE EYE

The fish eye is modified for the vision in air as well as in water. The eye in most of the fishes is lidless. But some sharks such as soupfin, *Galeorhinus* possess lid-like nictitating membrane, which clean the surface of cornea. However, some teleost have fatty eyelids of adipose tissue, which protects the eye (Gloria *et al.*, 2003).

Structure

The Cornea : It is anterior most part of the eye (Fig. 15.1). It is composed of corneal epithelium, corneal stroma and corneal endothelium. In most bony fishes the cornea is without pigment and hence is transparent. Some bony fishes such as *Perca* and *Amia* have green or yellow pigmented corneal epithelium membranes. Some species (*Solea*) have double layered cornea. Sometimes an specialized inner layer, i.e., autochthonous layer is found in teleostean cornea. The mudskipper which leaves the water for a considerable time, has a wetting device in the eye in the form of two layered cornea and fold of skin beneath the eye. The structure helps in retention of water and thus prevents drying of the eye.

The Sclerotic capsule : It is a tough and highly vasculated layer surrounding the eye-ball. It is supported by fibrous tissue in elasmobranch. In Chondrostei *(Sturgeon)* the sclerotic coat has a cartilaginous or sometimes bony support at its corneal border. The lobefin *(Latimaria)* has thick cartilage supported sclerotic coat in addition to calcified scleral plates. In teleosts, the sclerotic coat is fibrous and flexible inside the orbit and around the optic nerve, but there may be cartilaginous or even bony supporting structures in the outer part of the eye which can surround the entire cornea.

The Choroid layer : The choroid layer is present beneath the sclerotic coat. It is richly vasculated part. In most of the holostean and teleostean bony fishes, the blood vessels of choroid layer are arranged like a horse-shoe shaped structure known as 'choroid gland'. This lies between the choroid and sclerotic coat (Fig. 15.1) near exit of optic nerve. The choroid gland plays similar role for the secretion of oxygen as do the rate mirabile of gas bladder. The oxygen is secreted at higher tension than those found in the blood so as to meet high oxygen demand of the tissue. Generally, in these fishes the retina does not possess well developed circulation. Hence in these fishes choroid gland is large and well developed.

Thus choroid gland supplies the oxygen to the retina and also it acts as a cushion against compression of eyeball. In eel, however, the choroid is poorly supplied with blood capillaries and choroid gland is absent.

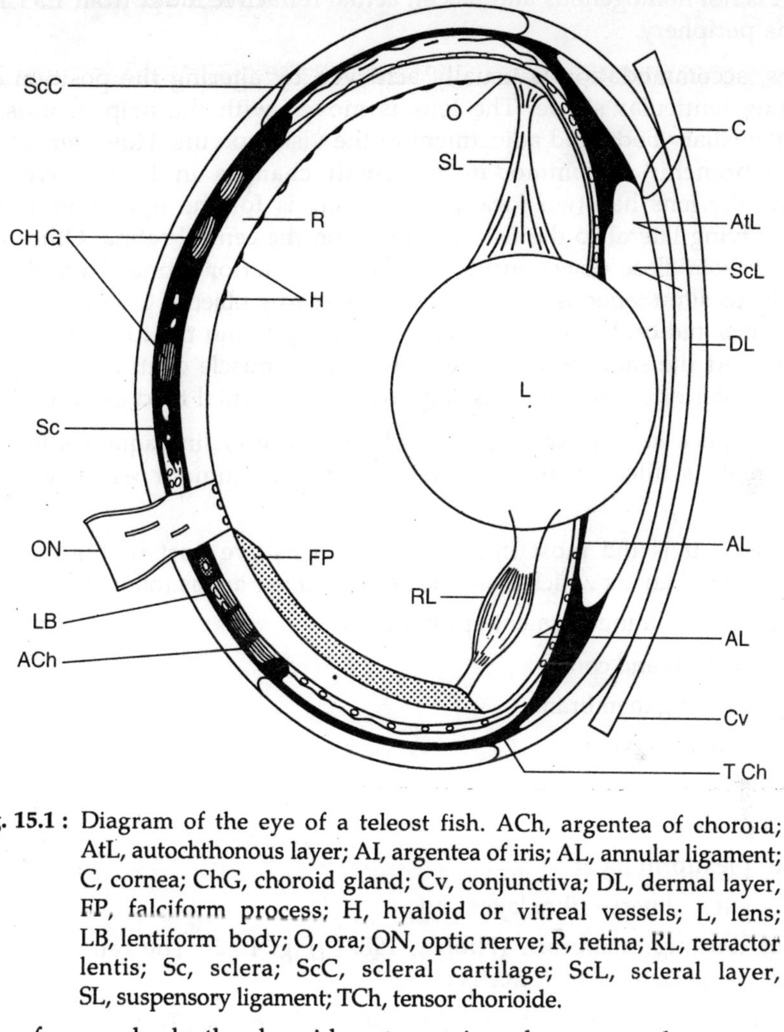

Fig. 15.1 : Diagram of the eye of a teleost fish. ACh, argentea of choroid; AtL, autochthonous layer; AI, argentea of iris; AL, annular ligament; C, cornea; ChG, choroid gland; Cv, conjunctiva; DL, dermal layer, FP, falciform process; H, hyaloid or vitreal vessels; L, lens; LB, lentiform body; O, ora; ON, optic nerve; R, retina; RL, retractor lentis; Sc, sclera; ScC, scleral cartilage; ScL, scleral layer, SL, suspensory ligament; TCh, tensor chorioide.

In the eyes of many sharks the choroid coat contains a lusterous substance, e.g., tapetum lucidum which collects considerable amount of light and reflects through guanine crystals for retina.

The Iris : The iris is a thin partition between the anterior and posterior chambers. It projects over the anterior surface of the lens with its free edge forming the pupil and controls the amount of light that reaches the retina. The iris of elasmobranch's eyes has muscular element, so they adjust the shape of pupil. Majority of fishes have fixed pupil, either circular or oval. Their iris has no muscles but some guanine and melanin are present.

The Lens : The lens is firm, transparent and ball-like and composed of non-collagenous protein. The lens is covered by the lens capsule and is filled by lens substance. Between them is present the lens epithelium, which plays an important role in metabolic processes of the lens. The lens substance is composed of lens fibres which are arranged in flat hexagonal prism. The lens fibres are highly modified epithelial cells. The bony fishes have roughly spherical lens. However, in sharks and rays the lens is horizontally compressed. Sometimes the lens is pyriform (Anablepidae) to provide for aquatic and aerial vision. In deep sea fishes, the eyes are projected with large lenses. Fish lens has a very high effective index of refraction. It is not homogenous and has an actual refractive index from 1.53 at the centre to 1.33 near the periphery.

In fishes, accommodation is usually achieved by altering the position of lens rather than changing lenticular shape. The lens is moved with the help of muscular papillae present on the ciliary body and adjustment of the vision occurs. However, some sharks and rays (Elasmobranchii) accommodate by small changes in lens convexity. In trouts (Salmoninae) the lens has two focal lengths, one is for the light rays reflected from a distant object lying lateral to the fish and focus on the central retina. Other is for the light rays reflected from close object and focus on the posterior retina. Thus these fishes have capacity to focus simultaneously the distant and nearby objects. Consequently, the distance between the lens and central retina remains unchanged, and the focus remains adjusted for distant vision. At the same time the retractor lentis muscle contracts and brings the lens closer to the posterior retina, which is concerned with frontal field of vision of the fish.

The small space in front of lens is filled with clear, saline aqueous humor. The main cavity of eyeball is filled by the transparent vitreous humour which is secreted by the ciliary body.

The retina : It is the most important and sensitive part of the eye. The retina is composed of several layers which from the outer to inner are as follows.

(i) Melanin containing pigment epithelium

(ii) Layers of rods and cones

(iii) Outer limiting membrane

(iv) Outer nuclear layer

(v) Outer plexiform layer

(vi) Inner nuclear layer

(vii) Inner plexiform layer

(viii) Ganglion and nerve fibre layer and

(ix) Inner limiting membrane (Hibiya, 1982) (Fig. 15.2). The retina has the following types of specialized nerve cells :

(1) **Visual cells :** They are of two types, rod and cone cells. The rod cells are concerned to detect the intensity of light, while the cone cells distinguish wave length, i.e., colour.

(2) **Horizontal cells :** They are present in the peripheral region of the inner nuclear layer. These cells give rise to some processes which extend horizontally near the outer nuclear layer and act as communicating lines between visual cells.

(3) **Bipolar cells :** These nerve cells are found in innermost part of the retina. They are the largest neurons of the retina. The axons from these cells assemble and form optic nerve.

(4) **Amacrine cells :** These cells are present between granular and inner plexiform layers and acts as horizontal lines of communication for visual stimuli.

The retina is well supplied with various capillaries which are of four types in the teleosts :

Type A : having falciform processes.

Type E : with vitreal vessels.

Type F : with retinal vessels.

Type G : unused instance.

Type A can be seen in the eye of rainbow trout which possesses falciform processes, containing the small blood vessels branches from the main vessel. The man vessel enters the eye from beneath the optic nerve papilla. The vessels are restricted to the falciform processes and are absent in the retina (Hanyu, 1959 a & b).

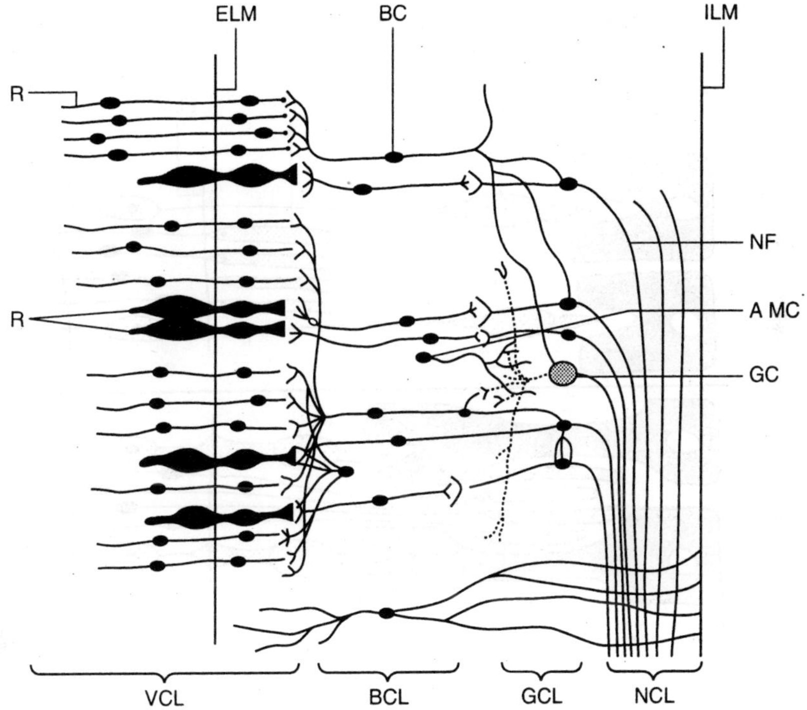

Fig. 15.2 : Diagram of nervous components of retina of Salmon. AMC, amacrine cell; BC, bipolar cell; BCL, bipolar cell layer, ELM, external limiting membrane; GC. ganglion cell; GLC ganglion cell layer; ILM, internal limiting membrane; NF, nerve fibre; NCL, nerve cell layer, R, rod; VCL, visual cell layer.

Type E blood vessels are present in the vitereo retinal boundary and are radially arranged from the centre, but they are absent in the retina. This type of intramolecular vascularization is found in carp and eel. In *Plecoglossus altivelis*, embryonic fissure is present in the retina. Therefore, the vascularization patterns in the eye of *Plecoglossus altivelis* and carp is further classified into type E-a and type E-b respectively (Hanyu, 1961)

The epithelium of retina contains melanin and borders on the choroid coat. Melanin is photosensitive and in bright light it spreads and shades the sensitive rods; in weak light, it aggregates near the choroid border, thus photosensitive cells are fully exposed to the amount of light available (Fig. 15.3). Simultaneously the contractile myoid element in the base of the rods and cones move the cell tips in such a manner that rods are moved away from the lens in the bright light to be covered by the epithelium melanin, whereas in dark the rod myoids contracts and brings the cells towards the lumen of the eyeball. Cones migrate in the opposite direction towards the lens in the bright light and towards the outer epithelium in dim light (Fig. 15.3). Such photomechanical or retinomotor reaction is much pronounced in fishes than other vertebrates.

The relative number of rods and cones vary considerably in different species. In sight feeders which are active during the day, the cones are more in number than rods, and the revenue is found in many crepuscular species which are more active in twilight.

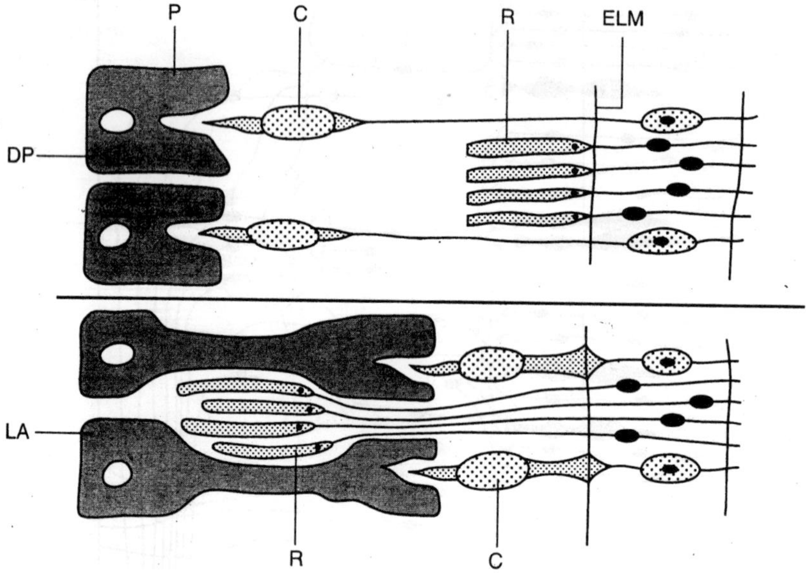

Fig. 15.3 : Diagram to show movement of rods and cones of retina of teleost fish in light and darkness. C, cone; DP, darkness adapted; ELM, external limiting membrane; LA, light adapted; P, pigment R, rod.

The retina of fishes contains two types of light sensitive pigments, rhodopsin and porphyropsin. Rhodopsin is purple coloured (marine Actinopterygii) while porphyropsin is rose-coloured (freshwater Actinopterygii). Migratory species like Atlantic salmon *(Salmo salar)*, eels *(Anguilla)* and lamprey *(Petromyzon marinus)* possess both rhodopsin and porphyropsin. More rhodopsin is present in the retina of lamprey migrating to the sea while porphyropsin predominates in the eye of spawners in freshwater. Rhodopsin and porphyropsin is synthesized by vitamin A in dark. In light there is found in the rods the yellow pigment retinine which can, however, be useful in resynthesizing the light sensitive purple or rose pigment.

The vision is photochemical process and involves reaction in the light sensitivity pigments of the rods and cones. But it is still not established that how these chemical changes become transformed into the electrical impulses that can be registered in retina nor how these ultimately converted into signals which is taken by optic nerve and travel to the brain.

THE INNER EAR

In fishes the middle ear apparatus is absent, only inner ear is present which is concerned with two senses, i.e., hearing and equilibrium. It is contained partly in the auditory capsule and partly in certain recesses outside and behind the capsule. It is made up of an upper portion, the pars superior and a lower portion, the pars inferior. The pars superior comprises three semicircular canals—anterior vertical, posterior vertical, and horizontal canal. Each semicircular canal enlarges at one of its anterior ends to from ampulla and a sac-like vesicle, the utriculus. In sharks the anterior vertical canal joins with horizontal canal to form a crus. The pars inferior which receives sound, is composed of two vesicles. The anterior sacculus and posterior lagena. The sacculus lies below the utriculus and is connected with utriculus through the recessus utriculi at its upper end. The cavity of inner ear is filled with lymph.

In bony fishes the utriculus, sacculus and lagena comprise lapillus, sagitta, and astericus respectively (Figs. 15.4a, b). These are calcified structures and are secreted by ectoderm and composed of calcium carbonate, keratine and mucopolysaccharides.

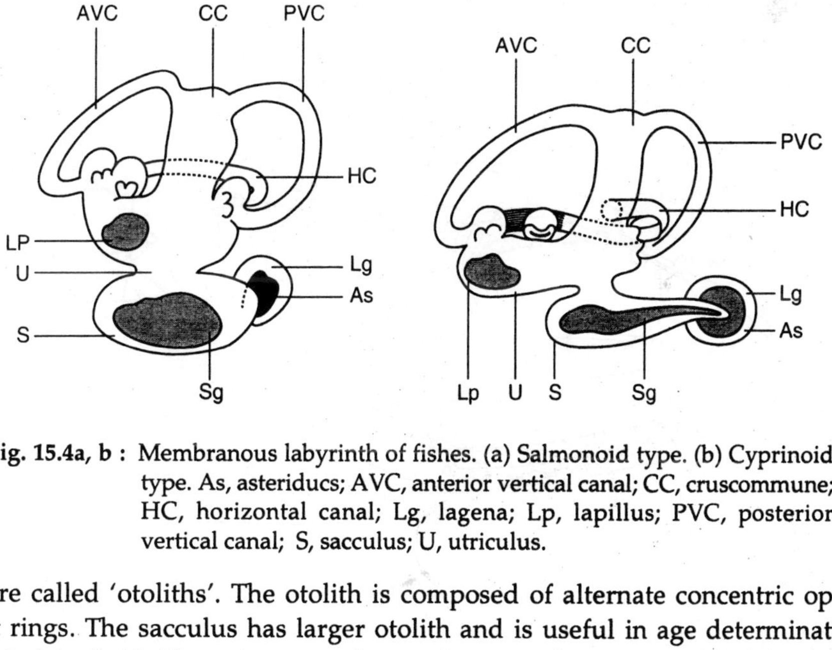

Fig. 15.4a, b : Membranous labyrinth of fishes. (a) Salmonoid type. (b) Cyprinoid type. As, asteriducs; AVC, anterior vertical canal; CC, cruscommune; HC, horizontal canal; Lg, lagena; Lp, lapillus; PVC, posterior vertical canal; S, sacculus; U, utriculus.

They are called 'otoliths'. The otolith is composed of alternate concentric opaque and translucent rings. The sacculus has larger otolith and is useful in age determination. They are suspended in fluid. There is a macular neglecta on the outer wall of the constriction between utriculus and sacculus. Internally the ampulla has receptor tissue arranged in form of patches, called 'cristae staticae' which possess sensory cells similar to those of lateral line. The cristae contain longer hair in their sensory cells than taste buds and they are covered by gelatinous cupula (Fig. 15.6).

The sensory units are supplied with VIII (i.e., acoustic) cranial nerve. The vestibular branch of auditory nerve is distributed in fan-like manner on the ventral surface of the utriculus and innervates it and of the ampullae of the semicircular canals. The saccular branch runs back and supplies the lagena, succulus, and ductus lymphaticus.

The sensory cells are covered by tactorial membrane, which transmit stimuli to the sensory hair by vibration (Hibiya, 1982). This apparatus is known as *organ of corti*.

The lapillus lies horizontally over the hair of sensory cells of the cristae utriculi, the lapillus is sensitive to the force of gravity and stimulates sensory cells of the cristae. To maintain the equilibrium of the body they work unitedly with the lower part of retina. Thus eye receiving light from above and utriculi receiving gravity from below push and pull keeping the fish in an upright position.

Besides the sound detection, the inner ear functions to 'orient' or 'balance' the animal, providing it feeling of direction in which gravity is acting when suspended in lightless, pelagic habitats. Having a greater density, water is a much efficient conductor of sound pressure waves than in air. Therefore, sound travel 4-8 times as faster under water.

The endolymphatic duct is a short and closed tube in teleosts. It is absent in *Salmo* and *Lampanyctus* (Jollie, 1968). In elasmobranchs an endolymphatic duct arises from the inner ear and extends outside. The sand particles enter the inner ear through this duct and reach the gelatinous cupulae covering the sensory cells. These particles acts as otolith.

Minnow, carp, catfishes and other teleosts connect the auditory system to the swim bladder with a chain of small bones called Weberian ossicles. The ossicles connect the pulsating swim baldder wall with Y-shaped lymph sinus, which abuts the lymph filled transverse canal joining the sacculi of the right and left ear.

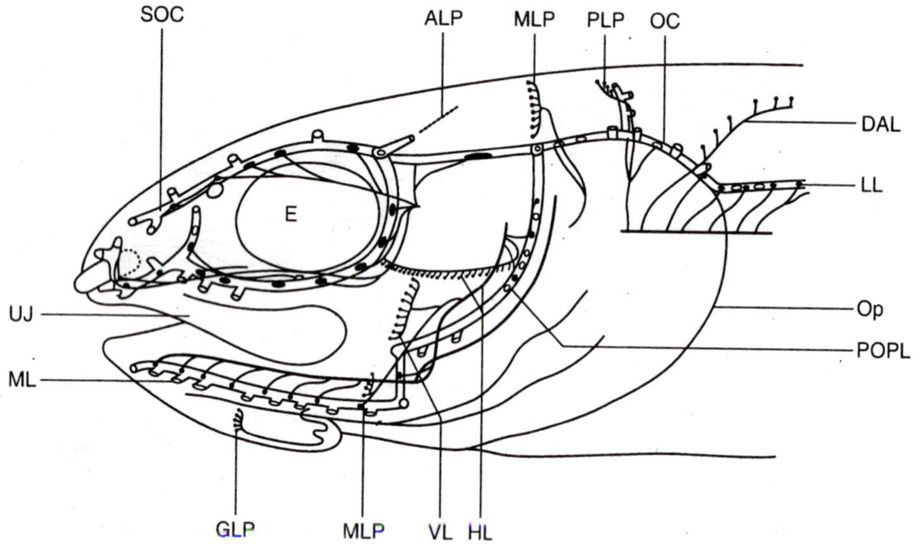

Fig. 15.5 : Sensory lines; organs and innervation of organs in bowfin (*Amia*). ALP, anterior pit line; DAL, dorsal sensory line organ; E, eye; GLP, gular pit line; HL, horizontal pit line; LL, lateral line; ML, mandibular line; MPL, middle pit line; OC, occipital canal; Op, operculum; PLP, posterior pit line; POPL, preopercular line; SOC, supra orbital canal; UJ, upper jaw, VL, vertical pit line.

There is a great diversity in the size and arrangement of the semicircular canals and vesicles among the fishes. The canals are selectively larger in the elephant fishes (Mormyridae) and in some blind cave fishes (Amblyopsidae). In the sharks (Squaliformes) the sacculus and the lagena are broadly united with one another than in bony fishes. The lampreys have only two semicircular canals and hagfishes have only one (Moyle and Cech, 1982).

The posterior and inferior parts of the inner ear differs in the degree of their development in rayfins (Actinopterygii) of different habits. The pars superior better developed in flying fishes and pelagic fast swimmers which need three dimensional orientation while pars inferior is highly developed in demersal species.

LATERAL LINE SYSTEM

It provides 'distant touch sense' of 'objects (Dijkgraaf, 1962). The object may create either mechanical disturbances or their presence may be made out from the reflected waves by echolocation. The lateral line system is an integral part of the acoustico lateralis system which includes the ear. It involves sensory lines distributed on the head and the body, pit organs and ampullae of Lorenzini. The basic pattern of lines (Fig. 15.5) are as follows :

(i) Supraorbital lying above the eye.

(ii) Infraorbital lying below and behind the eye.

(iii) Mandibular arch line lying in mandible part.

(iv) Hyoid arch line present in the hyoid region.

(v) Anterior and posterior pit lines present on the top of the head.

(vi) Dorsal, lateral and ventral lines present on the body (Jollie, 1968).

In addition to usual lateral line canals following sensory structures are also included in lateral line complex.

1. Ampullae of Lorenzini are present in the head region of sharks and rays (Elasmobranchii). They are small sac-like structures, which open outside by small pores. These structures are filled with jelly and have several diverticulae lines with sensory epithelium. The sensory cells of epithelium are supplied by branches of the facial nerve, i.e., (VH).

The ampullae of Lorenzini are sensitive to mechanical and weak electrical stimuli as well as changes in salinity. They are much sensitive to difference in electric potential. They are also considered to be electro-receptors which help in locating the prey (Sand, 1938).

2. The vesicles of Savi are found in rays (Rajiformes). They are arranged in lines and are gelatin filled follicles containing grey, granular amorphous material. The vesicles are supplied by branches of V (i.e., trigeminal) cranial nerve. Their function is still unknown.

3. Electrosensitive cutaneous cells are present in the elephant fish and eels.

4. The larvae of lungfishes (Dipnoi) have hair bearing sensory cells called organs of Fahrenholz resembling the lateral line organs of unknown functions.

The receptors of lateral line system are called 'neuromast' (Fig. 15.6). Each consists of individual hair cells with an attached 'cupula' which is composed of mucopolysaccharides. Water current bends the projecting cupula, which stimulate the hair cells by bending the attached cilia, i.e., sensory hair.

The lateral line proper may have the neuromast organs arranged variously.

1. They may be lodged in pits or arranged in many lines on the head or along the body (i.e., cyclostome and elasmobranchs) or in hands on ventral surface.

2. They may be free or grouped on the head and in lines on the body as in loach and cobitis.

3. They may be semi open or closed having various canals located.

(a) On the head with the infraorbital, supraorbital and hyomandibular canals.

(b) On the body having one canal only extending from shoulder girdle to base of caudal fin.

While equilibrium is maintained the neural impulse frequency is increased as the cupula are flexed in one direction, and decreased when fixed in the other direction. Thus the type of impulses from the free or canal neuromasts imparts a direction to the disturbance. Thus the movement of cupula produced by physical stimuli from the surrounding (pressure, vibration, touch and water current) strains the sensory hairs stimulating the hair cell. This stimulation is sensed by the nerve terminals of the hair cells.

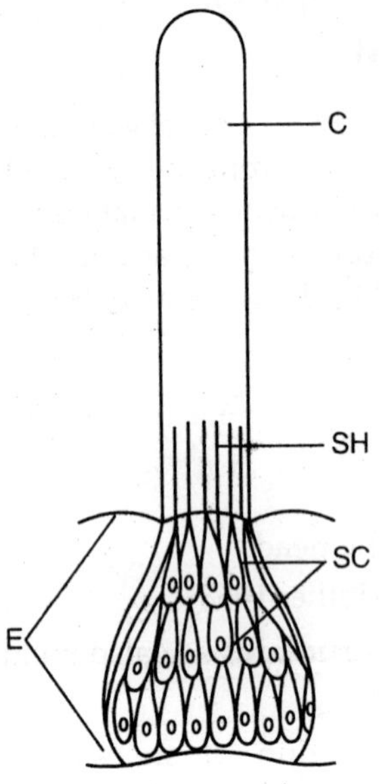

Fig. 15.6 : Superficial neuromast of bony fish. C, cupula; E, epidermis; SC, sensory cell; SH, sense hair.

Most of the lateral line organs of the head region are innervated by sensory fibres of the lateralis anterior root of cranial nerve VII (i.e., facialis). The rest of the organs of systems are innervated by the lateralis posterior root of the vagus (X). The fibres of both roots unite with those of labyrinth nerve VIII in the acoustic tubercle of the medulla oblongata (Moyle and Cech, 1982.)

The lateral line system is diversely modified and used by the fish having different habits. The canal neuromast systems are best developed in the roach *(Rutilus)* and the stoneloach *(Noemacheilus)* inhabiting the streams. On the contrary, canals are absent in those loaches that live in still water (Alexander, 1967). Generally the more active fishes have a greater number of canal neuromasts than free neuromasts. It is suggested that canal provides some protection to the laterally situated neuromasts against the water running over them. Thus canal based receptors help to detect weak local water displacement during rapid swimming (Dijkgraaf. 1962).

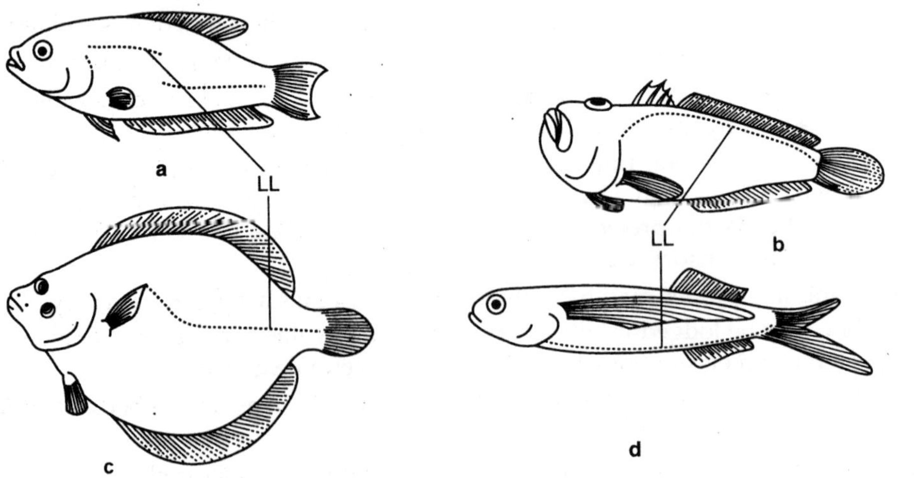

Fig. 15.7a-d : Diagram to show displaced lateral line on trunk due to dorsal shift of pectoral fin. (a) *Macropodus virdiauratus*. (b) *Pseudorhombus*. (c) *Uranoscopus scaber*, (d) *Exocoetus volitans*. LL. lateral line.

Lateral line canals on the body and head show several modifications. For example, the body canal may be extremely short in European bitterling *(Rhodeus promelas)*, intermediate in length as in the fathead minnow *(Pimephales premelas)*, interrupted in parrotfish *(Scarus)*, and strongly arched towards the back or decurved towards the belly (Lagler *et al.*, 1977).

In some fishes the lateral line displaced to dorsal side in the region of pectoral displaced to dorsal side in the region of pectoral fin (Fig. 15.7a) where this fin may drive water against the canal during locomotion. In these cases a straighter line would generate more 'noise' from locomotion alone and minimizes sensitivity to external water disturbances. The lateral line is displayed on dorsal side in *Uranoscopus* and on ventral side in *Exocoetus* (Fig. 15.7b), show selection towards the direction where water movements would be critical for survival.

The neuromast organs are said to be sensitive to low frequency vibration in the aquatic environment and help the fish to orient its body in respect to these vibration (Sand, 1938). They are also helpful in 'distant touch' and localizing the prey.

Experiments with adult saith *(Pollachius virens)* placed with opaque eye covers exhibited continued schooling behaviour, when they are placed among the non-blinded individuals as long as their lateral lines remained intact. Five blinded saith having their lateral lines cut at the operculum failed to schoo (Pitcher *et al.*, 1976).

THE OLFACTORY ORGANS

The organs of smell or olfactory organs in fish are pouch-like structure that opens to the water through incurrent and excurrent channels, i.e., naris or nares divided by flap of skin. The olfactory pit is internally lined with ciliated epithelium which is projected in multifolded olfactory rosette of receptor cells.

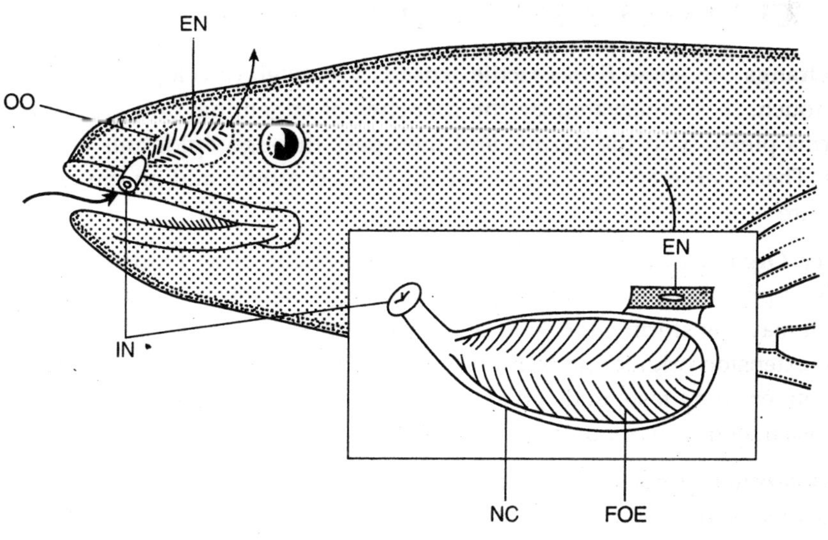

Fig. 15.8 : Diagram of olfactory organ of *Anguilla*. Box shows section of olfactory organ EN, excurrent naris; FOE, folds of sensory olfactory epithelium; IN, incurrent naris; OO, olfactory organ.

In teleosts the olfactory chamber is occupied with the olfactory rosette. The chamber receives opening of anterior naris and communicates out with posterior naris. The olfactory rosette is hollow oval sac, lies flat on upper side and convex on the lower side. It comprises connective tissue lined by the mucous membrane. The connective tissue is thickly developed on the lower side while it forms olfactory lobe in the one third of rosette. The mucous membrane of rosette is projected into large number of folds which are extended from one end of the sac to the other (Fig. 15.8). The folds are big and prominent in the middle and gets smaller towards the two ends of rosette. Each fold comprises a median core of connective tissue covered by a layer of ciliated columnar epithelium.

The cilia of olfactory pit drive the water and odours are perceived when the dissolved chemicals make contact with the olfactory rosette. Olfactory stimuli are conveyed to the olfactory lobe of brain via the first cranial nerve, i.e., olfactory nerve (Gloria *et al.*, 2003).

The fishes like eels, and sharks have big size olfactory lobes as they rely much on olfactory information. In sight feeders (Puffers) have much reduced olfactory lobes and olfactory organs and pits because these fishes probably rely more on sight for feeding.

In rayfin fishes (Actinopterygii) the paired olfactory organs are relatively smaller than those of elasmobranchs. The posterior nasal apertures are normally close to the anterior ones in many rayfins while they are relatively far apart in eels and morays. In sculpins *(Cottus)* and stickleback *(Gasterosteus)* only one nasa aperture and nasal pouch is found. They are alternately filled and emptied by the breathing movements.

Sense of smell is helpful in searching food and in orientation. Fishes that feed largely by odour, such as *Ichtalurus* and *Squalus* unable to detect their food when their nares are plugged. 'Pheromones' secreted by fish include glycerophospholipids. Some fishes like fathead minnows and eel can detect organic substances by smelling. Sutterlin (1975) found glycine and alanine to be specially attractive to winter flounder while alanine and methionine attracted Atlantic silversides effectively.

CUTANEOUS SENSES (TOUCH AND TASTE)

Tactile Orangs : In some fishes tactile sense is very well developed. The integument is sensitive to touch. It is innervated by the branches of trigeminal and facial nerves. These branches forms plexuses in the dermis and then pass through epidermis and terminate as free endings. These nerve endings are abundant around mouth and on the barbels which is tactile in nature.

Gustatory Organs : The taste buds are composed of three kinds of cells—these are receptor cells, supporting cells and basal cells. Among these cells some intermediate cells are present which may transform either into sensory or supporting cells. Each sensory cell has a thin hair like extension originating from taste bud and a nucleus. In barbels and lips of some fishes nerves innervating taste buds give of free nerve endings to the surroundings. The taste and touch sense may thus work unitedly.

Cyprinus carpio **:** It has greatly modified taste senses resembling to mammals, which can detect sweet, salty, acid, bitter stimuli. Bottom dwelling catfishes have relatively large numbers of taste receptors in their skin, fins and barbels. Taste buds have also been described on the palpatal organs in buccal cavities, gill arches and rakers of Cyprinids and Salmonids (Kapoor *et al.*, 1975).

REFERENCES

Alexander, R.M. (1967). Functional design in fish. London. Hutchinson Lib. 160 P.P.

Dijkgraaf, S. (1962). The functioning and significance of the lateral line organs. Biol. Rev. 38 : 51-105.

Gloria, A. Kapoor, B.G., Chardon, M. and Diogo, R. (2003). 'Fisheries' Vol. I Science Publishers. (USA).

Hanyu, I. (1959, a). On the falciform process, vitreal vessels and other related structures of the teleost eye-I, various types and their interrelationship, Bull. Jap. Soc. Sci. fish., 25 : 595-613.

Hanyu, I. Ditto. II. (1959, b) vascularization, *ibid.* 25 : 614-619,

Hanyu, I. (1961). Ditto. III. Formation of intraocular blood vessels, *ibid.* 27 : 504-509.

Hibiya, T. (1982). An atlas of fish histology, normal and pathological features. Lodansha Ltd. Gustav. Fisher Verlag.New York.

Jollie, M. (1968). Chordates Morphology, Reinbold Publishing Corporation, New York.

Kapoor, B.G., Evans, H.E. & Pervzner, R.A. (1975). The alimentay canal and digestion in teleosts. Adv. Mar. Biol. 13 : 109–239.

Lagler, K.F. Barsdach, J.E. & Miller. R.R. (1977). Ichthyology. John Wiley and Sons. Inc. N.Y.

Moyle , P.B. and J.J. Cech, Jr. (1982). 'Fishes an introduction to Ichthyology', Prentice Hall Inc. New Jersey.

Pitcher, T.J. Partidge, B.L., and Wardle, C.S. (1976). A blind fish can school. Science 194 : 963-65.

Sand, A., (1938). The function of the ampullae of Lorenzini with some observations on the effect of termperature sensory rhythms. Proc. Roy. Soc. London Ser. B. Biol. Sci, 125 : 524-553.

Sutterlin, A.M. (1975). Chemical attraction of some marine fish in their natural habitat, J, Fish Res. Bd. Canada 32 : 729-738.

Chapter 16

COLOURATION OF FISHES

Majority of fishes are vividly and brightly coloured. Colouration is one of the most common phenomena found among the fishes. The enormous range of colours and patterns that produced in fishes are generally related to their habits. Normally fishes are darker on the dorsal and lighter on sides or ventral side. This gives them protection from above and below. However, some fishes have uniform colouration as found in the gold fish, *Carassius*, which has brilliant colour all over the body. The bottom dwellers are often strongly and intricately coloured above and pale below. Variation in colour may be seen in a single fish. The trunk fish *(Ostracion)* has green body, orange tail and yellow belly with blue bands on the body. The pipe fish, Sea horse and angler living in weeds, often exhibit colour and pattern similar to weeds (Parker, 1943). Sometimes they also develop leaf like or filamentous *processes* on the body. Mahasheer *(Tor tor)* has dark grey colour on the back with golden or reddish on sides and silver on the abdomen. However, paird fins are yellowish or reddish. Colour differences in both sexes are quite marked in fishes. The males are generally brighter. Males of small million fish, *Lebistes,* are variously coloured while females are of a single colour. The variation of colours in males is due to genetic factor of Y-chromosome.

Another important feature is lack of pigment causes transparency in pelagic, free-swimming young's of many species. Similarly cave fishes living in total darkness, do not possess pigment and are colourless.

Sources of Colour

There are two main sources of colour production in fishes. These are chromatophores or biochromes and iridiocytes or iridiophores (Odiorne, 1957; Waring, 1963).

(I) The chromatophores — They are large and branched specialized cells. They are mostly present in the dermis, just beneath the epidermis or scales. They are also present around the brain and spinal cord. Chromatophores may be monochromatic, viz., and possess only one type of pigment, di or polychromatic. Cytoplasm of chromatophores contains different pigment granules, which are responsible for colour. These are flavines (yellowish green), carotenoides (yellow, red) and melanin (black and brown). Amalgamation of different chromatophores produces a wide range of colour, thus yellow and black (Fig. 16.1).

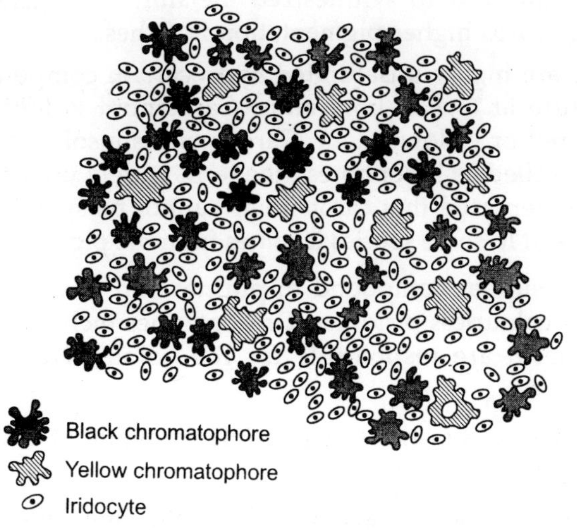

Fig. 16.1 : Coloration elements in the skin of the upper side of a flounder *(Platichthys)* (After Norman)

Chromatophores interspread among one another to produce green or brown colour.

Many fishes are able to change their colour of the body because of the migration of the pigment within the chromatophore. The pigment granules may disperse throughout the cell or aggregate in the centre to give different tone and pattern to the fish. There are four basic types of chromatophores based on colour of pigment granules present therein. These are erythrophores (red and orange), xanthophores (yellow), Melanophores (black or brown) and leucophores (white). The red and orange pigment granules of erythrophores and yellow pigment of xanthophores mainly consists of carotenoids. Fish obtain carotenoids through plant food. However, the black pigment of melanin is synthesized from amino acid tyrosine, under the influence of enzyme tyrosinase. Sometimes a brown pigment called eumelanin is also found in the chromatophores.

Pigments of chromatophore : Fish chromatophore contains following types of pigments.

Melanins : The melanin is the brown or black pigment derived from amino acid tyrosine. Melanin is synthesized usually in young melanophore and sometimes in adult melanophore. The first step of melanin synthesis is the oxidation of tyrosine to Dopa(3,4-dihydroxyp-

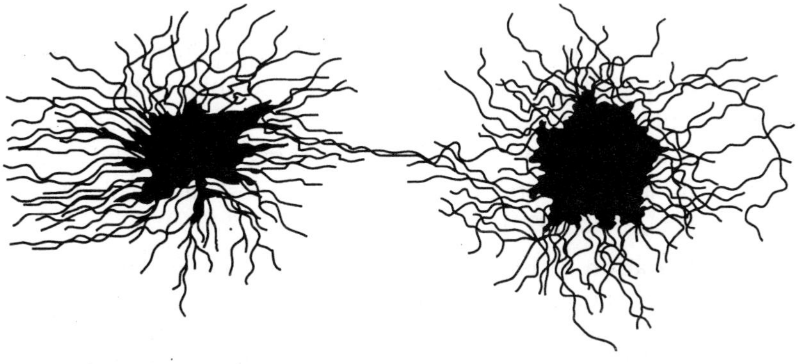

Fig. 16.2 : Nerves of the melanophores of a perch (From Ballowitz)

henylalnine) under the influence of enzyme tyrosinase. Dopa further oxidized to Dopa quinone, which is polymerized to synthesized melanin. It is generally considered that higher tyrosinase level causes higher pigmentation in fishes.

Carotenoids : These are highly unsaturated hydrocarbon compounds containing carbon chain with ring structure at one or both ends. Carotenoid is found in xanthophores or erythrophores causing red or yellow colour. Carotenoid is insoluble in water but soluble in organic solvents, hence called as 'lipophores', the term widely used to denote xanthophores and erythrophores. It is reported that carotenoids cannot be synthesized in body of fishes and it is derived from food. In some species it comes from the pigment found in the yolk.

Ptridines : This is a similar compound to purines and flavins. Fishes are reported to have both coloured and colourless ptridines. Drosopterines including drosopterin, isodroso-pterin and neodrosopterin are responsible for red colour. However, sepiapterins and isosepiapterins are yellow.

Purines : Guanine is a purine and is responsible for white or silvery tone in fishes. It is found in iridiocytes.

(II) The iridiocytes — They are also called as reflecting cells or mirror cells because they reflect light (Denton, 1970). The iridiocytes contain crystals of guanine, making them opaque and able to reflect light so as to produce either white or silvery appearance. This material is used in the manufacture of artificial pearls. Iridiocytes when present outside the scales, produce an iridescent appearance and when they present inside them, forming a layer called argenteum produce a white or silvery appearance.

Colour Changes : Colour change is both short and long term phenomenon due to pigment change. It is both physiological and morphological phenomena. A morphological change is a slow process as it involves formation of pigment granules in cells. Physiological change is rapid (for a short period within a few minutes) and exhibits rearrangement of pigment granules in the chromatophores. Both of these changes occur due to visual and non-visual stimuli. The later involve nerve and hormones.

Physiological or rapid colour change : In some fishes rapid change in colour occurs to match the changing surroundings. This type of colour adaptation is done by redistribution of pigment granules within the chromatophores. This type of colour change makes the fish inconspicuous over different backgrounds. The rapid colour change in fishes is known as cryptic or concealing colouration and may be of two types.

(i) Assimilation with background — In this type of colouration the fish harmonizes its colour to the background. The most common example of this type is pelagic laptocephalus larvae of eel, which is devoid of pigment. Sea horse and pipefish often have the colour that resembles the seaweeds. The green colour of 'tench' resembles to that of surrounding by assimilation. Another interesting example of rapid colour change is seen in the flat fish (Pleuronectiformes). These fishes have remarkable matching power. When they are kept on checkerboard, they will, after a short period, develop almost same colour and pattern as that of background (Nicol, *et al.*, 1973).

(ii) Disruptive colouration or breaking up the outline of the fish — The disruptive colouration is beneficial for concealment of fish. This is a sort of camouflage. In this type of colouration the continuity of body adapting different colour and tone disrupts surface or shape. The disruptive outline of the body helps fish to conceal. Various types of spots, stripes, lines and bands of brilliant colours on fish body, break up the outline making the animal less conspicuous. Sometimes disruptive colouration is used as a special camouflage, in which different parts of the body is concealed. Thus that particular part of the body is prevented from recognition on sight. In *Nassau grouper* a horizontal coloured line is present in continuation with the body, which makes the eye inconspicuous (Butcher, 1938). Similarly, a vertical line is present in the head of Jack-knife fish, to conceal the eye.

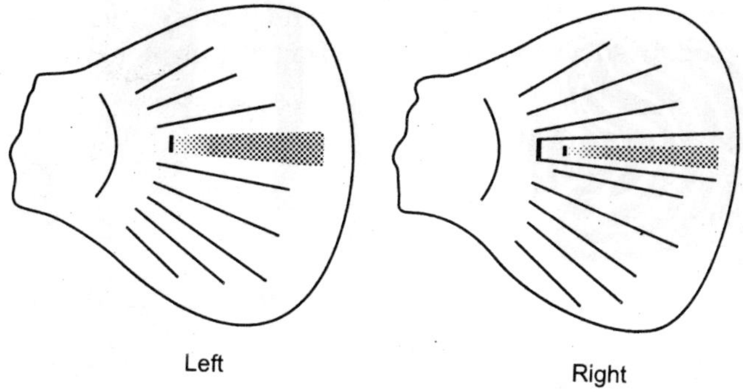

Fig. 16.3 : Diagrams showing, *left*, a cut in th tail of the fish *Fundulus* producing a band of dark melanophores; *right*, when the dark band had faded a second cut makes the melanophores again dark (After Parker, *Quart Rev. Biol*, 13)

(iii) Sematic or warning colouration — Besides concealing, another kind of colouration is sematic or warning colouration. In this type fish is usually adopt striking pattern and colour that reveals the animal then top conceal. This is of special significance for defence, as animal likely to attack are able to resemble the pattern and harmful effects previously associated with it. *Torpedo ocellata* has a prominent spot on electric organ for this purpose.

In some fishes obliterating colouration is adapted for concealment. The body of fish is counter shaded so that observer gets third dimension of the fish body, which reduces the visibility of fish.

Control of chromatophores — The regulation and coordination of colour change in fish is generally by interaction of nervous and hormonal control.

Nervous control : The chromatophores are supplied with nerve-fibres (Fig. 16.2), which are responsible for contraction of pigment granules, resulting the paling of skin colour. The nerve fibres are post-ganglionic sympathetic fibres . In some fishes the nerve fibres of any part of skin is cut, the chromatophores of that region expand making area darker.

Hormonal control : The colour change is also controlled by the action of posterior lobe of pituitary. It is evidently observed in the Atlantic minnow *Fundulus* that hypophysectomy result in the lighter body colour than normal individual, due to contraction of chromatophores (Abramowitz, 1953). The injection of pituitary extracts causes expansion of the chromatophores resulting dark colour of the body. It is believed that the two hormones of pituitary are responsible for colouration. The melanin dispersing hormone (MDH), i.e., intermedine causes darkening and the melanin-aggregating hormone (MAH) or W-substance causes paling of the body. It is evidently seen in *Scyllum*. Although the presence of MDH is found in many teleosts like *Anguilla* and *Fundulus* (Niell, 1940). In addition to pituitary

Fig. 16.4a-e : Colour patterns of various tropical fishes.
A, Muraena *(Gymnothorax)*; B, bat-fish *(Platax)*; C, butterfly fish *(Holacanthus)*; D, butterfly fish *(Chaetodon)*; E, perch *(Grammistes)* (From Norman)

hormones, adrenaline is also considered to control the action of chromatophore. It has chromatophore aggregating effect. Thyroxin is also believed to be responsible for colour change by effecting chromatophores.

Significance of Colouration : Colouration in fishes provides them power of adjustment with surroundings and also enables them to survive. The colour of fishes is used for concealment, communication, camouflage, sexual recognition and advertisement, warning or threat. Colouration also has taxonomic value. Different colour pattern in fish is often considered as character for distinction among species and subspecies. The specific pattern due to exact distribution of chromatophores is under genetic control. The colour pattern is also used in distinction of genera of some species like *Channa* and *Mystus*.

Factors Affecting the Colouration : There are various factors like temperature, light and stimuli which affect the functioning of the chromatophores. At low temperature chromatophores disperse causing darkening of the body while increase in temperature concentrates the chromatophores with substantial paling of the body. The light exerts its effect in two manners. In primary response the light affects the chromatophores by other sources then eyes. By secondary response chromatophores are affected by light through eyes.

External stimuli like tactile or psychic type also influence colouration of fish. The psychic type also influences colouration of fish. The psychic type of stimuli contribute much for change in colour during mating behavior of some fishes, when excited, shows psychological colour change in a short time, for example *Tilapia*.

Effect of Diet on Colouration

The colour of many fishes is also depends on their diets (Moe, 1982). Such types of diets may contain additional natural pigments to enhance colors of ornamental fishes. The carotenoid pigment found in most marine and a few freshwater invertebrates is astaxanthin. This pigment gives the characteristic colour to the flesh of salmon and is available in the diet of aquarium fish in shrimp and krill meals and salmon (fish) meal used as sources of protein in some feeds (Moyle, and Cech, 1988). Pure astaxanthin or canthaxanthin (synthetic astaxanthin) may also be added to fish feed to enhance red and orange coloration. These carotenoid pigments are often added to feeds for farm raised salmon and trout to give fillets a desirable red colour. Xanthophylls (yellow pigments) are found in corn gluten meal and dried egg that may be added to the diet to enhance yellows (Fujii, 1969). The ground petals of marigold flowers have also been used as a source of xanthophylls. The blue-green algae spirulina is a rich source of phycocyanin and may be added to a died to enhance blue coloration. The expense of supplementary pigments often limits the amount used in tropical fish feeds. These natural sources of pigments are in contrast to several methods routinely used to enhance colors of ornamental fish (Konings, 1993 Teichert-Coddington, Manning, and Eya, 2000).

Effect of Water Quality on the Colouration

Water quality may also play a support role in determining the colour of ornamental fish. Degraded water quality increases stress on captive fish and may dull fish colors. A high quality biological filter and routine at least bi-weeklywater changes will provide an environment enabling fish displaying their brightest colors. Feeding a varied diet rich in sources of pigments along with good water quality will ensure captive fish develop vivid colors.

REFERENCES

Abramowitz, A.A. (1953). Morphological colour change, *Fundulus xanthophores*. Proc. Nat. Acad. Sci, 21, 132-137.

Butcher, E.O. (1938). Relation of portions of eye to colour changes. Fishes J. Exp. Zool., 79, 275-297.

Denton, E.J. (1970). On the organization of reflecting surfaces in some marine animals. Phill. Trans. Roy. Soc. London. B, 178, 285-313.

Fujii, R. 1969. Chromatophores and pigments. pp. 301-353 in : W.S. Hoar and D.J. Randall (eds.). Fish Physiology. Volume III. Reproduction and Growth. Bioluminescence, Pigments, and Poisons. Academic Press, New York, NY.

Konings, A.(ed). 1993. Enjoying Cichlids. Cichlid Press.

Moe, M.A. 1982. The marine aquarium handbook-beginner to breeder. Green Turtle Publications, Plantation, FL.

Moyle, P.B. and J.J. Cech. 1988. Fishes an introduction to ichthyology. 2nd Edition. Prentice Hall, Englewood Cliffs, NJ.

Nicol, J.A.C., Arnott, H.J. and Best, A.C.G. (1973). *Tapeta lucidain bony* fishes (Actinopterygii): A survey. Can. J. Zool., 51, 69-81.

Niell, R.M. (1940). Colour change in young fishes; Colour change in *Anguilla*. J. Exp. Biol. 17, 74-94.

Odiorne, J.M. (1957). Colour changes. In the physiology of fishes, 2, 387-401. Academic Press,

Parkar, G.H. (1943). Vertebrates colour changes, review, Quart. Rev. Biol. 18, 205-227.

Teichert-Coddington, D., Manning, B. and Eya, J. (2000). Concentration of 17 alpha-Methyl-testosterone in hormone treated feed : Effects of analytical technique, fabrication, and storage temperature. Journal of the World Aquaculture Society 31 :

Waring. H. (1963). Colour change. Mechanism of cold-blooded vertebrates. Academic Press. New York.

Chapter 17

PARENTAL CARE

Parental care is one of the innate behaviors, which is found not only in higher vertebrates but also commonly seen in a large number of lower vertebrates including fishes. Fishes have evolved several means to provide care and protection to their eggs and youngs hatching out of them (Trivers, 1972). In this way fishes ensure survival of their young ones. Number of different behavioural strategies adopted by various groups of fishes have attracted attention of many biologists. According to them in some only males, while in some both male and female parents and in others only females afford care (Alexander, 1974).

Definition of Parental Care — "Parental care is protection and care provided to the eggs and young ones by one or both sexes among the fishes."

"Parental care may be defined as an association between the parents and the offsprings to increase the chances of survivorship of the offspring."

Parental care includes all the post spawning or post mating care afforded by the parents. Selection of sites, laying of eggs at suitable places, nest building, protection of eggs and young ones are some important aspects of parental care.

Additional Strategies

Many types of additional structures are present on the eggs to provide protection to them. In elasmobranches and holocephalans the shell gland of oviduct secretes a horny case around the egg. These cases protect them against predation and damage, and to certain extent help in osmotic control. An egg of some fishes like sculpin, trouts and minnows are coated with adhesive secretion, which keeps them held together and also helps them to attach to some objects so that they are prevented from being washed away (Wilson, 1975).

Selection of Spawning Site

Spawning at sheltered place is one of the most primitive types of protection afforded by the parents. Migratory fishes migrate to long distance in search of suitable place for spawning. The anadromous fishes like *Petromyzone marinus, Acipenser, Hilsa and Salmo* migrate extensively to spawn in suitable waters. The potadromous fishes like carps and trouts travel long distance within water in search of suitable spawning site and then return to their feeding grounds (Boisseau, 1968).

Parental Care by Nest Building

The most primitive is the habit of nest building. Many fish species build nest in one form or another, whether it is a simple pit dug into the gravel or the elaborate bubble nest. No special breeding setup is needed. When ready to spawn the fish constructs a nest by blowing bubbles, often using vegetation to anchor the nest. The male keeps the nest intact and keeps a close eye on the eggs. The female should be removed after spawning. Care is needed to raise the fry and the tank should have a glass cover to help keep the nest moist and warm. The Gouramis, Anabantids and some catfishes are the most common examples of this type of spawners.

The nest may be a simple trough or hollow cleaned out in the sandy bottom of a water body, as in case of *salmon*. Often one of the parents, usually the male, guards the eggs and undertakes the responsibility of incubating eggs and of rearing the fry till they are able to take care by themselves. The *Lepidosiren paradoxa* of South America constructs nest in the form of closed burrow in a swampy soil. It has been reported that during the breeding season male fish develops filamentous structures on their pelvic fins. Since these filaments are highly vascular, they secrete oxygen into water. According to some other workers, it is suggested that the filamentous gills supplement the oxygen requirement of the male fish, which during the time reduces its frequency to visit the surface.

The lungfish of Africa, the *Lapidosiren* scoops out a hallow depression in the mud of swamp, rich in aquatic weeds and grasses, which afford protection. The female lays eggs in a hollow nest. After that the male fish undertakes whole responsibility for nest building and subsequent care of the eggs and young ones. It swims around, chasing the predators and aerates the water in the nest for the eggs and young larvae.

The guarding parents usually stops feeding and utilize their body reserves during the period of care. According to Williams (1966), involvement of male parent in care and protection of young ones to make female parent free from burden of care so that they may feed properly and grow normally to maintain their fertility.

The bullhead, *Ictalurus natalis* either utilizes old tunnels of rodents or prepares a hole (Fig. 17.1). The male ventilates and guards thousands of eggs. The male looks after the young ones even after hatching till they grow up to two inches in size.

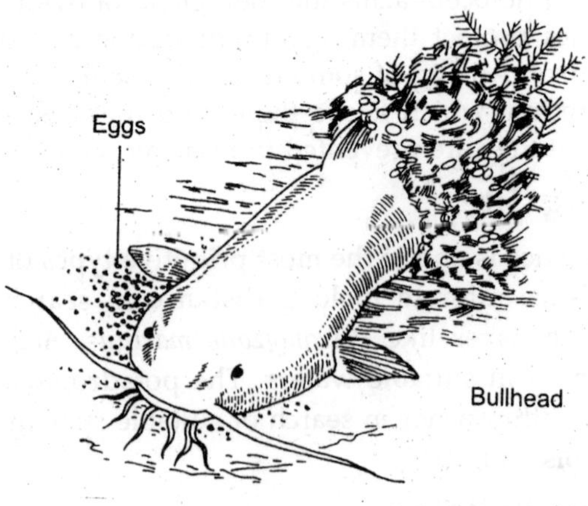

Fig. 17.1 : Parental care by male Bull head

Female *Trout, Tilapia* and the *Barbus tor* prepare their nest by scooping in gravel bottom (Fig. 17.2), although they have different egg laying habits and different breeding seasons.

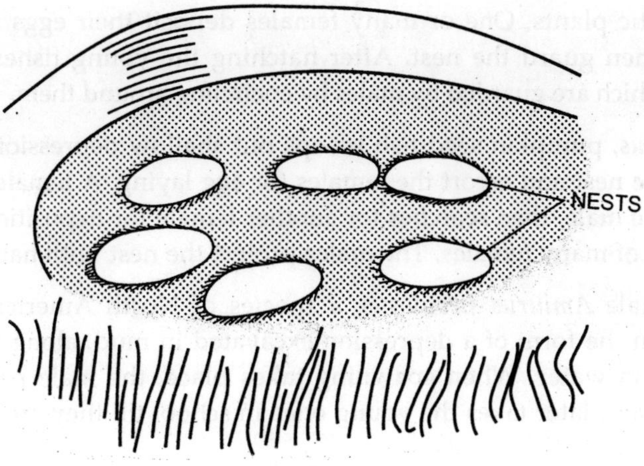

Fig. 17.2 : A drained pond showing nests of *Tilapia*

The African *Osteoglossid* (Clupisudis) makes a nest by cleaning a space in the aquatic vegetation. The nest is built in about two inches of water and it is about four inches across. The wall of the nest is several inches in thickness, made up of stems of grasses. The smooth bare ground constitutes the floor of the nest.

Females of Johney darter *(Etheostomal)* of Eastern North America build the nest in the form of shallow depression. The eggs laid are coated with mucus, so that they can stick on to stones lying around the nest.

The males of the three-spined stickleback, *Gasteroteus*, construct a much more elaborate nest. Their kidney secretes a special sticky and glue-like product. With the help of this glue males bind bits of plants and prepares the nest. Then male makes searches for a female, courts her and invites her to the nest, where she lay the eggs which he fertilizes. Soon after he again searches another female and the processes repeated till the eggs fill the nest. The male then undertakes the responsibility of guarding the eggs till they hatch. *Ophiocephalus* prefers weedy ponds to deposit their eggs. The male protects the eggs and later leads the youngs in school, till they grow about two inches long.

Male of giant kelp fish, *Heterostichus rostratus*, guards the nest carrying fertilized eggs. During the period of care the male gets aggressive against all predators .

Gymnarchus (Mormyrid) forms a large floating nest, which lies projected many inches above the water surface. The fishes enter in the nest from the end, which remains 6" below the water surface. Male of the Siamese fighting fish *(Betta splendens)* of family Anabantidae, builds nest made of air bubbles. They blow air bubbles in mucus of mouth, which are accumulated on the undersurface of some objects, floating in water. When enough bubbles are accumulated, courtship follows, and the eggs are fertilized. The male then relaxes and takes up position below the female. It picks up fertilized eggs in the mouth and swims upwards to adhere eggs to the nest by mucus. Males look after the nest, repair the nest and protect from predators. In these species the protective instinct in males is so strong that they often fight to enemies for defence of eggs till death.

Amia (North American) forms circular nests. They construct nests in swampy area of a lake, rich in aquatic plants. One or many females deposit their eggs at the bottom of the nest. The males then guard the nest. After hatching the young fishes leave the nest and swim in groups, which are guarded by males by encircling around them.

Lepomis gibbosus, pumpkin seed fish scoops out shallow depressions at the shores. The males construct the nest and escort the females for egg laying. A female after laying eggs is, driven away by the male, then searches for another female for depositing the eggs. Thus the nest contains eggs of many females. The males protect the nest until hatching takes place.

Male and female *Amiurus nebulosus*, a species of North American catfish, construct small crude nest in the form of a depression excavated in mud, along the bank of rivers or under any object in water. When spawning takes place, the male parent undertakes the protection of nest and later takes the young ones in school till they are shifted to safe water of shore.

Similar type of nest building habit is found in North American sunfish *(Centrarchidae)* (Fig. 17.3). The males of this fish scoop out a shallow and circular nest in the gravel bottom. The males prepare the nest and remove large stones, and only a fine sand layer is left. The females lay eggs, which are stick to sand. The male looks after the eggs until hatching takes place.

Fig. 17.3 : Male sunfish guarding eggs in nest

Egg Depositors

In this case, the eggs are either laid on a flat surface, like a stone or plant leaf or even individually placed among fine leafed plants like Java moss. The parents usually form pairs and guard the eggs and fry from all danger. The Cichlids are the best-known species for this. Some catfish and rainbow fish are also egg depositors. The set-up for these fishes will vary with the species, but usually a flat stone, broadleaf plant, cave or a broken flowerpot are to be provided. Sometimes you can remove the item that the eggs have been laid on to a separate hatching tank.

Paradise fish provides care to their eggs, which are lighter than water hence rise up and attach to bubbles in nest. Either of parents collects those eggs in mouth if falling down.

Parental Care by Other Means

One of the notable examples of parental care is found in a small carp *(Rhodeus)* commonly called Bitterling, inhabiting the rivers of Central Europe (Fig. 17.4). These fishes do not make nest, but discover tailor-made nest in the shells of the freshwater mussel. The

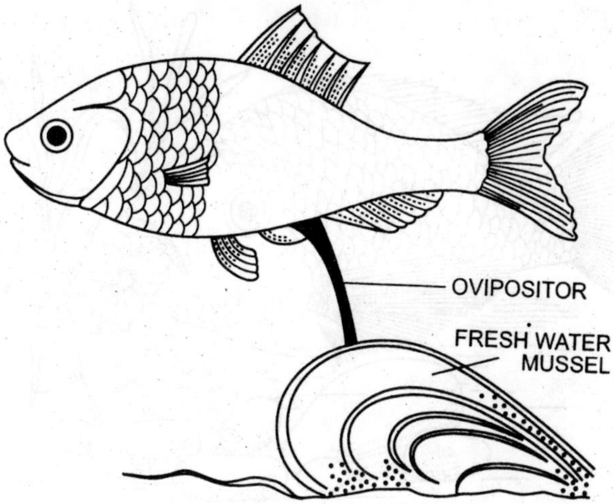

Fig. 17.4 : Female bitterling depositing eggs in the gill chamber of freshwater mussel

female fish draws out its oviduct to form a long tube just like an insect's ovipositor. With the help of this tube the females deposit their eggs in the bivalve shell, a safe place for eggs. Inside the shell the eggs undergo development and are aerated by respiratory current, which flows over the gills of mussel. It is observed that this period extends up to a month. It is amazing to note that the breeding season of the mussel coincides with the breeding season of the *Rhodius* fish. The glochidia larvae are temporary parasite on the fish and get dispersed wherever the fish swim about.

Another interesting type of parental care is seen in the Brazilian catfish *(Platystacus* (Fig. 17.5). The belly of female of this species becomes soft and spongy during the breeding

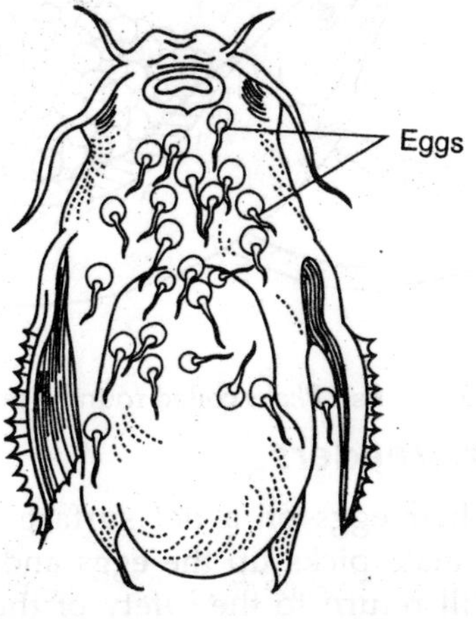

Fig. 17.5 : Parental care by female obstetrical cat fish

season. The female lies on the fertilized eggs, which get attached to the skin by small stalked cup-like structure. The cup is vascular and nourishes the developing embryo until they hatch (Williams, 1966).

Some species provides parental care by modification of certain structures. The lower lip of male of a Brazilian fish enlarges in the form of a pouch. The fertilized eggs are incubated

Fig. 17.6 : *Tilapia mossambica* : The young one taking shelter in the female buccal cavity when danger apprehended

in this pouch. However, mouth itself is used as an oral incubator in two catfishes, viz., *Bagre marinus and Geleichthys jelis*. The males of these species *carries in his mouth* 10-30 developing eggs. The African fish *Tilapia mossambica* is known as 'mouth brooder' as hatchlings come in the oral cavity at the time of danger (Fig. 17.6).

The Butterfish (*Pholis*) makes a ball of its spawn and roll around carry in their mouth to protect the eggs (Fig. 17.7). Both the parents carry out this movement alternately.

Fig. 17.7 : Butter fish *Pholis* coiled round a mass of eggs

Parental Care by Mouthbreeders

The females usually lay their eggs on a flat surface where the male then fertilizes them. After fertilization the female picks up the eggs and incubates them in her mouth. Even after hatching the fry will return to the safety of their mothers mouth if danger is near. Brood numbers are usually small, since by the time the fry are released they are well formed and losses are minimal. The best-known mouthbreeders are the African lake *cichlids*. The female fish keeps the eggs in her buccal cavity and provides shelters to the young ones.

The males of marine catfish *(Aridae)* collect the fertilized eggs in the mouth and keep them till their complete development takes place. During this period they do not feed.

Parental Care by Modifying Special Parts

Some fishes attribute parental care by modifying their organs in special manner.

Male of Florida pipefish (*Syngnanthus*) develops a pair of flaps on under side of body (Fig. 17.8). During the breeding season these flaps fuse to form a brood pouch. The female extends its oviduct and deposits the eggs into the brood pouch. The development takes place into brood pouch.

The Australian *kurtus* (*Kurtoidei*) offers parental care to the fertilized eggs in a different manner. The male fish produces its dorsal fin and forwards into a spinous hook, to which the eggs adhere to in clusters.

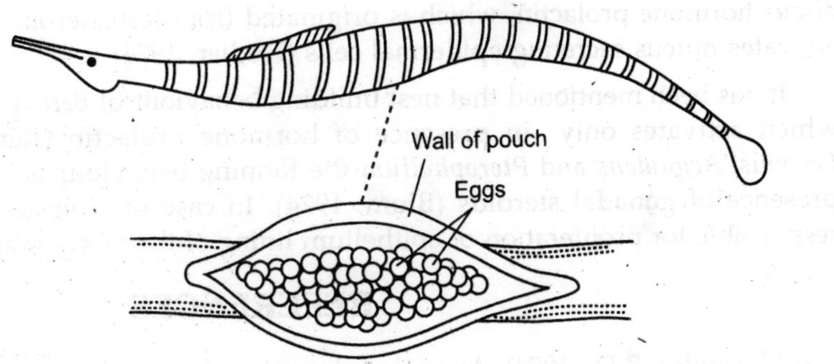

Fig. 17.8 : Male pipefish *Syngnathus* and its brood pouch dissected and opened to show eggs

In the *Hippocampus* (Sea horse), the male possesses brood pouch beneath the tail (Fig. 17.9). At the time of breeding season this pouch becomes thick walled and receives much blood vessels. The cloaca of the female becomes extended to form a genital papilla, which serves as an intermittent organ for the transfer of eggs to the brood pouch of male. When the eggs hatch into embryos, the vascular wall of the pouch nourishes them. At a time at least 6-7 miniature seahorse may be accommodated in the pouch.

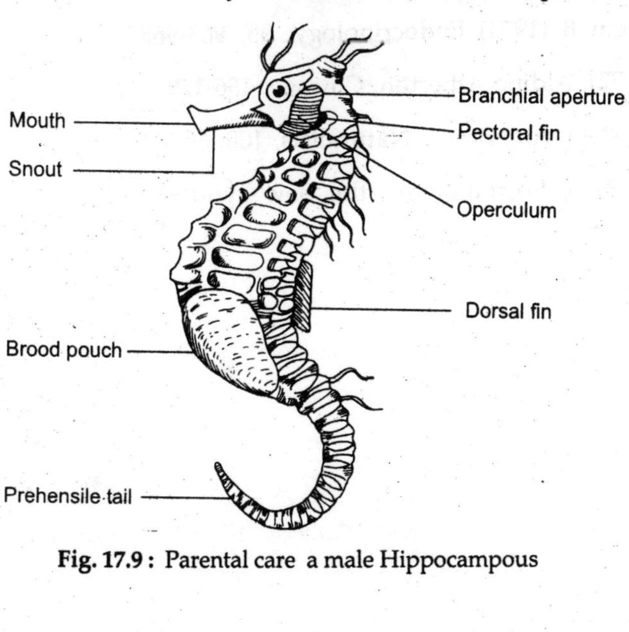

Fig. 17.9 : Parental care a male Hippocampous

Parental Care by Egg Buriars

The annual Killifish are known for this method of parental care. As the pools where they live dry out, the fishes spawn pressing their eggs into the substrate. The pools dry out completely and the adults die, but the eggs remain in the dried mud. When the rains return and the pool refills, the eggs hatch and the cycle is repeated. Killifish eggs can stay viable for many years in the dried out mud.

Influence of Endocrine Glands on Parental Behaviour

It has been reported that more than one hormone regulate the reproductive behaviour of the fishes. Fiedler (1974) reported that prolactin hormone accelerates the secretory activity of buccal epithelium of *Betta* at the time of breeding season. Thus the fish becomes enable to secrete air bubbles in mucus for building of nest. In another fish *Symphyrodon discus* hormone prolactin, which is originated from testosterone and is called as paralactin, activates mucus secreting epidermal cells (Fiedler, 1974).

It has been mentioned that nest building behaviour of *Betta* is controlled by testosterone, which activates only in presence of hormone prolactin (Ketz and Eckstein, 1974). In *Lepomis, Acquidens* and *Pterophyllum* the fanning behaviour is induced by prolactin in the presence of gonadal steroids (Blum, 1974). In case of *Hippocampus* it is prolactin that is responsible for proliferation of epithelium lining of the brood pouch (Boisseau, 1968).

REFERENCES

Alexander, R.D. (1974). Annual review of ecology and systematics, 5, 325-383.

Blum, V. (1974). Die Rolle des. Polaktins. Bei der. Cichlidenbrutpflege. Fortschr. Zool., 22, 310-333.

Boisseau, J.P. (1968). Polactine et incubation chez 1 hippocampe Coll. Int. CNRS. No. 177, 205-214.

Fiedler, K. (1974). Hormonale Kontrolle des Verhaltens bei fischen. Fortscrh. Zool., 22, 269-309.

Ketz, Y. & Eckstein, B. (1974). Endocrinology, 95, 963-968.

Trivers, R.L. (1972). Aldine Atherton. Chicago, 136-179.

Williams, E.C. (1966). American Naturalists, 100, 687-690.

Wilson, E.Q. (1975). Sociobiology; the new synthesis, Harvard University Press, Cambridge.

Chapter 18

BIOLUMINESCENCE

A number of fishes especially marine species are known to produce characteristic light through their special organs called luminous organs. These organs are commonly found in fishes living in deep-sea where the sunlight ceases to enter. The luminous organs are absent in freshwater fishes (Harvey, 1957). The most important function of bioluminescence is to illuminate surroundings for the purpose of camouflage, schooling and for recognition of movement of predators in the water (Denton, *et al.*, 1972). The luminous organs or photophores are special gland cells of the epidermis. Their distribution on the body type and adaptive value may vary in different species of fishes.

Structure of luminous organ or photophore — On the basis of anatomy of photophores they may be categorized in two types.

1. **Simple photophore** — They are small in size, about 0.1 to 0.34 mm in width. It consists of light generating cells called as photocytes. Simple type may be provided with or without mantle of pigment. The lenses are formed by grouping of cells known as lenticular cells. The distal part of photocyte is provided with acidophilic granules. A layer of melanophres surrounds the photophore. Simple type of photophores is present in sharks. In *Stomias* the luminous organs are logged in gelatinous corium of the epidermis.

2. **Compound photophore** — This type of photophores consists of additional structures like reflectors, pigmented mantle and sub-ocular organs. The latter one is a large organ deeply

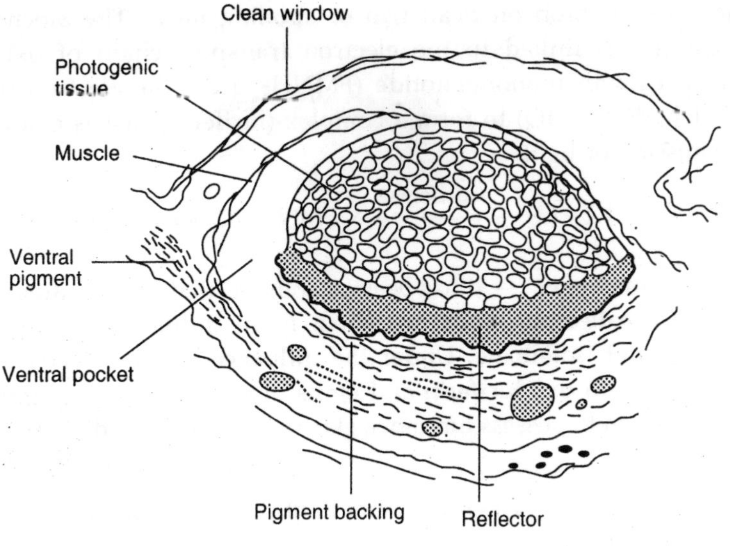

Fig. 18.1 : T.S. through a sub-ocular light organ of *Astronesthes richardsoni*

embedded in dermal tissue. The photocytes are arranged in the form of cords and bands. Photogenic tissue, pigment and reflector layers are provided with nerves and blood vessels (Denton, *et al.*, 1972) (Fig. 18.1). The photogenic tissues are found in the centre of the photophore and consist of two types of glandular cells.

The mechanism of light production is peculiar in fishes and takes place the special sets of muscles present around the photocytes. When these muscles contract, they pulls the outer surface of photophore downwards causing brighter surface to be concealed. In contrast the relaxation of these muscles exposes bright surface of the photophores. In some species, movement of pigmented layer carries out concealing and rotating of photophores (Fox, 1957).

Types of Luminescence

On the basis of source of illumination it may be classified as follows :

1. **Extra cellular luminescence** — Light may be generated by luminous secretion from the glandular tissues. Extra cellular luminescent organs are found in a very limited species of fishes. Certain fishes like rat tails and searssids emit light by secreting extra cellular slime. Rat tails possess special glands near its anus, which secretes slime of sufficient luminosity.

2. **Intracellular luminescence** — In this type the light is produced within the glandular cell or intrinsic photocyte. These luminous organs developed from the epidermis. Fishes ornamented with this type of luminous organs belong largely to the family of teleosts, i.e., *Sternoptychidae* (hatchet fish), *Myctophidae* (lantern fish), *Halosauridae* (Halosaurid eel), *Stomiatidae* (scaly dragonfishes), *Brotulidae* (Brotulus), *Lophiidae (anglerfish)* and *Zoarcidae* (eel pouts) (Lawry, 1974).

3. **Bacterial luminescence** — In this type, symbiotic bacteria present in the photophore or luminous cell discharge light (Nicol, 1969). Many different species are recognized particularly the genus photobacterium and achromobacterium have been isolated and grown in cultures. They are common on dead fish or spoiling meat. The biochemical step in bacterial luminescence is linked to the electron transport chain of oxidative phosphorylation, in which flavin mononucliotide ($FMNH_2$) from the electron transport chain reacts with an aldehyde (RCHO) to form a complex (luciferin) that is oxidized to an acid (RCOOH) with emission of light.

$$FMNH_2 + RCHO + O_2 \xrightarrow{\text{luciferase}} FMN + RCOOH + H_2O + 0.1/hv$$

4. **Chemical luminescence** — It has been established that the glandular tissue secretes a chemical substance called as luciferin, which is an indole derivative consisting of tryptamine, arginine and isoleucine. Under the influence of the enzyme luciferase, this substance is converted into oxy-luciferin and emits blue or blue-green light. Apogon, the Parapria-canthus is known to possess luminous glands containing crude form of luciferin and luciferase.

Control of luminescence — The function of light producing organs is controlled by the nervous or endocrine system.

Nervous control — Several workers have reported that light production by the luminous organs is controlled by the nervous system, probably by the peripheral sympathetic system.

The nerves innervate the phagocytes. The efferent nerves enter the photogenic cells and activate them.

Hormonal control — It has been reported that some fishes have hormonal control on the photophores. Endocrine gland like supra renal activate them. Adrenalin or noradrenalin is known to control light emission from the photophores.

Mechanical control — The muscles present beneath the photophores contract and rotate the photophores in such a way that they get concealed. Thus fish is prevented from illumination specially when in danger. In *Photoblepharon palpebratus* the ventral part of luminous organ has a fold of black tissue (Fig. 18.2). This fold can be drawn over the photophores and conceal the light. In some fishes the light production is also supposed to be influenced by the movement of pigment in the chromatophores.

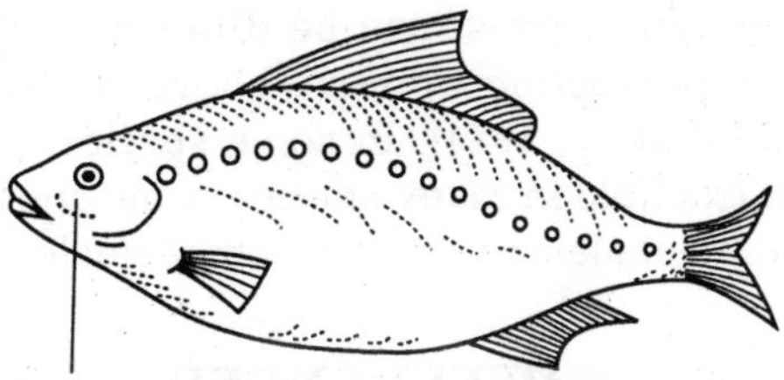

Fig. 18.2 : Light producing organs of *Photoblepharon*

Biological Significance of luminescence — This is useful in variety of ways in marine fishes specially in deep-sea fishes.

1. **Illuminates surroundings** — Some fishes utilizes their luminous organs to illuminate their surroundings in the event of dimness. Thus they become able to search their prey in the dark waters. Some species (stomiatoid) are able to emit beam of light from the specially designed luminous cheek organ to catch the small creatures like planktons. The cheek organs of *Anamalops* produce light like a torch.

2. **As defensive device** — Many fishes produce sudden flash of light from their luminous organs, which helps in diverting the attention of their predators. The emission of light facilitates an escape of fish by puzzling the enemy. *Alepocephalidae* produce a glowing spark, which confuses the predator for a spur of moment, and help the fish to escape. However, some fishes uses luminous organs to enable them inconspicuous. In doing so they illuminate their ventral surface that makes them inconspicuous against lighted background above.

3. **As a warning signal** — A number of fishes uses its luminous organ to warn the predators. For instance, the midshipman *Porichthys* that possesses, a toxic sign, flashes light when it is attacked by a predatory fish and avoids the danger (Fig. 18.3).

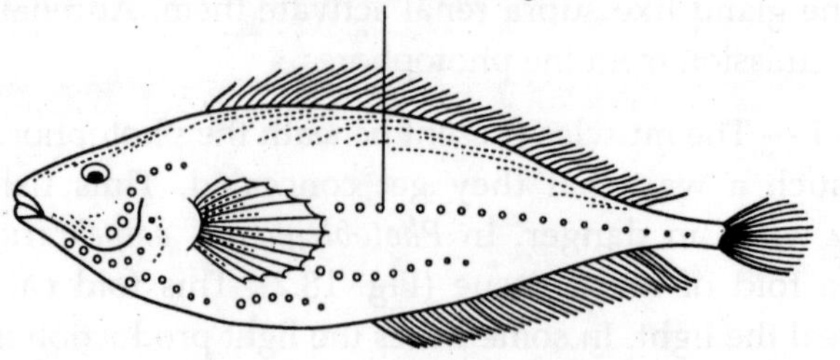

Fig. 18.3 : Hight producing organs of *Porychthyes*.

4. **Recognizing own species** — Every species has a unique arrangement and distribution of photophores on their body, which help the fish to recognize species of same type and thus help in schooling behaviour. The luminous organs are also helpful in recognizing the mates for courtship, as the light organs may be different both in male and female. Male lanternfish has one or many photophores present above but both in the female possesses it below the caudal peduncle (Lawry, 1974). In some species the size of luminous organ is different in both sexes. For example in many species of melanostommiatidae, the postorbital luminous organ is larger in the male and smaller in the female.

REFERENCES

Denton, E.J., Gulpin-Brown, J.B. and Wright, P.G. (1972). The angular distribution of light produced by some mesopelagic fish in relation to their camouflage. Proc. Roy. Soc. London B, 182, 145-158.

Fox, D.L. (1957). The pigment of fishes. In the physiology of fishes, 2; 367-385. Academic Press, New York.

Harvey, E.N. (1957). The luminous organ of fishes. In the physiology of fishes, 2; 345-366. Academic Press, New York.

Lawry, J.V. (1974). Lanternfish compare down dwelling light and bioluminescence. Nature, 247, 145-157.

Nicol, J.A.C. (1969). Bioluminescence. In the fish physiology, 2; 355-400. Academic Press, New York.

Chapter 19

ENDOCRINE GLANDS

Classically the endocrine glands have been defined as being ductless glands, since they release their secretory product directly into the blood or lymph. The components of endocrine system can be classified on the basis of their organization, which is as follows :

(A) Discrete endocrine glands : These include pituitary (hypophysis), thyroid and pineal (Fig. 19.1).

(B) Organs containing both endocrine and exocrine functions. In fishes, it is kidney, gonads (Fig. 19.1) and intestine. Kidney contains heterotopic thyroid follicles, interrenal, and corpuscles of Stannius.

(C) Scattered cells with endocrine function : They are known as diffused neuroendocrines. They are present in digestive tract (Fig. 19.1). They are generally called as paracrines (e.g., somatostatin). There are gastrointestinal peptides whose definite classification as hormone or paracrine agent has not yet been established, these are designated as putative hormones (Ross and Reith, 1985).

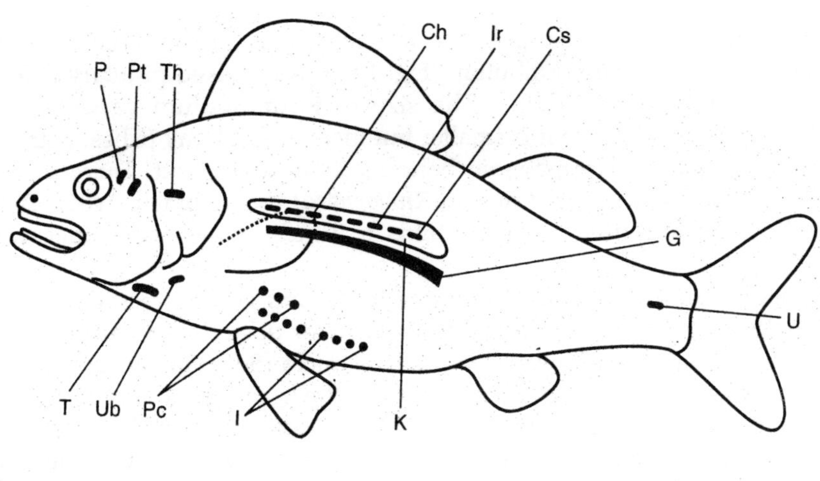

Fig. 19.1 : Schematic diagram to show position of various endocrine glands in fishes. Ch. chromaffin tissue; Cs, corpuscles of Stannius; G, gonad; I, intestinal tissue; Ir, interrenal tissue; K, kidney; P, pineal; Pc, pancreatic islets; Pt, pituitary; T, thyroid; Th, thymus; U, urohypophysis; Ub, ultimobranchial.

Chemically, hormones can be divided into three classes

(I) Steroid hormones (testosterone & estradiol)

(II) Protein (peptide) hormones (e.g., insulin) and peptide hormones are secreted by hypophysis, thyroid, internal tissue and pancreatic tissue.

(III) The amino acid analogues are norepinephrine and epinephrine, collectively called catecholamines.

Endocrine glands of increasing complexities are found in cyclostomes, elasmobranchs and osteichthyes. Elasmobranchs (sharks) possess well developed endocrine glands but these show some interesting differences from those of higher chordates. However, osteichthyes (bony fishes) have endocrine glands rather more similar to higher chordates. The difference between fish and mammal endocrine glands is probably due to the development and modification of various body systems in these two classes, and also due to exigencies of an aquatic mode of life. Mammalian endocrine glands are well advance and well studied but fish endocrinology is limited to the work on its influence on chromatophores, action of sex cells, function of pituitary and thyroid and control on migration. The detailed study of histochemistry and immunocytology of endocrine in fish is further needed.

Unlike nervous system, the endocrine system is basically related to comparatively slow metabolism of carbohydrate and water by adrenal cortical tissue, nitrogen metabolism by adrenal cortical tissue and thyroid glands and the maturation of sex cells and reproductive behaviour by the pituitary gland and gonadal hormones.

THE PITUITARY OR HYPOPHYSIS

Origin : The pituitary gland occupies the same central part in the endocrine signalling system of fish that it has in mammals. This master endocrine gland originates embryologically from the two sources. One as ventral downgrowth of a neural element from the diencephalon called the infundibulum to join with another, an ectodermal upgrowth (extending as Rathke's pouch) from primitive buccal cavity. These two outgrowths are thus ectodermal in origin and enclose mesoderm in between them, which later on supply blood to the pituitary gland, originating from the interrenal carotid artery.

Location : The pituitary gland is located below the diencephalon (hypothalamus), behind the optic chiasma and anterior to saccus vasculosus, and is attached to the diencephalon by a stalk or infundibulum (Fig. 19.2). The stalked pituitary is found in *Barbus stigma* and *Xiphophorus maculatus*. The size of infundibulum varies according to the species. Usually in cyclostomes it is smaller but increases in bony fishes, with prominence in groove or depression of parasphenoid bone receiving the gland. There is no sella turcica comparable to that found in mammals in *Xiphophorus* (Schreibman, 1964). The short, thick-walled, hollow infundibular stalk contains a lumen, which continues with the third ventricle.

Shape and size : The pituitary is an oval body and is compressed dorsoventrally. The size of sexually mature platyfish has a mean anterior posterior length of 472.9 micra, with mean width of 178 micra and mean depth of 360 micra (Schreibman, 1964). Male glands are smaller than those of females. On ventral aspect the gland gradually tapers caudally from rounded anterior end. The dorsal surface of the pituitary of platyfish is concave, ventrally it is slightly convex. The pituitary gland is completely enveloped by a delicate connective tissue capsule.

Anatomy of the gland : Microscopically, the pituitary gland is composed of two parts.

(i) Adenohypophysis, which is a glandular part originated from the oral, ectoderm.

(ii) Neurohypophysis, which is a nervous part originated from the infundibular region of the brain. Both parts are present in close association.

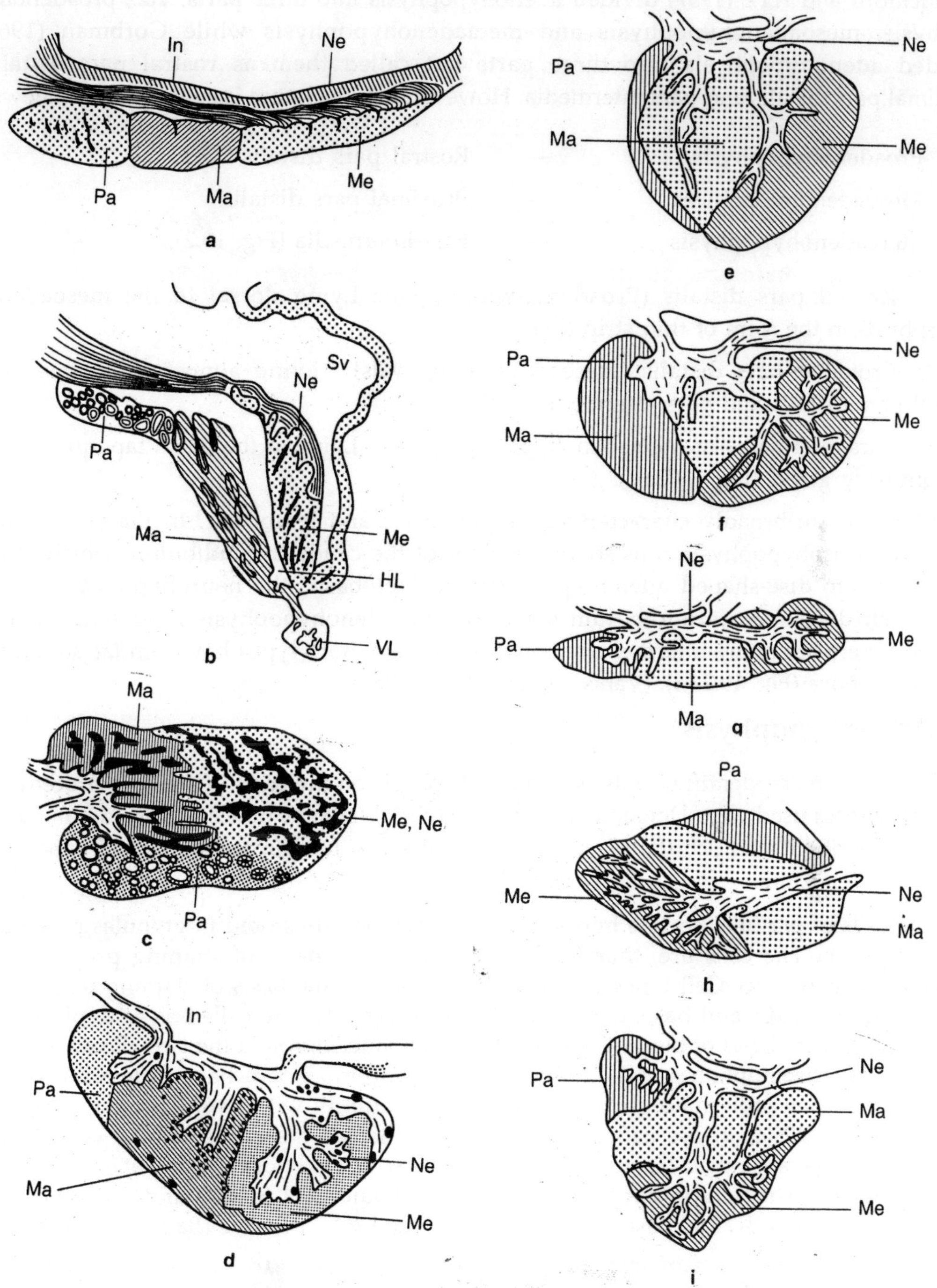

Fig. 19.2a-i : Diagrams of pituitary of various fishes. (a) *Petromyzon*. (b) Dogfish shark (*Squalus*). (c) Trout (*Salmo*). (d) Perch (*Perca*). (e) Rainbow trout. (f) Ayu (*P. altivelis*). (g) Eel. (h) Carp. (i) Yellow tail. HL, lumen of hypophysis; In, Infundibulum; Ma, mesoadenohypophysis; Me, metaadenohypophysis; Ne, neurohypophysis; Pa, proadenophypophysis; SV, saccus vasculosus; VL, ventral lobe; (Source : a-d, Pickford and AtZ, 1957; e-i, Hibiya, 1982).

Pickford and ATZ (1957) divided adenohypophysis into three parts, viz., proadenohypophysis, mesoadenohypophysis and metaadenohypophysis while Gorbman (1965) divided adenohypophysis into three parts but called them as rostral pars distalis, proximal pars distalis and pars intermedia. However, nomenclature is synonyms as follows :

Proadenohypophysis	—	Rostral pars distalis
Mesoadenohypophysis	—	Proximal pars distalis
Metaadenohypophysis	—	Pars intermedia (Fig. 19.2).

1. Rostral pars distalis (Proadenohypophysis) : Lying dorsal to the mesoadenohypophysis in the form of thin strip (Fig. 19.2a-i).

2. Proximal pars distalis (mesoadenohypophysis) : Lying almost in between the rostral pars distalis and pars intermedia.

3. Pars intermedia or metaadenohypophysis, viz. : Lying at the distal tapering end of the pituitary gland (Fig. 19.2a-i).

Pituitary are broadly characterized as platybasic and leptobasic. In platybasic form (Eel), the neurohypophysis consists of flat floor of the caudal infundibulum which sends processes into disc-shaped adenohypophysis. In leptobasic, the neurohypophysis has a fairly well developed infundibulum stalk and the adenohypophysis is globular or egg shaped. There are many intermediate between the two. Both types have similar structures described above (Fig. 19.2a-i). (Vanoordt and Peute, 1983).

I. Adenohypophysis

Earlier workers identified cells of adenohypophysis on the basis of staining procedures. The procedures used were Heidenhain's azon method, Masson's poncean acid fuschin anilin blue, the periodic acid Schiff's reaction (PAS), aldehyde fuschin technique (AF) and then they made cell counts.

The cells of pituitary secrete hormones and hormones are stored in granules present in the cytoplasm. The cells are, therefore, classified on the basis of staining properties of granules of these cells. Cell types of adenohypophysis, on the basis of staining reaction, to the mixture of acidic and basic dyes with secretory granules are called as acidophilic and basophilic. On the basis of binding affinities with ribonucleoprotein the two classes are also classified as chromophobes and chromophils. The chromophobes have little affinity with dye while chromophils stain strongly as they have affinity with dye. Chromophilic cells which take acidic stain are called as acidophils whereas the chromophilic cells which bound basic dye are called basophilic and the cells which do not take any stain are called chromophobes. The acidophilic cells are PAS (periodic acid Schiff) and AF (aldehyde fuschin) negative cells. The basophilic cells are AF and PAS positive (Ball & Baker, 1969).

Recently on the basis of immunocytochemistry, the cells are classified according to hormones released by the proadenohypophysis. For example, the cells which take basophilic stain but produce adrenocorticotropic hormones, they are called ACTH cells but if secrete thyroid stimulating hormone, these cells are called thyrotrophs and if they secrete FSH hormones they are called gonadotrops although they are basophilic in nature. The cells of adenohypophysis when stained with periodic and Schiff (PAS) and aldehyde fuschin methods, if do not take stain, they are PAS and AF negative.

The teleost hypothalamopituitary system is unique amongst vertebrates, as there is direct innervation of pars distalis by neurosecretory neurons of hypothalamus and there is loss of modification of the typical vertebrate hypothalamohypophysial portal vascular system for transport of neurohormones to pars distalis (Peter and Fryer, 1983).

(a) **Proadenohypophysis :** It contains cells which secrete prolactin and corticotropin (ACTH) exclusively in addition to other hormones.

(b) **Mesoadenohypophysis :** The mesoadenohypophysis (proximal pars distalis) contains cells which produce gonadotropin (GTH) and growth hormone (GH). Thyrotropin cells may occur in either or both in rostral pars distalis and proximal pars distalis. Acidophils rounded or oval or sometime pyrimidal shaped. They are coarsely granular, and give the cytoplasm a splotched appearance. They have round to oval peripheral nuclei. Basophils (cyanophilic) are spherical with large, round, centrally located nuclei. Their cytoplasm is finely granular, chromophobes cells are similar in structure as they are found in proadenohypophysis. The basophilic (cyanophilic) cells are PAS positive and AF positive.

(c) **Metaadenohypophysis :** It also encompasses more neurohypophysial tissue than any other region. The meta-adenohypophysis basophilic cells are PAS positive. However, the granular cells do not show consistent staining reaction with PAS and AF stains.

II. Neurohypophysis

The neurohypophysis occupies considerable portion of the gland and possesses many interesting and distinctive features. The neurohypophysis comprises connective tissue, neuroglia cells and loosely tangled network of nerve fibres. These nerve fibres are scattered horizontally along the dorsal part of the adenohypophysis and run vertically, which are generously interspread with granular meterial, large irregularly shaped amorphous masses and large nuclei. They are located in the middorsal region. The amorphous masses are called "Herring bodies", which have an intimate relation with the diencephalic neuro-secretory cells called nucleus preopticus by means of a fibre tract known as the preoptic neurohypophysial tract (Knowles & Vollarth, 1966). The diencephalon another parts of the brain contain a group of neurons and each group is called nucleus. The NPO and NLT are important as their axons are in association with both adenohypohysis and neurohypophysis (Fig. 19.3). These possess neurosecretory cells. The nucleus preopticus (NPO, preoptic nucleus) is

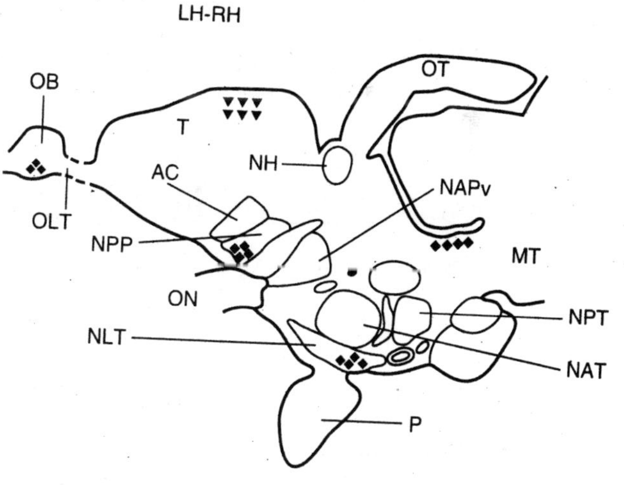

Fig. 19.3 : Diagram to show NPO and NLT in the brain of platyfish. AC, anterior commissure; MT, mid-brain tegmentum; NAT, nucleus anterioris tuberis; NAPv, nucleus anterioris periventricularis; NH, nucleus habenularis; NLT, nucleus lateral tuberis; NPO, nucleus preopticus; NPP, nucleus prepoticus periven-tricularis; NPT, nucleus posterioris tuberis; OB, olfactory bulb OLT, olfactory tract; ON, optic nerve; OT, optic tectum; P, pituitary; T, telencephalon.

situated on either side of the optic recessus slightly in front of the optic chiasma. The preoptic nucleus (NPO) is further subdivided into two parts, I, Pars parvocellularis, it is located anteroventrally and consists of relatively small cells, II, pars magnocellularis, it is situated posterodorsally and comprises relatively larger cells (Ball, 1981).

The preoptic nucleus (nucleus preopticus, NPO), their axons and nerve endings in the pituitary are stainable by neurosecretory stains. The neurons with Gormori's chrome alum haematoxylin, aldehyde fuschin and alcian blue, can differentiate the NPO from other nuclei in preoptic region as they are neurosecretory in nature (Graus Stenson, 1977).

Blood supply : Vascularization of the pituitary has been studied in variety of species. In brook trout, *Salvelinus fontinalis* and Atlantic salmon, *Salmo salar*, there is a separate blood supply to the neurointermediate lobe from the caudal hypothalamic artery and to the combined rostral proximal pars distalis from hypophysial arteries that branch off the anterior cerebral arteries. According to Follenius (1963), there is no separate blood supply for rostral proximal pars distalis and pars intermedia in *Salmo gairdneri* but the entire blood supply originated anteriorly from the hypophysial arteries. However, in teleosts, the rostral proximal pars distalis receives blood supply from the extensive loopings of arterioles which are found near the interface with the pars distalis (Fig. 19.4). These vessels

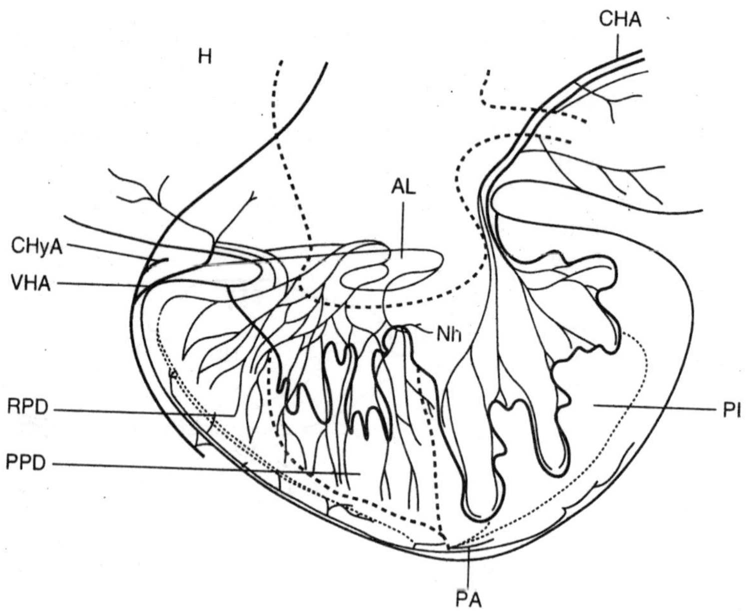

Fig. 19.4 : Para sagittal view of pituitary gland of brook trout showing blood supply. AL, arterial loops; CHA, caudal hypothalamic artery; CHyA, caudal hypophysial artery; H, hypothalamus; Nh, neurohypophysis; PA, peripheral artery; PI, pars intermedia; PPD, proximal pars distalis; RPD, rostral pars distalis; VHA, ventral hypothalamic artery.

are invaded into the pars distalis together with the interdegitations of the anterior neurohypophysis. It has been considered that these arterior loops are the rudiment of the hypothalamohypophysial portal system (Henderson, 1969). However, there is no neurovascular connections with these blood vessels, as are typically found in the median *eminence* of various vertebrates. This hypothesis is argued as a portal system. The function of the hypothalamohypophysial portal system as a means of transport of neurohormones to the pituitary has become redundant and pure vascular in function, probably because the pituitary cells have direct innervation by neurosecretory endings.

In spite of this, a typical but small hypothalamohypophysial portal system has been described in variety of teleosts (*Channa punctatus*, Belsare, 1965 a,b; *Heteropneustes fossilis*; Sathyanesan and Haider, 1971, a, b; and *Clarias batrachus*, Sathyanesan, 1972).

According to Sathyanesan and colleagues, the branches of hypothalamic arteries form "primary capillary plexus" located in the meningeal tissue and the adjacent neural tissue of hypothalamus anterior to the pituitary stalk. This plexus converges into vessels that enter the pituitary or proximal pars distalis or pars intermedia. Thus in teleosts this portal system is the only and primary source of blood for the pars distalis. It has been considered that among teleosts the Cypriniformes or Siluriformes have reduced portal system. It is clear, however, that teleosts have neurohormones secreted more or less directly to the pituitary, and that some have the potential for vascular transportation of neurohormones by a portal system as well.

HORMONES OF PITUITARY

There are (seven) various hormones secreted by pituitary but it is generally agreed that one cell type-one hormone concept, is correct. The different hormones secreting cell are not localized in specific region but are spread over in part of the adenohypophysis (Fig. 19.5).

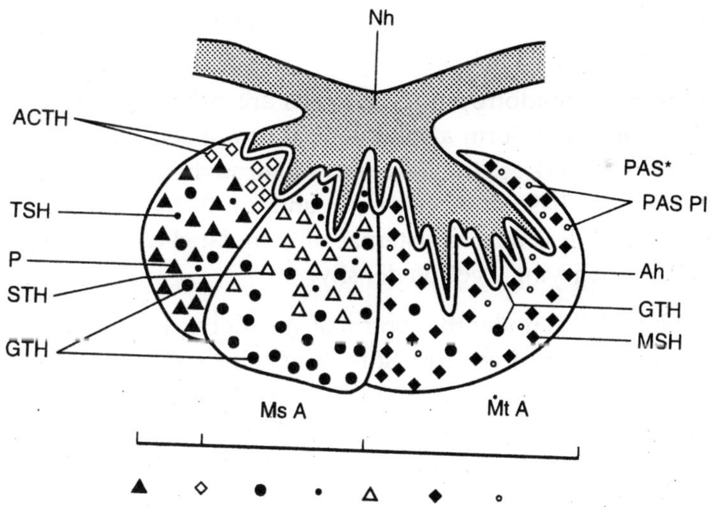

Fig. 19.5 : Section of the pituitary to show various hormone secreting cells in the adenohypophysis (Ah). ACTH, adeno corticotropic cell; GTH, gonado tropic cell; MsA, mesoadenohypophysis MtA, metaadenohypophysis; MSH melanotropic cell; Nh, neurohypophysis; PAS, periodic acid Schiff-positive cell in pars intermediate; P prolactin producing cell; STH, somatotropic cell; TSH; thyrotropic cell.

All hormones secreted by the pituitary are necessarily proteins or polypeptides. There exists a slight difference in the pituitary hormones of the different groups of fishes. The pituitary hormones of fishes are of two types (I) one which regulates the function of other endocrine glands. Such hormones are called tropins or tropic hormones. These are :

1. Thyrotropin activates thyroid.
2. Adrenocorticotropic hormones activate adrenal cortex.
3. The gonadotropin FSH and LH (Leuteotropins, various steroid hormones).
4. Growth hormones, somatotropin (actually they are not tropic).

(II) Second which directly regulates the specific enzymatic reactions in the various body cells or tissues. These hormones are melanin hormones (MH) and melanophore stimulating hormone (MSH), etc. Thyrotropin hormone is secreted from proadenohypophysis (rostral pars distalis) and stimulate activity of thyroid hormones (Prasad Roa, 1972). The TSH is secreted under the influence of (TRH), thyroid releasing hormones from diencephalon in fishes. It is proved that TRH influences the TSH cell activity and thyroid activity in fish. In *Carassus auratus,* crude extract of the hypothalamus or goldfish results decreased radioiodine uptake by the thyroid, which indicates the presence of TRH activity in hypothalamus (Peter & Mekeown, 1975).

In teleosts, the TSH cells have direct innervation by neurosecretory endings, which are adjacent to the cells having no synaptic contact or the endings and may be separated from the TSH cells by a basement membrane. In *Tilapia mossambica* (Bern *et al.*, 1974) and *Carassius auratus* (Kaul and Vollarth, 1974) the TSH cells have direct contact with endings containing elementary neurosecretory granules, and with endings containing vesicles having dense granules (Breton & Weil, 1973).

(ii) **Gonadotropin** : Gonadotropin (GTH) cells are richly found in the proximal pars distalis (PPD), where they may form a solid ventral rim of cells. Such situation is found in Cyprinoide. In salmonids and eel they are spread throughout rostral pars distalis (RPD) and PPD. Gonadotrops are basophilic cell types and are PAS and AB positive. These cells have irregular and more or less dilated cisternae of granulated endoplasmic reticulum (GER) containing granules with varying electron density.

The gonadotroph (GTH) is under the control of gonadotropin releasing hormone. In many teleosts, unlike mammals, neurosecretory stimuli may pass along the nerve fibres piercing the laminae that separates the neuro from adenohypophysis, and penetrating into the endocrine parenchyma of pars distalis (Ball, 1981). There are two types of nerve fibres designated as A and B types. The fibres of A type remain in contact with hormone producing cells, including gonadotrops and even terminate with synapse on these cells. B type fibres form synaptic contact with a large granular vesicle of 60-100 nm diameter, while the A synapse have granules of 100-200 nm diameter (Knowles *et al.*, 1967; Fridberg and Ekengren, 1977).

The gonadotropin (GTH) releasing hormone (GnRH) of teleost is similar to luteinizing hormone releasing hormone (LH-RH) is localized in ventral lateral nucleus preopticus periventricularis (NPP) and posterior lateral nucleus lateral tuberis (NLT) as well as other areas. In hypothalamus, localization of immuno reactive fibre tracts from cells in the NPP and NLP to the pituitary gland suggests that these areas are the origin of endogenous releasing hormone.

Studies of Peter and Crim (1978) on *Carassius auratus* indicate that the nucleus lateralis tuberis (NLT) pars posterior and the NLT pars anterior which are situated in the pituitary stalk, are actively take part in regulation of GTH secretion for gonadal recrudescence (Fig. 19.6). In several fishes GTH secretion is associated with ovulation. In *Carassius auratus*, GTH level becomes higher on the day of ovulation. However, in sockeye salmon, *Oncorhynchus nerka*, high level of GTH found during spawning.

In fishes there is only one functional gonadotropin is found, which is often regarded as piscian pituitary gonadotropin (PPG). This single gonadotropin has similar properties of two hormones. LH and FSH of mammals. Mammalian luteinizing hormone (LH) promotes release of gametes from nearly mature gonads in fishes and stimulates appearance of secondary sexual characters. This indicates that there must be a similar hormone in fishes also. Salmon pituitary secretes gonadotropins which resembles LH. Furthermore, the gonadotropins from human chorion and urine of gravid mares, have LH like properties which hasten the release of eggs in female fishes.

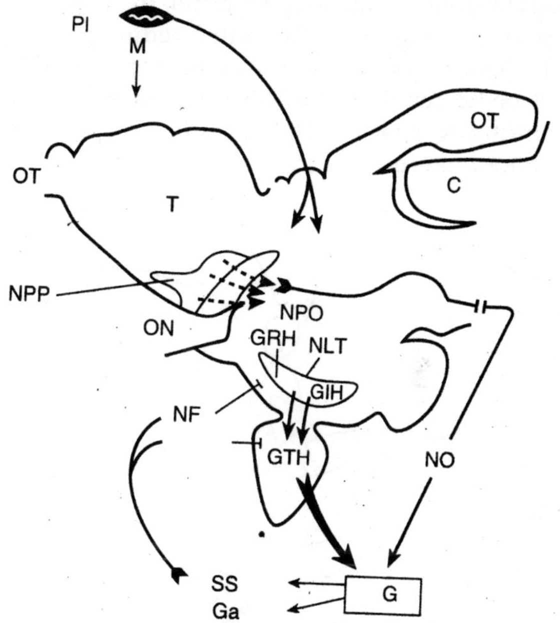

Fig. 19.6 : Diagrammatic representation of the neuroendocrine regulation of gonadotropin (GTH) secretion by releasing hormone (GRH) and one inhibitory hormone (GIH) in gold fish brain. C, cerebellum; G; gonad; Ga, gametes; M, melatonin; NF, negative feedback; NLT, nucleus lateralis tuberis; NPP, nucleus preopticus periventricularis, NPO, nucleus preopticus; NO, neural output; ON, optic nerve; OT, olfactory tract; PI, photoperiod input; SS, sex steroid; T, telencephalon. (Source; Peer and Fryer, 1983).

The presence of follicle stimulating hormone in fishes (FSH), which is the second gonadotropin hormone found in mammalian pituitary gland, is still not confirmed.

Recently, prostaglandin, which is hormone-like substance has been isolated from testis and semen of bluefin tuna, *Thynnus thynnus* and flounder (*Paralichthys olivaceus*).

(iii) **Adrenocorticotropic hormone (ACTH) :** It is secreted by ACTH cells located between the rostral pars distalis and the neurohypophysis. Secretion of ACTH from pituitary is stimulated by the hypothalamus through corticotrophin releasing factor (CRF) (Fig. 19.7). Hypothalamic and telencephalic extracts of *Carassius auratus* and longnose suckers, *Catostomus catostomus* stimulated secretion of ACTH in *Carassius auratus* in vivo. The nature of this telencephalic hypothalamic CRF is unknown. However, it shows similarity with mammalian CRF.

In *Carassius auratus*, ACTH cells are innervated by aminergic like type B fibres, which originate from nucleus lateralis tuberis (NLT).

In teleosts, neurohypophysial peptides may regulate the ACTH secretion. Implantation of cortisol pellets in *Carassius auratus* shows that corticosteroids exert negative feedback effects on the brain to suppress ACTH secretion. Cortisol added to the medium inhibits the activity of ACTH cells and release of ACTH which also suggest direct negative feedback effect of cortisol on the ACTH cells.

(iv) **Prolactin :** It is a similar hormone that influences lactation in mammals and is released from proadenohypophysis. In some fishes like mummichog *(Fundulus heteroclitus)*, prolactin along with the intermedin enhances the laying down of melanin in the melanophores of the skin. Among the several hormones the prolactin is also involved in electrolytic regulation in teleosts but its importance in maintaining homeostasis varies according to species. The secretion of prolactin from the teleost pituitary is under an inhibitory neuroendocrine control of hypothalamic origin (Nagahama *et al.*, 1975).

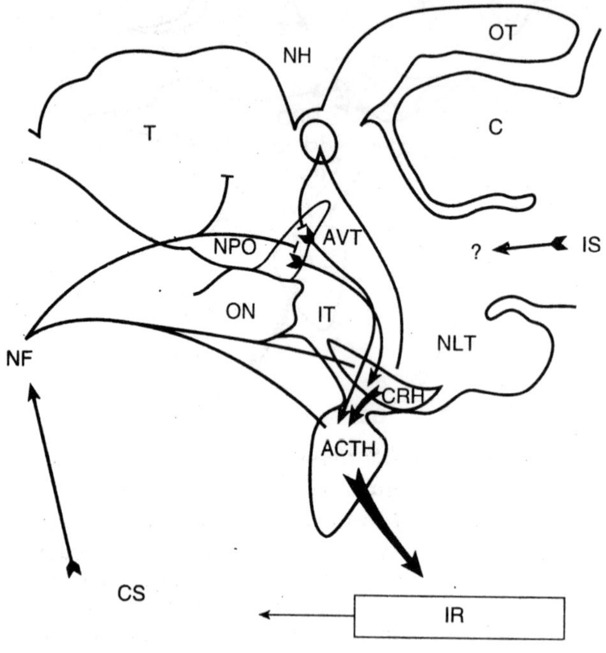

Fig. 19.7 : Diagrammatic illustration of neuroendocrine regulation of (ACTH) by a releasing hormone (RH) and (NPO)→ arrow indicates stimulatory influences; Question (?) indicates unknown pathways. AVT, arginine vasotocin; IT, isotonin; IR, inter renal; IS, input of stress; C, cerebellum; CS, cortico steroid; NF, negative feed back; NH, nucleus habenularis, NLT, nucleus lateralis tuberis; NPO, nucleus preopticus; ON, optic nerve; T, telencephalon.

(v) **Growth hormone (GH) :** Mesoadenohypophysis secretes a growth hormone which accelerates increase in the body length of fishes. Very little is known concerning its control, mode of action on cell division and protein synthesis in teleosts. It has been reported that teleost GH cells are capable of some spontaneous activity and continue to synthesize and secrete GH in vitro (*Anguilla anguilla, Salmo gairdneri;* Baker and Ingleton, 1973).

It is evident that the GH secretion may be influenced by osmotic pressure, as release of GH from cultured *Salmo gairdneri* and *Anguilla anguilla* pituitaries greater in a medium containing low sodium than in a high sodium medium, relative to plasma sodium levels. However, GH release from *Poecilia latipinna* pituitary has no effect of osmotic pressure.

Recently a zone of the hypothalamus has been recognized, which is believed to be responsible for control of GH in *Carassius auratus*. In this fish nucleus anterior tuberis (NAT) and sometimes nucleus lateralis tuberis (NLT) forms an area which stimulates GH secretion and is perhaps the origin of a growth hormone releasing hormone (GRH).

The hypothalamic control of GH secretion is revealed by ultrastructural studies on the pituitary. In teleosts GH cells of pars distalis have direct synaptoid contact with type B endings as in *Carassius auratus*, which have direct contact without the synaptoid appearance in *Tilapia mossambica*. Very few species like *Oryzias latipes* contain synaptoid contact of type A endings on GH cells (Davis and Northcutt, 1983); in other teleosts, type A fibre may have direct contact with GH cells, but generally the endings are separated from the cells by a basement membrane. Thus it is clear that on neuroendocrine factor reaching the GH cells and probably the GH cells are regulated by a dual hormone.

(vi) **Melanocyte Stimulating Hormone (MSH) or Intermedin :** MSH is secreted from the meta-adenohypophysis and acts antagonistically to melanin hormone (MAH). MSH expands the pigment in the chromatophores, thus takes part in adjustment of background. It also stimulates the melanin synthesis. Pars intermedia of teleost pituitary comprise two kinds of secretory cells, which can be identified by their staining properties. One cell type is PAS^{+ve} periodic acid Schiff positive and PbH^{-ve} (lead hematoxylin negative). However second cell type is PbH^{+ve} and PAS^{-ve} (Holmes and Ball, 1974). Salmonoid seems to have only PbH^{+ve} cells. These cells are source of melanocyte stimulating hormone (MSH) which stimulates melanin despersion in the melanocytes and darkening of skin. Neurointermediate lobe of *Salmo gairdneri* appears to have a melanin concentrating factor. Several authors have demonstrated the occurrence of MSH and/or its precursor ACTH in PbH cells of several species of teleosts by immuno fluorescence techniques (*Tinca tinca;* Romaine *et al.*, 1974; *Anguilla anguilla*, Olivereau *et al.*, 1976). Thus these observations confirm earlier correlations of body colour or background adaptation with activity of the PbH cells.

In teleosts, the neuroendocrine control of pars intermedia varies according to species. In fishes like *Cymatogastes aggregate, Anguilla anguilla* and *Salmo gairdneri*, neurosecretory axons do not enter the pars intermedia but terminate in extravascular channels bordering the pars intermedia or terminate at the basement membrane. However, other teleosts such as *Carassius auratus*, and *Gillichthys mirabilis* have direct innervation from neurosecretory axons.

Secretion of MSH in teleost may be supressed by catecholaminergic mechanism. Treatment of 6 OHDA to destroy catecholaminergic nerve terminals also causes activation of the MSH cells in *Gillichthys mirabilis*, darkening of skin or activation of MSH cells in *Anguilla* anguilla, catecholamine inhibits the release of MSH directly, when the former is MIH. Also the catecholamine may affect MSH secretion indirectly by promoting the release of an MIH, or by inhibiting the secretion of an MRH nerve terminals within the pars intermedia.

Several histological investigations demonstrate a stimulation of the pars intermedia associated with reproduction. During spawning period pars intermedia of clupea becomes intensely active. In *Carassius auratus* the number of PbH^{+ve} cells increases after spawning in both number and activity during oogenesis and breeding season.

(vii) **Oxytocoin and vasopressin hormones :** In fishes the neurohypophysis secretes two hormones, i.e., oxytocin and vasopressin, which are stored in hypothalmic neurosecretory cells. These endocrine substances have well known effect on mammalian metabolism. Vasopressin and antidiuretic (ADH) hormones are responsible for the constriction of blood vessels in mammals and thus stimulates retention of water by their action in kidney. Oxytocin stimulates mammalian uterine muscles and increases the discharge of milk from lactating mammals. The fish pituitary hormones are capable to produce such effects in higher vertebrates but presumably the target organs are specific site of their action in fishes and probably is different from those of higher vertebrates. In fishes they control osmoregulation by maintaining water and salt balance.

Use of pituitary hormones in induced breeding : The pituitary hormones have practical applications by injecting and implanting to force or stimlulate spawning of certain fishes of great economic value, such as trouts (Salmoninae), catfishes (Ictaluridae), Mullets (Mugliliade), and sturgeons (Acipenseridae). The synthesis of sex hormones in the gonad is controlled by pituitary gonadotropin. Hence pituitary extract containing GTH are taken from sexually mature male or female fish than injected to the same species for inducing and hastening and spawning. For the preparation of pituitary extract closely related species may also be used as donor.

THYROID GLAND

Location : In many teleosts the thyroid gland is situated in the pharyngeal region in between the dorsal basibranchial cartilages and ventral sternohyoid muscle. The thyroid surrounds anterior and middle parts of first, second and sometimes third afferent branchial arteries of ventral aorta, as found in *Ophiocephalus* species (Fig. 19.8). In *Heteropneustes* it occupies almost the entire length of the ventral aorta and afferent arteries. In *Clarias batrachus* the thyroid gland is concentrated around the ventral aorta, middle ends of two pairs of afferent arteries and the paired inferior jugular veins.

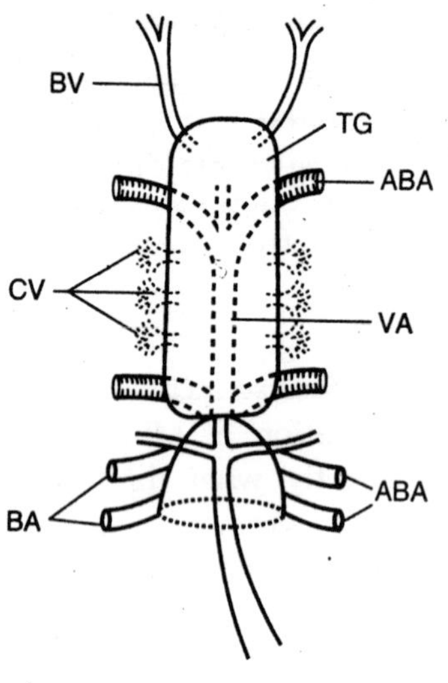

Fig. 19.8 : Ventral view of thyroid gland and their blood vessels in *Heteropneustes*. ABA, afferent branchial artery; BA, branchial vessel; BV, buccal vein; CV, commissural vessel; TG, thyroid gland; VA, ventral aorta.

Shape and size : In majority of teleosts the thyroid is unencapsulated and thin follicles are dispersed or arranged in clusters around the base of afferent branchial arteries. It is thin-walled, sac-like, compact dark brownish and enclosed in a thin-walled capsule of connective tissue in these fishes might be correlated with the air breathing habit because thyroid gland acts here as thermoregulator to adapt the fish to a semiterrestrial environment of low thermal capacity. In *Heteropneustes* the thyroid gland is an unpaired thin-walled brownish but cylindrical in shape. In *Clarias batrachus* the thyroid gland is not covered by definite wall, i.e., uncapsulated and is elongated in shape.

Histology of the thyroid gland : In teleosts, histologically the thyroid gland consists of a large number of follicles, lymph sinuses, venules and connective tissues. The follicles are round, oval and irregular in shape. Each follicle contains a central cavity surrounded by a wall composed of single layer of epithelial cells. The structure of epithelium varies according to its secretory activity. Less active follicles generally have thin epithelium. Epithelial cells are of two types.

(i) Chief cells which are columnar or cuboidal in shape, having oval nuclei and clear cytoplasm.

(ii) Colloid cells or Benstay's cells. They possess droplets of secretory material. The follicles are supported in position by connective tissue fibres, which surrounds them. The central lumen of follicle is filled with colloid containing chromophilic and chromophobic vacuoles.

Blood supply : The thyroid gland is highly vascularized and is generally well supplied with blood. A single buccal vein and two pairs of commissural vessels supply blood to thyroid gland. From its posterior end a pair of veins arise, which merge immediately to form the posterior *inferior* jugular vein sending blood to the heart. In *Ophiocephalus,* the thyroid gland also receives blood from same vessels. The buccal vein collects blood from the buccal region and after running for a short distance beneath the anterior end of pairs of the thyroid gland opens into it. The two commissural blood vessels are highly branched and bring blood from the floor of pharynx. One pair opens at the anterior end while the other at the middle of the gland on either side. In *Heteropneustes* the commissural vessels are more than two pairs.

Hormones of Thyroid Gland

Thyroid hormone is synthesized in the thyroid gland, for which inorganic iodine is extracted from the blood. These inorganic iodine combines with tyrosine. The thyroid hormones of fishes appear to be identical with those of mammals, including, mono- and di-iodotyrosine and thyroxin. These hormones are kept stored in the thyroid follicles and are released into blood stream on metabolic demands. The release of the thyroid hormone from the follicle is controlled by the thyrotropic hormone (TSH) of pituitary which in turn is influenced by genetically determined maturation process along with certain factors like temperature, photoperiod and salinity. The thyroid glands in sharks and higher teleosts are diffused in nature. Therefore, it is difficult to remove or inactivate. In spite of the certain deficiency, studies have been made by physiological blocking or radiothyroidectomy using. In teleosts, there is no respiratory stimulation by thyroxin, which is best known in mammals. Physiologically used small quantity of thyroxin and tri-iodothyroxine result in thickening of the epidermis and fading of goldfish *(Carassius auratus).* Induced thyroid hyperactivity accelerates the transformation into juvenile smolt stage in salmon but high thyroid the titer retards growth of the larva in the same genus. Induced thyroid hyperactivity in mud skipper *(Periophthelamus)* shows morphological and metabolical changes in response to the more terrestrial existence of fish living mostly outside the water. Thyroid gland of salmons and sticklebacks is known to influence osmoregulation. In salmon the thyroid gland becomes hyperactive during their spawning migration. It has been considered that thyroid influences the growth and nitrogen metabolism in goldfish, as indicated by high ammonia excreted by them. Thus the action of thyroid is conjugated with other vital processes including growth and maturation and also the diadromous migration of fishes.

Adrenal Cortical Tissue or Interrenal Tissue

Location : In Lamprey (Cyclostomata) the endocrine interrenal cells are present throughout the body cavity close to the postcardinal vein. Among the rays they lie in more or less close association with posterior kidney tissue, including some species possessing interrenal tissue concentrated near the left and in other near the right central border of that organ. In sharks (Squaliformes) they are present between the kidneys. In teleosts the interrenal cells are multilayered and situated along the postcardinal veins as they enter the head kidney (Fig. 19.9).

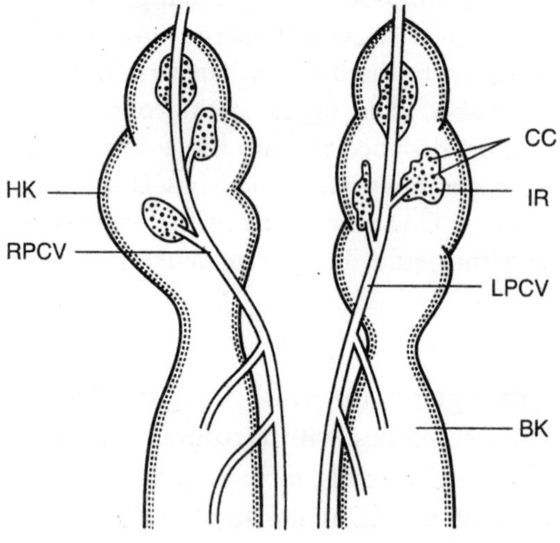

Fig. 19.9 : Digram to show location of interrenal glands in fishes. BK, body kidney, CC, chromaffin cells; HK, head kidney; IR, interrenal tissue; LPCV, left post cardinal view; RPCV, right post cardinal vein. (Source : Habia, T., 1982).

Anatomy : In some fishes like *Puntius ticto* interrenal cells are arranged in form of thick glandular mass while in others like *Channa punctatus* they are present in form of lobules. Each interrenal cells is eosinophilic and columnar with a round nucleus.

Adrenal Cortical Hormone

Adrenal cortical tissue or interrenal tissue secretes two hormones. These are (i) mineral corticoids concerned with fish osmoregulation, (ii) glucocorticoids, which regulate the carbohydrate metabolism, particularly blood sugar level. *Salmo gairdneri* treated with mineral corticoid excretes higher than normal amount of sodium ions through its gills but conserve more than normal amount of sodium in the kidneys and osmoregulation in the body.

Intramuscular injection of corticosteroid compounds to the oyster toadfish causes increase in blood sugar level thus showing control on carbohydrate metabolism.

The cortisone level of blood plasma of salmons rises during spawning period and declines during the more sedantary stages. During the spawning phases 60% of total body protein is catabolised in *Oncorhynchus*, which is correlated with the six fold increase in plasma corticosteriods and rises in liver glycogen. Like higher vertebrates administration of adrenal cortical hormones stimulates lymphocyte release in *Astyanax* and antibody release in European perch. Corticosteroids structurally similar fo androgens and produce androgens side effects.

Secretion of adrenocortical hormones is under control of the adrenocorticotropic hormone (ACTH) of hypophysis.

Chromaffin Tissue or Suprarenal Bodies or Medullary Tissue

In lamprey (Cyclostomata) the chromaffin cells are present in the form of strands along the dorsal aorta as in the ventricle and the portal vein heart. In sharks and rays (Elasmobranchii) these tissues are found associated with the sympathetic chain of nerve ganglia while in bony fishes (Actinopterygii) the chromaffin cells have wide variation in their distribution. They are elasmobranch like, distributed as in flounders *(Pleuronectus)*. On the other hand they have true adrenal arrangement as in sculpins *(Cottus)* where chromaffin and adrenal cortical tissue are joined into one organ, similar to the mammalian adrenal gland (Figs. 19.10a, b, c, d).

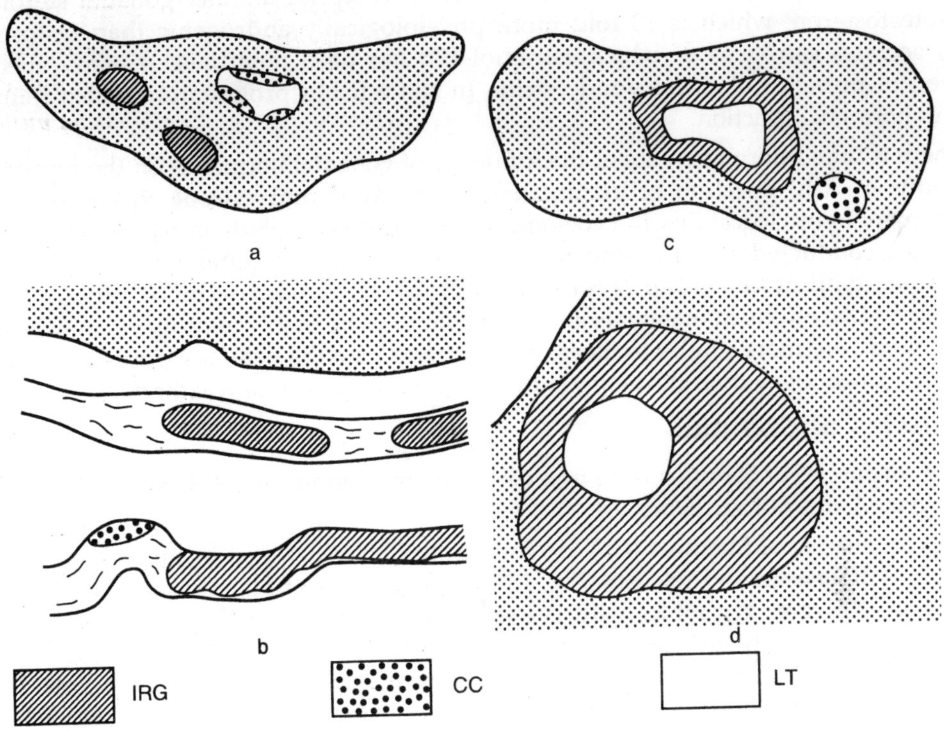

Fig. 19.10a-d : Diagram showing chromaffin tissue in fishes (a) Rainbow trout. (b) Eel. (c) Carp. (d) Yellow tail. IRG, interrenal gland; CC, chromaffin cell; LT, lymphoid tissue. (Source : Hibiya, T., 1982)

Chromaffin tissue of fishes richly contain adrenaline and noradrenaline. Injection of adrenaline and noradrenaline causes changes in blood pressure, bradycardia, branchial vasodilation, diuresis in glomerular teleosts and hyperventilation.

The Ultimobranchial Gland

Typically the gland is small and paired and is situated in the transverse septum between the abdominal cavity and sinus venosus just ventral to the oesophagus or near the thyroid gland (Fig. 19.1). Embryonically the gland develops from pharyngeal epithelium near the fifth gill arch. In *Heteropneustes*, the gland measures 0.4 × 1.5 mm in diameter in average adult of 130 to 150 mm body length (Belsare, 1974, b). Histologically, it consists of parenchyma, which is solid and composed of cell cords and clumps of polygonal cells covered by capillary network. The gland secretes the hormone calcitonin which regulates calcium metabolism. Calcitonin is said to be related with the osmoregulation. Eel calcitonin causes decrease serum osmolarity, sodium and choloride in Japanese eels. The ultimobranchial gland is under the control of pituitary gland.

The Sex Glands as Endocrine Organs

The sex hormones are synthesized and secreted by specialized cells of the ovaries and testis. The release of sex hormones are under the control of mesoadenohypophysis of pituitary. In fishes these sex hormones are necessary for maturation of gametes and in addition secondary sex characteristics such as breeding tubercles, colouration and the maturation of gonopodia. In elasmobranch (Raja) and in salmon the blood plasma contains male hormone testosterone with a correlation between plasma level and the reproductive cycle. *Oryzias latipes* (medaka) and sockey salmon comprise another gonadal steroid, i.e., 11-ketotestosteron, which is 10 fold more physiologically androgenic than testosterone. Ovary secretes estrogens of which estradiol-17B has been identified in many species in addition of presence of estrone and estriol. In some fishes progesteron is also found but without hormonal function.

There is little information about the influence of gonadial hormones on the reproductive behaviour of fish. Injection of mammalian testosterone and estrone to lamprey causes development of its cloacal lips and coelomic pores, which contribute in reproductive process. Such tests conducted for rays and sharks (Elasmobranchii) could not give any results whereas ethynil testosterone (pregnenolone) which produces mild androgenic and progesteron like effects in mammals and birds, found to be highly androgenic in fishes. Male sex hormones are more similar to those of vertebrates than the ovarian hormones, the former strongly influences ovarian development in a loach, the Japanese weatherfish.

Corpuscles of Stannius

The corpuscles of Stannius were first described by Stannius in 1939 as discrete gland like bodies in the kidney of sturgeon.

The corpuscles of Stannius (CS) are found attached or lodged in the kidneys of fishes particularly holostean and teleosts (Fig. 19.11). Corpuscles of Stannius are asymmetrically distributed and often resembles with cysts of parasites but lie different from the latter by higher vascular supply and dull white or pink colour. Histologically, they are similar to the adrenal cortical cells. Their number varies from two to six according to species. The CS. may be flat, oval as in goldfish, trout, salmon. It is made up of columnar cells which are covered by a fibrous capsule. The columnar cells are of two types (i) AF-positive and (ii) AF negative. They are filled with secretory granules. The parenchyma of CS comprises vasculo-ganglionic units consisting of a bunch of ganglion cells, blood vessels and nerve fibres.

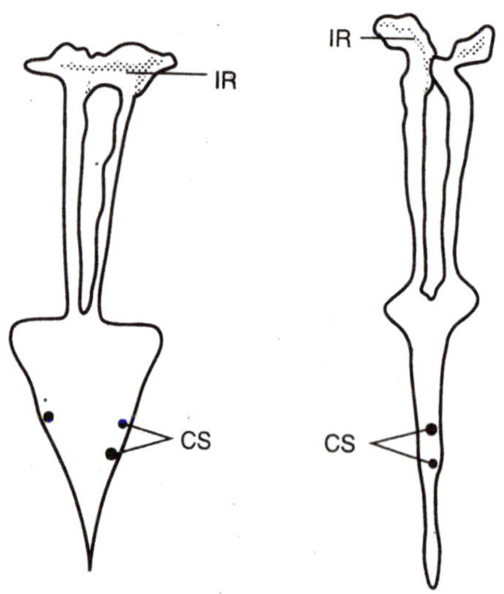

Fig. 19.11a, b : Diagram of kidney of fishes showing corpuscles of Stannius. (a) *Cirrhina mrigala*. (b) *Labeo rohita*. CS, corpuscles of Stannius; IR, interrenal corpuscles.

The number and position of CS vary in different species. Their may be single CS as in *Heteropneustes setani* and *Notopterus notopterus* (Fig. 19.11a, b), while as many as ten CS are present in some species like *Clarias batrachus*. In other species their number varies from one to four. According to Garrett (1942), there is a gradual reduction in number of CS that has occurred during evolution of Holostei and Teleostei.

In salmonids the CS is located near the middle part of mesonephros but in majority of fishes they are situated at the posterior end of kindney. Garrett (1942) pointed out that CS moves progressively backward during the course of evolution as a result of body cavity rather than a migration of CS. In *Notopterus notopterus* the CS are present in anterior end of kidney perhaps because most of the body cavity is occupied by the air bladder and also other organs are compactly arranged in limited space. The presence of the CS at the extreme anterior end of the kidney in *Heteropneustes setani* is probably because this species has wide body space and long archinephric duct. Therefore, the variation in number and position of CS in teleost species seems to be an embryological speciality.

There is only one-cell type present in the CS of pink salmon. However, two-cell types are found in pacific salmon.

The corpuscles of Stannius reduces serum level in the *Fundulus heteroclitus,* which have environment containing high calcium, such as sea water. Recently, it has been shown that corpuscles of Stannius work in association with pituitary gland, which exerts hypercalcemic effect, in order to balance relatively constant level of serum calcium.

Intestinal Mucosa

The intestinal mucosa produces secretin and pancreozymin, which are controlled by nervous system and regulate pancreatic secretion. Secretin affects flow of enzyme carrying liquids from the pancreas, whereas pancreozymin accelerates flow of zymogens. These hormones are usually synthesized in anterior part of the small intestine. In carnivorous fish these hormones are brought into the stomach, containing acidified homogenate of fish flesh or by injection of secretin into gastric vein which stimulates the secretion of pancreas.

Islets of Langerhans

In some fishes like *Labeo, Cirrhina,* and *Channa* small islets are present which are separate from pancreas and are found near gall bladder, spleen, pyloric caeca or intestine.

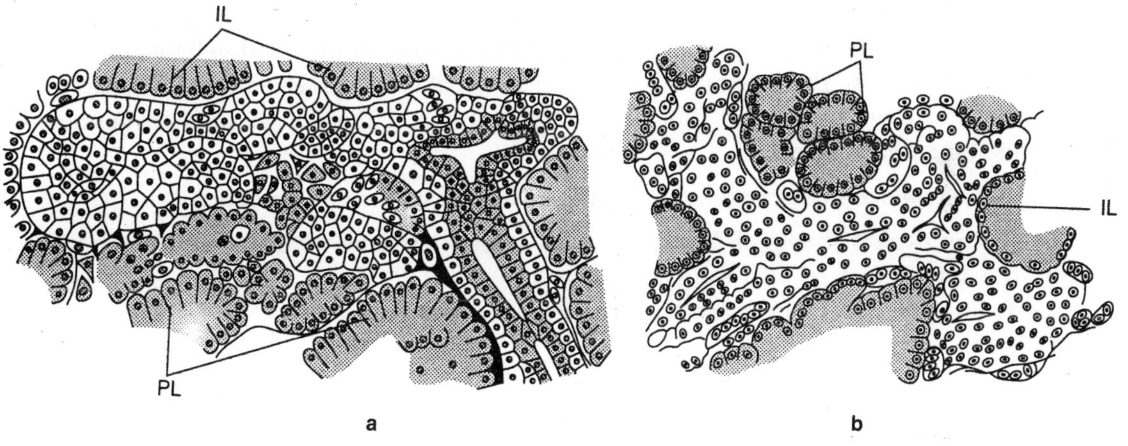

Fig. 19.12a, b : Diagram of pancreas showing endocrine components. (a) *Torpedo marmorata.* (b) *Mustelus laevis.* IL, islets of langerhans; PL, pancreatic lobule.

Such islets are often referred to as principal islets. But in some species like *Clarias batrachus* and *Heteropneustes fossils* the number of large and small islets are found to be embedded in the pancreatic tissues, similar to the higher vertebrates. In fish the islets are big and prominent and consist of three kinds of cells (Fig. 19.12a, b). (i) The beta cells which secrete insulin and take aldehyde fuschin stain. (ii) Another type of cells are alpha cells, which do not take aldehyde fuschin stain and have two types, A_1 and A_2 cells, which produce glucagon. The function of the third type of cells is not known. Insulin is secreted by beta cells and regulates the blood sugar level in fishes.

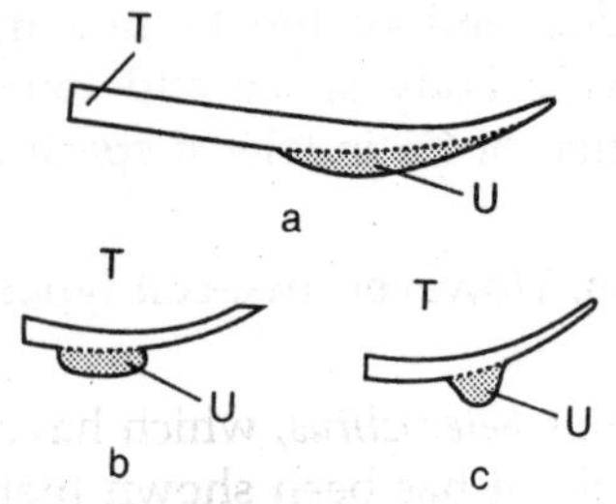

Fig. 19.13a-c : Diagram of urohypophysis of teleosts. (a) rainbow trout. (b) carp. (c) Yellow tail. T, tail; U, urohypophysis (Source : Hibiye, T., 1982)

Pineal Organ

It is situated near the pituitary (Fig. 19.1). Inspite of being a photoreceptor organ the pineal organ shows endocrine nature of doubtful function. Removal of pineal from *Lebistes* species causes reduced growth rate, anomalies in the skeleton, pituitary, thyroid and corpuscles of Stannius. It has been reported that thyroid and pituitary glands influence the secretion of pineal.

Urophysis

Urophysis is a small oval body, present in the terminal partofspinal cord (Fig. 19.1, Fig. 19.13a, b, c). It is an organ deposits, which releases materials produced in the neurosecretory cells situated in the spinal cord. These cells together with the urophysis are called the caudal neurosecretory system. This neurosecretory system is found only in elasmobranchs and teleosts but it corrosponds to the hypothalamo neurosecretory system present in vertebrates.

In caudal neurosecretory system, neurosecretory cells are diffused in terminal part of spinal cord. Axon terminals of these cells assemble at the ventral side of the region and form urophysis with blood capillaries.

The neurosecretory cells is a large nerve cell and has basophilic cytoplasm and a polymorphic nucleus.

In Ayu (*P. altivelis*) the urophysis is extended like a bow. In carp and yellow tail it is a conspicuous oval body. The urophysis is made up of spinal cord elements like neurosecretory axon, glia, and ependymal and glia fibers and meningeal derivative such as vascular reticulum and reticular fibres.

The caudal neurosecretory system is said to be related with osmoregulation.

Urophysis extract shows ability to contract smooth muscles of ovary and oviduct of guppy (*Poecilia reticulata)* and the sperm duct of goby (*Gillichthys mirabilis*), suggesting the possibility of involvement in reproduction and spawnning.

REFERENCES

Baker, B. I. and Ingleton P. M. (1973). Factors affecting the release and synthesis of teleost pituitary polypeptide hormones. J. Endocrinol. 59 × 1 (abstr.).

Ball, N.J. (1981). Hypothalamic control of the pars distalis in fishes, amphibians and reptiles. Gen. Comp. Endocrinol. 44, 135-170.

Ball, N.J. and Baker, B.I. (1969). The pituitary gland : Anatomy and histophysiology. In W.S. Hoar and D.J. Randall (eds), Fish physiology, Vol. II, p.p. 1-11, Academic Press, New York.

Ball, N.J., Baker, B.I. Olivereau, M. and Peter, R.E. (1972). Investigation on hypothalamic control of adenohypophysial functions in teleost fishes. Gen. Comp. Endocrinol, Supll. 3 : 11-21.

Belsare, D.K. (1959). Thyroid gland in some Indian air breathing fishes. Proc. first India Zool. Congr. 5. 75-80.

Belsare, D.K. (1965). Vascular supply of the pituitary in *Channa punctatus* Bloch. Nature. (Lond.) 206-211.

Belsare, D.K. (1974,a). Morphological types of the preoptico-neurohypophysial neurosecretory pathways in some teleosts. Z. Mikrosk. Anat. Forsch, Leipzig 88.6S 1029-1035.

Belsare, D.K. (1974,b). Histomorphology of the ultimobranchial gland in *Heteropneustes fossilis* Bloch. 'Mikroskopie' Bd. 305. 31-33.

Bern, H.A., D. Zambrano, and Y. Nagahama (1974). The relationship between nerve fibres and adenohypophysial cell types in the cichlid teleost *Tilapia mossambica*. Rech. Biol. Contempt. 179-94.

Breton, B. and C. Weil (1973). Effects due LH/FSH-RH sythetique e.d' extraits hypothalamiques de carpe surla secretion d' hormone gonadotrope in vivo chez lacarpe (*Cyprinus carpio. L.)* C.R. Hebd. Seances Acad. Sci. Ser. D : Sci Nat. 277 : 2061-64.

Davis, R.E. and R.G. Northcutt (1983) : Fish neurobiology, Vol. 2. The university of Michigan Press. Ann. Arbor.

Follenius, E. (1963). Etude comparative de la cytologic fine du noyau preoptique (NPO) et du noyau lateral du tuber (NLT) chez la truite *(Salmo irideus Gibb)* et chez la perche *(Perca fluivilitis)* Comparison des deux types de neurosecretion. Gen. Comp. Endocrinol 3 : 66-85.

Fridberg, G. and Ekengren, B. (1977). The vascularization and the neuroendocrine pathways of the pituitary gland in the Atlantic Salmon, *Salmon salar*. Can. J. Zool. 55 : 1284-96.

Garrett, F.D. (1942) : The development and phylogeny of the corpuscles of Stannius in ganoid and teleostean fishes. J. Morph. 70, 41-67.

Gorbman, A. (1965) : Endocrine terminologic usage. Gen. Comp. Endocrinol 5. 129-130.

Grau, E.G. and Stenson, M.H. (1977). Pituitary autotransplants in *Fundulus heteroclitus* : Effect on thyroid function. Gen. Comp. Endocrinol. 32 : 427-31.

Henderson. N.E. (1969). Structual similarities between the neurohypophysis of brook trout and tetrapods. Gen. Comp. Endocrinol. 12 : 148-53.

Hibiya, T. (1982). An Atlas of Fish Histology, Normal & Pathological features. Gustav Fischer verlag.

Holmes, R.L. and Ball, N.J. (1974). The pituitary gland—a comparative account. Cambridge Univ. Press, London.

Kaul, S. and Vollrath, L. (1974). The goldfish pituitary. II. Innervation. Cell Tissue Res. 154 : 231-249.

Knowles, F. and Vollrath, L. (1966). Neurosecretory innervation of the pituitary of the cells *Anguilla* and Conger. I. The structure and ultrastructure of neuro-intermediate lobe under normal and experimental conditions. Philos. Trans. R. 50c, Lond. B : Biol. Sci. 250 : 311-27.

Knowles, F. Vollrath, L., and Nishioka R.S. (1967). Dual neurosecretory innervation of the adenohypophysis of *Hippocampus*, the sea horse. Nature (Lond) 214-309.

Nagahama, Y.R. Nishioka, S. Bem, H.A. and Gunther, R.L.; (1975). Control of prolactin secretion in teleosts, with special reference to *Gillichthys mirabilis* and *Tilapia mossambica*. Gen. Comp. Endocrinol. 25 : 166-88.

Nazar Ahsan, (1966). Effect of gonadotropin hormones or male hypophysectomised lake chule, *Collesius plumbeus*. Canadian J. of Zoology 44 : 703-717.

Olivereau, M. Bugnon, C. and Fellmann D. (1976). Localisation cytoimmunologique de MSH et d' ACTH dens les cellules hypophysaires colorables avec I' hematoxyline an plomb (HPb) Chezl' anguille. C.R. Hebd. Seances, Acad. Sci. Ser. D. Sci. Nat. 283 : 1321-23.

Peter, R.E. (1970). Hypothalamic control or thyroid gland activity and gonadal activity in the goldfish, *Carassius auratus*. Gen. Comp. Endocrinol. 14 : 334-56.

Peter, R.E. (1973). Neuroendocrinology of teleosts. Ann. Zool. 13 : 743-55.

Peter, R.E. and J.N. Fryer (1983). Endocrine functions of hypothalamus of Actinopterygians. In fish neurobiology Vol. 2. Higher brain areas and functions, Ed. R.E. Davis and R.G. Northcutt. The Univ. of Michigan press. Ann arbor.

Peter, R.E. and L.W. Crim. (1978). Hypothalamic lesions of goldfish : Effects on gonadal recrudescence and gonadotropin secretion. An. Biol. Anim. Biochem. Biphys. 18 : 819-23.

Peter, R.E. and R. Billard (1976). Effects of third ventricle injection of prostaglandins on gonadotropin secretion in goldfish. *Carassius auratus*. Gen. Comp. Endocrinol. 30 : 451-56.

Peter. R.E. and Mckeown, B.A. (1975). Hypothalamic control of prolactin and thyrotropin secretion in teleosts, with special reference to recent studies on the goldfish. Gen. Comp. Endocrinol. 25 : 153-65.

Pickford, G.E. and ATZ, J.W. (1957). The physiology of the pituitary gland of fishes. New York Zoological Society. New York.

Prasad Rao. P.D. (1972). Identification of gonadotropins and thyrotropin secreting cells, the hypophysis of the catfish, *Heteropneustes fossilis* (Bloch). Z. Mikrosk. Anat. Forsch. 86 : 21-38.

Romaine, P.; Bugnon, C. & Fellmann, D (1974). Etude cytoimmunologique des cellules HPb Positive diadenohypophysis de la tanch (Tinca tinca L.) C.R. Seances SOC. Biol. 168 : 1245-1348.

Ross, M.H, and Reith E.J. (1985). Histology—a Text and Atlas. Harper and Row, Publishers, J.B. Lippincott Company New York.

Sathyanesan, A.G. (1972). Hypothalamo-hypophysial vascularization in a teleost fish with special reference to its tetrapodan features. Anat. 81 : 349-66.

Sathyanesan, A.G. and Haider, S. (1971, a). Presence of hypothalamo-hypophysial portal system in the teleost *Heteropneustes fossilis* (BI) Experientia (Basel) 27 : 462.

Sathyanesan, A.G. and S. Haider (1971, b). Tetrapod like features of the hypothalamo hypophysial vascularization in the air breathing teleost *Heteropneustes fossilis* (BI) Gen. Comp. Endocrinol. 17 : 360-70.

Schreibman, M.P. (1964). Studies on the pituitary gland of *Xiphophorus maculatus* (the platyfish) Zoolgica, New York, Vol. 49. (3) 217-243.

Vanoordt, P.G. W.J. & Peute, J. (1983). The cellular origin of pituitary gonadotropins in teleosts. In : Fish physiology (Eds. Hoar, W.S. Randall, D. J. & Donaldson, E.M. Vol. IX. Academic Press, London. pp. 137-186.

Chapter 20

REPRODUCTIVE SYSTEM

The reproduction in fishes is bisexual, hermaphrodite or parthenogenic. The most predominating process is bisexual reproduction. In such fishes the sexes are separate, e.g., the individuals are dioecious. In a few fishes, both sexes are present in same individual, i.e., such fishes are hermaphrodite. The examples are *Perca, Stizostedion, Micropterus.* In some fishes juvenile hermaphroditism has been noticed. In *Poecilia formosa* parthenogenesis occurs, actually the processes is gynogenesis, *i.e.*, the development of young without fertilization.

The gonads of elasmobranchs developed by the migration of primordial germ cells from the extraembryonic endoderm which originate from peritoneal epithelium and interrenal analoge (Chieffi, 1967). In teleosts also the primordial germ cells originate from the genital ridge (as swelling). The primordial germ cell is quite large and slightly oval in shape with a large spheroidal nucleus containing one large nucleolus. The genital ridge develops into cortex and medulla in vertebrates. But gonads in teleosts and cyclostomes consist only cortex

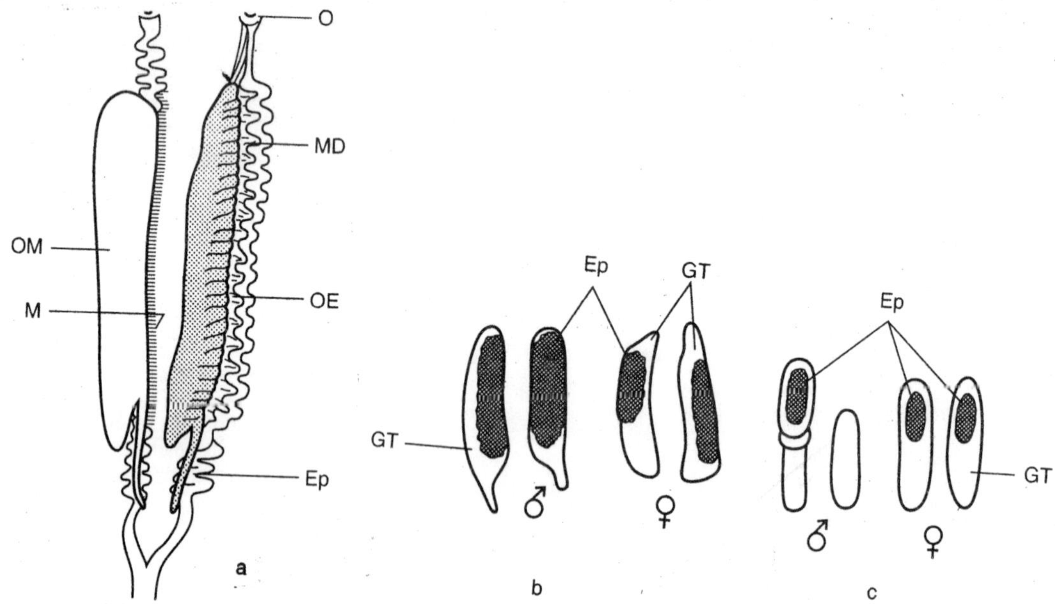

Fig. 20.1a-c : Reproductive organs of lungfish. (a) Sagittal section of *Protopterus.* (b) and (c) Position of epigonal organ in gonadal tissue. Ep, epigonal tissue; GT, gonadal tissue; M, Mesovarium; MD, Mullerian duct; O, ostium; OE, ovary with eggs; OM, ovarian membrane.

and lack medullary tissue. In elasmobranch it develops into medulla, but in females they develop into the cortex (Chieffi, 1959). Later the primordial germ cells underago meiosis and proliferate and ultimately differentiate into male and female gonads.

The fishes have one pair of bilateral gonad. They are generally symmetrical. They are suspended from the dorsal portion of the body cavity by mesenteries across the roof of the body cavity in close association with kidneys. The mesentery in female is called mesovarium. They are also present in *Neoceratodus* (Fig. 20.1a, b, c). While in male it is known as mesorchium. It is richly supported with blood vessels and nerve fibres.

HAEMOPOIETIC EPIGONAL ORGAN

In all elasmobranchs gonads are associated with haemopoietic epigonal organ. The testis like ovaries are also associated with epigonal organs (Fig. 20.1b, c). They extend back to cloaca, it may be relatively small in *Heptanchus* or absent in *Squalus*. In *Scyllium*, the testis are fused with epigonal organs. It is paired in *Cetorhinus*, the one of the right side lying posterior and slightly dorsal to the ovary and suspended by a posterior extension of mesovarium. It measures about 60 cm in length and has diameter of 20 cm. The anterior region is intimately fused with the posterior one third of the ovary. The left epigonal organ in female is similar in size and shape although not associated with a gonad. In female sharks of many genera only one ovary develops (*Scyliorhinus, Pristiurus, Galeus, Mustelus*) and, therefore, in accordance the female eipgonal organ is often unpaired, so they are better developed on one side than the other (Fange, 1984).

FEMALE REPRODUCTIVE ORGAN

The female reproductive organs comprise the ovaries, oviducts and in some fishes pseudocopulatory papilla (Fig. 20.2 and Fig. 20.3). The ovary in relation with oviduct and transport of ova is distinguished into cytovarian type, semicytovarian type and gymnovarian type. In cytovarian type the lumen of the ovary is connected with oviduct and released in water through genital pore.

In semicytovarian type, the oocytes in place of oviduct pass through a funnel shaped transparent groove, which opens into genital pore. Such condition is present in Notopteridae, Osteoglassidae, etc, where oviduct degenerate partially or completely and, therefore, ova are shed into the coelomic cavity and then carried through pores or funnel. Gymnovarian type of ovary is not continuous with the oviduct. The ovary hangs down like curtain and the oocytes are ovulated into the coelomic or peritoneal cavity and carried to oviduct by cilia.

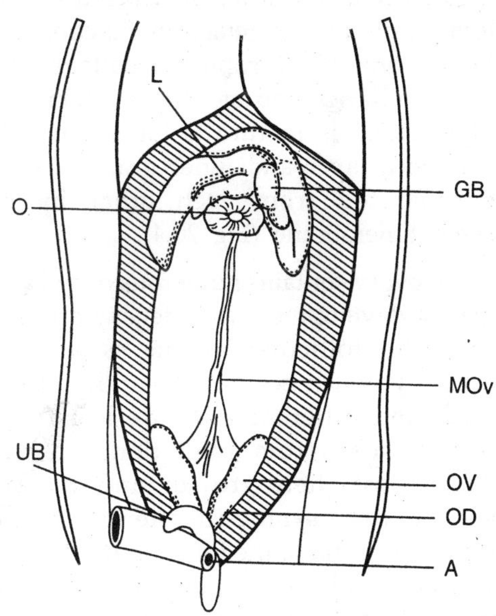

Fig. 20.2 : Female reproductive organs of *Wallago attu*. A, anus; GB, gall bladder, L, liver, MOV, mesovarium; O, oesophagus; OD, oviduct; OV, ovary, UB, urinary bladder.

Normally ovary are sac like and paired and symmetrical structure both in teleosts and elasmobranchs and are functional. However, in some elasmobranchs, the right ovary become becomes functional (Mathews, 1950) while in some cases the left ovary only develops (Babel, 1967).

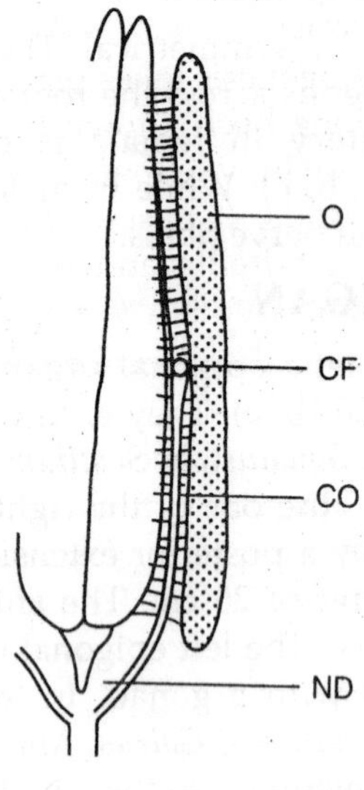

Fig. 20.3 : Reproductives system of female *Polypterus*. CF, coelomic funnel; CO, coelomic oviduct; ND, nephric duct extended; O, ovary.

Histology : The ovary contains supporting tissue known as stroma, which contains oogonia and oocytes. The oogonium (germ cell) after multiplication develops into primary oocyte. The oogonium cell in the early stages of development is a large cell with a large nucleus having prominent nucleolus with chromophobic cytoplasm. The size of the cell becomes smaller after multiplication, but is not important. The important difference is in the chromosomes. The chromosome becomes thread like in the nucleus (leptotene stage). The chromosomes undergo zygotene stage followed by pachytene and ultimately diplotene stage. At this stage of development to oocyte is surrounded by follicle cells. The follicle cells are multilayer in elasmobranch and amniota but single layer in teleosts.

In mammals, the follicular cells give rise to stratified epithelium, the stratum granulosa surrounding the oocyte and ultimately proliferate to form granular cells. The stromal cells immediately surrounding the follicle, form a sheath of cells, the theca follicle which is distinguished into theca externa and theca interna. In elasmobranch, the theca interna lies immediately below the externa and consists of 4 to 6 rows of flattened and elongated cells, the walls of which are extended outward into proceses (Dodd, 1983). Theca externa cells are secretary in function and contain organelles such as ER and mitochondria. It contains fibroblasts, collagen fibres and blood capillaries and steroid producing cells. The ooctye which is mature is having two layers, an outer theca layer and inner granulosa layer, both are separated from each other by a distinct basement membrane (Fig. 20.4).

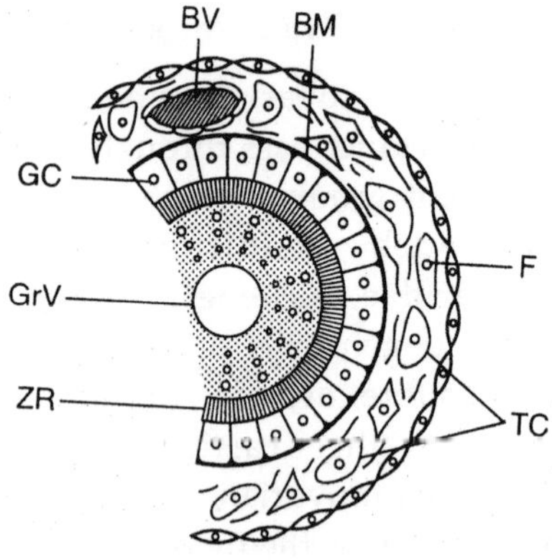

Fig. 20.4 : Diagrammatic illustration of the follicle layer surrounding an early vitellogenic oocyte of salmonid fish. BM, basement membrane; BV, blood vessel; F, follicle GC, granulosa cell; GrV, germinal vesicle; TC, theca cell; ZR, zona radiata.

The ovary contains numerous oocytes in different stages of development and degeneration (Saidapur, 1978). The follicular atresia is a degenerative process by which oocytes in various stages of their growth and differentiation are lost from the fish ovary as also from ovulation (Guraya, 1993). Follicles lost their growth and ditferentiation. They do not synthesize steriod hormones and carry out any endocrine function (Guraya, 1993). The dengerating follicles are of two types, the corpora lutea and corpora atretica. In corpora lutea, the enclosed oocytes are removed from follicles by the invading granulosa cells and ingrowing theca cells. Corpora lutea and corpora atretica are present in holocephalon ovary (Hissaw and

Hissaw, 1959). In oviparous and viviparous species they are usually present in late summer. They probably represent yolky oocytes which for some reason have failed to be ovulated. In *Squalus acanthias* corpora atretica is absent in the ovaries.

Vitellogenesis : Egg in the beginning is without yolk but vitellogenesis takes place later on (atresia). In oviparous and ovoviviparous vertebrates it has long been known that the blood of mature female contains a calcium-binding lipophosphoprotein, vitellogenin, which is synthesized in the liver under stimulation of female sex steriods. It passes from the plasma into oocytes in which it gives rise to the yolk protein lipovitellin and phosvitin (Wallace, 1978). Vitellogenesis in the dogfish is similar to that found in other vertebrates.

Craik (1978) measured the rate at which vitellogenin is synthesized and converted into yolk granules.

The vitellogenesis takes place, in which accumulation of yolk substance occurs. The conspicuous change noticeable is that the cytoplasm which was basophilic, now becomes acidophilic. There are three essential yolk substances—the yolk vesicles, yolk globules and oil droplets. The yolk vesicle is provided with glycoprotein they stain slightly red with eosin but are PAS positive. They take deep red stain. The yolk vesicles later become the cortical alveoli and take part in the formation of perivitelline space.

The yolk globules are made up of mainly lipoprotein with some carbohydrates and other substances. They are weakly positive to PAS. The old droplets contain generally glycerides and small amount of cholesterol.

With the completion of vitellogenesis movement of germinal vesicle, fusion of yolk granules and grouping of oil droplets occur.

After the movement of germinal vesicle to an animal pole, the first meiotic division occurs and the first polar body is released. Subsequently the second meiotic division starts and an egg in which division arrested at the metaphase is ovulated. Chieffi (1962) believes that true corpora lutea (i.e., in the mammalian functional sense) are found in both oviparous and viviparous elasmobranch. In the former (*S. stellaris*) they are believed to develop from ovulated follicles (corpora lutea) and in the latter (*T. marmorata*) by follicular atresia (corpora atretica).

Steroids are present in the ovary of elasmobranch. 17 beta estradiol and traces of progesterone and estrogen were present in the ovary of *Squalus*. It is demonstrated in *Squalus acanthias* that progesterone can be synthesized from (HC) pregnenolone. The enzymes required for steroidogenesis in the ovary of *Squalus acanthias* such as glucose-6 phosphate dehydrogenase (G-6PDH), 3 beta-hydroxysteroid dehydrogenase (3 beta HSD), 3 alpha-hydroxysteroid dehydrogenase (17 beta HSD) and 20 beta hydroxysteroid dehydrogenase (20 beta HSD) are present in the ovary.

3 beta HSD is found in corpora lutea but not in corpora atretica and 3a-HSD was weakly present in the granulosa of late corpora atretica (Dodd, 1983, Chieffi and lupo di Prisco, 1983).

MALE REPRODUCTIVE ORGANS

Testes Morphology : The testes in fishes, generally, are paired structures situated on either side somewhat lateral to and below the kidneys. The testis has two major functions, the production of spermatozoa (spermatogenesis) and another function is the production of steriods (Steroidogenesis). The size of testes in the breeding season, become enormous. The testis may be elongated, sac-like or lobulated, may be rounded or triangular in cross sections. In mature gobies, *Acanthogobius fluviatilis* the testis is small and thread-like. In *Notopterus* and *Amphipnous* testis is unpaired. In Syngnathids the testis is a simple tube. The testes occupy different position in the abdominal cavity. They are suspended from the

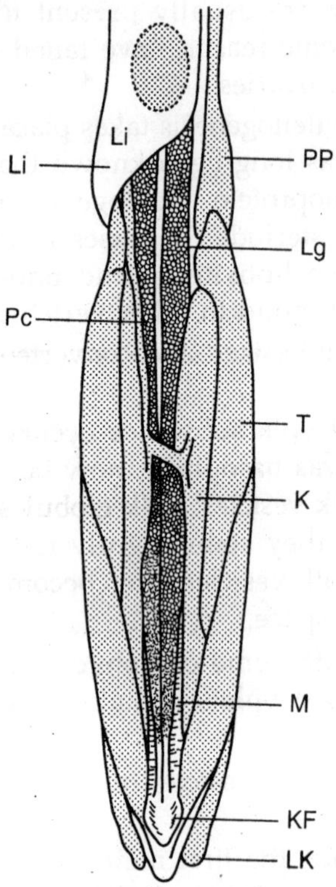

Fig. 20.5 : Urinogenital system of a lungfish, *Protopterus* with gut removed. K, kidney, KF, fused area of kidney,; Lg, lung; Li, liver, LK, lymphoid part of kidney; M, mesentery; Pc, postcaval; PP, progonial part; T, testis.

dorsal body-wall by mesentery known as mesorchia, which is vascularised and contain nerve fibres. In dipnoi fishes e.g. in *Neoceratodus* and *Protopterus* the right testis is attached anteriorly to the tip of the liver, while the left gonad extends forward to the region of the ductus Cuvieri (Fig. 20.5). The right is usually larger to the left. The testis of *Protopterus* extends the entire length of the body as a thick band closely and broadly bound to the kidney. The testis of *Lepidosiren* is an elongated mass and embedded in fatty tissue.

The testes of teleosts are long and rounded in section. In *Latimaria* the right testis is two to three times larger than the left but both produce sperms. The right testis extends through the middle of quarter of the body cavity ventrolateral to the swim bladder. In Chondrichthyes the position of testis varies from far anterior back to about the middle. In some sharks the testis is associated with an epigonal organ which is lymphoid in nature.

From each testis a sperm duct or vas deferens originates. In some fishes mesonephric ducts unite with testis to form vas deferens and vasa efferentia (epidydimous). In some fishes the sperm duct is shared with kidney and often called nephric duct (Wolffian duct) (Fig. 20.6a, b).

In lungfishes many efferent ducts extend from the central canal into the kidney and connected with the capsules of the nephric tubules. In teleosts the sperm duct or vas deferens is a modified nephric duct and may be from one or several ureters carrying the excretory fluid to the cloaca (Jollie, 1962) (Fig. 20.6c-d).

The sperm duct : The sperm duct opens into another chamber known as seminal vesicle. The seminal vesicle is thickened and often has more diameter than the sperm duct. The vas deferens and seminal vesicle are clearly visible in the sharks. There sperms are stored for short period of time. It is unanimously accepted that the seminal vesicle is absent in teleosts, because it does not store sperms and are not the part of the Wolffian duct system, hence it is not homologous to the seminal vesicle of higher vertebrates (Nagahama, 1983). However, in goldfish, a median wall of the sperm duct forms inconspicuous protrusion where the sperms are stored (Takahashi and Takano, 1972). This structure is considered as homologous to the seminal vesicle, although it is not glandular in nature. A seminal vesicle as that of cartilaginous fishes are noticed in Chondrostei, Holostei and Dipnoi. The sperm duct from each testis often join to form a common duct and opens outside through genital pore. In *Squalus*, the nephric ducts join at urinary sinus, which also serves as the ureters and opens into cloaca. From the cloaca the spermatophore (bundle of sperms) are forced by the action of siphons, muscular sacs underlying the skin and extending forward from the cloaca, through the grooves of the claspers into the cloaca of the female. In some sharks a rudiment of the Mullerian duct enters. In Muraenids the urinary sinus opens into urinary bladder, in salmon and perch it opens into the excretory sinus. In *Lota,* it opens into the anal margin. It opens into the jointed anal and excretory opening in, synganitids (i.e. into a cloaca).

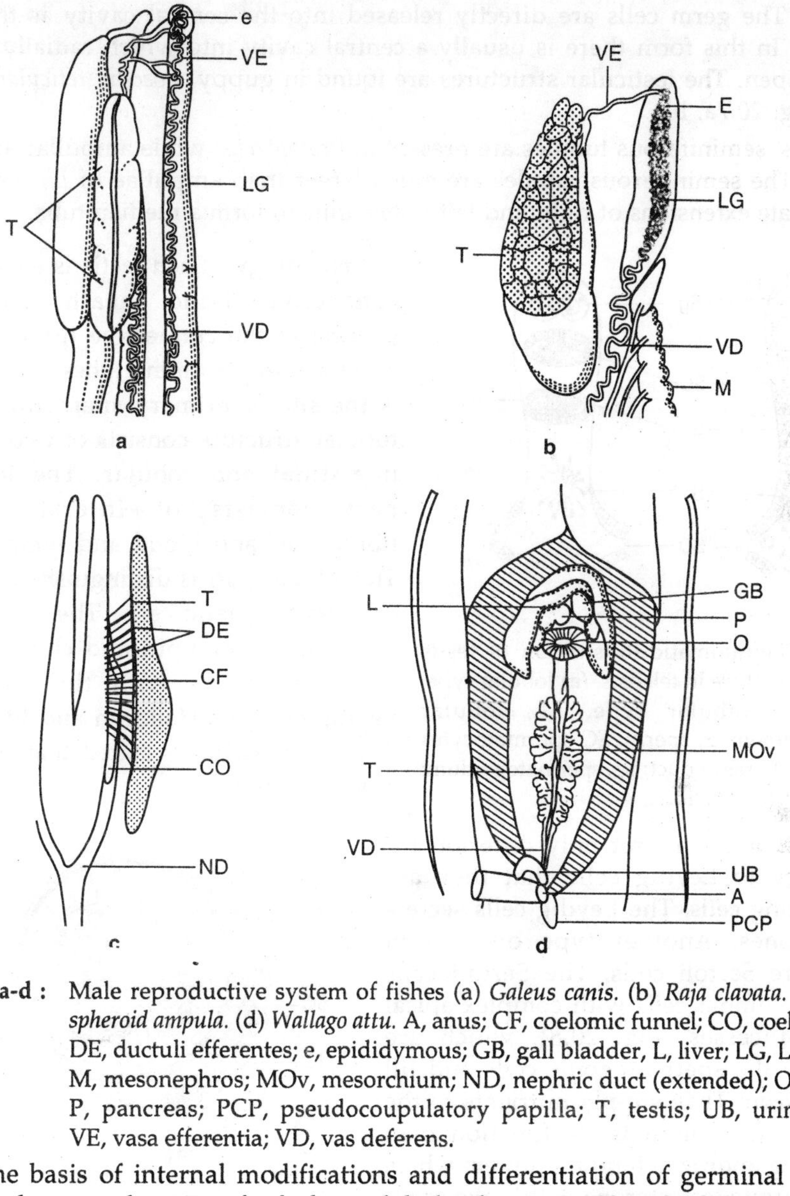

Fig. 20.6a-d : Male reproductive system of fishes (a) *Galeus canis.* (b) *Raja clavata.* (c) *Acipenser, spheroid ampula.* (d) *Wallago attu.* A, anus; CF, coelomic funnel; CO, coelomic oviduct; DE, ductuli efferentes; e, epididymous; GB, gall bladder, L, liver; LG, Leydig's gland; M, mesonephros; MOv, mesorchium; ND, nephric duct (extended); O, oesophagus; P, pancreas; PCP, pseudocoupulatory papilla; T, testis; UB, urinary bladder; VE, vasa efferentia; VD, vas deferens.

On the basis of internal modifications and differentiation of germinal tissue or their release in the central cavity of tubular or lobular forms. In Chondrichthyes, the wall of the testis forms spheroid ampullae or follicles and does not acquire tube-like structure. During developmental stages, the formation of ampullae starts from ventral free margin, where new follicles are germinating, and progresses towards the dorsal attachment. At the time of breeding, the ripe ampullae are connected through narrow ductuli recti (Dodd, 1983) with the central canal which lies along the attached margin of the gonad. The ampullae are provided with germ cells and they undergo spermatogenesis inside the ampullae and release sperms into the central cavity through narrow ducts. The ampullae originate in ampullogenic zone.

In teleosts, both ampullae and tubular forms are observed. The structure of ampullae is similar to that of elasmobranchs. The tube-like structure are from external tunica propria to the central cavity. The germ cells or spermatogonia are located in these tube like arrangements. The germ cells are directly released into the central cavity as there is no lobular lumen. In this form there is usually a central cavity into which radially arranged short tubules open. The testicular structures are found in guppy, *Poecilia reticulata* (Billard *et al.*, 1982) (Fig. 20.7a, b).

In dipnoans, seminiferous tubules are present in *Protopterus* while ampullae are present in *Lepidosiren*. The seminiferous tubules are much larger than ampullae of *Lipidosiren*. The canalicular or rate extensions of right and left testes unite to form a median tube.

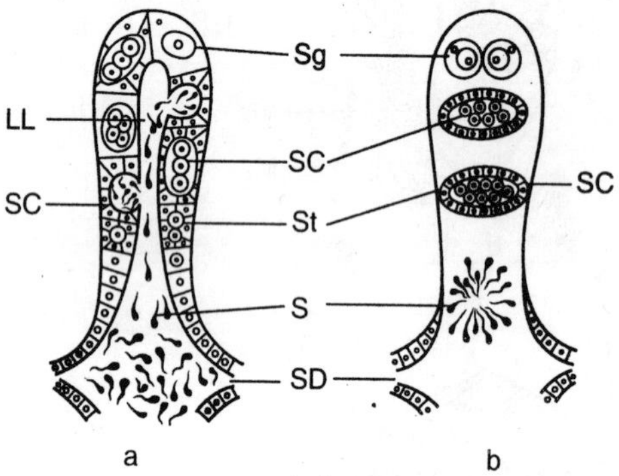

Fig. 20.7a, b : Diagrammatic illustration of testis structure in teleosts. (a) lobular type. (b) tubular type. LL, lobular lumen; S, sperm; SC spermatocytes SD, sperm duct; Sg, spermatogonium; St, spermatid.

Histology : Each testis is enclosed in a connective tissue sheath. The tunica propria of the connective tissue projected into the lumen forming tubes. The blind end is the site of primary spermatocytes. The tubular structure consists of two parts, the interstitial and lobular. The interstitial part consists of interstitial cells, fibroblasts, and blood and lymph vessels. The lobular part is distinguished into germ cells and somatic cells. The somatic cells are positive for lipid and cholesterol and are homologous with the mammalian Leydig cells (O' Halloran and Idler, 1970). A few investigators stated that cholesterol positive lipid.

The nature of cell is not only criterion for their homology to Leydig cells that they are steroid producing cells. The Leydig cells secrete steroid hormones. Another type of cells in interstitium are Sertoli cells. The Sertoli cells are basically columnar cells with complex apical and lateral processes (Fig. 20.8), which are surrounded by the spermatogonic cells and fill the space between. They supply nutrients to the germinal cells. In mammals its function may include secreting androgen binding protein which serve to concentrate testosterone within the seminiferous epithelium and proximal part of the genital duct system. Its secretion helps in regulation of mitosis, meiosis and steroidogenic functions of the Leydig cells and the release of gonadotropins from pituitary.

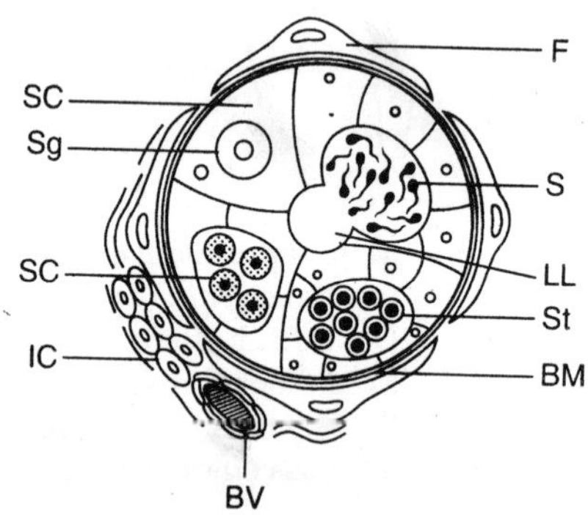

Fig. 20.8 : Diagram of tubular type testis showing a close relationship between germ cells at various developmental stages. BM, basement membrane; BV, blood vessel; F, fibroblast; IC, interstitial cell; LL, lobular lumen; S, sperm; SC., spermatocyte; Sg, spermatogonium; St, spermatid.

Grier and Linton 1977) found sertoli homolgous cells as they take Sudan black stain. Regarding the terminology whether these cells

cells should be called Sertoli cells, cyst cells or Sertoli homologous cells, there is a unanimity that they may be termed as Sertoli cells in teleosts (Nagahama, 1983).

SPERMATOGENESIS

Spermatogenesis is the developmental process through which the spermatogonic cells are transformed from undifferentiated diploid (2n) spermatogonia into highly specialized haploid (n) spermatozoa. There are generally three principle phases :

(i) The spermatogonial phase or spermatocytogenesis.

(ii) The meiotic phase or meiosis.

(iii) Spermatid phase or spermiogenesis.

The spermatogonium in the early stages in fishes is a large oval cell with one large round nucleolus. It is distinguished into primary spermatocyte then to secondary spermatocyte by maturation (or meiosis) division and then ultimately develops into spermatid. The spermatid then develops into a spermatozoan.

REFERENCES

Babel, T. S. (1967). Reproduction, life history, and ecology of the round sting ray, *Urolophushalleri* cooper, Calif, Fish, Bull. 137 : 1-104.

Billard, R., Fostier, A., Weil, C. and Breton, B. (1982). Endocrine control of spermatogenesis in teleost fish. Can. J. fish. Aquat. Sci. 39 : 65-79.

Chieffi, G. (1959). Sex differentiation and experimental sex reversal in elasmobranch fishes. Arch. Anat. Microsc. Morphol. Exp. 48 : (Suppl.) 21-36.

Chieffi, G. (1962). Endocrine aspects of reproduction in elasmobranch fishes. Gen. Comp. Endocrinol. Suppl. 1 : 275-285.

Chieffi, G. (1967). The reproductive system of elasmobranchs : Development and endocrinological aspects. In "Sharks, scates and Rays" (Eds. Gilbert, P.W., Mathewson, R.F. and Rall, D.P.), Johns Hopkins Press, Baltimore, Mary land. pp. 553-580.

Chieffi, G. And Lupodi Prisco, (1963). Identification of sex hormones in the ovarian extract of *Torpedo marmorata* and *Bufo vulgaris*. Gen. Comp. Endocrinol. 3 : 149-152.

Craik, J. C. A. (1978). Kinetic studies of vitellogenin metabolism in the elasmobranch *Scylorhinus canicula* L. Comp. Biochem. Physiol. A. 61A : 355-361.

Dodd, (1983). Reproduction in cartilagenous fishes (chondrichthyes). In Fish physiology (Eds. Hoar.; W. S.; Randall, D. J. & Donaldson, E. M.) Vol. IX A. Academic Pres, N. Y. pp. 31-95.

Fange, R. (1984). Lymphomyeloid tissue in fishes. Vidensk, Meddr Dansk. Foren. 145 : 143-162.

Grier, H. J. and Linton J. R. (1977). Ultrastructural identification of the Sertoli cell in the testis of the northern pike, *Esox lucius* Am. J. Anat. 149 : 283-288.

Guraya, S. S. (1993). Follicular (or oocyte) atresia and its causes and functional significance in fish ovary. In 'Advances in Fish Research' (Ed. B. R. Singh). Vol. 1 Narendra Publishing House., Delhi. pp. 313-332.

Hissaw, F. L. Jr. and Hissaw, F. L. (1959). The corpora lutea of elasmobranch fishes. Anat. Rec. 135; 269-278.

Jollie, M. (1962). Chordata Morphology. Reinhold Publishing Corporation, New York.

Mathews, L. H. (1950). Reproduction in the basking shark, *Cetorhinus maximum* (Gunner). Philos. Trans. R. Soc. London. Ser. B. 234 : 247-316.

Nagahama, Y. (1983). The functional morphology of teleosts gonads. In : 'Fish physiology'. (Eds. Hoar, W. S.; Randall D. S. and Donaldson, E. M.) Vol. IX A. 223-276.

O'Halloran, M. J. and Idler, D. R. (1970). Identification and distribution of the Leydig cell homology in the testis of sexually mature Atlantic salmon (*Salmo salar*) Gen. Comp. Endocrinol. 15 : 361-364.

Saidapur, S. K. (1978). Follicular atresia in the ovaries of non-mammalian vertebrates. Int. Rev. Cytol. 54 : 225-244.

Takahashi, H. And Takano, K. (1972). Morphogenesis of accessory reproductive organs in male goldfish, *Carassius auratus.* Bull. Fac. Fish Hokkaido Univ. 23 : 53-64.

Wallace, R. A. (1978). Oocyte growth in non-mammalian vertebrates. In : "The ovary" (Ed. Jones., R. E.) Plenum, New York. pp. 469-502.

Chapter 21

DEVELOPMENT OF FISH

The embryonic development starts with the penetration of sperm in the egg. The process is called as impregnation. The sperm enters the egg through micropyle. In some fishes, the micropyle is funnel-shaped. As soon as sperm penetrates, there occurs a cortical reaction which prevents further entry of sperms. Even in those cases where polyspermy occurs, only one sperm fuses with the egg nucleus. After the completion of the cortical reaction, the vitelline membrane is known as fertilization membrane (Gloria *et al.*, 2003).

The fertilization in teleost is external, taking place in water outside the body. So in these eggs, water hardening takes place, when the water uptake completes, the egg is turgid.

The gametes of fishes have the fertilization capability even after leaving the body. The ability could be artificially maintained by using modern techniques of cryopreservation i.e., by deep freezing at – 196°C. The preservation of gametes will help in breeding fish and also for improving stock for market and for selective purposes. Amongst elasmobranchs, 12 families of Squaliformes are entirely viviparous, 2 are oviparous and 2 are mixed (Wourms, 1977; Budker, 1971). The dogfish, *Squalus canicula,* is oviparous species, which lay shelled egg. *Latimeria chalumnalis* the only living representative of lobe finned fishes, Coelocanthine, gravid female contains advance young. Each has a large yolk sac with no apparent connection to the surrounding oviduct wall (Smith *et al.*, 1979).

During the fertilization the pronuclei of sperm and ovum unite together with the fusion of cytoplasm. At this stage the egg contains yolk in the centre and cytoplasm occupy the periphery. The quantity of cytoplasm is slightly more where the egg nuclear material is present. The cytoplasm is completely separated from the yolk in *Cyprinus carpio. Ophiocephalus punctatus* and *Gasterosteus aculeatus.* The vitelline membrane is double layered.

Fish spermatozoa is divisible into head, middle piece and tail (Fig. 21.1a-g). The teleost spermatozoa lack a head cap, the acrosome. The holostean spermatozoa are also devoid of acrosome. The fishes in which head is present in sperm, the heads are often ovoid and the middle piece is small. The tail is relatively long and contains microtubules and forms cytoskeletal framework. The microtubules have 9 + 2 arrangement (Fig. 21.2). However, a few investigators in Anguilliformes and Elopiformes, have reported that the central microtubule shows 9 + 0 pattern. In viviparous fishes, head and middle piece are elongated (Gardiner, 1978).

The middle piece contains a substantial number of mitochondria as found in *Poecilia reticulata* (Stoss and Donaldson, 1982). Biflagellate cells are found in *Porichthys notatus,* and *Ectalurus punctatus* and *Poecilia retulata.* The mobility of the sperm is limited to a period of second to minute in freshwater spawners because of lysis (Billard, 1978). Motility is considerably lònger in saltwater spawner. It is 15 minutes in Atlantic cod (Devenport *et al.*, 1981) or several days in Herring (Yanagimachi, 1957). There is little doubt that

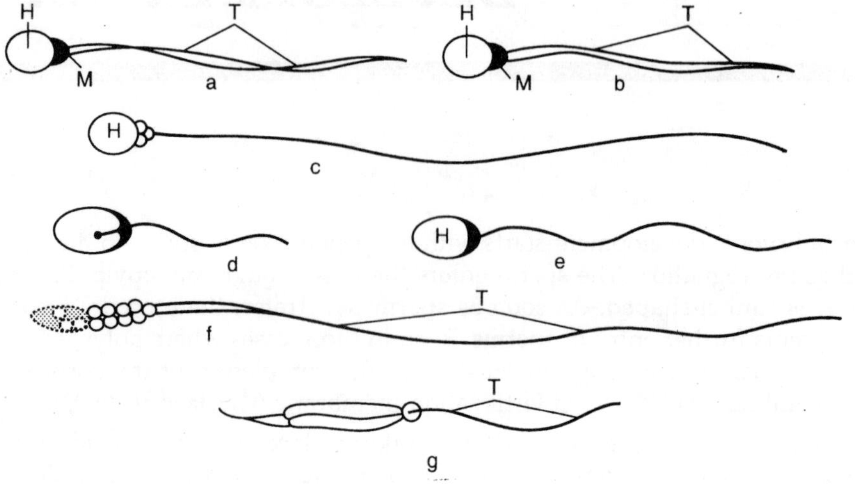

Fig. 21.1a-g : Structure of spermatozoa of bony fishes. (a) *Esox leucius.* (b) *Pirca fluviatilis.* (c) *Clupea harengus.* (d) *Salmo gairdneri.* (e) *Salmo trutta.* (f) *Lebistie reticulatus.* (g) Anguilla anguilla. H, head; M, middle piece, T, tail.

K^+ ions from seminal plasma blocks motility and the spermatozoan motility can be enhanced in proper physiological solutions by controlling pH and dilution of K^+ and proper fish Ringers solutions. This technique is used in the preservation of the spermatozoa (Cryopreservation).

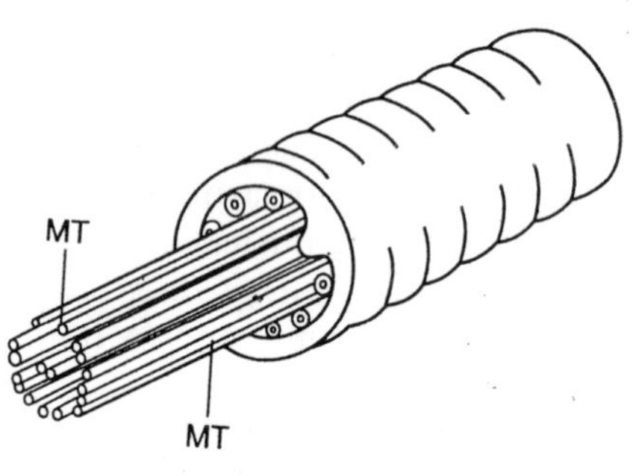

Fig. 21.2 : The diagrammatic sketch showing 9 + 2 arrangement of microtubule in tail of spermatozoa. MT, microtubule.

Structure of Ovum : The egg is surrounded by tough layer called chorion, next to chorion is the plasma or vitelline membrane or pellicle. This layer surrounds the yolk and cytoplasm (ooplasm).

The yolk is present in considerable quantity in lungfish, *Neoceratodus and Lepidosiren*. The quantity of yolk is more in cartilaginous fishes such as *Acipenser*. Egg size and yolk contents are independent variables. The chorion of teleost fish arises entirely from oocytes and made up of proteins and polysaccharide.

The unfertilized teleost eggs are generally opaque and heavier than water. According to Swarup (1958), the eggs of newly caught female are light orange but if the female stickleback have been previously kept in aquarium and fed on tubifex, they are colourless. The eggs of *Cyprinus carpio* are yelllow (Verma, 1970, 1971). The eggs of some culturable variety of fishes are classified as non-floating and floating. The non-floating eggs are further divisible as non-adhesive and adhesive. The eggs of *Catla catla* are light red, *Cirrhinus mrigala* are brownish and *Labeo rohita* are reddish, but the eggs of *Labeo calbasu* are bluish. These eggs are non-floating and non-adhesive. The eggs of *Clarias batrachus* and *Heteropneustes fossilis* are adhesive and non-filamentous and are green in colour. The eggs of *Notopterus notoptorus* and *Notopterus chitala* are yellowish. The floating eggs are of *Channa punctatus* and *C. striatus*. Their colour are amber (Srivastava, 1964 & Jhingran, 1988).

FERTILIZED EGGS

The fertilized eggs become gradually more and more transparent and the vitelline membrane separates from egg proper and develops a space known as perivitelline space, which is filled with fluid (Fig. 21.3a).

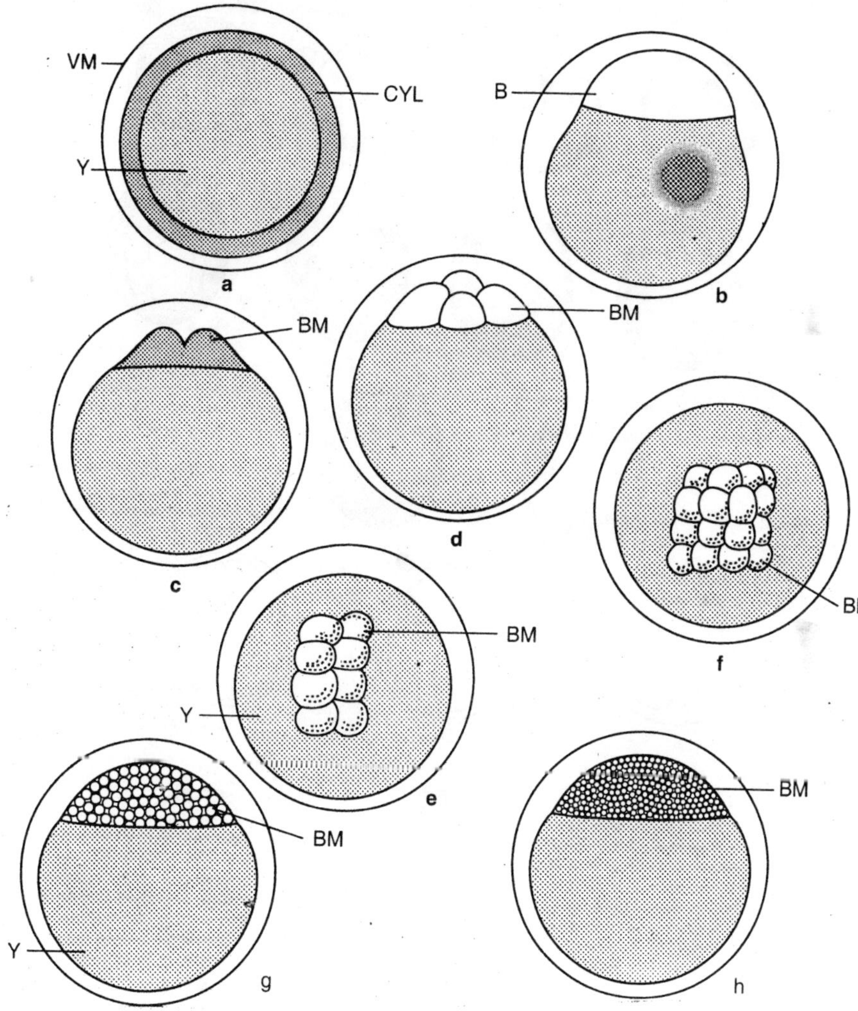

Fig. 21.3a-h : Diagrammatic representation of developmental stages. (a) Fertilized egg. (b) Blastodisc stage. (c) 2-cell stage. (d) 4-cell stage. (e) 8-cell stage. (f) 16-cell stage. (g) Early morula stage. (h) Late morula stage. B, Blastodisc; BM, blastomere; CYL, cytoplasmic layer, VM, vitelline membrane, Y, yolk. Source, Verma (1970).

Formation of Blastodisc : Just after fertilization, the cytoplasm which is present on the periphery, starts flowing towards the area where the sperm has probably entered the egg. The accumulation of cytoplasm is due to the contraction wave which is set in vegetal pole, pass through the equator and at the animal pole. The polarity at this stage is set up. The completion of one contraction cycle takes about 2 minutes. About twenty of these cycles follows one another, each cycle adding more and more cytoplasm at the animal pole and soon forms a cap-like structure, the blastodermic cap or blastodisc (Fig. 21.3b). The blastodisc in teleost is disc shape.

Most of the eggs have two principal regions in common, a centre, which is relatively stable to centrifugation and an endoplasm which is displaceable containing yolk and other inclusion.

Cleavage : The further development in major teleosts are almost identical, followed by the process of cleavage.

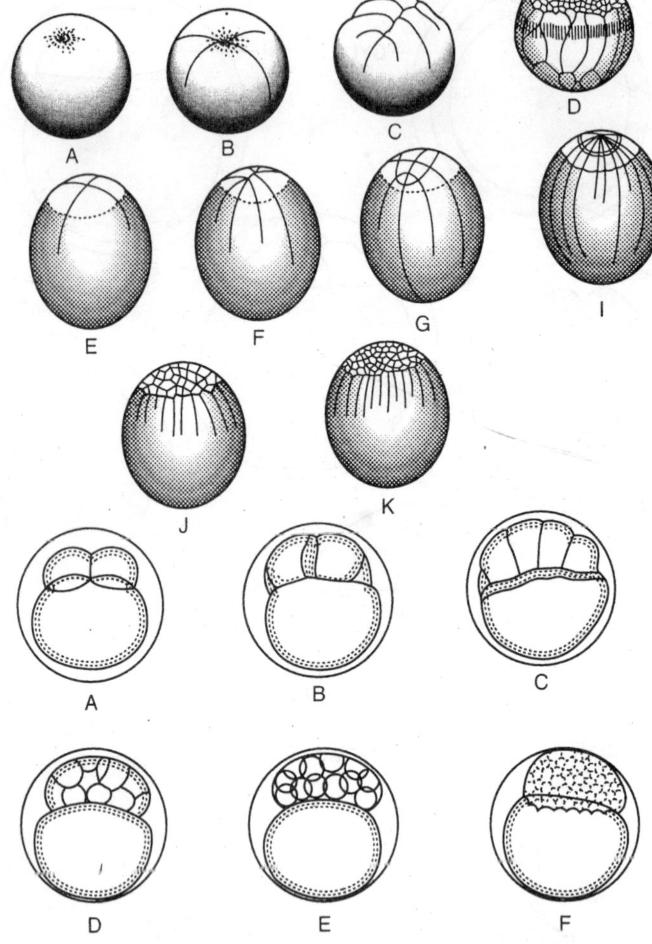

Fig. 21.4 : Superficial view of the early developmental stages. (A-K) *Acipencer.* (A-F) *Crenilabrus* pavo. Source : Gresse (1957)

The teleost egg has blastoderm in the form of blastodisc, the cleavage is meroblastic, i.e., limited to blastodisc, the entire zygote is not divided. The segmentation starts from 1 to $1\frac{3}{4}$ hours after fertilization. The factors which bring the cleavage are many but the major changes are the orientation of nuclear spindle and viscosity visible. They are parallel and on either side of the second cleavage plane and right angles to the first and third. In this way 16 cells are formed (Fig. 21.3f).

32-Cell Stage : The 16-cell stage undergoes further division, but from now, the cleavage furrows are horizontal as well as vertical. The four central cells are divided by a horizontal division into 8 cells which are arranged in two layers of four cells each. With the exception of the four corner cells in which the division is more or less diagonal, the rest of the cells are divided by vertical division. These are either parallel to the first or to the second cleavage furrow. In this way the 32-cell stage is formed.

Early Morula : At the end of cleavage, a ball of cells the morula is formed. The total area occupied by the cells of the early morula is more or less the same as that of the original blastodermic disc (Fig. 21.3g). The superficial view of egg showing cleavage and formation of morula is given in diagrams (Fig. 21.4 A-K, A-F).

Late Morula : The cells of morula divide further and becomes smaller in size. In a side view, this stage appears as a mass of cells with prominent hemispherical projection and the convex base which rests in hollow concavity of the yolk (Fig. 21.3h). A large number of oil granules pass out from the cell mass into the yolk, where they combine to form bigger globules.

The cells of morula loose and they become separated from one another under slight pressure. A syncytial layer is formed between yolk and the convex base of the cell mass. This syncytium is called periblast.

Cleavages result in the formation of two kinds of cells, blastoderm or periblast. The blastoderm cells are distinct and produce the embryo. The periblast or trophoblast cells lie between the yolk and cells of blastoderm and cover the entire yolk mass, having originated from the most marginal and outlying blastomeres. This syncytial layer helps in the mobilization of yolk reserves.

The nuclei arise at the edge of the blastoderm from division of marginal cell nuclei, each resultant nuclei is drawn into the divided yolk protoplasm or the periblast. The periblast is syncytial i.e., multinucleated cytoplasm. These nuclei resemble the nuclei of the blastomeres and since spindle, asters and chromosomes were observed, it is concluded that they divide mitotically.

According to Beer's law, percent transmission of a nucleus would be inversely proportional to the number of absorbing molecules in that nucleus, and the relationship would be logarithmic rather than linear.

Blastula : The cleavages or segmentations results in the formation of two kinds of cells, the 'blastoderm' and 'periblast' (Fig. 21.5a-g). The embryo is formed by the blastoderm, while the periblast or trophoblast cells, which lie between yolk and cells of blastoderm, which is syncytial in nature helps in mobilization of yolk reserves. There are substantial cohesive forces between developing blastomeres and the surrounding periblast which are important in the subsequent morphogenetic movement. It is suggested that periblast acts as an intermediary between two 'non-wettable' components—blastoderm and yolk. When the blastoderms diameter is 4/5 of the egg diameter, it is converted into blastula (Riley, 1974).

The hemispherical mass of cells of morula project from the yolk. The cells then flattened and extend outwards. The periphery of the blastodisc lines with the periphery of the yolk. The marginal or peripheral cells remain in close contact with the periblast. Whereas the central cells of the floor of the blastodisc are raised. As these cells are raised, a space is developed. This space is called as segmentation cavity or blastocoel (Fig. 21.5a). Soon blastocoel becomes well developed. The blastula formation starts 15 hours after fertilization in stickleback while it takes 8 hours after fertilization in *Cyprinus carpio.*

At the end of segmentation, the blastodisc becomes radially symmetrical. The radial symmetry changes to bilateral şymmetry because the flattening of cell mass is expressed in one sector and thus this region becomes thicker. The thicker sector is very important because

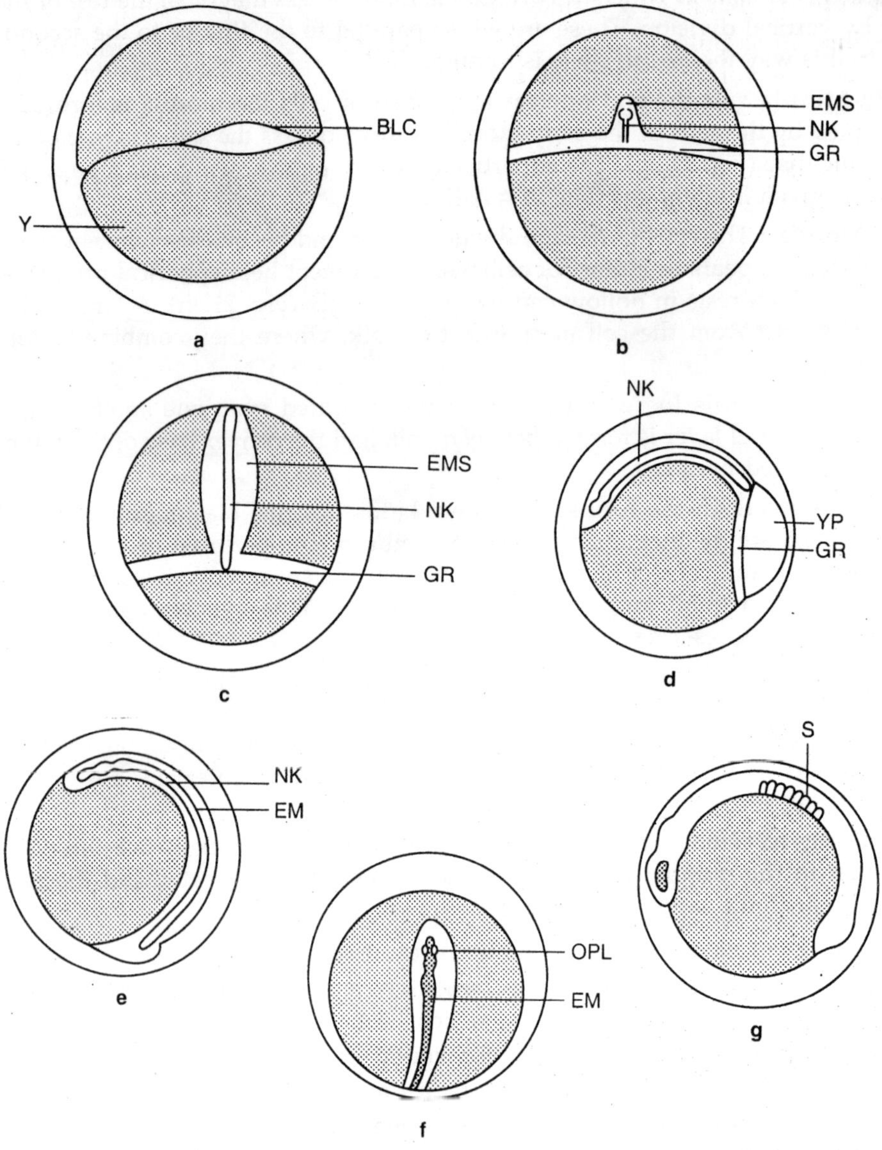

Fig. 21.5a-g : Diagrammatic representation of developmental stages. (a) Blastula stage. (b) Gastrulation stage, germ ring descend to the equator of the egg. (c) Three quarter of the egg surface covered. (d) showing small yolk plus. (e) Showing closure of blastopore, side view. (f) Same, dorsal view. (g) Formation of somites. BLC, blastocoel; EM, embnyo; EMS, embryonic shield; GR, germ ring, NK, neural keel; OPL, optic lobe; OPV, optic vesicle; S, somite; T, tail; Y, yolk; YP, yolk plug.

it is embryonic material and future embryo develops from it, and its median plane becomes the median plane of the embryo. At this stage the anterior and posterior sides of future embryo are also fixed. The distal part of the thicker sector is the prospective posterior end of the embryo and its central part corresponds to the prospective anterior end of the embryo.

Gastrula : The appearance of distrinct primitive streak on the embryonic shield is the beginning of gastrula. The gastrulation generally ends with the closure of blastopore. According to Riley (1974), this distinction is arbitrary. Both epiboly and emboly actively take part in the formation of gastrula.

Invagination or emboly : It takes place at about 21-26 hours after fertilization, the cells of the thicker sector invaginate at the limit of cytoplasm and yolk. This marks the beginning of gastrulation (Fig. 21.6a-d). The invagination which originally starts at one point, then extends laterally around the edge of the blastoderm and soon spread to the periphery of the entire blastodisc. The invaginated layer does not extend over the floor of the subgerminal cavity but is confined to the edges of the blastoderm thus forming a prominent ring, known as 'germ ring' (Fig. 21.5b). The only part which shows further invagination is the region of the germ ring which is formed by the thick sector of blastoderm disc.

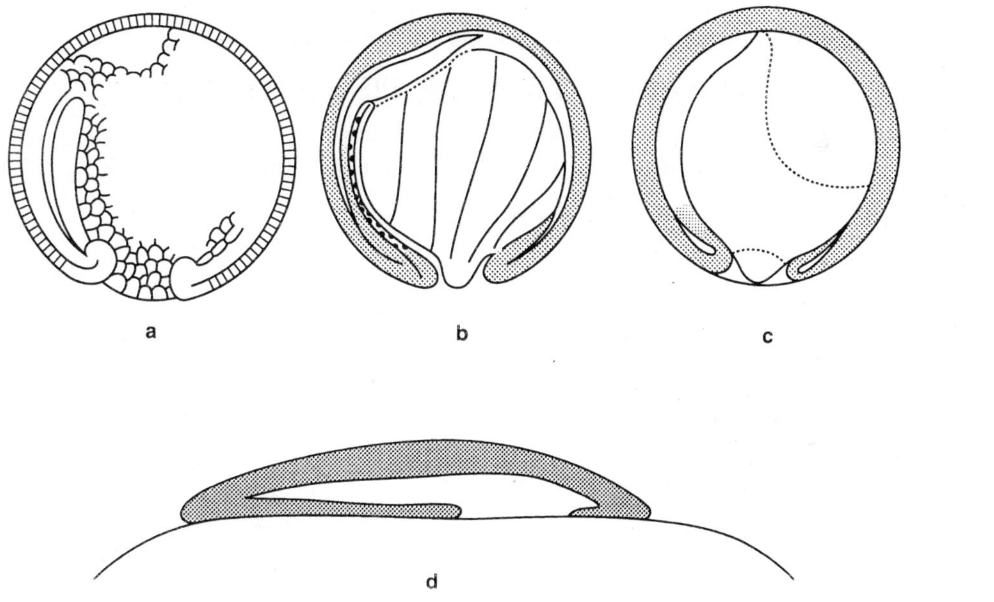

Fig. 21.6a-d : Gastrula. (a) *Acipenser,* (b) *Amia calva,* (c) *Perch,* (d) Archenteron (Source: Grasse, 1958)

As soon as the germ ring becomes established, it moves towards the yolk of the egg (Fig. 21.5b). The width remains constant but it increases in its circumference. Regarding further invagination, it advances more rapidly at one place than round the rest of the periphery of the blastoderm making it in triangular shape (Fig. 21.5c). The apex is pointing towards the animal pole of the egg. The embryo loses its triangular shape and becomes elongated. If blastoderm is seen from above, the posterior pole is roughly triangular which is thicker than the adjacent area. This makes the embryonic shield more prominent. The embryonic shield has been differentiated as embryonic and extraembryonic area. The embryonic shield is made up of an epiblast of polygonic cells covered by epidermic stratum and a complex lower layer which is known as entochordamesoblast. This entochordamesoblast is the analogue of mesoderm and endoderm. The thickened median portion will become the prechordal plate and chorda, while somewhat loosely arranged cells will form entoderm.

In the extraembryonic region an elongated subgerminal cavity bounded laterally by germ ring extends between the periblast and epiblast. The presumptive mesoderm in the meantime has coverage towards the dorsolateral edges of the blastodisc where it involutes, passing to the inside between the entoderm and ectoderm. The mesoderm becomes arranged on either side of the median notochordal material in developing embryo.

The notochord, prechordal plate and mesoderm which are differentiated but continue with entochordamesoblast, later ectoderm differentiates and all three germinal layers are distinguished ectoderm, mesoderm and entoderm.

With the advancement of development notochord, Kuffer's vesicle and neural plate are differentiated.

Epiboly : Simultaneously with emboly, the epiboly also starts and the cells overgrow the yolk and at the same time migrate at its periphery. The blastoderm become flattened. The flattening of blastoderm causes it to spread over yolk. Eventually the rim of blastoderm converges at or near the opposite side of the yolk and the opening closes by the contraction of rim. The rim of the blastodisc corresponds to the lip of the blastopore. Later before the blastopore closes, a yolk plug is seen projecting from the blastopore (Fig. 21.5d).

ORGANISATION OF FISH EMBRYO

Presumptive areas can be mapped out in the wall of the gastrula. The fate map for gastrula of *Cyprinus carpio* has been given by Verma (1971) (Figs. 21.7A-F & Fig. 21.8).

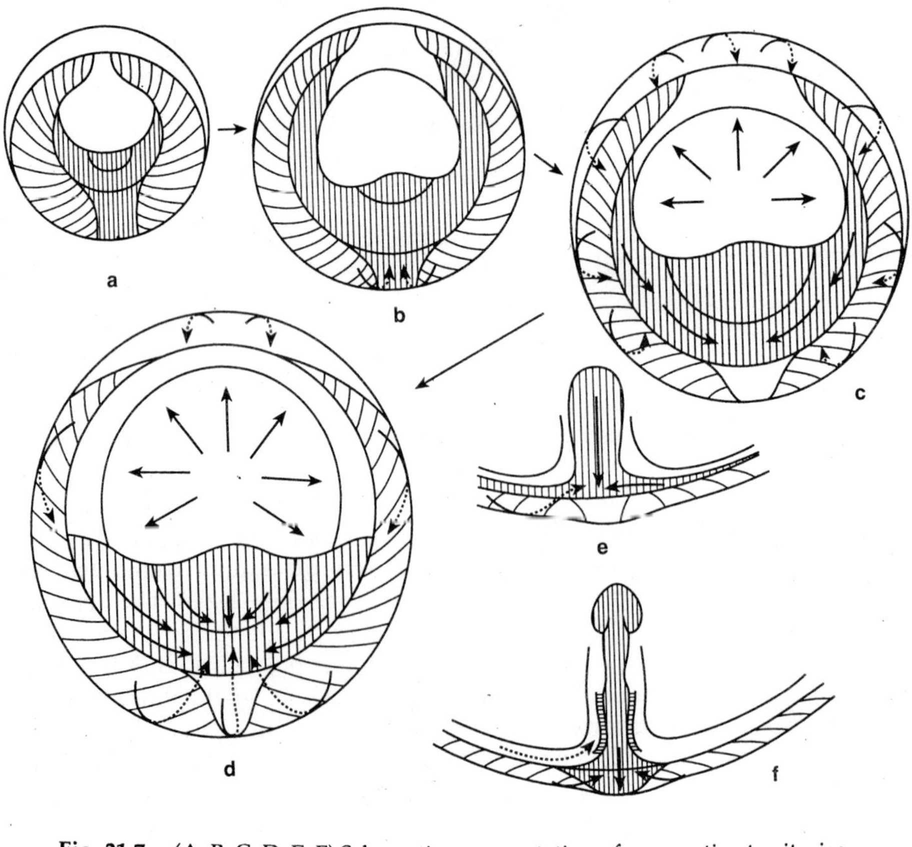

Fig. 21.7 : (A, B, C, D, E, F) Schematic representation of presumtive territories (Source: Gresse, 1957).

Organogenesis : About 27 to 50 hours after fertilization, due to further contraction of the lips, the blastopore closes.

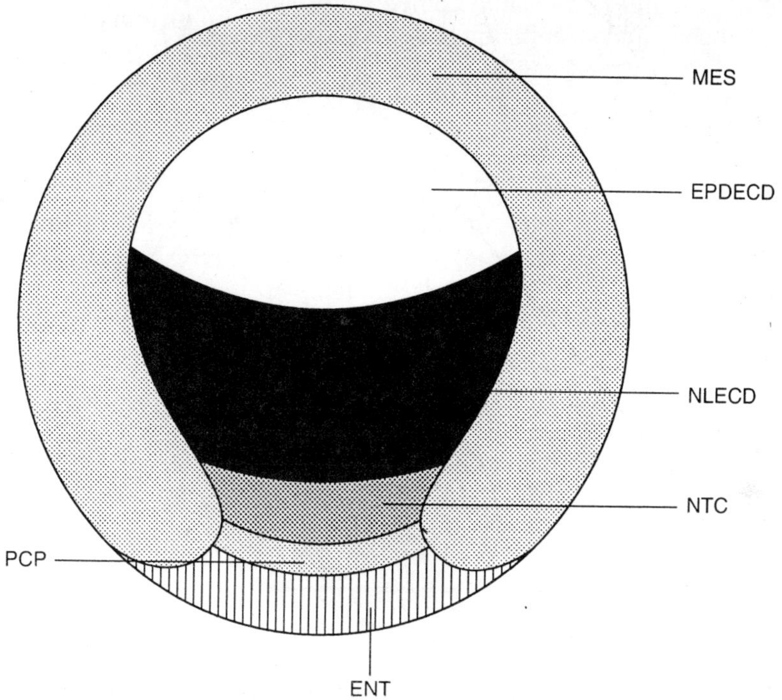

Fig. 21.8 : Fate map. ENT, entoderm; EPDECD, epidermal ectoderm; MES, mesoderm; NLECD, neural ectoderm; NTC, notochord; PCP, precaudal plate.

Notogenesis : The presumptive notochordal cells migrate inwards and roll up along the posterior edge of the blastoderm and thus form a solid string like notochord.

Neurogenesis : The presumptive neural plate sinks down to the space vacated by the inwardly migrated notochordal cells. The edges of neural plates rise up and fuse with each other at the middle line enclosing a cavity, 'neurocoel'. Thus a hollow tube like neural tube is formed just above the notochord. The anterior part of neural tube swells to form the brain whereas the part behind it remains as such and forms the spinal cord. By the two consequtive invaginations in the brain, it is differentiated into fore, mid and hind parts. The optic lobes appear by lateral outgrowth from the forebrain. The unsegmented parts of the embryo converge towards the embryonic axis. This convergence along with the development of the central nervous system causes a thickening of the embryo proper, which now protrudes from the surface of the egg extending approximately half way round the circumference of the yolk sphere.

In a few hours, after further contraction of lips, the blastopore closes 60 hours after fertilization in stickleback and 21 hours after fertilization in *Cyprinus carpio*, 5-10 parts of somites appear in the middle of embryo on either lateral side of the nerve cord (Fig. 21.5g). Each pair of somite is formed by lateral mesoderm. Later the somites give rise to the muscle of the trunk, appendages and their skeleton. Simultaneously blastopore closes, the whole germ ring fuse with embryo proper which now appears elevated and well demarcated from the yolk. By the development of central cavities in optic lobes, they become optic vesicles (Figs. 21.9a-e).

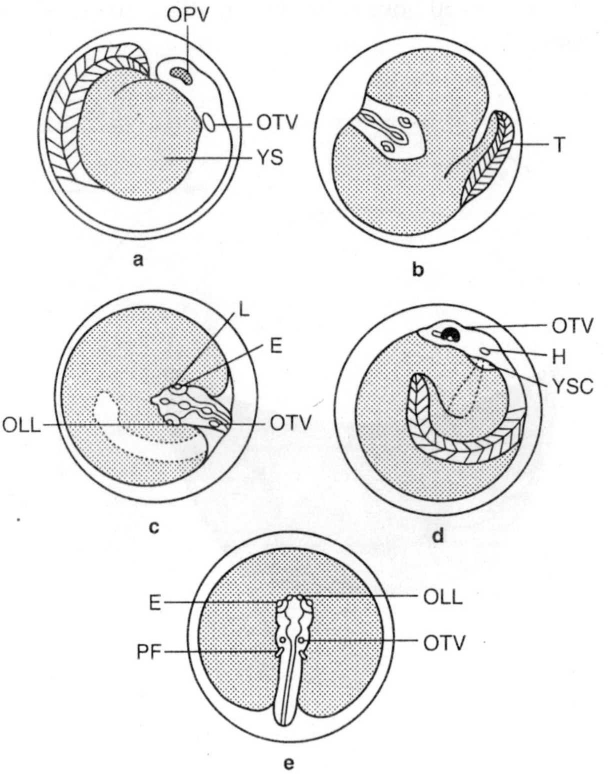

Fig. 21.9a-e : Diagrammatic representation of development stages. (a) Appearance of otic capsule & Kupffer's vesicle. (b) Appearance of lense, brain ventrili. (c) Appearance of heart & olfactory lobes. (d) Pigmented eye & blood circulation. (e) Appearance of pectoral fin, closure of ventrcles of fore brain. E, eye, H, Heart; L, lens; OLL, olfectory lobe; OPV, optic vesicle; OTV, otic vesicle; PF; pectoral fin; T, tail; YS, yolk sac; YSC, yolk sac circulation (Source : Verma, P. ; 1970, Redrawn).

The appearance of optic capsule and Kuffer's vesicle developed after 30 hours of fertilization in *Cyprinus carpio*. The head of embryo is differentiated further. The optic vesicles are converted into optic cups and the lenses are also formed. The brain develops as a median dorsal furrow which widens in fore and mid brain to ventricle which opens dorsally. Lateral to the hind brain a pair of optic capsules can be recognized. Ventral to the posterior part of brain the pericardium appears although heart is not yet visible. The somites increase in number, Kuffer's vesicle is now visible ventrally at the posterior end of the body. Heart and tail develop at 88 hours of development in stickleback and 55 hours in *Cyprinus carpio*. Before this at 70 hours of development the pectoral fins and ventricle are formed and forebrain closes.

Hatching : After the development of the various organs in the embryo, its body becomes cylindrical and bilaterally symmetrical. The connection between the body and yolk sac gradually narrows to form a stalk. The yolk sac gradually decreases in size as the embryo grows. Finally embryo hatches into a small free swimming larva.

Larval development : The freshly developed larva of *Cyprinus carpio* measures about 4.5 mm in length which is characterised by (a) the head slightly bent on the yolk, (b) mouth is open but no alimentary canal, eyes are large, pectoral fins are rudimentary and the tail is heterocercal (Fig. 21.10a-e).

One day old larva increases to 5.5 mm in length. The head becomes straight than the preceeding stage where the head is slightly bent. Eyes become dark black, heart enlarges in size and the alimentary canal differentiated above the yolk sac. Mouth is bounded by the jaws but covered by a thin membrane. Gill arches with rudimentary gill filaments are developed which are not yet covered over by an operculum.

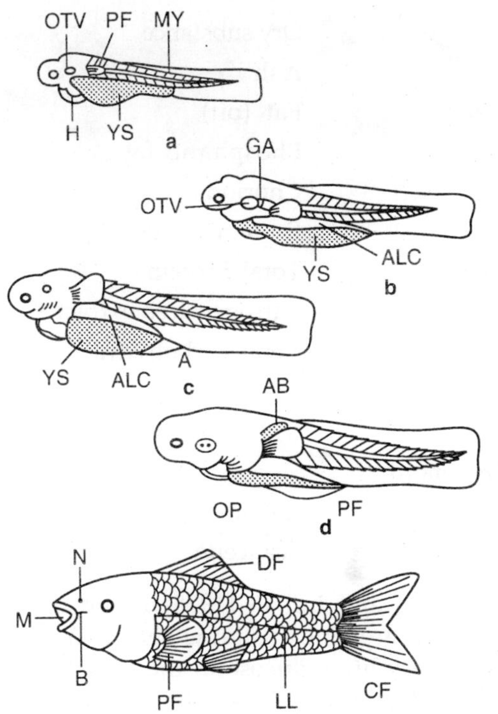

Fig. 21.10a-d : Diagrammatic representation of developmental stages. (a) Freshly hatch larva. (b) One day old larva. (c) Two day old larva. (d) Four day old larva. (e) One month old larva. (Source : Verma, 1970, Redrawn). AB, air bladder ALC, alimentary canal; CF, caudal fin; Df. Dorsal fin; GA, gill arch; H, heart; LL, lateral line; M, mouth; MY, myotome; N, nostril; OTV, optic vesicle; PF, pectoral fin; YS, yolk sac.

In two days the mouth of larva opens and becomes slit like, alimentary canal opens through anus. At this stage larva starts respiration with gills and feeding with mouth. There is complete absorption of yolk at 7 mm stage larvae which is almost 4 days old. At 10 days, the larva assumes the shape of a fish with a convex dorsal profile.

Feeding, starvation and weight change of early fish larva of *Tilapia sparmanii* and *Paralichthys oliyaceus* (marine) have been studied by Ishibashi (1974). According to him, incubation time for *Tilapia* was 48 hours at 27 °C. The total length was 4.2 mm at hatching, the larva being inactive and without a functional mouth. After two days mouth opens, caudal fin starts to move actively, and after three days the larva begins to swim. Weight increases rapidly after hatching. The weight on third day was 0.65 mg heavier than at hatching. At this stage the larva began to take food and weight of 8.80 mg was increased and on ninth day, the unfed larvae were inactive and weight was 1.24 mg, 25% less than that on third day. Only 1% of the unfed larvae survived to day 12.

Factors influencing early survival : Light, oxygen, temperature and feeding are some important factors which are responsible for survival during embryonic development. According to Pinus (1974), tiulka (*Culpeonella delicatula*) is most abundant fish of the sea of Azon with catches amounting to 40-50% of the total fish landed from this sea. He found optimum condition for the survival or eggs of this fish, when the temperature reaches up to 15-18 °C.

BIOCHEMISTRY OF EGG

The fish eggs with its relatively bulky yolk is the most difficult subject for chemical analysis. Information has been sought from the techniques of cytochemistry and more refined modern methods of electrophoresis and chromatography. The analysis of 100 mg egg of *Salmo irideus* is as follows (Hartmann *et al.*, 1947).

Water	66.18
Dry substance	33.82
Ash	1.29
Fat (oil)	3.61
Phosphatic fat	3.71
Chorion	2.69
Protein	20.16
Total Protein	23.19
Nitrogen free extract	2.32
Total carbohydrate	0.18-0.20
	(Smith, 1952)

Carotenoid

Lutein	1.2 g
Astacene	7.12 g

Young and Inman (1938) found 0.4% of ash and about 4.38% of volatile material. However, arginine, histidine and lysine present in respective ratio of 4 : 1 : 3. Hayes (1930) found very little glucose in the salmon egg, only 0.049 mg for 100 mg of egg. The rest of carbohydrate probably is bound to protein (Smith, 1952).

Amino acid composition of *Salmo gairdneri* egg during development (Suyamar, 1958)

	g/mg nitrogen
Arginine	7.0
Cystine	1.5
Histidine	2.4
Isoleucine	7.2
Leucine	9.8
Lysine	8.0
Methionine	2.9
Phenylalanine	4.8
Threonine	5.7
Tryptophane	1.0
Tyrosine	4.8
Valine	7.3

ENZYME

The chorion contains an enzyme known as chorionase. The chorion also resists digestion by trypsin and pepsin, it possesses pseudo-keratin. The hardening of chorion is due to the enzyme in the perivitelline fluid. Ca++ affects the enzyme rather than chorion itself. Hardening resulted from oxidation of SH to SS groups by means of aldehydes produced by polysaccharide with a glycol groups. Hatching enzyme works best under alkaline condition, pH 7.2-9.6 and temperature of 14-20 $^{\circ}$C (Blaster, 1969).

Acetylcholinesterase, the enzyme associated with nervous stimulation of the muscles, has been detected on 10th day after fertilization in *Salmo gairdneri* eggs. In the eye stage increase in activity is almost coinciding with the development of excitable tissue (Uesugi and Yamazoe, 1964). Ornithine transcarbamylase and arginase of the five O-U cycle enzymes were reported in the egg of *Salmo* which were capable of excreting nitrogen (Rice and Stoke, 1974).

METABOLISM OF NITROGENOUS WASTES

Metabolism and nitrogenous wastes in the eggs and alevins of Rainbow trout, *Salmo gairdneri* were studied by Rice and Stoke (1974). Ammonia, urea, uric acid and total protein and free arginine were recorded in eggs in alevines. According to them, ammonia and concentrations through hatching and greatest concentrations were found in alevines. Ammonia level increases for the first few days, then declines as the yolk was absorbed. Uric acid concentration did not change dramatically during development, but concentration before and after hatching period were lower than just shortly after fertilization and when the yolk was nearly absorbed. Urea and free arginine both increased in concentration while the developing embryo was still within chorion. Peak concentration of urea and arginine occurred soon after hatching but thereafter concentration of both compounds decrease.

Protein concentrations were relatively constant during first 25 days of embryonic development, steadily decline thereafter.

Production of ammonia in salmon embryo is expected even though metabolism of fat is the predominant source of energy. The proteins in the yolk are broken down to amino acids. Before they are resynthesized into protein in the embryo (Monroy *et al.*, 1961), an amino acid pool is available for catabolism. However, most amino acids are resynthesized into protein rather than catabolised, because total egg protein values remain stable until the day 21, after which catabolism leads to a decrease in total protein. Urea appeared to be synthesized when from yolk proteins were degraded by arginine. Rice and Stoke (1974) supported the hypothesis that O-U cycle enzymes can function to produce intermediates in other metabolič pathways.

Abnormal development : Swarup (1958, 1959 a, b, c, d) found twin forms if the freshly fertilized egg of ***G. aculeatus*** were subjected to heat and cold (32.5 to 37°C and 0 to $\frac{1}{2}$°C). The malformation includes synophthalmia, monophthalmia, microphthalmia and anophthalmia. Not only above mentioned changes occur but abnormality in blastodisc also occur due to high and low temperature.

REFERENCES

Billard, R. 1978; As reported in Reproductive Physiology of Fish. C. J. J. Ruchter and H.J. Th. Goods Centre for Agriculture Publishing and Documental Wageningen.

Blaster, J. H. S. (1969). Development : Eggs and Larvae. In : W.S. Hoar and D. J. Randall (eds.), Fish Physiology, 3, pp. 177-252. New York, Academic Press.

Budker, P. (1971). The Life of Sharks. Columbia Univ. Press, New York.

Devenport, J. S. Lonning & E. Kjorsvik (1981). Osmotic and structural changes during early development of eggs and larvae cod, *Gadus morhu,* L. J. Fish Biol. 19 : 317-331.

Gardiner, D. M. (1978). Fine structure of spermatozoon of; the viviparous teleost *Cymatogaster aggregata*. J. Fish Biol. 13 : 435-438.

Gloria A., Kapoor, B. G., Chardon, M. and Diogo, R, (2003). Fisheries, Vol. II, Science Publishers (USA).

Gresse, P. P. (1957). Triate De Zoology, T. B. Agnathes et. Poissons.

Hartmann, M. Meden, F. C. Kuhn, R. and Bielig, H. J. (1947). Investigation on the fertilizing material of rainbow trout. Z. Naturf. 2B : 330-349.

Hayes, F. R. (1930). The metabolism of developing salmon eggs. II. Chemical changes during development. Biochem. J. 24, 735-745, 71-72.

Ishibashi, N. (1974). Feeding, starvation and weight changes of early fish larva. Ed. J. H. S. Blaster Spring Verlag, New York, 339-344.

Jhingran, V. G. (1988). Fish and fisheries of India. Hindustan Publishing Corporation. Delhi, India.

Monroy, A., Shida, M. and Nakano, E. (1961). The pattern of transfer of yolk material to embryo during the developmnt of the teleostean fish, *Oryzias latipes*. Embryol. 6 : 151-158.

Pinus, G. N. (1974). Some factors influencing early survival and abundance of *Chipeonella* in the sea of Azon. In : The early life history of fish Ed. J. H. S. Blaster Spring, Verlag, 81-85.

Rice, S. D. and R. M. Stoke (1974). Metabolism of nitrogen wastes in the egg and alevins of rainbow trout, *Salmo gairdneri* Richardson. In : The early life history of fish. Ed. J. H. S. Blaster. Spring Verlag, New York, 325-337.

Riley, J. D. (1974). The distribution and mortality of sole eggs (*Solea solea* L.) In inshore areas. 39-52. In the early life history of fish. Ex. J. H. S. Blaster Spring Verlag.

Smith, S. (1952). Studies in the development of the rainbow trout (*Salmo gairdneri*) II. The metabolism of carbohydrates and fat. J. Exp. Bial. 29, 650-666.

Smith, C. L., Charles, S. Rand, Schaeffer, B. and James W. (1979). *Latimeria* the living Coelocanth is ovoviviparous (Ed. Milton S. Love and Gregor M. Cailliet). Prentice Hall of India Pvt. Ltd., New Delhi).

Srivastava, R. S. (1964). Study of normal morphological development of *Ophiocephalus punctatus* (Bloch). J. Vik. Univ. Ujjain, 61-69.

Stoss, J. and E. M., Donaldson (1982). Preservation of gametes. In Reproductive physiology of fish. Proc. of Int. Symp. on Wageningen, the Netherlands.

Suyamar, M. (1958). Changes in the amino acid composition of protein during the development of rainbow trout eggs. Bull. Jap. Soc. Scient. Fish 13 : 789-792, 70, 76.

Swarup, H. (1958). Stages in the development of the stickleback *Gasterosteus aculeatus* (L.). J. Embryol. Exp. Morph. 6 : 373-383.

Swarup, H. (1959, a). Abnormal development in the temperature treated eggs of *Gasterosteus aculeatus* (L.) III. Twinning. J. Zool. Society of India, II : 7-10.

Swarup, H. (1959, b). Abnormal development in the temperature treated eggs of *Gasterosteus aculeatus* (L.) I. Cleavage Abnormality. 108-113.

Swarup, H. (1959, c). Abnormal development in the temperature treated eggs of *Gasterosteus aculeatus* (L.) IV. Microcephaly 2 : 102-108.

Swarup. H. (1959, d). Abnormal development in the temperature treated eggs of *Gasterosteus aculeatus* (L.) II. Gadulation 2 : 1-6.

Smiths, (1974). Early development and hatching. In the early life history of fishes. E.D. J. H. S. Blaster. Spring Verlag, New York.

Uesugi, S, Yamazoe, S. (1964). Acetylcholineserase activity during the development of the rainbow trout eggs. Gunma J. Med. Sci. 13 : 91-93.

Verma, P. (1970). Normal stages in the development of *Cyprinus carpio* var. Communis L. Acta. Biol. Acad. Sci. Hung. 21(2) : 207-218.

Verma, P. (1971). The early development of *Cyprinus carpio* var. Communis (Linn.) Acta Anat. 80 : 388-417.

Wourms, J. P. (1977). Reproduction and development in Chondricthyan fishes. American Zoologist, 17 : 379-410.

Yanagimachi, R. (1957). Studies of fertilization in Clupea pallasii. Parts l-111. Zool. Mag. (Tokyo) 66 : 218-233.

Young, E. G. and Inman, W. R. (1938). The protein of the casing of Salmon eggs. J. Biol. Chem. 124 : 189-193.

Chapter 22

CRYOPRESERVATION OF GAMETES

The cryopreservation is a storage technique with practically no time limits. Preservation of fish gametes has been tried with this technique. The cryopreservation of sperms is a well established procedure in animal husbandry. Now cryopreserved sperms are used successfully in insemination of cattle, horse, pig, sheep and poultry breeding. It is interesting to note that great efforts have been put to transfer this technology to the preservation of sperms, ova and embryos of fish. This technique will help the fish breeders in the following ways.

(1) The problem of non-coincidence maturation of male and female can overcome, if sperms or ova are held in storage.

(2) The programme of selective breeding can be undertaken which will improve the indigenous stock and will help the third world to breed their indigenous species and to produce special strains of superior stalk.

(3) The technique of cryopreservation of gametes will play an important role in conservation of indigenous germplasm as many of indigenous Indian fish cannot compete with exotic fishes. By this technique the gametes from threatened species are banked as a hedge against dwindling genetic variability caused by environmental disturbances.

(4) It can help in production of monosex culture.

(5) If gametes can be successfully preserved we can develop fish throughout the year as per our requirement and can help to establish gene bank to preserve genetic originality of population and keep them available for re-introduction when general conditions for survival have improved.

The sperms, eggs and embryos can be cryopreserved, but the preservation of egg and embryos are difficult because of unavailability of cryoprotectant to be absorbed sufficiently in these cells because of their large size in comparison to spermatozoa (Ashwood Smith, 1980; Harvey and Kelley, 1984). The cryopreservation of spermatozoa in fishes is successful in many species. These species are *Oncorhynchus, Salmo* (Stoss and Donaldson, 1982), *Salvilinus fontinalis, Hucho hucho, Thymallus thymallus, Esox lucius* and *Cyprinus carpio* (Bayrle, 1980), *Serothodon mossambicus* (Harvey, 1983 & Harvey and Kelley, 1984); *Labeo rohita, Catla catla* and *Cirrhinus mrigala* (Rao, 1989; Kumar, 1989).

The following steps are to be followed in preserving spermatozoa.

(1) Collection of milt (semen).

(2) Preparation of extender solution.

(3) Selection of cryoprotectant.

(4) Freezing.

(5) Success of fertilization.

Collection of Milt

The first step is to collect sperms. Normally the sperms are collected from healthy fishes by stripping method. The cathetor may be used, which is better alternative to stripping. The male brooders with best characteristic traits were selected and washed with Ringers solution and genital papillae region is made dried. The brooder may be anaesthetized or not anaesthetized. The anaesthesia generally used is 0.3 ml/l of phenoxyethanol. According to Kumar (1989), an injection of suston 250 (organon) prior to stripping gives better result in Indian carps. Majority of workers in this field recommended unanaesthetized fishes for stripping for obtaining milt.

The milt is collected in syringes or haemolysis tubes and later preserved in glass ampules (sealed or unsealed), plastic straws and often in plastic bags. The colour, volume, density, pH and motility of sperms are to be noticed. The sample should be free from urinary and faecal matter. A smear of the sample is to be examined under microscope to notice abnormalities in sperms. If sample is OK, then put for further processing, otherwise fresh sample from new brooders may be taken.

During the period of collection and preservation of spermatozoa, the syringe/ampules/ straw may be kept in pond water to maintain constant temperature. Legendary and Billard (1980) collected sperms in haemolysis tubes in melting ice and then stored on a refrigerated surface at 4 °C.

The preparation and selection of extender, a solution in which milt is collected, is most critical for cryopreservation. The extender prevents depletion of sperm energy and maintain sperm in quiscent condition but alive.

There are two basic extenders. They are called Mounibs medium (M) and Menezo medium (Me). In these two media bovine serun albumin (BSA) and tellurite egg yolk are added.

For Indian fishes, several extenders were tried by modifying in their following constituents.

The common constituents are sodium chloride, KCI, $CaCI_2$, $NaHCO_3$, Na_2HPO_4, $MgSO_4$. In addition to these, nutrients & glycine are also added to improve cryopreservation. Antibiotics such as gentomycin or streptomycin are also added to the extender solution if needed.

Common Extender

Sodium Chloride	75 gm	for ionicity
Sodium Bicarbonate	500 gm	for buffer
Fructose	500 gm	nutrient
Mannitol	500 gm	nutrient
Veg. Lecithium	750 gm	for protection of membrane.
Pancryozyme C	1000/ml	

Cryoprotectant

The cryoprotectant's main function is to penetrate in the cell and may help spermatozoa to tolerate freezing temperature. The most important and used widely is DSMO (Dimethyl sulfoxide) and glycerol. Harvey (1983) used methanol, DSMO and glycerol in combination with milk powder or egg yolk.

Cryoprotectant and extender solution in a fixed ratio, generally one part of cryoprotectant and nine parts of extender solution is known as diluent. In this diluent the sperms are collected for further processing for freezing.

The diluent (cryoprotectant + extender + sperms) are taken in polyvinyl straws and both ends of straws are sealed and now they are ready for freezing or chilling.

The milt can be chilled, the temperature is maintained between 0-5 °C. While in freezing, CO_2 and liquid nitrogen are used Chilled storage of teleost milt (0-50 °C) has been studied extensively in Salmonids.

In freezing procedure many problems such as cold shock, formation of intracellular ice and osmotic shock should be controlled. Cryopreservation generally means storage of cells or tissue at –196 °C, the temperature of liquid nitrogen (Fig. 22.1).

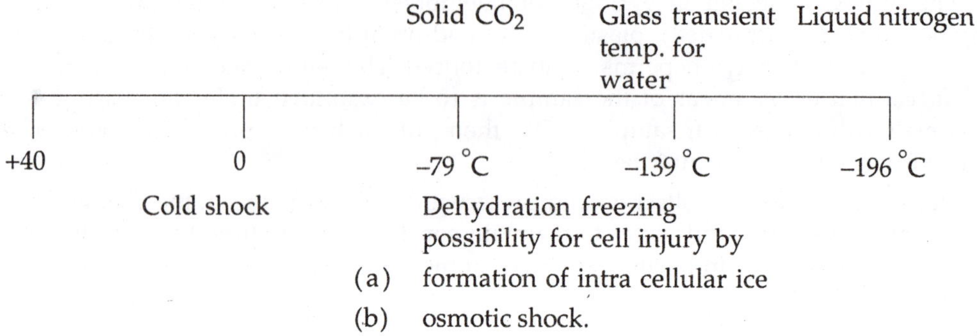

Fig. 22.1

Cell injury should be controlled. Three types of cell injury generally occur. The first is cold shock, which is caused by cooling above freezing temperatures. This is more important for ova, not very much importent for sperms. This occurs up to 0 °C. When the temperature goes from 0 °C and–80 °C the osmotic shock and formation of intracellular ice in ova and sperm takes place. For successful cryopreservation these things are controlled.

In short, the straws containing diluent, are transferred into cannisters. The cannisters holding straws after allowing varying equilibriation period are dipped first into liquid nitrogen vapours and then into liquid nitrogen at the temperature –196 °C. Legendra and Billard (1980) stored the sperms of rainbow trouts in dry ice (solid CO_2–79 °C) or in liquid nitrogen up to 6 months. In a number of fresh water species freezing has been carried out by pelleting suspend sperms into dry ice (Nagase and Niwak, 1964). The transition from 0 °C to –70 °C lasts approximately 2 minutes resulting in a rate of 35 °C/min. Since the rate decreases above – 60 °C.

The best results are obtained in liquid nitrogen.

Thawing : The thawing is reverse event of freezing. Thawing is done in water bath at temperatures ranging 10 °C to 60 °C (Horton and Ott, 1976). The success of whole procedure depends upon the motility of spermatozoa and capability of fertilizing ova. Stoss and Holtz (1982) have increased the motility of pink salmon spermatozoa from 30 sec to 10 minutes by activating with an 120 nm $NaHCO_3$ solution to which 1 BMX (3-isobutyl-l-methylxanthium) had been added. Kumar (1989) stated that length of storage period is about 20 days in *L. rohita* and *H. molitrix* under cryogenic condition.

The eggs of *Salmo gairdeneri*, *Onchorhynchus kisutch* and *O. keta* were fixed by removing through the process of stripping and sealed in plastic bags together with coelomic fluid and kept under oxygen at 1 °C. The eggs are then allowed to spread to form a layer one or two eggs deep (Harvey, 1983, and Ashwood-Smith, 1980).

The eggs then mixed with extender solution and cryoprotectant almost in similar way as described earlier for sperms. According to Harvey and Kelley (1984), unfrozen (chilled) storage of ova has been moderately successful in salmonids, with storage times in order of a few weeks obtained through refrigeration of oxygenated eggs in coelomic fluid.

REFERENCES

Ashwood-Smith, M.J. (1980). Low temperature preservation of cells tissue and organs. In : Low temperature preservation in Medicine and Biology. (Eds. Ashwood-Smith. M. J. and Farranb, J.) Pitman Medical, Tunbridge wells. pp. 39-43.

Bayrle, H. (1980). Untersuchungen Zur optimicring der Befruchlung fahigikeit Von gefrierkonserviertem Fischsperma, Dissertation. Tech. Univ. Munchen.

Harvey, B, (1983). Cryopreservation of *Serotherodon mossambicus* spermatozoan. Aquaculture 32 : 318-320.

Harvey, B; and Kelley, R.N. (1984). Chilled storage of Sarothercdon mossambicus milt. Aquacultures 36 : 85-95.

Hortan, H.F. and Ott, A.C. (1976). Cryopreservation of fish spermatozoa and ova. J. Fish Res. Bd. Canad. 33 : 995-1000.

Kumar, K. (1989). Studies on cryogenic preservation of carp spermatozoa. In. "Fish Genetics in India" (Eds. Das. P. and Jhingran, A.G.). Today and Tomarrow's Printers and Publisher, New Delhi. pp. 199-206.

Nagase and Niwak (1964). Deep freezing bull semon in concentrated pellet form. 1 factors affecting survival of spermatozoa. proc. 4th intern, congr. Anim. Reprod. Artif. insemin. Trento. 3 : 410-415.

Legendre, Par M. & Billard, R. (1980). Cryopreservation of rainbow trout sperm by deep-freezing Reprod. Mutr. Develop 6 : 1868.

Rao, K.G. (1989). Cryopreservation. In : Fish Genetics in India (Eds. Das P. & Jhingran, A.G.) Today and Tomarrow's Printers Publ. New Delhi. pp. 193-198.

Stoss, J. & Holtz (1981). Cryopreservation of rainbow frout *(Salmo girdneri)* sperm II, effect of pH and presence of buffer in the diluent. Aquaculture 25 : 217-222.

Stoss, J. and Donaldson, E.M. (1982). Preservation of gamets. In : Reproductive physiology of fish. (Eds. Richter C.J.J., & Goos, H.J.Th.) Centre of Agricultural Publishing and Documentation Wageningen. pp. 114-127.

PART II

APPLIED FISHERIES

Chapter 23

AQUACULTURE

Aquaculture is defined as farming of organisms under water under controlled conditions. Depending upon the species, Aquaculture amongst animals can be done in freshwater (Inland water), brackish water (estuaries) and marine water (sea). It is like farming of agriculture products on land and can also be compared to animal husbandry.

After independence, the nation has made green (agriculture), white (milk & milk products) and blue (fish culture) revolutions. Blue revolution is obtained through fish culture/aqua farming or aquaculture. Aquaculture amongst animals principally deals with fishes, crustaceans and molluscs, because their culture is commercially viable. Blue revolution is the effort made by both the state Government and the central Government. A large number of Fish Education centers were established by the Ministry of Agriculture, in addition to imparting education of Aquaculture in universities and colleges. CIFE and CMFE institutions of Indian Council of Agriculture Research made a project 'Lab to land' for the benefit of fishermen and rural population. *Due to these efforts, the total fish production increased from 4.2 million tons in 1992 to 5.6 million tons in 1996 (Anon,1996,1997&1999). During 1999, the share from marine and inland sources was 3.03 and 2.57 million tons respectively. This has been achieved due to capture and culture fisheries.*

Though, in India there has been a great success in Fish Culture, there is further need for its expansion as aquaculture industry involving hightech technology and sustainability to produce protein to meet challenge of food for our one million population.

What is the need of aquaculture when such species are available in nature?

It is an established fact that there is limitation in tapping the natural resources like availability of only a particular species in particular environment during a particular season. There is always a mixing of the non-preferential species while tapping the resources from nature. There is always unpredictable per capita income per effort per period in the natural environment, necessitate the need to go for aqua farming. Instead, the farming in water bodies under scientifically supervised conditions enables the farmer to produce the selective aquatic flora or fauna for an anticipated harvest of planned quality and quantity so as to meet the increasing demand for aquaculture product (Sinha,1999; Jana & Jana,2003; Biswas, 2004).

Aqua farmers must have in mind some of the important issues. The intensive agriculture practices have led to environmental problems both for terrestrial and aquatic ecosystem. Extensive use of pesticides has brought adverse effect on capture fisheries in India. The fish development and its growth cannot be left on nature. Fish culture on planned manner will conserve the nature resources.

Importance of Fish Culture

A. If fish culture is properly taken up and proper awareness amongst people is made. This will improve economy and social culture of the nation but it will help in getting nutritious food to the population. The fish contains about 14 to 21 percent of protein, which is significantly higher than milk. In milk the percentage of protein is only 3 to 5 %. It is higher than cereals also where the % of protein is only 7 to 9.

B. The fish has unsaturated fatty acid which is essential for food of those persons who have undergone bypass surgery and are prone to coronary heart disease. Omega 6 (linoleic acid) and Omega 3 (linoleic acid) are essential fatty acid and are mainly present in fish and shell fish.

C. Ecosopentanoic acid and decosohexanoic acid are present in fish oil. They have been shown to prolong platelate aggregation. The vitamine E is present in more quantities in fish and vitamin E is essential as an antioxidant.

Biotechnology in Fish Culture

Using gene transfer techniques, scientists have created genetically engineered variety of Atlantic salmon that grows to market weight in about 18 months, compared to 24 to 30 months that it normally takes for a fish to reach that size (Carol Lewis, 2001). For fish farmers, raising these so-called transgenic fish could be faster and cheaper because it takes less food and saves about half the time to raise for sending the name ultimately to the market.

Breeders can now use the tool of biotechnology to introduce new characteristics into the fish.

REFERENCES

Anon, 1996. Sewage treatment through aquaculture. Central Institute of Freshwater Aquaculture, Bhubaneswar, Orissa, India.

Anon, 1997. Aquaculture Development. FAO Technical Guidelines for Responsible Fisheries. No. 5 Rome, Italy.

Anon, 1999. Pisciculture, marine fisheries and aquaculture. Pages 517-567 in Indian agriculture. Ministry of Agriculture, Government of India, New Delhi, India.

Biswas, KP. 2004. Industrial Fishers. Days Publishing House, New Delhi.

Carol Lewis, 2001. U.S Food and Drug Administration, FAD Consumer Magazine. January-February year.

Jana, B.B. & Jana, Santana (2003). The potential and sustainability of Aquaculture in India. Journal of Applied Aquaculture: 13, 283-316.

Sinha, V.R.P. 1999. Rural Aquaculture in India. RAP Publications, 21, Bankok, Thailand.

Chapter 24

ANIMAL USED FOR AQUACULTURE

(Field identification and breeding habits of cultivable species)

As stated previously, Aquaculture amongst animals can be done in freshwater (Inland water), brackish water (estuaries) and marine water (sea). The most important point is that those species are to be selected, whose culture will be profitable.

Selection of Freshwater Animals for Culture

There are about **22 species** of fishes and **2 species** of prawns, which are suitable for fresh water aqua farming as they are economically viable (Srivastava, 2005; Sandhu, 2007). The most useful species are **major carps**; they are called major carps because they grow large and fast utilizing food available in water. The major carps can be categorized into two major categories, **indigenous and exotic**. The indigenous major carps are *Catla catla, Labeo rohita* and *Cirrhinus mrigal*, apart from these varieties *C.reba* and *Labeo calbasu* are also economically viable. *Catla catla* is the fastest growing Indian major carp amongst fin fishes. It is palatable and is liked by the Indian population (Figs. 24.1 and 24.2a-c).

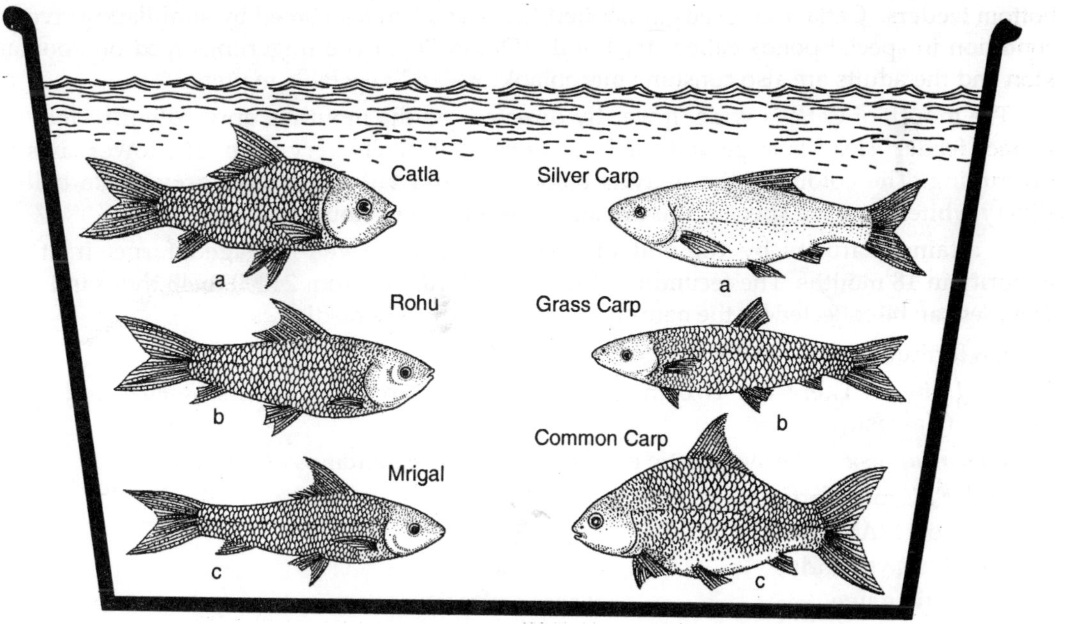

Fig. 24.1a-c : Common names of Indian Major Carps : (a) Catla, (b) Rohu, (c) Mrigal.

Fig. 24.2a-c : Common names of exotic Major Carps : (a) Silver Carp, (b) Grass Carp, (c) Common Carp.

The exotic varieties are *Cyprinus carpio* (Common carp), *Ctenopharyngodon idella* (Grass Carp), *Hypopthalmichthys molitrix* (Silver carp) and *Tilapia mossambica*. Fig. (24.2a-c).

Common carp brought in India in 1939 and was successfully acclimatized in Ooty lake in Tamil Nadu and since then it is well adopted in Indian water and is now available in almost all provinces in India.

The exotic and indigenous major carps are suitable for culture, as they could breed in stagnant water through the process of induced breeding by pituitary injection or by HCG or synthetic analogs.

The brief classification, important field identification, characteristics and breeding habits are given as under (Winfield Nelon, 1441) (Dhandapani, 1995 Jyoti & Sharma, 2006) :

Distribution, Ecology and Seed Availability: Indian Major Carp

1. *Catla Catla*

It belongs to order : Cypriniformcs, (Division : (A) Cyprini) and family : Cyprinidae. It is called in Hindi speaking regions as Catla. In Oriya, it is known as Bhakur whereas in Tamil it is called as Thoppa meen, in Telugu it is named as Bacha.

In nature, the fish is found in lakes, ponds, rivers and reservoirs. It breeds only in running waters that is in rivers. It breeds in natural habitat during southwest monsoon and seeds are available from May to August and could be collected from the breeding grounds. Due to the success in induced breeding, this fish now can easily breeds in stagnant waters in fish farms. It is a highly suitable fish for composite culture along with fishes that are column and bottom feeders. Catla too breeds in flooded fields and can also breed by stimulating riverine condition in special ponds called dry bunds (Dubey,2003).The fingerlings feed on zooplankton and the adults are also consumc meroplanktons and vegetable matters.

The mouth of *Catla* is wide upwardly directed. As the mouth is upwardly directed, it is surface feeder. Head is large and the body is comparatively very deep. The lower jaws are protruding. The colour of the body is black, fins and tails are dusky grey while belly is silvery white. In some species, fins are also dark black in colour.

It attains maturity by the end of second year. In well-managed farms it attains maturity in 18 months. The fecundity of female fish ranges from 24000 to 42 thousand eggs. The fries can be collected by the natural sources or induced breeding.

Fin formula: D, 3-4/14-15; A.3/5: P, 21: V, 9; C, 19-20.

2. *Labeo* : There are two main species. They are *Labeo rohita (Rohu)* and *Labeo calbasu* (Kalbasu).

Labeo calbasu : The systematic position in animal kingdom is as follows:

Order : Cypriniformes
Division : (A) Cyprini
Family: Cyprinidae

Scientific name	*Labeo calbasu*
Vernacular names	
Hindi	Kala abenese
Tamil	Kaka meen
Telugu	Kalla Kundumeen
Oriya	Kala abenese

The body of *Labeo calbasu* is stumpy, stout and head is large. The colour of the body is blackish green but the fins are black. **The lower lip is fringed and is a distinguishing character**. It is found in rivers and can survive well in ponds. As far as breeding habits are concerned it does not breed in stagnant water but by induced breeding methods it can breed well in stagnant water too. Fingerlings and adult can feed on vegetable debris, *Daphnia* and decomposed bottom material.

Labeo rohita

It is common in North India, Orissa and Bengal and is called Rohu but in Assam it is known as Rohiti. The body is elongated; dorsal profile is more convex, the colour of the body is bluish above and silvery at the flanks. Eyes are redish. In juvenile stage, it posseses dark band at the caudal peduncle. It has redish tinge at the dorsal, pelvic, anal and caudal fins. **The distinguishing feature is that the lips are fringed**. The maxillary barbels are prominent.

As far as the distribution is concerned, it is available throughout nation in rivers and ponds. It is most delicious and, therefore, there is a great demand of this fish. It breeds in monsoon in rivers. Like other Cypriniformes, it does not breed in stagnant water. During breeding season the fish before breeding moves along with current of water, then take a turn and moves against the current of water and breeds in shallow water. It can successfully breed by induced methods. At 10 mm stage it has prominent concentration of chromatophores at the base of caudal fin, nape and at the base of all fins. The barbels are conspicuously black in colour.

It is widely distributed and inhabits in rivers and ponds of India. It grows fairly large in size but in comparison to *Catla catla* the growth is less. The fish becomes sexually mature in one year or so. The fecundity is 1.5-2.00 (lac/g body weight). The fry and fingerlings could be obtained from the breeding grounds. Induced breeding in fish farms can successfully breed it.

Adult as well as fingerlings feed on vegetable debris, *Daphnia* and decomposed water material.

Other species of *Labeo* could also be cultured but from the economic point of view *Labeo rohita* and *Labeo calbasu* are profitable.

3. *Cirrhinus mrigala*

The common Indian species are *Cirrhinus mrigala, C. cirrhosa, C. latia, C. reba* and *C. fulungee.*

The systematic position of *Cirrhinus mrigala* is as follows:

Zoological name. *Cirrhinus mrigala*

Order : Cypriniformes

Division : Cyprini

Family : Cyprinidae

Vernacular name

Hindi : Mrigal

Oriya : Mirikali

Tamil : Mirikali

The body is streamlined; the abdomen is round with deeply forked caudal fin. The snout is depressed, mouth is broad, transverse and obtusely round and the upper lip is entire. Barbels are two in number. The body colour is coopery, the flank is white with tinge of golden colour. The colour of eye is golden. It attains an average length of about 40 cm. It is found in rivers and lakes of northern India. It is also well acclimatized in South India for aquaculture.

It breeds during monsoon months. It is most suited for induced breeding and now available throughout India.

Remarks : More than five inter-generic hybrid fries are available for culture. The fingerlings and adult feed more on animal protein. Both male and female mature at the age of two years. It is said that the induced breed fish mature only at the age of one year.

C. reba breeds from June to September. Seeds are cultured in large quantities in any freshwater body due to rapid growth of the species. Food is composed of mostly vegetables matter. Induced breeding is very successful for maintaining good population of this species.

C. mrigala breeds during monsoon. The fingerlings are available from natural grounds from July to November. The fish breeds naturally in rivers or induced riverine conditions due to the effect of pituitary hormone or other synthetic hormones. The number of ova/g body weight is 32-280, the number of ova/g ovary weight is 880-1830 and with fecundity per kg body weight is 1.5-2.00 lakh.

Exotic major carps which are well adopted to Indian water are as follows:

1. *Cyprinus carpio* (Common carp)
2. *Hypopthalmichthys molitrix* (C.&V.)
3. *Ctenopharyngodon idella* (Val.)(Grass Carp).
4. *Tilapia mossambica*
5. *Osphronemus goramy* :

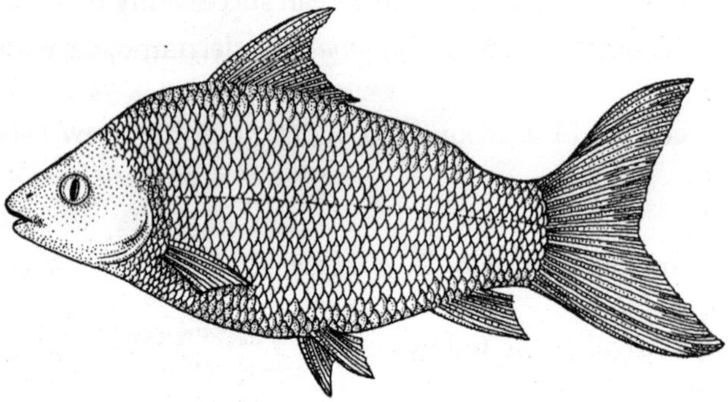

Fig. 24.3 : *Cyprinus carpio*

Zoological name *Cyprinus carpio*

Vernacular name is Common carp. One of the world's important cultured fish.This fish was brought from China and was put in Ooty Lake. In 1947, it was introduced in Nainital and Kumanun hill. This was a cold-water fish. This fish is now well acclimatized in Indian waters and now found throughout the nation. Induced breeding in controlled conditions successfully breeds it.

It has three main varieties.

1. *Cyprinus carpio specularis* : Common name is mirror carp. The body is covered with large unevenly bright yellow scales. Due to unevenly arrangement of scales a large area of the body remains scaleless. Scales are mostly restricted to lateral line.

2. *Cyprinus carpio communis* : The common name is scale carp. The body is covered with regularly arranged small scales. Asians prefer wild-type scale pattern.

3. *Cyprinus carpio nudus* : The common name is leather carp. The body is devoid of scales except for a single row of somewhat degenerate scales along the base of the dorsal fin. Sometimes it extends from head to tail. Carps with reduced scale pattern are highly valued in the European market (Beaumont and Hoare, 2003).

The body is round and deep, mouth is anteriorly directed, and lips are thin. The fins are short. It is a bottom dweller fish. It is omnivorous and consumes all material available in the bottom of the tank. In India, it becomes adult in six months time. But in cold water it gets maturity in one year. In China and Japan they mature in two years and it is reported that this fish attains maturity only in one and half years in India. It grows 800 g in a year. In India, after first and second years it acquires weight 1 to 2 kg respectively (Dubey, 2003).

The number of ova/g body weight is 8.3-37.3, the number of ova/g ovary weight is 437-1793.This is the carp that can easily breed in stagnant water (Piska, 1999).

Fecundity : Eggs quantity is 39,500 to 16,60,000. It is reported that it can breed four to five times in a year. Breeding habit is similar to carps.

Hypopthalmichthys molitrix. Silver carp

It is also introduced in Indian waters from China. The fingerlings of this fish was brought from Japan in 1955 and cultured in fish farm at Cuttack. It is the fastest growing carp amongst indigenous and exotic fishes. The head is small and pointed, the lower lip is protruded interiorly but the upper lip is slightly bifurcated. The jaws are of equal length, the lower one is provided with tubercles, and upper jaw is slightly notched. The origin of dorsal fin is behind that of ventral fin. It is non-predatory fish like phytoplankton and zooplankton. It is the fastest growing exotic carp. It atttains 2.0, 3.5 and 5 kg in first, second and third year respectively. The eyes are small. It feeds on protozoan and rotifers, vegetable are also consumed. It controls the algal bloom. It is surface feeder on account of mouth position. Barbels are absent. The fecundity per kg body weight is approximately 0.8-1.0 (lakhs).

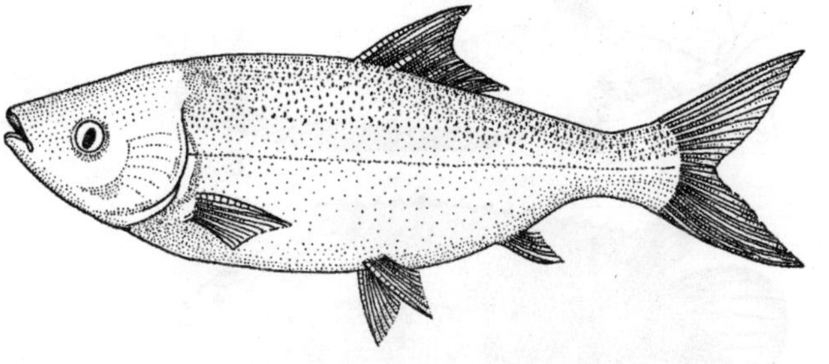

Fig. 24.4 : *Hypopthalmichthys molitrix*

Ctenopharyngodon idella. (*Grass carp*) (Fig. 24.5). It is natural inhabitant of rivers of China. This fish was introduced in India first time in 1959 from Hong Kong and is now well acclimatized in India and distributed to various states in the country. Its growth is also fine and is a tasty fish and liked by Indians.

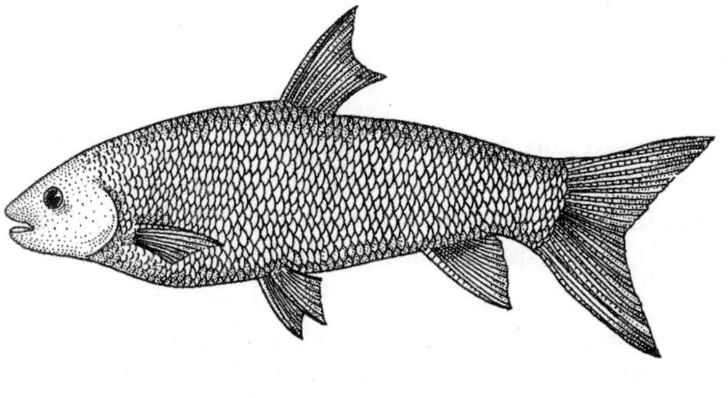

Fig. 24.5 : *Ctenopharyngodon idella (Grass Carp)*

The body is oblong and fairly compressed, head is small and pointed. Mouth is upturned. Breeding habit is similar to carps.

Air Breathing Fishes

The air breathing fishes are considered as a medium quality fish for meat purposes but have well liking by the people of the various part of the country. They are brought to the market in good living condition due to the presence of air breathing organs. These fishes are also used as recreational fishes in some part of the world (Gloria *et al*, 2003). The culture of these fishes is done as a profitable business. These fishes have high protein contents between 18-22% along with calcium, phosphorus and iron. The fishes are *Clarius batrachus, Heteropneustes fossilis, Anabas testudineus, Channa* striatus, *Channa marulius* and *Channa punctatus* (Fig. 24.6a-f).

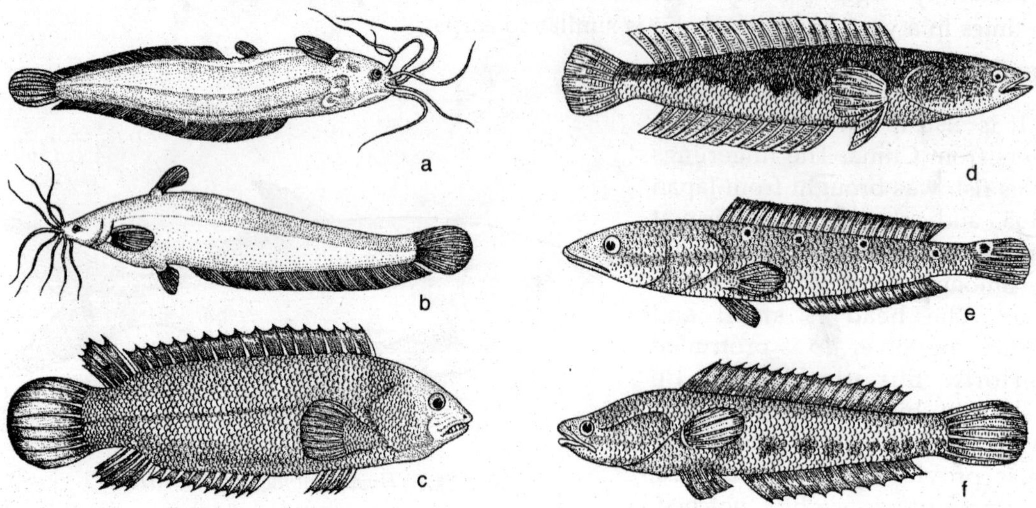

Fig. 24.6a-f : (a) *Clarias batrachus,* (b) *Heteropneustes fossilis.* (c) *Anabas testudineus,* (d) *Channa striatus,* (e) *Channa marulius* and, (f) *Channa punctatus.*

Clarias batrachus. The systematic position is as follows:

Order : Cypriniformes

Division : Siluri

Family : Channidaeas

Vernacular name

Hindi & Assamese — Magur

Telugu — Manguro

Oriya — Marpula

The body of the fish is laterally compressed and devoid of scales. The colour of the body is brownish green. Head is dorso-ventrally flattened. Four pairs of barbels, the maxillary barbels, are well developed. The pectoral fins are long and serrated. Juvenile fish has broad anal fin. A spot near the pectoral fins exist. The fishes inhabit fresh- and brackish water in India.

Breeding season is July to August. It breeds in marshy pools, ponds and lakes. It is nutritious and good for convalescing patients. The fish gets maturity by the end of the year. The spawn is kept in nests, which contains about 2000 to 5000 eggs. The eggs are blue in color. The species can breed in cisterns or small ponds. During heavy rains the fishes migrate to shallow marginal areas of ponds (Jhingran, 66th session of ISCA).

Heteropnuestes fossilis

Order: Cypriniformes, Division : Siluri, Family : Heteropnuestidae

Vernacular names Bengali : Singhi, Hindi: Bitchu, Oriya: Singhi, Tamil: Thelimeen, Telugu : Mapujella)

The body of *Heteropnuestes fossilis* is elongated, head is dorso-ventrally flattened, and the body is blackish brown. They are present throughout nation. Pectoral fins are strongly serrated spine. The juvenile has narrow anal fin, maxillary barbels are well developed. It is the fish which breed in confined water during monsoon. Induced breeding is highly successful. Its breeding habits is similar to *Clarias batrachus*. It has also nutritious value and its farming is highly profitable.

Channa (formerly it was named as *Ophiocephalus*) : They belong to order Channiformes. They are also known as murrels. There are five important species of *Channa* for culture purpose; they are *Channa marulius, C. striatus, C. gachua and Channa stewarti*. In Indian species, scales on head moderate, 9 or 10 between orbit and angle of preopercular, 15 or 16 between dorsal fin and snout. The shape of the body is long and cylindrical in all Indian species.

Channa marulius.

Body is elongated and round. The head of these fishes is similar to the head of the snake, hence also commonly classified as snake headed fish. Fins are without spines. Air breathing organs are present. The colour is greenish gray dorsally, the belly is yellowish orange with spots scattered all over the body. There are six blotches along the flank. The fish is inhabitant of freshwater—commonly found in ponds and stream, reservoir rivers. It is tasty having fine flavour and highly preferred. The caudal fin possesses a spot of black colour, which is surrounded by an orange ring. It is carnivorous bottom feeder.

Murrels breed in ponds, tanks and swamps. It breeds during April to October.

Channa striatus.

The head is dorso-ventrally flat, the body is laterally compressed. Four pairs of barbels provide the fishes. A spot is present on the pectoral fin. The fishes inhabit fresh and brackish waters of India. The fishes spawn during July to August in marshy pools, ponds and lakes. It is found in freshwater and brackish waters, and during monsoon they migrate to low laying area and paddy field and breed there also. It is also reported that it starts breeding at the outset of southwest monsoon. It has the ability to adapt to freshwater and brackish water and also with low oxygen content, i.e., poor environment conditions. It is highly beneficial fish for large scale rural farming in India. It breeds twice during monsoon and post monsoon months, whereas O. punctatus breeds almost throughout the year (Kohli, 2003) .

Channa gachua. Body is long and cylindrical and head is depressed. The dorsal surface is dark black; laterally the colour is deep blue and provided with orange and blue strips.

Clarias batrachus : The culture of magur (vernacular name of *Clarias batrachus*) is done either by the collection of fry or fingerling from natural sources from nests found in ponds/race way/marshy places or spawn can be generated by induced breeding, i.e., controlling endocrine function by giving appropriate injection of pituitary extract of fish pituitary or by giving injection of HCG (Human Chorion Gonadotropin) or by giving injection of synthetic compounds available in the market.

The eggs of this fish are demersal, spherical and small in size varying in diameter from 1.5-2.00 mm. Normally, the fecundity of this fish is 3000 to 40,000.

The temperature of water suitable for spawning is 26-29 °C, pH should be 7.0, Ammonia 0, Nitrite 0, Nitrate 0, GH, 100 ppm and KH 70 ppm.

Hatching from the fertilized eggs takes about 6-8 days. In hatching the fry comes out to combat.

C. striatus and C. gachua are similar so far as the breeding is concerned.

Notopterus notopterus (Razorfish)

There are two species of this genus, *Notopterus notopterus* (Razorfish) *and N. chitala* (Figs. 24.7A and B). They are inhabitants of freshwater, very tasty and liked by the people. The breeding in captivity is not known.

Body is silvery white with minute brown spots. Dorsal profile is plain, body is without markings.

Juvenile is similar to adult, is found all over India particularly in lakes and ponds. It inhabits in brackish water also.

Seeds are available in large lakes and ponds from July to August.

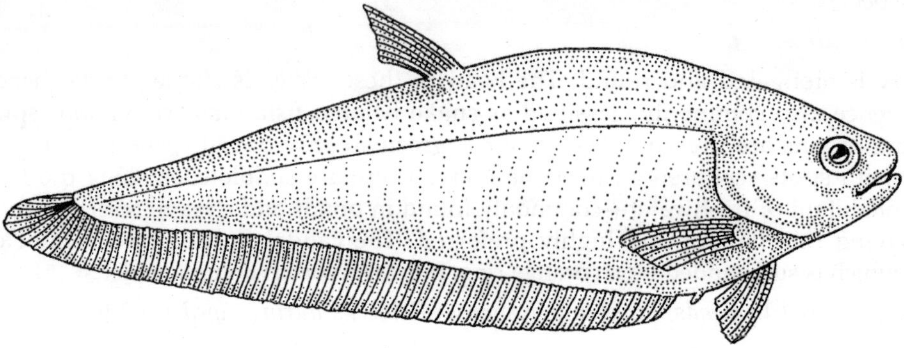

Fig. 24.7A : *Notopterus notopterus*

Notopterus chitala (Hump backed razor fish)

The peculiar feature of this fish is that they have hump dorsally. The color is brown. It has vertical oval patches running almost at equal distance from anterior to posterior end. The oval patches being prominent in juvenile fish otherwise resembles with adult. Its marketable size is 30 cm. It is tasty and sport fish.

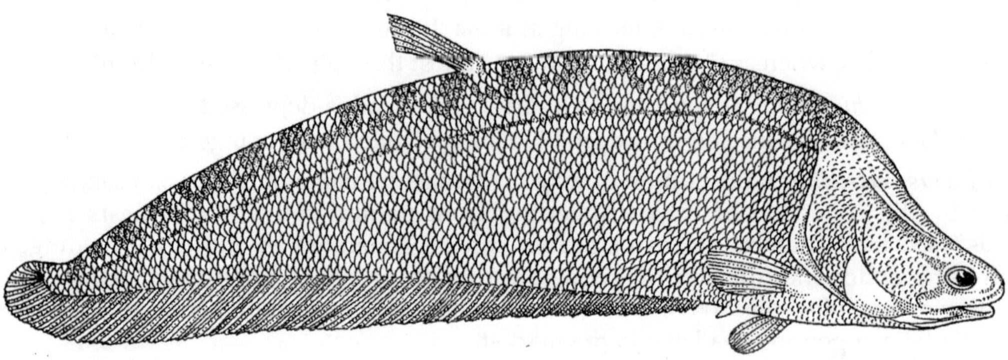

Fig. 24.7B : *Notopterus chitala*

It inhabits in large freshwater rivers of northern and central India. It is found in reservoirs in huge quatities (Kulshrestha, 2003). The juveniles in considerable quantities could be collected from fresh swamps. Spawns are available from August to October.

Mastacembelus armatus

Order : Mastacembelliformes

Family : Mastacembelldae

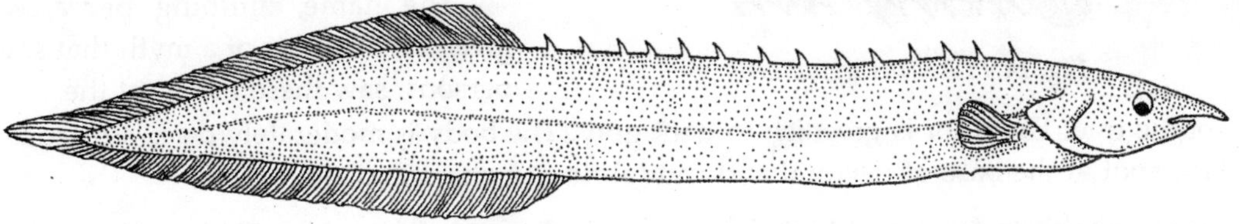

Fig. 24.8 : *Mastacembelus armatus*

Vernacular name : Assamese : Bami, Bengali : Bami, Oriya : Bummi, Hindi : Bamm, Tamil : Aaraah, Telugu : Mudibommidal

The body is cylindrical and laterally flattened. Snout is pointed with rostral appendage. The colour of the body is rich brown with zigzag lines. The pectoral fins are spotted.

Could be cultured in impounded waters with good aeration. It attains the size of 50 *cms.*

Anguilla begalensis *Indian log fin eel*

Order : Anguilliformes

Family : Anguillidae

This family includes two species, *Anguilla begalensis and Anguilla bicolour*, both are tasty and profitable. The vernacular name in Bengal is Kuchia, in Oriya it is called as Thumbi while in Tamil it is known as Porivelangu.

Body is elongated and the fish is eel like. The body is naked or with minute scales which are embedded in skin. Dorsal and anal fins are long and narrow. The body colour is yellowish brown. The skin is shining.

It inhabits in rivers and associated lakes and ponds throughout India. The juveniles in large numbers are collected by nets from estuaries. They migrate in rivers at night only. The culture is also very successful.

Anguilla bicolour bicolour

Vernacular name : Andaman : Jeetahadah, Tamil : Vilangu, Telugu : Malagu pamu.

It differs from *Anguilla bengalensis* in body colour. The body is bluish brown dorsally and lighter ventrally. It is also long and snake-like in outline.

The juveniles are known as elevers and look like earthworms. They inhabit freshwater rivers, ponds and associated lakes. Seeds are available in huge quantities and can be collected for culture during night at the time of migration. Elevers can be successfully cultured.

Anabas testudineus (Climbing perch)

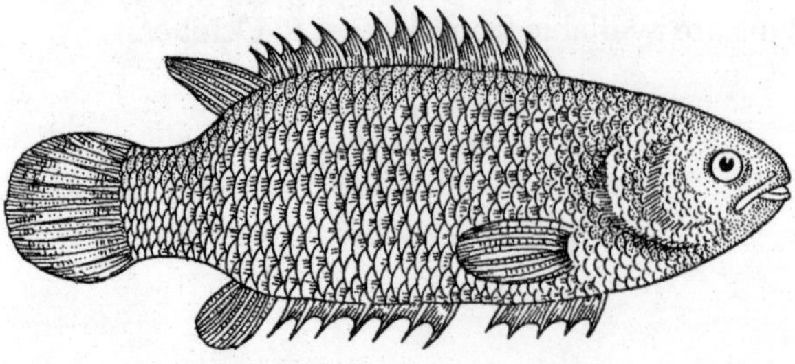

Fig. 24.9 : *Anabas testudineus*

The classification is as follows:

Order : Perciformes

Family : Anabantidae

Vernacular name : Hindi: Kowai, Tamil : Panayeri kendai,Telugu : Thattimata, Oriya : Kou.

The name climbing perch was given on the basis of a myth that some persons saw these fishes at the top of the tree. The color is greenish to dark gray. The body is moderately deep. Presence of a black spot at the base of pectoral and caudal fin is a characteristic feature.

They inhabit in fresh and brackish waters and are also found in ponds and mostly in swamps. It is also a tasty fish. Breeding period is rainy season.

Colour of the female and non-breeding males is watery gray to yellowish. The dorsal fins are black with red margins. Pectoral fins light red.

They are found inhabiting in lakes, pond, and estuaries, highly adapted to salinity. It is a mouth breeder. This fish could be nuisance competitors with other fishes, hence should be treated as a separate community.

Tor tor (Mahaseer). The systematic position is as follows:

Order : Cypriniformes

Division : Cyprini

Family : Cyprinidae

Its body is stout, fairly deep. The colour is grayish green dorsally. Flanks are pinkish green. A mixture of gold and yellow colour shades spread over the head and sky blue colour on the operculum. It is found on montane and sub-montane area of Northern India. Breeding season is August to December.

Farming In Brackish Water

There are some animals which inhabit in brackish water. Some fishes and prawns live in sea water but their larval stages, post larval/ juvenile and adult migrate to the brackish water and stay there till they become adult but they again migrate to sea for the maturation. Similarly, there are some fishes which breed in freshwater but their larvae and post larval stages migrate to brackish water till they become adult. Some of them completely dependent on brackish water for their larval development. About 12 species of finfishes, four species of prawns and one species of crab is described as under (Handbook of aquafarming).

	Name	Distribution	Breeding ground	Development
1.	*Lates calcarifer*	Seas, estuaries & rivers	Coastal waters & Brackish water	Sea brackish water
2.	*Chanos chanos*	Costal water & estuary	Brackish water	Brackish water (Fry: March to June)
3.	*Mugul cephalus*	brackish water	Breed in Chilka and Pulit lakes.	
4.	*Liza macrolepis*	Coastal, estuary & backwater	Breed during monsoon	Induced breeding successful Pulicat, Chilka Kayankulam lakes
5.	*Liza parsia*	Coastal, estuaries, lagoons	Seed available in brackish water	Malta estuary in Orissa
6.	*Valamugil cunnesius*	Coastal water, estuaries, back-water frequently enters fresh water	Breed in natural water (sea/brackish) west monsoon.	Cultured and varying salinities
7.	*Etropius suratensis*	Brackish water and coastal regions close to rivers.	Breeds in fresh-water/coastal/brackish water. Delicious brackish water fish of high market value.	Seeds available in brackish water, ponds and lakes. It can be cultivated in brackish water.
8.	*Eleutheronema tetradactylum*	Inhabits sandy shores and muddy estuaries.	Seed available in brackish water lagoon and back water, Bree-ding grounds.	
9.	*Polynemus threadfin* (paradise)	Inhabits shallow sandy inshore areas.	Breeding season is March to June.	Juveniles are collected from estuaries. Highly suitable of cage culture.
10.	*Megalops cyprinoids*	Highly tolerant to salinity variation, hence found in coastal waters, lagoons, backwater of eastcoast of India	Breeding ground and seed collection.	
11.	*Sillago sihnama*	Coastal fish, migrate to estuaries from February to May	Plenty in Chilka lake.	Farming could be possible in sea.

Prawns which are inhabitant of Brackish Water

Vemband lake prawn, *Macrobrachium idella*. It is endemic in Kerala backwater. Breeding occurs in the sea and larvae are available in estuaries, backwater and in surf of intertidal region seed are produced in hacteries also. Seeds are available in plenty in all the backwater of India. Culture is profitable and also exported.

Penaeus indicus (white prawns) : Available in coastal areas. Breeding occurs in sea and larvae are available in estuaries, backwater and in the surf of inter-tidal region. Seeds are produced in hatcheries. Cultured in coastal regions.

Penaeus merguiensis : This is present in North of Karwar in west coast and North of Chilka in the East coast of India.

Seed. Breeds in sea and juveniles are produced in hatcheries also.

Crabs. *Scyila serrata* (Mud crab). Plenty of seeds are harvestable in the estuary regions by dragging from November to March. Culture can be done in baskets.

Farming in the sea is also possible and if successful it will create a blue revolution. Amongst fin fishes which can be cultivated in cages known as cage fisheries, they are 1, *Lutjanus fulviflamma, Lutjanus malabaricus, Lutjanus sebae, Lutjanus argetimaculatus, Epicephelus malabaricus.* Sea is a great source of obtaining fishes for sale and can be categorised as capture fisheries. Recently great progress has been done to culture fresh as well as sea fishes which have both economic as well as nutritive value and be cultured in cages and great profit may be obtained by cultivating in sea water.

Crab Culture

Spiny lobster is present close to shores. Rearing them is of great commercial importance. The seeds are available in inter-tidal region, in mangrove swamps and also in coral reefs. The important lobsters are as follows :

Panulirus homarus, P. Longipes, Panulirus ornatus., P. penicillatus, P. polyphagus, P. versicolor (Silas,1992).

Bivalves Molluscs

It is not useful for major human population, but could be well used for poultry food. These are grown on ropes suspended on floating raft.

REFERENCES

Beaumont, A.R & Hoare, K., 2003. Biotechnology and Genetics in Fisheries and Aquaculture. Black Well Publishing. pp. 158.

Dhandapani, P 1995. Handbook of Aquafarming, Shrimp Hatchery. Anaswara Printing & Publishing Co, SRM Road, Kochi, India.

Dubey, Kiran, 2003. Biology, reproductive biology and embryonic development of carps. 12-29. In: Carp and catfish breeding and culture. Published by CIFE, Versova, Bombay.

Gloria, A., Kapoor, B. G. Chardon, M & Diogo R 2003, Cat fishes Cat fish, Cat fish book.

Jhingran, V.G. Presidential Address, Indian Science Congresss Association. 66th session.

Jyoti; M. K. & sharma, A. 2006. Fishes. Aid to collection, preservation and identification. Daya Publishing House, Delhi.

Kholi, M.P.S. (2003) : Breeding of Air breathing fishes. In : Carp and Cat fish breeding and culture, Published by CIFE, Versova, Bombay. PP. 177-181.

Kushrestha, S.D., 2003. Fish and Fisheries (in Hindi). ABD Publishers, Jaipur, India.

Piska, R. S. 1999. Fisheries and Aquaculture. Sri Uday Printers, Hyderabad.

Sandhu, G. S. 2007. Applied fisheries in India : issues and concern, concept publishing company India.

Silas, E.G. 1992. (Ed.) Freshwater Prawns. Proceedings of the national symposium on freshwater prawns. Kerala Agriculture University, Trissur,India. EMJAY Offset Printers, Kochi.

Srivastav, CBL 2005. A textbook of Fisheries and Indian Fishers, Oslar Publishing 302, Nitika Tower- II, Complex, Delhi.

Winfield, I., J. & Nelson, Joseph, S. (Eds.) 1991. Cyprinid Fishes : Systematics, Ecology and Exploitation. Kluwer's Fish and Fisheries Series. University of British Columbia.

Chapter 25

CONSTRUCTION OF AQUA FARM

Construction of Aqua Farm / Fish Farm and its Maintenance

The major concern of the commercial fish farm is the profitability. The farmer's profit is calculated on the cost of construction of fish farm, the production of fish (spawn, fry, fingerling and table fish) per unit of the water area and profit obtained by selling the product in the market. Appropriate site is the site which fulfil the social need of the community (Anon, 1995).

The size of the pond depends upon the type of stages to be kept and normally they are classified as follows:

1. *Nursery pond* may be 0.4 hectare (one acre), it is for rearing spawn to fry (2.5-4 cm).
2. Rearing pond may be about 2 hectares (5 acres) used for rearing fry to fingerling.
3. Stocking pond may be five hectares in area for rearing from fingerling to table fish (Zaidi, 2000 Langur, 2002).

The followings points are to be kept in mind for the construction of aqua farm.

1. The **initial capital** construction and operating cost of artificial aqua farm should be as low as possible. The farm will give high lèvel of the yield. Water source should be available with sufficient quantity of water. So, daily exchange of water is also possible. Minimum facilities for aeration of water are available. Govt. of India has also provided guidelines. The main points are as follows (Shang, 2005) :

1. Topography
2. Hydrology
3. Soil condition.

1. TOPOGRAPHY

A proper survey is to be made for the site selection. The ideal site for artificial fish farm is the land that is flat or nearly flat with at least about three metres deep soil. The water table should be 30-50 cm from the bottom of the pond and the maximum distance should not be more than 5-6 metres. The best soil for the construction of the aqua farm is clay loam to sandy clay loam; such soil comes under moderately textured soil category. Flooding possibilities are to be avoided, because the flood will increase detritus effects, erosion potential and for the introduction of undesirable species into the water of the fish pond. The site should not be hilly or stony. The site should be close to the road for proper transportation of fish to market and also the transportation of construction material at the site.

The gravitational flow may be checked because the gravitational flow is the cheapest mean to fill the pond. The slope should be more than 2%. It will avoid additional cost, if the water is to be removed between two harvests. A slope of 5% is not desirable because of the extra construction cost associated with back filling and increased possibility of run off problems. There should not be thick forest and high trees. Other infrastructural facilities such as electricity should be available.

2. **HYDROLOGY**

Fish cannot survive without water. Hence water is most important and crucial for the maintenance of the aqua farm. Sources of freshwater are several. The ground water resources are wells, tube wells, and springs. Ground water is better source for aquaculture as it is more uniform, free from predatory insects, less pollutant and having constant water temperature. The correct assessment of water for the use of fish pond depends on its physical, chemical and biological properties and, therefore, the correct analysis of water is required for any fish culture activities.

Physical Properties

Dissolved Oxygen

Surface water comes in contact with atmosphere; it absorbs oxygen, depending upon temperature and pressure of that place. Water should not have less oxygen and more carbon dioxide. Dissolved oxygen (DO) is the most important parameter for water quality to maintain good fish population in the fishpond, the (DO) should be around 6.0 mg/lit.

Temperature plays an important role for the maintenance of dissolved oxygen. The higher the temperature, lower will be the solubility of oxygen. Therefore, in summer the condition is worst. At low temperature the oxygen solubility is more, so DO is more in winter. Low DO in monsoon may be due to increased turbidity and high growth of bacteria, bacteria utilized oxygen for their metabolic activities.

Sufficient dissolved oxygen (DO) maintains aerobic condition in the water column and also helps to avoid septic conditions. In anaerobic condition decomposition of organic material takes place, which leads to the formation of toxic gases like CO_2.

Gases such as hydrogen sulphide, methane should not be present. Excess quantity of iron and other metals are estimated.

There are numbers of new equipments which are available in the market and analysis can be done by the farmer himself or the sample can be sent to Laboratory for knowing the correct picture (Jackson,1967; Pollingher, 1981; Wetzel & Liken, 1991). Winkler's method is supposed to be most convenient and popular method for the measurement of DO (dissolved oxygen).

Colour

Colour of water is not a parameter in water analysis but would be useful to know the humic substances in water. Humus is formed by the microbiological decomposition of organic matter in the soil.

Pure water is colourless, pure water is not found in natural condition. Water in the nature contains foreign substances, dissolved or suspended which gives colour.

Suspended Solids

The solids can remain in suspended or dissolved forms. The solids suspended in the water are inorganic and organic particles (Segall & Alpert, 1990). They come under the head 'chemical parameters.' The inorganic solids are clay, silt and other soil constituents mixed in the water. Organic substances are generally biological solids (algal cells, bacteria, etc) which are common constituents of surface water and may cause diseases. Some of the strains of algae produces toxin, which causes damage to the fish life.

Turbidity

Turbidity is defined as the measure of the extent to which the light is either absorbed or scattered by suspended material in the water. Turbidity of water is mainly caused by sand, clay, silt, discharge effluents, decomposed organic matter and microscopic organisms that interfere with the penetration of light in ponds/rivers. The minimum values of turbidity are in winter and summer seasons. The maximum turbidity is in monsoon period.

The main cause of turbidity in the fish ponds is due to the erosion of colloidal material like silts fragments and metal oxide from the soil. The main cause of turbidity in fishpond is vegetable fibres and microorganisms. Turbid meters are available in the market. Transparency of water is ranging from 20.0 to 58.5 ms. Sacchi disc is also used for the measurement of turbidity.

Temperature

Temperature of water is the most important factor in an aquatic environment. The survival of fish and biological species depends on temperature. Temperature also has a pronounced effect on the solubility of gases such as dissolved oxygen (DO) and carbon dioxide. Both are essential parameters for culture of fishes. At 7 °C, 14 mg/lit of DO is to be dissolved in water. At 20 °C (temperature) DO of water is 8 mg/lit.

pH

The biochemical activities depend on the pH of the surrounding water.

The disinfections, water softening and hydrogen ion concentration is generally expressed in term of pH, which is defined as the negative logarithm of the hydrogen ion concentration. It is simply a means of converting unwieldy number into a more convenient system.

Very small changes in the hydrogen ion concentration cause important alterations in the living cell. It affects metabolic and physiological processes of the fish. Protein and enzymes work on a specific pH. pH is a factor in chemical coagulation. It is generally agreed that unpolluted river has a pH 7.7 to 8.7.

The pH from 6.5 to 9 are generally regarded as most suitable for pond fish production

pH 4.5-5.0 is toxic to warm water fishes.

pH 5.5-6.5 causes low reproduction.

pH 10.0-10.3 is not good for health.

pH 10.5-11.0 is lethal to fish.

Water with heavy algal growth often has a pH value as high as 9-10. pH can be maintained by liming the farm.

Liming of the Fish Farm

Liming in the pond is done only for the adjustment of pH. It is an old age practice and is used for correcting the acidity of soil and water; it also helps in speeding up decomposition of organic matter. The decomposition of organic matter results in the conversion of nutrient of the soil available for the fish to grow. The liming also helps in keeping the sanitation of the pond.

Due to the presence of lime, the contents of carbonate are increased and avoid in the lack of carbon dioxide in the water. If there is lack of carbon dioxide, it will not act as limiting factor for photosynthesis. It also adds calcium in the water and soil. After liming and with a gap of 20 to 30 days, the next step is to fertilize the pond. The pond is to be fertilized by the fertilizer which contains P_2O_5 and nitrogen. The ratio should be 1:3 for the freshwater fish farm and 1:1 for brackish water pond. The fertilizer should be applied @ of 5-10kg/ha at 2-4 weeks interval. High rate of nitrogen fertilizer is also recommended but not essential. Urea and ammonia fertilizers are also recommended because they are cheap. Nitrate fertilizer is more recommended as nitrogen is not toxic and is fully oxidized. In fish culture, the effect is brought about through longer chain of conversion consisting of soil water column fertilizer bacterial-aquatic plants-zooplankton and zoobenthos fish. The objective of fertilizing the fish pond is to increase growth of fish food organisms by providing the good nutritional status of the soil as well as water of the pond.

Total Dissolved Solids : The material remaining in the water after filtration for the suspended-solids analysis is considered to be dissolved. The water is the universal solvent. The inorganic substances are minerals, metals and gases, while the organic material comes from the decaying vegetation, from organic chemicals, and the organic gases are common dissolved organic constituents of water. For finding out the TDS it is desirable to know the *ion balance.*

The ions usually accounting for the vast majority of TDS in natural water is presented in the table.

Biochemical Oxygen Demand (BOD) : It is the amount of oxygen required by the microorganism to decompose biological organic matter under aerobic condition and at a standard temperature and time. BOD determination is most useful technique to assess the level of the organic pollution in the river system. Rhyner, *et al.*, 1995 suggested that water having BOD value of 3.0 mg/l or more is of doubtful quality and that with more than 5 mg/l is bad. The following material is needed for proper nutrients.

1. Proper nutrient supply like $CaCl_2$, Sodium phosphate (Na_2PO_4), Magnesium sulphate ($MgSO_4$) and Ferric chloride ($FeCl_3$).
2. pH should be 7-8.
3. Heavy metals and toxic substance such as Cl_2 should be nil.
4. Maintenance of temperature and time (Rhyner, *et al.*, 1995).

Chemical Oxygen Demand (COD) : It is the amount required for the chemical oxidation of organic matter. Chemical oxygen demand represents chemically oxidized load of organic matter in water. Normally, COD values range between 7.2 mg/l to 68.3 mg/l. The upper limit for COD for drinking and irrigation is 20 mg/l and 150 mg/l respectively.

Biological oxygen demand and chemical oxygen demand (COD) have great importance in water quality assessment since these are directly related to organic matter content in the water and the quantity of oxygen required for stabilization. BOD gives a fair idea of the quantity of biodegradable matter present in water which is aerobic decomposition by micro organisms (Young,1984).

3. SOIL CONDITION

The selection of soil is important point for consideration for the construction of pond. The major soil groups (of aquaculture) available in India are black, red, lateritic and alluvial soils. The occurrence and development of microorganisms and higher plants depend upon the chemical environment and that is why the study of soil reaction is so important (Antipchuk, 1979, Segall, & Alpert, 1990). Soil is classified as follows (The word 'loam' is used as it became popular).

1. Fine texture soil may have clay, silty clay and sand clay.
2. Moderately fine texture soil contains silty clay loam, clay loam and sandy clay loam.
3. Medium textured soil is loam, silty loam, or silt.
4. Coarse texture soil is sandy loam or loamy sand.
5. Very coarse texture soil contains sand alone.

The main consideration for the selection of aqua farm is based upon two important points as under:

1. Water retention capacity of the soil
2. Fertility of the soil

The permeability and fertility of soil depend upon the texture and the structure of the soil. The retention capacity of water of the soil is known as permeability. The methods, both scientific and conventional for the estimation of permeability, are known.

Best soil for the construction of the aqua farm is clay loam to sandy clay loam. Further, clay is impermeable to air, water and plant roots. The clay also contains aluminum hydroxide as one of the constituents. Such types of soil encourage the development of benthic community. Nitrogen and phosphorus should be available in the soil.

Acidity and alkalinity of the soil is also important factors and so the redox potential, nitrogen and phosphorus should be available in the soil.

Fertility is important as it affects the cost of feeding. No external feed is required if fertility is good. The cost may be reduced if the fish could derive food from the pond under natural condition. The soil with large amount of clay may cause difficulties because clay is sticky when wet, dried out slowly and becomes hard.

Conventional Methods for testing Soil

1. The **simple test** and non-conventional method is to make a ball of the soil with the help of water and then throw this ball up to 60 cm above and then take it in the palm of the hand. If the ball is broken into pieces then the soil is not fit for the excavation of pond as the soil contains more sand.

2. **Another method** is to take the soil of the site, to make a ball by adding water and place it in the sun. If on drying the ball is broken down, then the soil is not suitable. If the space of the ball remains almost the same, then it shows that the soil contains sufficient quantity of clay suitable for the excavation of pond (Singh, 1997).

The consistency of the soil can be measured by **Attenberg limit**. The liquid limit and plastic limit should also be ascertained. If on the palm the ball is having almost the same shape, the soil is suitable of the excavation of pond as it contains the required quantity of clay. Soil of high contents of organic matter should be avoided in the level construction because of settling from decomposition could cause creaking and seepage.

Coefficient of Permeability

The permeability can be measured by finding out the co-efficient of permeability and by calculation of seepage rate. The coefficient of permeability should be less than $K= 5x\ 1x10^{-6}$, such soil is good for the bottom construction. The seepage should not be more than 125 cm per year.

Conventional Methods for checking Permeability

A simple technique for the measurement of permeability is to dig a hole, which is one metre deep, one metre broad and one metre in length. This is to be filled with water up to the brim and the level of the top is maintained by filling with water. If on the second day in the evening, it contains 80% of water, the soil is fit for the excavation and Constructions of pond as it has a good retention capacity.

The second and more reliable method is to have a hole by bucket augur up to one metre deep. The circumference and depth should be correctly measured. The filling of water up to the surface is to be continued till half an hour with 5-minute intervals. With the help of one metre scale and stopwatch note the water level till becomes stationary. The co-efficient of permeability K could be calculated by the following formula :

$$K = \frac{r.I_n\,(h_1 \div h_2)}{2(t_2\text{-}t_1)}$$

r = circumferance of the hole; I_n = Neparian logarithm; h_1 and h_2 level of two reading of water; t_1-t_2 = time in second between two measurements. K is obtained in milli second. $2(t_2\text{-}t_1)$

REFERENCES

(Anon) 1995., American Public Health Association, Standard methods for the examination of water and waste-water. 18th ed. Washington, New York.

Antipchuk, A.F., 1979. Microbiological control in fish farming ponds. Pishevaya Promishlenostov, Moscow. 145 pp. (in Russian).

In : Carp and catfish breeding & culture. C.I.F.E, Versova, Mumbai.

Jackson, M.L., 1967. Soil chemical Analysis. Prentice Hall of India Pvt. Ltd., New Delhi.

Langur, R.K., 2002. Management of carp rearing ponds. 62-65. In: Carp and catfish breeding & culture. C.I.F.E, Versova, Mumbai.

Pollingher, U,1981. The structure and dynamics of phytoplankton. Res. 3, 93-104.

Rhyner, C.R.; Schwart, L.J.; Wegner, R.B & Kohrell, M.J, 1995. Waste Management and Resource Recovery. Lewis Publisher, BocaRaton, New York, London, Tokyo.

Segall, L & Alpert, J., 1990. Compost market strategy. Biocycle, 31(2): 38.

Shang,Y.C., 2005. Chapter 24. Interlink. Economic aspects of aquafarm construction and maintenance. pp. 1-14.

Singh, C 1997. Muchli Palan (Fish Farming; in Hindi), Prakashan Nidashalai,Govind Vallabh Pant Agriculture & Technical University, Pant Nagar (U.P), India.

Wetzel, R.G. &, Liken, G.E., 1991. Limnological Analyses.

Young, J.C., 1984. Waste strength and water pollution parameters. Water Analysis,. Vol.111. Academic Press, New York.

Zaidi, S.G.S. 2000. Carp hatcheries, commercial utilities and operation. 44-53.

Chapter 26

AQUATIC WEEDS

Aquatic Weeds and Management in Freshwater Aquaculture

It is unanimously accepted that the main concern of the commercial fish farm is the profitability. One of the prime difficulties is uncontrolled growth of aquatic weeds. The control of aquatic weeds in almost all the systems of aquaculture is a problem noticed all over the world. The weeds problems are more in aquatic systems of tropical and semitropical countries like **India** and then in the **western countries.**

The weeds can be defined as unwanted and undesirable plants that are adopted to grow and reproduce under aquatic conditions. Limited growth of aquatic plants may be useful in the maintenance of water quality and may serve as shelter and substrate for food organisms in water but their uncontrolled growth may cover the surface of the water of the pond resulting large numbers of problems, which are mentioned below :

1. The penetration of the light will not be adequate and thus affect the productivity of fish production.
2. The nutrients in the water, which is generated by fertilizing the pond, will be consumed more by plants, and fish will not get the sufficient nutrients, hence it will badly affect the growth and production rate of the cultured species.
3. Blooms of the algae often results in the depletion of oxygen (as a result of dead and decaying algal mass that quantity of oxygen production will be reduced) causing anoxia (less quantity of oxygen) for the fishes cultured in the pond. Dissolved oxygen (DO) is the parameter for water quality and reflects the physical and biological processes prevailing in the water. To maintain good fish population in the fish-pond, the DO should be around 6.0 mg/lit.
4. Fishing with nets will be difficult. It will be a problem for cage fisheries also as the algae will stick to the cages.

Common aquatic weeds comprises both filamentous and single cells algae. They have been divided according to look, 1996 & Naithani, 1990 as follows :

A. Floating weeds : They are free floating with their leaves over the surface of water while roots are under water. The most common are *Eichornia, Azolla* and *Pistia* (Fig. 26.1).

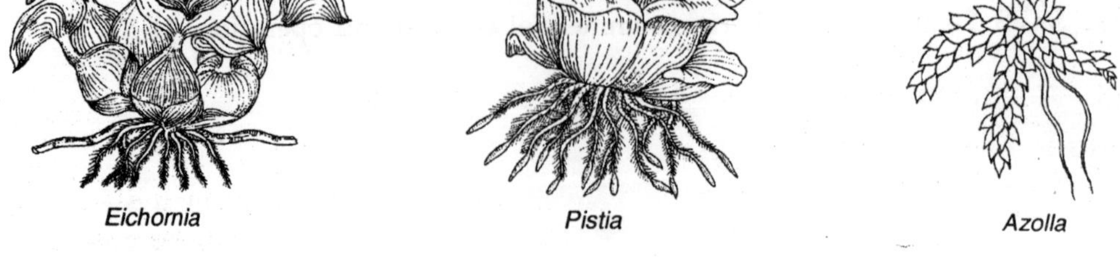

Fig. 26.1 : Floating plants

B. Emergent weeds : They possess roots which are present in the soil of the water while their leaves or shoots are present over the surface of water, e.g. *Utricularia* Nymphaea, Trapa, Myriophyllum, Otella, Vallisneria etc. (Fig. 26.2).

C. Submerged weeds are those weeds, which are completely submerged under water, but their roots are present in the soil of the pond. The common examples are *Hydrilla* and *Naja*. The *Ceratophyllum* and *Utricularia* also come under the category of submerged weeds but their roots are not present in the soil and the roots are free floating (Fig. 26.3).

D. Marginal weeds, grow on the sides of the pond and are rooted in the water-logged soil. The common marginal weeds are *Typha* and *Phargmites* (Fig. 26.4).

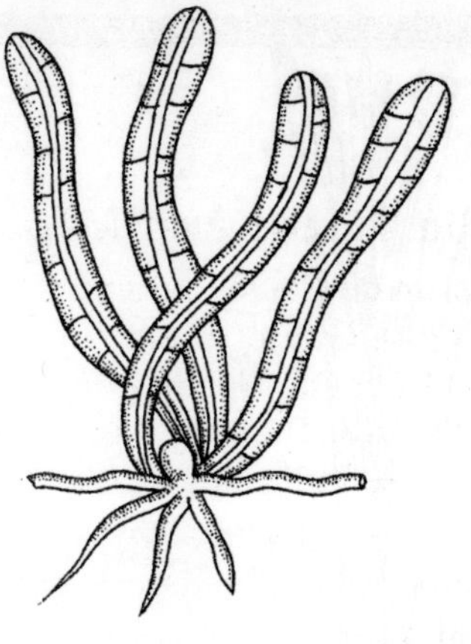

Fig. 26.2 : Anchared Submerged Hydrophytes

E. Filamentous algae, form mats in the marginal area or scum in the main water body. The most common filamentous algae in the ponds are *Spirogyra* and *Pithophora*.

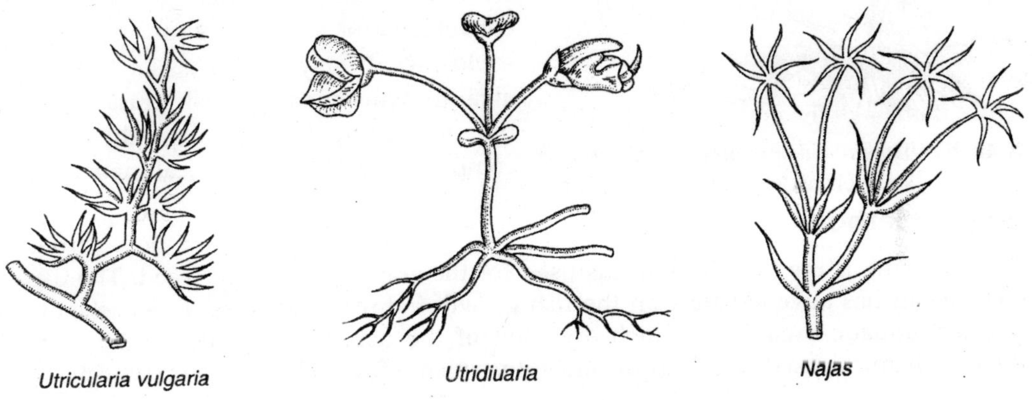

Fig. 26.3 : Submerged Weeds

F. Planktonic algae proliferate very rapidly and form algal bloom in water. The most common example is *Microcystis* and *Anabaena* (Sharma Sangappa, 1993) .

Control Management *:* Following are the methods for the control of weed :

1. Manual
2. Mechanical
3. Chemical
4. Biological
5. Control by intra-specific competition.

Manual Method

The manual method is the most convenient, cheap and easy in India and in the developing nations, since there the labour is cheap. It is not long lasting because the weeds grow again as it is difficult to eradicate the weeds completely.

The emergent and marginal weeds are removed by pulling them with hand. They are kept under check by cutting their floating roots repeatedly. The floating weeds are removed either by hand or wire-coirs nylon-nets. For the clearing of rooted submerged weeds simple methods like hand-pull bottom rakes, bamboo poles with toothed pronged can be used.

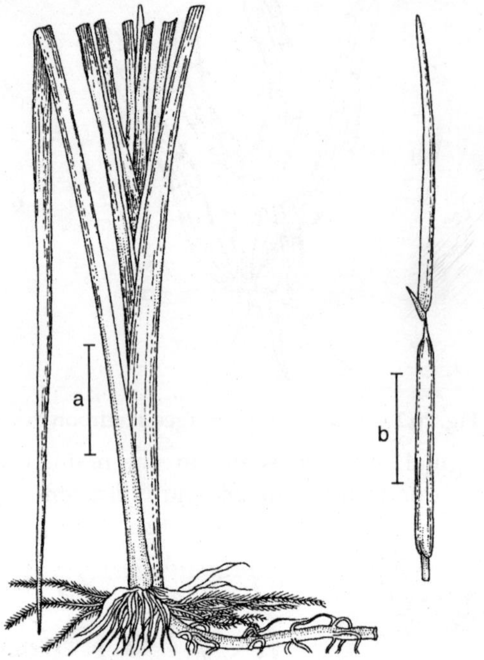

Fig. 26.4a, b : *Typha dominengesis* (a) sterile shoot (4 cm), (b) inflorescence (8 cm)

Repeated removal, combined with biological or chemical method, may be continued to control the weed.

Mechanical Methods

The mechanical methods are many; their application depends upon different size of lakes. They are different if they are used for extensive fish farms. For moderate fish farm of 100 ha, the most common method is the weed cutter. It is used for removal of submerged and emergent weeds. The weed cutters are either flat-bottomed boat fitted with cutting beams or other cutting devices. Amphibious boats fitted with weed-cutting devices are especially convenient for use in shallow ponds and enclosures. It may appear easier to remove algal growths such as musk grass (Charra) but it is difficult to remove the clumps completely by this mechanical method.

Chemical Methods

The following is the list of chemicals used for the eradication of the weed. The use of these chemicals has adverse effect on the fish growth indirectly because these chemicals badly effect aquatic organisms which are food of the fish. Some of the herbicide and weedicide, chemical used are Sodium arsenite, xylene, 2,4 dichlorophenoxy acetic acid (2,4 D), etc.

Biological Control of Aquatic Weed

There are large numbers of biological control methods. The use of herbivorous fishes and other aquatic animals has been used for biological control of weeds. The biological control of aquatic weeds is easy and popular and is achieved by fishes those are phytophagus or herbivorous fishes. The most common fishes are *Ctenopharyngodon idella* (grass carp) *Carassius carassius, Carassius auratus, Tilapia mossambicus*. These fishes consume aquatic weeds as their food. These fishes are having strong pharyngeal teeth and, therefore, could macerate plant material. *Hypophthalmichthys molitrix* feeds on phytoplankton. It is reported that grass carp could consume about 19.9 metric tons of water weed and could produce about 195 kg of fish flash. It can thus control growth of aquatic weed very efficiently.

The aquatic weed can be used as feed for fishes. It can also be used as protein for human consumption.

The following parts of the weeds are used as food for human use.

Ipomea aquatica, young leaves and stem, leaves.

Marsilea species, flowers, leaves and rhizome.

Nymphia species, seeds

Euryale ferox are used for human consumption.

Weeds are also used as the source of energy and biogas.

The common weeds are water hyacinth. The cow dung and water hyacinth are mixed in the ratio of 1:1.

Waste Water Treatment

Some aquatic plants such as *Scirpus lacustris, Ceratophyllum demersum, Spirodella polyrhiza and Lamina minor* are used in waste-water treatment.

Aquatic Weeds are used in Pulp

Some aquatic weeds are used in pulp, paper and fibre industry. There are some other uses of weeds. They are used for the construction of huts, etc.

REFERENCES

Cook, C.D.K. 1996. Aquatic and wetland plants. Oxford University Press, Oxford, New York. Delhi.

Naithani, H.B., 1990. Flowering plants in India, Nepal and Bhutan, not recorded in Sir J.D. Hooker's Flora of British India. I-VIII-I-711.

Sharma, B.D. and Sanjappa, M. 1993., Flora of India. Botanical Survey of India, Calcutta. 1-639.

Chapter 27

INDUCED BREEDING

Major carps are most important species from the point of view of their high food and nutritive values. Hence they have kept attention of scientists and aqua farmers. They have peculiar habit of breeding in running waters of rivers and streams where they have large space for movement. During breeding season, water is sufficiently provided with minerals, O_2 and food contents. This friendly aquatic environment provides stimulus for spawning. Carps do not breed in confined water of captivity even if their gonads are matured and ovulation might have taken place in natural environment.

For the increasing production of carps it is necessary that they should be made to breed in confined water so that increasing demand of good quality fish and their seed could be available. This can be done by adopting induced breeding technique by which ripe or mature fishes breed in confined water when stimulated by injection of pituitary hormone. The pituitary hormone is an important gonadotropin, which is extracted from the hypophysis of a mammal or a mature fish.

Induced breeding of carps in captivity by the use of pituitary hormone injection has been successfully done. Many scientists are involved in doing research to improve induced breeding technique. Several attempts are being made to establish pituitary banks to meet out the need of pituitary hormones throughout the year.

Induced breeding technique has following steps:

1. **Collection of pituitary extract** — Pituitary gland is collected from a mature fish, which is called as a donor fish. Most widely used donor fish is the common carp, *Cyprinus carpio*. The best time for preparation of pituitary extract is May and June. The Indian major carps like *Catla catla* , *Labeo rohita, Labeo kalbasu, Labeo gonius* and *Cirrhinus mrigala* do not breed in confined water and have need to be subjected to induced breeding.

To remove pituitary gland, the head of the fish is dissected and brain is exposed. The gland is immediately removed from the brain and stored in a refrigerator. It may be preserved in absolute alcohol at room temperature. The gland then is homogenized in distilled water. The homogenate is centrifuged and clear supernatant is used as source of hormone to which 0.3% sodium chloride solution is added to it. This extract is ready for the immediate use. If pituitary extract is to be stored for a longer period the glycerin or tri-chloroacetate acid may be used instead of sodium chloride.

2. **Selection of Breeders** — Medium sized fully ripe and healthy fish of around 2 to 4 years of age is preferred for induced breeding. The weight should be 1 to 5 kg. Healthy male

and female breeders should be identified and netted out before the breeding season and should be kept in spawning pools. They should be provided with supplementary food.

3. **Injection of Pituitary extract** — To ensure higher success rate of fertilization it is important to coincide time of ovulation with the release of milt of male fish. For this purpose usually ratio of female and male 2:1 is maintained in every set. Dose of pituitary extract to be given is decided according to age, sex, weight size and state of maturity of both donor and recipient. A dose 2 to 3 mg of gland per kg body weight is given to female breeder. There is no need of injecting dose to the male breeder if it in a state of milt oozing. After 6 hours of the first dose of injection another dose of 5-8 mg of gland per kg of body weight may be given to female if needed. However, a dose of 2-3 mg per kg body weight is recommended for the male breeder. More than 2 injections should not be given. The injection given may be intra-macular at caudal peduncle or shoulder or intra-peritoneal at the bases of paired fins. The first injection should be given at the early hours of the day, while the second one in the evening. Weather should be rainy or cloudy for easy and early spawning. The fishes should be transferred to the breeding hapa after injecting the pituitary hormone.

4. **Spawning in breeding hapa** — A pair of breeder is released into the breeding hapa for spawning after injection of pituitary extract. The breeding hapa is a rectangular case of fine netting. For larger fishes its size is 8′ x 3′ x 3′ , but for the smaller fishes it is 5′x 3′x 3′. It is held on four bamboo poles, one at each corner of the rectangular case. The roof of the hapa may be open or closed. The hapa is made of mosquito net cloth through which laid eggs and milt cannot escape out. Three-fourth part of hapa is summarized in water whereas upper one-fourth part remains in air. After 3 to 6 hours of injection of pituitary extract spawning takes place. The fertilized eggs are white and opaque whereas unfertilized eggs are transparent and bead-like. A hatching hapa is also rectangular and made of muslin or malmal cloth and is open from above. The mosquito net hapa is present inside the hatching hapa.

5. **Precautions for Induce breeding** —

(1) To avoid diseases and parasitic infections, breeders should be properly washed with $KMnO_4$ solution (0.5 g in 100 litres of water) for a few minutes. After this they should be kept in formalin (200 mg/lt of water) for one hour.

(2) Breeder should be protected from mechanical injuries during handling.

(3) Water condition should be favourable having temperature about 24 to 31 °C and turbidity about 100 to 1000 ppm. Flowing water with higher O_2 content is of great use. The intensity and duration of light also affect the induced breeding and spawning. Pituitary glands taken from the same or related species as the recipient species are said to be more effective.

Advantages of Induced Breeding (Jain, 2003 & Srivastava, 2005)

- The seed spawn is timely available, its availability from natural sources is quite uncertain.

- A pure spawn of a desired species is made available. The spawn obtained from the rivers are not pure. They are mixed with the spawns of other species and sorting of pure seed from the mixed spawn is not possible.

- Any quantity of pure spawn can be made available.
- Several carps attain sexual maturity in ponds but they do not breed in confined water. Such fish can be subjected to induced breeding and spawn can be collected.
- It is economical to obtain a spawn from induced breeding experiments in comparison to its collection from the riverine sources.
- The induced breeding technique is very simple and can be learnt even by a layman.

REFERENCES

Jain, K.K. 2003, Induced breeding of carps by hypophysation. In : Carp and Cat fish breeding & culture CIFE. Publication, Versova. Mumbai.

Srivastava, C.B.L. 2005, A TextBook of Fisheries and Indian Fish.

Chapter 28

AIR BREATHING FISH CULTURE

Introduction

Extensive research work has been carried out during the last century for the culture of air breathing fishes [*Clarias batrachus, Heteropneustes fossilis, Channa spp., Anabas testudineus* Fig. 28.1(a-d)]. Now culture of these air breathing fishes are practised also in India. The habit, habitats and the breeding behaviour have been thoroughly worked out (Marshall, 2001; Kohli, 2003).

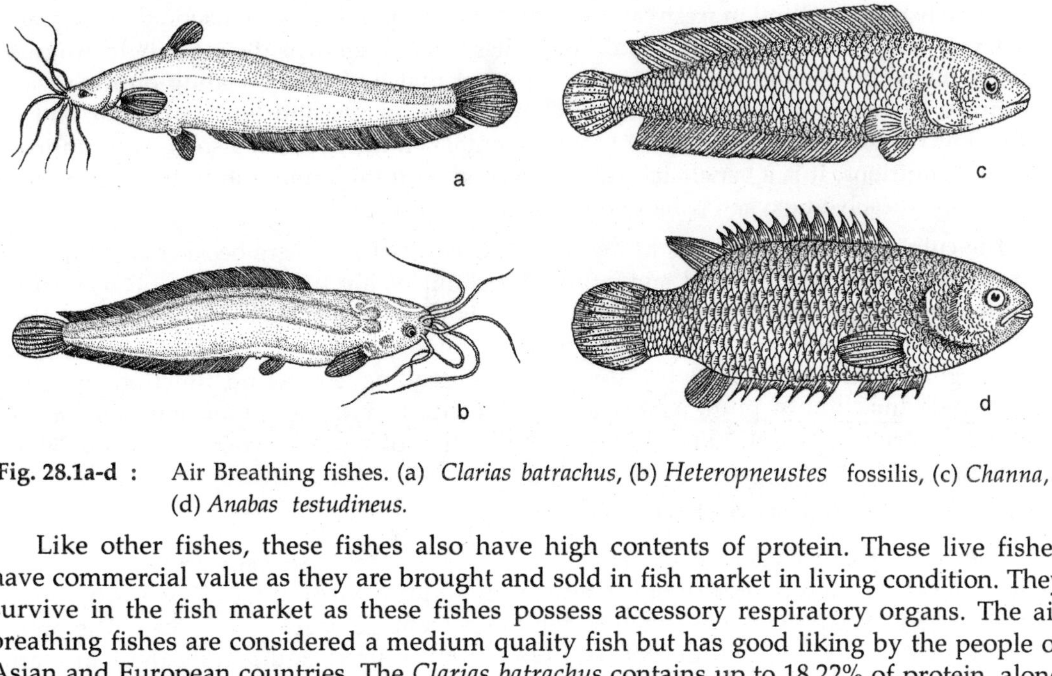

Fig. 28.1a-d : Air Breathing fishes. (a) *Clarias batrachus*, (b) *Heteropneustes* fossilis, (c) *Channa*, (d) *Anabas testudineus.*

Like other fishes, these fishes also have high contents of protein. These live fishes have commercial value as they are brought and sold in fish market in living condition. They survive in the fish market as these fishes possess accessory respiratory organs. The air breathing fishes are considered a medium quality fish but has good liking by the people of Asian and European countries. The *Clarias batrachus* contains up to 18.22% of protein, along with such high contents of protein it also contains calcium and phosphorus. The liver of fishes also contains vitamin A; therefore it is an important natural source. It is highly valuable as nourishing food for convalescing patients.

These fishes are also considered as recreational fishes in some parts of the world and amateur fishermen are allowed to fish in these water against fees based on quality of the fishes caught during fishing.

Amongst catfish, *Clarias batrachus* is the most valuable fish. Other Asian catfishes of commercial importance are *Clarias macrocephalus* and *Clarias lazera* (African catfish or

Nile catfish) and *Pangasius sutchi* (Lee, J.S 1973 : Kloke and Potaros, 1975 Chaturvedi, 2003 & Jain 2003) Other catfishes, such as *Chrysicthys spp.*, are also being experimented for pond and cage culture.

The culture of these fishes can be done by the following methods (Colman 1982; Hogendoorn, 1983.

(1) By the capture of fry and fingerling from the natural breeding ground and rearing them in ponds.

(2) By hypophysation. The hypophysation technique used is similar to the one used in carp culture. Later, spawn, fry or fingerlings and market size fishes are obtained on scientific basis.

The culture of these fishes can be done by the following methods :

A. Pond culture.

B. Race way culture.

C. Floating bamboo cage culture.

D. Pan culture.

Culture of Catfishes, Clarias batrachus

These fishes are found in fresh- and brackish waters and during monsoon they migrate to low laying areas and paddy fields and breed there also. It spawns during July to August in marshy pools, ponds and lakes. It is also reported that it started breeding on the onset of southwest monsoon. It has the ability to adapt to freshwater and brackish water and also with a low oxygen contents, i.e., poor environmental condition. It can also grow in improved swamp conditions. It is a beneficial fish for large-scale rural farming in India because it can sustain low dissolved oxygen (Chaturvedi, 2003 & Jain 2003).

The culture of catfish, *Clarias batrachus*, has been discussed here because it is a popular fish in many parts of the world and cultured techniques are well mastered. The culture of magur (vernacular name of *Clarias batrachus*) is done either by the collection of fry or fingerling from natural sources from nests found in ponds/raceways/marshy places, or spawn can be generated by induced breeding, i.e., controlling endocrine function by giving appropriate injection of pituitary extract of fish pituitary or by giving injection of HCG (Human Chorion Gonadotropin) or by giving injection of synthetic compounds available in the market.

Collection of Fry from Natural Sources

The fry are collected from natural sources. The natural sources are rice fields and irrigation canals. Natural breeding is common in paddy fields during rainy season but the output is poor. The fry could be obtained but the percentage of survival is poor. The process for culture starts by the collection of fry from natural sources and rearing them in rearing ponds. Another method is to obtain spawn and fry by hypophysation in pond under controlled conditions. *Clarias batrachus* lays egg in nests and fry are hatched out in the nests. Incubation period in Indian species such as in *C. batrachus, H. fossilis and A. testudinus* vary between 18-24 hrs while in murrels it is between 26 to 65 hours.

The fry are collected from the nests, which are situated on the margin of water bodies. The nests are present or built about 50 cm below the water surface (Fig. 28.2). The nests are small and round hollow holes which the female constructs. The sizes of the nests are about 25-30 cm in diameter and are approximately 5-8 cm deep. The bottom of the nest is generally grassy. The eggs are deposited in the nest and being adhesive stick to the surrounding grass present at the bottom of the nest. The male guards the nests and the female stays nearby.

From one nest about 2000-15000 fry can be procured by keeping watch over the courtship behaviour during spawning season. In these fishes one the male courts with one female, male becomes U-shaped in texture and the female touches the genital of the male (Fig. 28.2a). During this process eggs and milt protruded out and fertilization occurs. Fertilized egg takes six to ten days in hatching. About eight days are needed for the complete absorption of yolk sac. Fingerlings take about seven months for harvesting.

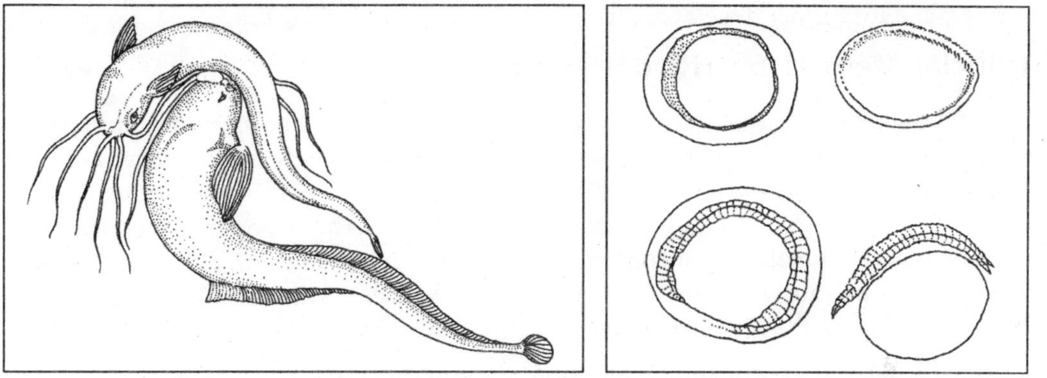

a b

Fig. 28.2a, b : Courtship & fertilized egg

Clarias batrachus and *Clarias macrocephalus* will easily be bred in ponds and other confined waters, if necessary environmental conditions are available. *Clarias lazera, the African* catfish can mature in ponds but breeds in the flooded rivers. They can be successfully bred in confined water with pituitary extract injection. Collection of fry of both these

Fig. 28.3 : Cat fish

fishes from natural breeding grounds is identical as natural spawning of *C. macrocephalusis and* C. *batrachus* is same. Even though the methods of pond breeding have been developed, farmers still depend to a larger extend on fry collection from natural waters, irrigation canals, rice fields, etc. Fry collection from water bodies is done about 50 cm below the water surface. About 2000-15000 fry can be found in a nest. The fry are then transferred to nursery pond for rearing.

The spawn is also found in the rice field and is procured during rainy season. The female makes nests in the form of hollow holes about 30 cm in diameter and 5-8 cm deep in the grassy bottom. The fry are collected from the nest by hand nets and then transferred to rearing pond for further development.

Artificial Method by Hypophysation

For obtaining the spawn/fry by artificial method, the brooders are collected. The brooders were collected either from wild or captive stock. These brooders are fed on high protein diet every day. The diet is 90% trash fish and 10% rice bran. For this spawning, rearing and stocking ponds are constructed separately. These ponds are constructed on scientific basis. Normally these conditions are available in fish farm, either in Government (mostly) or in private fish farm. The first point is to identify the healthy male and female brooders for giving the pituitary extract injection/HCG/synthetic analog.

In artificial/induced breeding, injection of pituitary extract is given in required dose to the brooders and spawn could be obtained. If small quantity of spawn is needed, then this exercise could be done in aquarium. If more quantity of spawn is needed then pond is to be used (De Kimpe & Micha, 1974).

Fig. 28.4 : Artificial nests used for spawning European catfish in Hungary (from M. Huet, 1989)

The aquarium should be 100-200 litres capacity of water and it must have continuous flow of water. The doses vary from country to country depending upon temperature, salinity and chemical composition of water.

After the penetration of first dose if the spawn are not delivered after 24 hours a subsequent injection is to be applied. If the fish fails to deliver spawn, then the same procedure is to be repeated.

If the eggs hatched out in the aquarium/spawning pond or in a pen, the fry are collected from the artificial ponds where nests are prepared, after driving out the male and transferring them to suitable fry rearing pond. This artificial hatching technique in pan spawning is more convenient as the nest can be reutilized quickly and the farmers can know more accurately the number of eggs hatched. The eggs can be hatched in jars, troughs and baskets also before transferring then to rearing tanks.

A. Identification of Male and Female Fish

Normally magur breeds once in a year. The fish is heterosexual. The sexual dimorphism is quite prominent during breeding season. In *Clarias batrachus,* the genital opening in a male is conical and prominently long while in a female it is oval and slit-like. On exercising of a little pressure on the abdomen, eggs come out in female while in male milt oozes out. The genital papillae are short, oval or of button shape. The colour of the vent is pink in female. In Indian species, the colour of the mature fish becomes grayish during breeding season. The female becomes darker in colour. The distinguishing features of other air breathing fishes during breeding season are given by Huet 1989.

B. Pituitary Extraction/HCG/Synthetic

The inject doses of various extract is as follows:

(1) Pituitary extract: 13 mg/kg body weight; pituitary of same species or *Clarias batrachus* preferred.

(2) Acetone dried carp pituitary at 4 mg per kg body weight is also suitable and is commonly recommended for *C. batrachus.* In Indian species 10-15 ml/ 1000 g body weight injection intramuscularly at two points just above the lateral line in the posterior half of the body and another injection above the cloacal opening of the female. The dose results in releasing the milt/spawn in breeding container, i.e., plastic pool. Glass aquaria, troughs, glass fibre tanks and small earthen ponds.

However, in *Pangasius sutchi,* pituitary extract is injected between the dorsal fin and the pectoral fin or at the base of pectoral fin. The male is given one injection and the female two. In *C. lazera* only one dose of injection is sufficient, injection to the male is not successful on account of position of testis. The male is to be scarified for milt. The female is stripped after 11-16 hours of injection.

(3) The HCG dose is 1760 mg/kg body weight;

(4) Synthetic analogs are given as per the recommendations of the company.

Structure of Spawn

The eggs of this fish are demersal, spherical and small in size varying in diameter from 1.5-2.00 mm. Normally, the fecundity of this fish is 3000 to 40000.

The temperature of water suitable for spawning is 26 °C to 29 °C, pH should be 7.0, Ammonia 0, Nitrite 0, Nitrate 0, GH 100 ppm and KH 70 ppm.

Hatching from the fertilized eggs takes about 6-8 days. After hatching the fry comes out to combat.

Pond

Ponds may be of three major types, —spawning, nursery and stocking. The size of pond varies from 0.05 to 0.01 ha. The depth should range from 0.75 to 1.0 m but bigger farm can also be used if properly managed. Apart from the ponds, Bheel and swampy marshes and low line areas may also be used. In foreign countries the spawning ponds are usually 8000-16000 m square in area. Nests are in the form of 30 cm deep hollow (20-30 cm in diameter) are made on the banks about 20-30 cm below the water line, to resemble nests built by the species in the natural waters for spawning. If the brood stock has to be held in the spawning pond, a section of about 20-30 percent of the pond is dug deeper by about *1m. Brood fish are stocked at this rate.*

The nursery ponds should be 400-800 acres of area with a depth of 0.8-1 m. It is essential that the nursery ponds should be manured by chicken manure and rice bran to produce adequate quantity of food. Stocking is generally done at the rate of 100-300 fry/ m sq. but if it is planned to grow fry for a longer time, a lower rate of density is preferred. The rearing period is about 15-35 days depending upon the fry required. Steamed poultry are considered to be the best food for fry after the yolk sac is absorbed. For first two days they are fed twice a day. It is suggested that 10 eggs are sufficient for 100000 fry. After that they are fed on trash fish twice a day at the rate 1 kg per 1000000 fry. The quantity is increased by about 1 kg every two or three days. No addition or reduction of water is needed. Fry of 3-5 cm in length or larger in size are in greater demand by the fishermen.

Raceways with adequate flow of water produce more fishes in the ponds. With a flow of 10 litres of water about 4000-50000 fingerlings can be stocked. Stocking can be done in spring and early summer.

Cages are installed in open waters rather than in ponds (Fig. 28.5). The stocking rate is 65-70 fingerlings per m. cube. Floating fed is supplied.

Fig. 28.5 : Channel Catfish spawning check in a pond

Diseases : Septicaemia disease is an important bacterial disease of the catfish. Protozoan such as ichthyophthiriassis or ich are common diseases (Mukerjee, 2003).

Harvesting and marketing are to be taken care of.

REFERENCES

Chaturvedi, C.S., 2003. Biology, culture and breeding technique of Indian Magur. In : Carp and Catfishes breeding and culture. Central Institute of Fisheries Education, Mumbai Publication, pages 74-79.

Colman, J.A. 1982. Pond management, water environment and fish grow-out performance relationship in *Clarias* culture trail. National Inland Fisheries Institute, Bangkok.

De Kimpe, P. and Micha, J.C., 1974. First guidelines for the culture of *Clarias lazera* in Central Africa.

Hogendoorn, H., 1983. The African Catfish (*Clarias lazera* C & V. 1840—a new species for aquaculture). Dissertation. Agriculture University. Wageningen.

Hogendoorn, H. (1983) Growth and production of African Catfish, *Clarias lazera* (C & V) 11. Effects of body weight, temperature and feeding level in intensive tank culture. Aquaculture, 34.265-85.

Huet, M. (1989). Textbook of fish culture, 2nd edition, Fishing News Books. Oxford.

Jain, K.K., 2003. Identification of fish seed (Indian major carps and Expotic carps and catfishes. In : Carp and catfishes breeding and culture. Central Institute of Fisheries Education, Mumbai. Publication, pages 66-73.

Kholi, M.P.S (2003) : Breeding of Air-Breathing Fishes, In : Carp and catfishes breeding and culture. Central Institute of Fisheries Education, Mumbai. Publication, pages 117-181.

Kloke C.W. and Potaros, 1975. The technology and economics of catfish (*Clarias* spp.) farming in Thailand. IPFC. Occasional papers.

Lee, J.S.,1973. Commercial cat farming. The Interstate Printers and Publishers, Danville.

Marshall, D., 2001. The climbing Perch, *Anabas testudineus*. web. site of Ryedale Society. Pickering North Yorkshire. England.

Mukerjee, S.C. (2003). Common Disease Problems in Fish and their Health Management. In : Carp and catfishes breeding and culture. Central Institute of Fisheries Education, Mumbai. Publication, pages 161-172.

Chapter 29

COLD WATER FISHERIES

Introduction

Cold water Fisheries occupy an important place amongst the freshwater fishes of India. The cold water fisheries deal with fisheries activity in water where temperature of water ranges from 5 to 25 degrees centigrade. The water temperature under cold water fisheries should not be more than 25 °C even in summer. Such conditions in India occur in Himalayan and peninsular regions. India has significant resources in terms of upland rivers/streams, high and low altitudes of natural lakes in addition to man made reservoirs existing both in Himalayan regions and Western Ghats. There are 16 big and small rivers in Himalayan and peninsular regions having an area of about 3885 km. Apart from these rivers area of natural lake's is 20500 hectares. The man made reservoirs are spread over an area of 265000 hectares. Much research work has been done in connection with cold water fisheries. Fish culture has been successfully carrying on in certain lakes and reservoirs of J & K, Himachal Pradesh, Uttaranchal, Nagalad, Meghalaya, Arunachal Pradesh, Tamil Nadu and Kerala. Government of India established National Research Centre on Cold Water Fisheries (NRCCWF). It has played a significant role in the improvement and conservation of indigenous as well as in the establishment of exotic species in this region.

The biological functions of fishes at high altitude depend upon several factors resulting in different adaptations. The torrential streams of water in mountains contain rich oxygen (7.8 to 10.2 ppm) and low carbon di-oxide quantity, the parameters of which are most important for the growth of fishes. The gills of cold water fish are greatly reduced and the gill opening being smaller in size is the main adaptation.

The fish fauna of mountain region contains the following indigenous species (Khanna, 1992; Kulshrestha, 2003).

(1) **Mahaseer :** There are 4 species of mahaseer. It has a big head hence it is known as Mahaseer. It is a migratory fish running in the main rivers for spawning. Two of them are edible fishes and are of commercial use, they are *Tor putitora* and *Tor tor*. They exist from Kashmir to Assam. They are about 1.5m in length and breed from July to September. *Tor mosal* is available in Kashmir, Sikkim and in rivers of Assam. *Tor khudree* (Deccan Mahseer) are found in the rivers of southern part of India. It is found at an elevation of 875 below. Normally they are found between pitted rocks and stone with filamentous algae.. It requires DO(ppm) 7.8-8.4 and temperature of water should be between 18.5-27.6. (Figs. 29.1a-c).

(2) **Snow Trout :** Two genera are present; *Schizothorax* Spp. and *Schizothoraichthys* are endemic to Kashmir and Ladakh while *Schizothorax prograstus* occurs in Eastern Himalayas up to certain stretch. *Schizothorax richarsonii* is distributed almost all along the Himalayas. *Schizothoraichthys* has some species, namely *Longipinnis curvifrons*,

planifrons, ecocinus and *micropogon,* all are endemic to Ladakh and Kashmir (Fig. 29.2a, b). They are illiophagic and herbivorous. The fecundity of these fishes per kg body weight ranges between 8000-23000 and the peak breeding session is from March-April. It has lacustrine habitat.

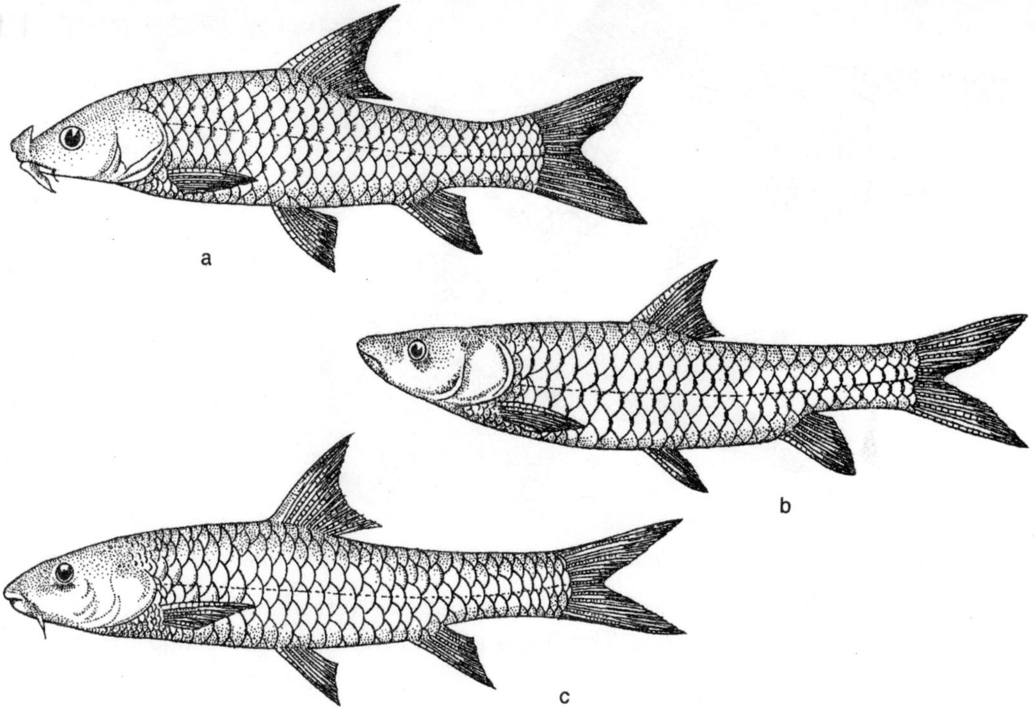

Fig. 29.1a-c : Important Mahaseer fishes of found in Cold Water : (a) *Tor tor,* (b) *Tor putiora ,* (c) *Tor Khudree*

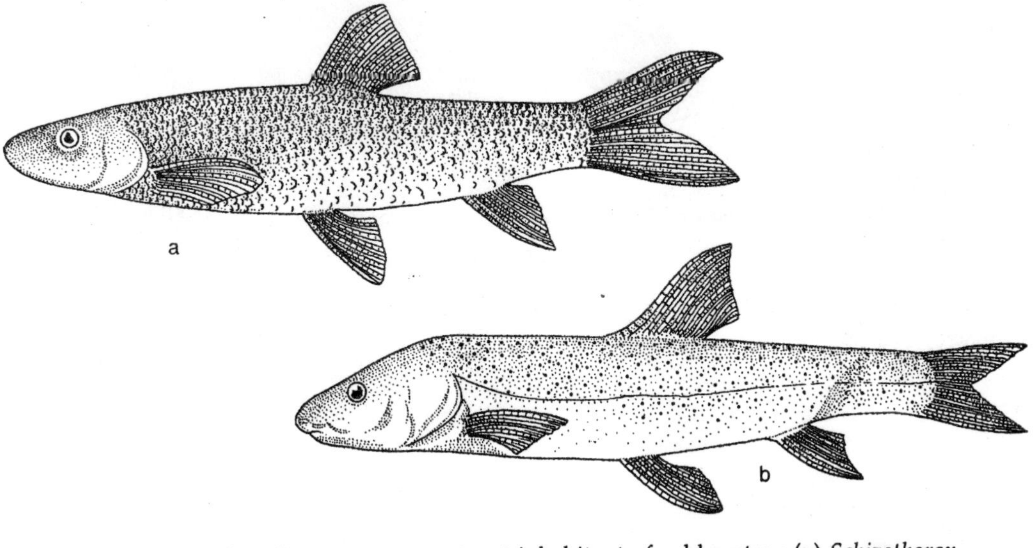

Fig. 29.2a, b : Important snow trout inhabitant of cold water : (a) *Schizothorax,* (b) *Schizothoraichthys ecocinus*

(3) ***Barilius*** **:** It is used as manure. Found in Assam, Punjab, Kumaun, Nilgiri, and Nepal. *Barilius* vagra, B. bendelisis and, *B. bola* these fishes are also called as Mountain trouts (Fig. 29.3a, b).

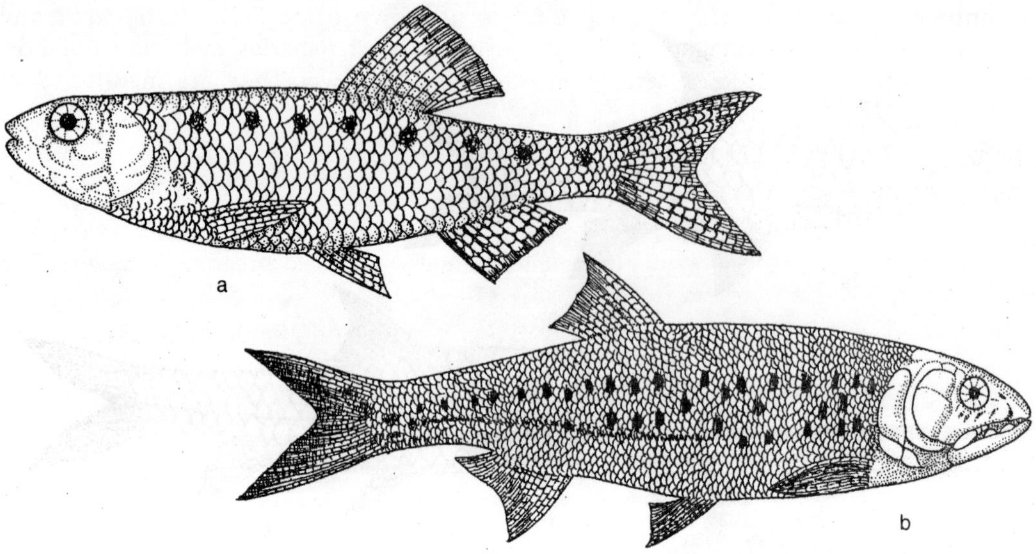

Fig. 29.3a, b : The Indian hill trout : (a) *Barilis bendelisis,* (b) *Barilis bola*

(4) ***Labeo*** **:** The fishes which are present in mountain rivers are minor carp and have no significance as commercial fisheries. The fishes are *L. dera, L. dyocheilus* and *Crossocheilus latius* (Fig. 29.4a, b).

Fig. 29.4a, b : Fishes present in mountain rivers : (a) *Labeo dora,* (b) *Carassius auratus*

(5) ***Garra*** **:** There are two species found in lakes of the Himalayas as they are not edible fishes but are relished as food of fishes. They are *Garra lamta* and *Garra gotyli*.

Exotic Fishes

1. ***Trouts*** **:** The following trouts are now acclimatized in the streams, lakes and reservoirs in Indian waters, particularly in Nilgiri, Kodai Kanal, Kashmir and Himachal Pradesh. The fishes are now available for food and farming. The management practice is important for myskii promoting angling tourism of brown trout, *Salmo trutta fabrio* and *Oncorhynchus* are two main genera. They constitute trout fisheries in Indian uplands. *Salmon* has three main species while *Oncorhynchus* has two. They are mentioned as follows:

Salmo trutta fabrio, Salmo eredius, Salmo gairdinarii

Oncorhynchus mykiss (rainbow trout), Oncohrynchus nakrii

2. ***Cyprinus carpio*** **:** This fish has been well cultured in Kashmir and in the north-eastern region. The culture is being successfully done by induced breeding methods. The methods are described in carp breeding.

The culture of indigenous snow-trouts endemic to Kashmir inhabiting lakes, river/ streams has declined but conserved by artificial propagation in this part of India.

Field Incubators : On account of difficulty in handling culture practices at the breeding grounds in hilly tracts, field incubators are designed which are easy to handle. In the absence of full fledged hatchery, a field incubator was designed for production of seed at streamside. In this incubator the water enters from below through an inlet and leaves through an outlet at the top so as to maintain continuous flow of water. A feeble current is maintained inside so that eggs are not damaged and agitated. The fertilized eggs of *Schizothorax* spp. a cold water fish endemic to Kashmir are gently placed in hatching troughs inside the incubators with top covered. The hatching period of the eggs is about 8-15 days, later newly hatchlings can be transferred in the nursery tanks. The advantages of this incubator are:

1. It is handy and economical.
2. It can conveniently be installed in the field at the side of the stream and breeding ground.
3. Oxygen depletion inside the incubator is controlled.
4. The size of the incubator can be changed depending upon the availability of the brood stock and quantity of seed to be produced. This miniature hatchery has greatly helped in seed production activity of the snow-trout. This has helped in the rehabilitation of this species by ranching it at suitable site in the streams.

Flow through hatchery : This is another device for rearing seed up to the fry stage. It has been designed and fabricated by National Research Centre on Coldwater Fisheries at Bhimtal in Kumaun hills. It has the capacity of incubating 0.225b million eggs. It is reported that about 0.2 million fry and using this hatchery produced 0.15 million advance fry. The capacity of flow through hatchery further increased with installation of more troughs/ trays and nursery facilities with substantial increase in the quantity of water through overhead tanks. This gives 30% more survival rate from traditional pattern.

REFERENCES

Kulshrestha, S.D. (2003). Fish and Fisheries (In Hindi). A, B, D Publishers, Jaipur India.

Khanna, S.S. (1992). An Introduction to Fishes. Central Book Depot, Allahabad. India.

Chapter 30

SEWAGE FEED FISHERIES

Sewage and Sewage Feed Fisheries

The disposal of sewage or effluents from sewage treatment facilities is a problem of global importance. In some part of the world, fish culture has been adopted as an aid to such disposal through actual utilization of sewage. Thus fishponds have served as ancillary sewage treatment facilities as well as yielding a high production of fish per unit area. The sewage fisheries, the fish culture in sewage, will play an important role for the production of animal proteins like fish, and at the same time help in the improvement of water quality. The proper treatment of sewage and removal of pollutants from sewage by sewage fish culture is becoming widespread. The utilization of this technique is, therefore, of interest of pollution control agencies, fish culturists, ecologists, conservationists, agriculturists and urban planners.

Recently, due to environmental protection laws, the industrial waste is first treated in the factory premises itself, then only it is allowed to outlet from the factory under safety limits. The domestic waste in many big cities are also collected by the Municipal, and the sewage is treated by primary, secondary and tertiary ways and then only the wastewater of the sewage is used for aquaculture and other purposes.

It is essential to understand the broadly acceptable definition of Sewage so that we can know the constitution of it. If composition of the sewage is known, then it will be easy to made sewage wastewater free of pathogen, containing only nutrients useful for culture of fishes and other aquatic organisms on profitable basis.

The sewage is regarded as a turbid liquid, may be purely domestic or it may partly consists of industrial or agricultural wastewater. Depending upon the source, it contains different percentage of suspended solids, dissolved solids, semi-decomposed organic matters and a few essential minerals. Sewage sludge is defined as a semi-liquid substance consisting of suspended sewage solids combined with water and dissolved material in varying amount. It is also essential to understand the sharp definitions of sullage and sludge (Allen, 1969 & Rhyner *et al.*, 1995).

Sullage is wastewater resulting from personal wash excluding faecal matter and urine, laundry, and cleaning of kitchen utensils.

Sludge is any solid, semi-solid, or liquid waste generated from a municipal, commercial, or industrial wastewater treatment plant or air pollution control facility.

The wastewater contaminants are suspended solids, biodegradable organics, pathogens, nutrients, refractory organics, and heavy metals and dissolved inorganic solids. The most significant components of wastewater are usually suspended solids, dissolved solids and pathogens. Suspended solids are removed as sludge. Removal of suspended solids is essential prior to discharge or reuse of wastewater.

Detailed Constitution of Sludge

The raw sewage contains 90% of water and the rest include fat, grease, suspended solids organic matter, bacteria, protozoan and fungi. The common gases and silt are present. The common gases are CO_2, H_2S, NH_3 and PO_4. It emits foul and obnoxious gases, which are due to high rate of microbial decomposition. Other heavy metals present in traces are Zn, Cu. chromium, magnasium, nickle and lead. The dissolved oxygen contents are almost nil or insignificant. Fresh sludge is muddy or black clouds in colour due to suspended colloidal particles. It is the solid substance settled at the bottom of the water body containing waste.

Activated sludge contains single cell protein (ASCP 39-46 protein) obtained by the bio-oxidation of crude domestic sewage has been evaluated as a supplementary protein and calorie in the diet, which is regarded as efficient diet. It is regarded as an efficient diet of trout in Germany.

The sewage water is also rich in inorganic nutrients such as Ca, Na, K, P and nitrogen that is essential for the production of phytoplankton. Nitrogen, phosphorus and carbon are essential food materials/nutrients for the phytoplankton.

Phytoplanktons are first stage of the pyramid of food chain necessary for fish growth. Nitrogen source in the aquatic system is from animal waste and fertilizer. Minerals, inorganic and soluble organic substances are directly used by the phytoplankton and for the photosynthesis and they are eaten by zooplankton as food source. Carbon and nitrogen should be available in the ratio of 1:3, which are the good sources of nutrition for fishes. Municipal waste contains rich quantity of phosphate. CO_2 can immediately be used in the photosynthesis process. But this water is on the other hand contains rich quantity of infection materials such as pathogen and microbial fauna, which are harmful for both fish and human being if the fish is consumed.

However, this problem could be solved by giving proper treatment to fish before it is marketed. The most important point for the consideration of wastewater is the proper dissolved oxygen. Fish cannot survive if D.O. is not sufficient. Further insufficient D.O. causes anaerobic decomposition of organic materials, which lead to formation of gases such as H_2S, CH_4, CO_2 and the formation of gases causes sludge formation.

The proper treatment causes reduction in B.O.D. and NH_3, H_2S, CO_2.

Micro and macro particles of sewage form direct food source to zooplankton, benthos and fishes.

The dissolved solids are first converted to suspended solids and subsequently removed as sludge. So maximum objectionable material initially present in the sewage is first concentrated in the form of sludge and the disposal of wastewater reduces constitution of sewage in acceptable manner. **This wastewater could be used for aquaculture and agriculture.** (Blackman, 1993).

The fishes, which are recommended for sewage culture, are carps, and air breathing fishes. The indigenous carps are *Labeo rohita*, *Catla catla* and *Cirrhinus mrigala*. The combination of Indian and Chinese carp could also be cultured in the sewage water. The combination of 1:2:1 is suggested as the best combination because the whole water columns are utilized. The domestic sewage is commonly used for rearing fry and fingerlings in the rearing ponds.

REFERENCES

Allen, G.H., 1969. A preliminary bibliography on the utilization of sewage in fish culture. In cooperation with inland Fishery Branch, Fishery Resource and Exploitation division. Department of Fisheries, FAO.

Blackman, W.C., 1993. Basic Hazardous Waste Management. Lewis Publisher, Boca Raton.FL.

Rhyner, C.R., Schwartz, L., Wegner, R.B. & Kohrell, M.G. 1995. Waste Management and Resource Recovery. CRC Press, Inc. USA.

Chapter 31

CAGE CULTURE

Introduction

Cage culture involves the rearing of aquatic organisms such as fish or shellfish to fingerlings/juveniles or to commercial size in a volume of water enclosed on all sides with cage netting material including bottom, while permitting the free circulation of water and waste removal through the mesh of cages. This cage is supported by a rigid framework made by bamboo, wood, metal and set in sheltered portion of reservoirs, lakes, rivers, and estuaries. In general, it consists of rearing of fish from fry to fingerlings or raising fish from juvenile to a table size fish (Tamot *et al.*, 2008).

Fish pond culture has its tradition of about 4000 years , cage fish culture is a recent origin. It seems that as earlier as 1920, cage culture has developed independently in at least two countries such as Kampuchea and Indonesia. In Kampuchea where fishermen in and around the great lake region would keep *Clarias* spp. and other commercial fishes in bamboo cages and baskets; and in Indonesia when bamboo cages were used to grow *Leptobarbus hoeveni* fry as early as 1922. Now cage culture has rapidly propagated throughout the world to about 40 countries in Asia, Europe, United States, USSR, etc.

Modern cage culture began in the 1950s with the advent of synthetic materials for cage construction. In the United States universities did not begin conducting research on cage rearing of fish until the 1960s. Cage research has been limited mostly because large scale open pond culture was more economically viable and, therefore, received most of the research focus.

Today cage culture is receiving more attention by both researchers and commercial producers. Factors such as increasing consumption of fish, some declining wild fish stock, and poor farm economy have produced a strong interest in fish production in cages. A large number of small or limited resource farmers are looking for alternatives to traditional agricultural crops. Aquaculture appears to be a rapidly expanding industry and one that may offer opportunities even on a small scale. Cage culture also offers the farmer a chance to utilize natural water resources for aquaculture, which in most cases have only limited use for other purposes (Ariyaratna, 2001).

Therefore, it is necessary to take start up for *in situ* rearing of large size fingerlings in these reservoirs. The fish may be reared preferably up to a size of 100 mm or above and then released into the reservoir. This will reduce loss generally occurred during transport and advanced fingerlings would be available for stocking in sufficient quantity. This technology has advantage of ruling out possibility of fish mortality during transport. This technology is an ecofriendly technology and poses no hazard to surrounding environment or flora-fauna. Cage aquaculture also plays a very significant role for increasing production in reservoirs, lakes, ponds and riverine fishery.

Advantage of Cage Culture

For taking maximum output from the aquatic resources, it is necessary to develop command areas of these resources by application of biotechnological tools. The principal cause of unsatisfactory productivity of the aquatic resources is stocking of fry or small size fingerlings, which are not recommended for large size water bodies. Under current practices the stocking of large fingerlings becomes uneconomic. Generally, seed producers prefer to rear fry or fingerlings up to maximum 50 mm size. Growing of large size fingerlings takes duration of 3-4 months and greater quantity of feed is consumed. The growth and survival from fry to fingerling is less than 30%. Therefore, the fingerlings of above 50 mm cost very high. Further, transportation of large size fingerlings is cumbersome and costs high as frequently mass mortality is observed during their transportation. Therefore, fish farmers prefer to stock fry or fingerlings measuring between 25-50 mm size which are not grown well to compete with the fishes of the large water body and consequently low rate of production is obtained. To overcome these constraints cage culture technology may be adopted in which fingerlings are grown in the floating cages of fine mesh net in the reservoir itself and all hazards discussed above are ruled out.

Following are the main advantages of the technology :

- Large water resources can be used which could not otherwise be used for aquaculture.
- Easier handling, less maintenance and complete harvesting of fish.
- Better control of fish, faster growth, and higher production.
- Protection from predators and competitors.
- Simple observation and sampling of fish.
- Efficient use of feed.
- High stocking density.
- Less manpower required.
- Minimum supervision.
- Total harvesting and immediate return on investment.
- This system allows the flexibility of raising at least two different species of fish; one inside the floating cages and another outside or in the water body for different sizes of the same species.

Methodology

Basic need for cage aquaculture are strong stable and good quality net or webbing, a framework support preferably with working platform, buoys for floating the cages and sinkers for keeping them in a fixed location (APHA, 1995).

i. **Types of cages**

Generally, two types of cages are mainly used, such as fixed and floating. The fixed cages is suitable for a water depth of 1 to 5 metres. The floating fish cages are suitable for water depth of more than 5 metres.

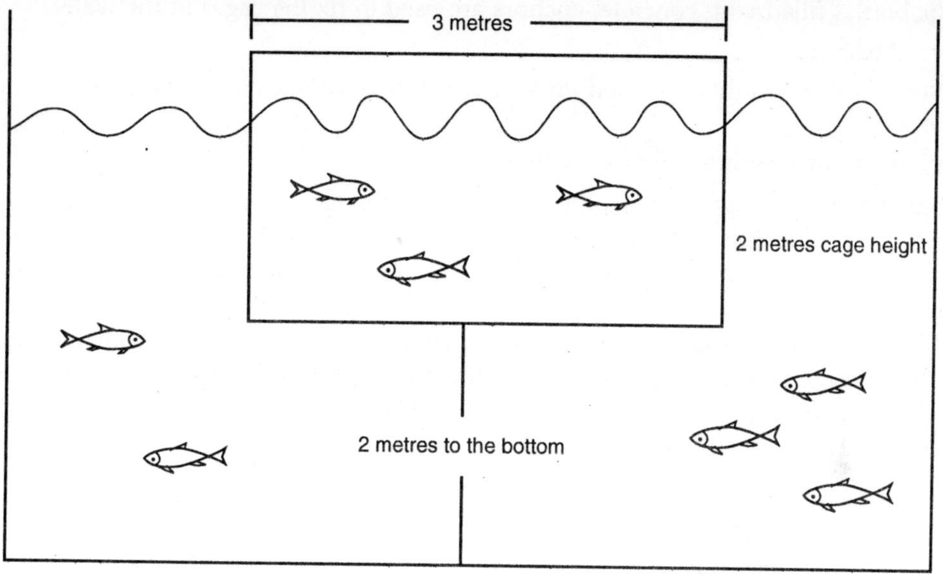

Fig. 31.1 : Floating cage system in pond/reservoir

ii. **Site selection**

Site selection is one of the very important parameters because all water bodies are not suitable for cage culture.

Some of the important features for location of cage installation in a water body are as under :

- Good water circulation.
- Depth of water should be at least 10 to 15 feet.
- Cage site should be away from coves and weed beds, which can reduce wind circulation.
- Cage should be away from frequent disturbances from people and animals.

Site should be protected from the winds and waves.

- Availability of electricity and aeration devices.
- The site must be accessible to land and water transportation.
- The water body should be free from problems of aquatic weeds, surface scum, over population of wild fish, or oxygen stress.
- Good water quality such as adequate dissolved oxygen, stable pH, low turbidity and absence of pollutants.

iii. **Cage material**

The body of the cage consists pieces of netting made of Polyamide (PA), Polyethylene (PE), Polypropylene (PP) and other synthetic material or wire. The material is selected depending on availability, rot resistance, water absorption, tendency to foul, strength & cost.

iv. **Floats, sinkers and anchors**

For flotation of cages inner tubes, waterproof foam rubber, capped PVC pipe, air tight PVC tank, wood, bamboo, old oil barrels/drums, etc. Floats can be selected on the basis of site and the weight of platform. Sinkers are made of ceramic, steel pipes, stones or concrete blocks or plastic bottles filled with concrete, anchors are used to fix the cages in the water.

v. **Mesh size**

Mesh size is mainly depended on size of fry/fingerlings. Normally 4-15 mm mesh size can be used.

vi. **Cage size, design and arrangement**

Small and large cages can be prepared. For 4 metre deep ponds the size of cage should be 3 m × 3 m × 2 m (length × width × depth) and for deeper water bodies 3 m × 3 m × 3 m (length x width × depth) or more can be used. For carp culture, rectangular cages are suitable and easily manageable. Set of four or more cages can be arranged depending upon the space available in reservoir and direction of wind. Two rows of four cages parallel to each other can be arranged perpendicular to the direction of water current. This type of cage arrangement provide mass oxygenated water supply.

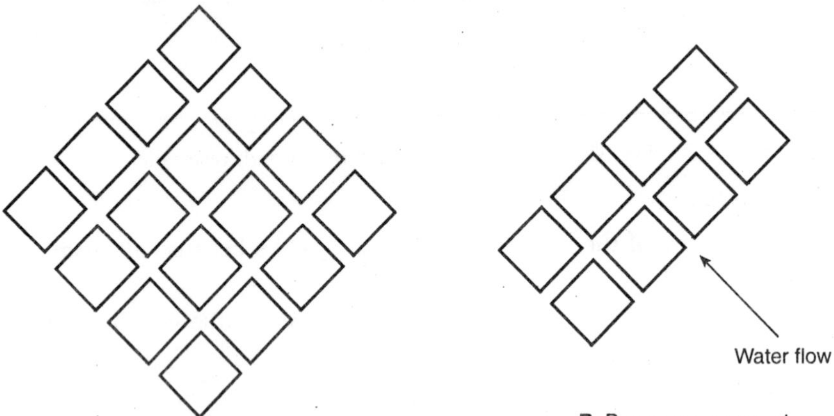

A. Improper arrangement

B. Proper arrangement

Fig. 31.2 A, B : Arrangement of cages in reservoir for maximum water exchange.

Choice of fish

For cage culture some of the desired characteristics of fish species are:

- Fast growth rate.
- Tolerance for crowded condition.
- Native to the region.
- Has a market value.

Stocking & culture period

50 to 200 fry/m^3 of 25 to 65 mm length is recommended. For raising table size fish in cage, stocking size should be 60 to 100 mm length and stocking density should be 25 to 100 fingerlings/m^3. For raising fry to fingerlings of carp 30-60 days rearing is sufficient and for raising table size fish a period of 6-8 months is required.

Maintenance

The webbing is to be cleared at weekly interval by brushing biofouling organisms such as algae, sponges and detritus material. Water quality monitoring from time to time is also required. Monitoring and repairing of floating platform, sinkers, anchors, etc. are also important. Before the onset of predictable bad weather are to be loosened anchor ropes, all the units towards the sheltered area should be low.

Feed & feeding tray

For higher production, good quality fish feed containing economical feed ingredients should be provided and made locally available. Feed is given to the fish using 3-5% of body weight at least twice a day. Feed can be prepared in pellets form in 2-6 mm diameter using rice bran, soy cake, fish meal, molasses, soy oil, wheat flour, maize, vitamin mix, etc. For effective and proper utilization of artificial feed, feeding tray can be used.

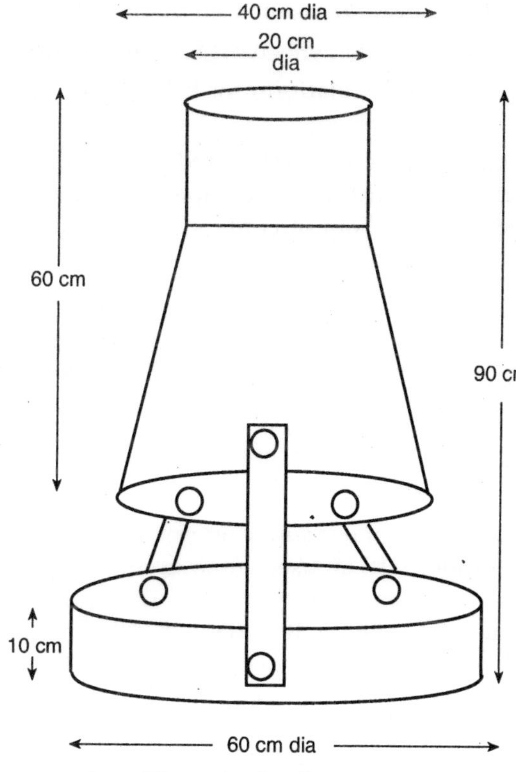

Fig. 31.3 : Feeding tray

REFERENCES

APHA. (1995) Standard methods for examination of water, 19th Edition. Published jointly by American Public Health Association and Water Pollution Control Federation, Washington, DC, USA.

Ariyaratna, M.H.S. (2001). Performance of cage reared fingerlings of commonly cultured fish species in response to different feeds. In : Reservoir and Culture Based Fisheries : Biology and Management (Ed. De Silva, SS : 359-366).

Tamot, P, Mishra, R, Somdutta (2008). Proceeding of Taal, 2007 : In 12th Lake Conference : 318-324.

Chapter 32

CRAFTS AND GEARS

Fishing Craft and Gear

The use of crafts and gears in fishing technology plays very important role and help enhancing the production commercial bases. The success of fishing largely depends on to how and which types of nets are used to capture the fish (Jyoti & Sharma, 2006). There are two main types of devices used to capture fishes in both marine and inland fisheries.

(1) Nets or gear — these are instruments used for catching fish.

(2) Crafts or Boats — It provides platform for fishing operations, carrying the crew and fishing gears.

There are various types of gears and crafts used in different parts depending upon the nature of water bodies, the age of fish and their species. Some nets are used without craft, however, others are used with the help of crafts. Generally, locally made gears and crafts may be non-mechanized or mechanized.

Crafts and Boats

There are many types of fishing crafts being successfully made and used for marine and inland fisheries.

Marine Fishing Crafts

Different crafts are used due to different conditions of sea on the east and west coasts.

Crafts used on the East Coasts

(1) Catamaran —The word *catamaran* is originated from a Tamil word *Kattumaram* which means 'lashing timber'. It is used mainly on the east costs of Orissa from Kanyakumari. It is also used on northeast cost of Kerala. It is the most primitive, traditional, economical and efficient craft. It is made by tying many wood logs in such a way that it takes the shape of a canoe, which consists of two main logs and two side logs cut into boat shape and held together with rope. The logs are held in position by loose rubber called *Teppa*. Generally the Catamaran is 5-10 metre long, 0.5 metre wide and 0.3 metre deep.

Types of Catamarans

(i) Orissa and Ganjam type
(ii) Coromandal type
(iii) Andhra type
(iv) The boat-Catamaran

Orissa and Ganjam type — It is made by five logs pegged with wood. The logs are cut in boat shape and are not tied with rope.

Coromandal type — It is used in Tamil Nadu to capture flying fish of Nagapattanam. It is made by 3-5 logs. A modified type of Coromandal type is called *Kolamaram,* which is made by 7 logs.

Andhra type — It is the modified form of Orissa type, larger in size—about 5-7 metres long, hence made by nine heavy side wood logs that are fitted with a median logs.

The boat-Catamaran —It is made of three wood logs tied in boat shape. It is used in Mandapam and Mukkun coastal regions.

Masula boat — It is a weakly constructed boat of about 8-12 metres long (Fig. 32.1). It is used in clear weather near the shore. The masula boat is keel less and frameless made by mango planks, which are stitched with palm leaf fibres. There are several variations of the musula boat. In Orissa it is called 'Bar boat' and in Andhra called *Padava* or *Padgam*.

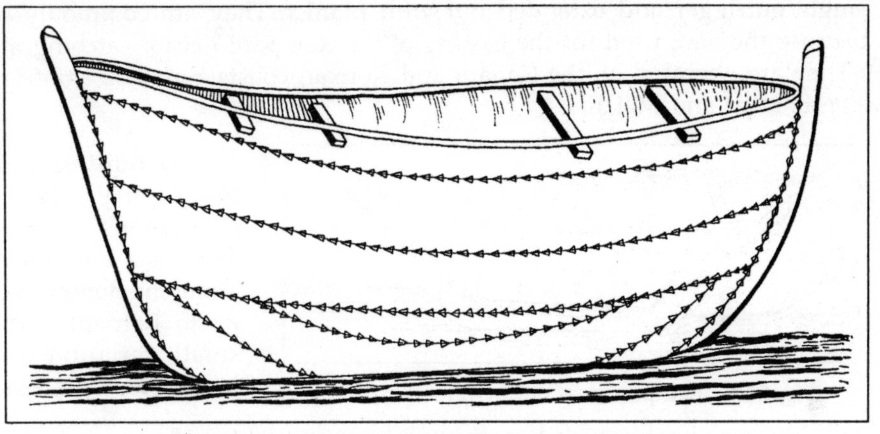

Fig. 32.1 : Padava—Masula boat of Andhra coast

Nauka and Dinghi — These boats are with carvings and are operated in West Bengal and Orissa. These are well-designed large boats measuring about 11-13 m × 2-3 m × 2 m in size.

Tuticorin boats or fishing luggers — These are operated in inshore waters and are used as cargo boats. They are carved boats of 11 m × 2 m × 1 m in size (Fig. 32.2).

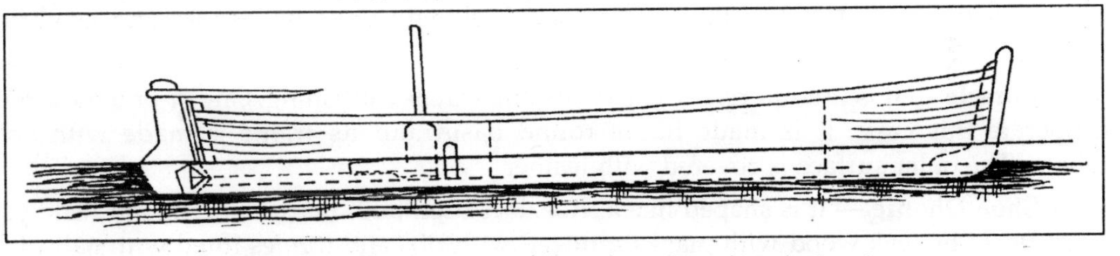

Fig. 32.2 : Tuticorin-type fishing boat (Tamil Nadu State)

Crafts used on West Coasts

(1) **Dugout Canoes** — These are constructed from large wood logs (Fig. 32.3). These logs are hollowed by scooping inner part. Their bottom is thicker than sides. They are operated on Kerela and Konkan coasts. The small boats of 5-10 m long are called 'Thonies', which are

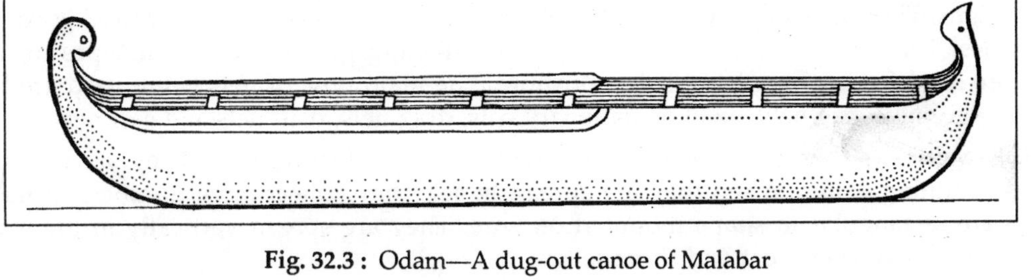

Fig. 32.3 : Odam—A dug-out canoe of Malabar

used for gill nets or drift fishing and for seining. The large boat of 10-22 m long is called *Vanchi* or *Odam* and is used for operating variety of nets on Malabar Coast. Dug-out canoes are also operated on the west coast from Colachal to Kathiawar.

(2) **Plank-built Canoes** — It is a kind of dug-out canoes. It is extended with planks on sides. They are popularly used on coast of Kerala, Karnataka and North Bombay.

(3) **Outrigger Canoes** — These are large size canoes of about 15 m long. It has narrow keel and single outrigger and extended out with planks. They are commonly called as *Rampani,* because they are used for the casting of the Rampani net for catching mackerels (Fig. 43.4). They are operated on the Kanara and Konkan coasts. Small size outriggers are used between Bhatkal and Majali.

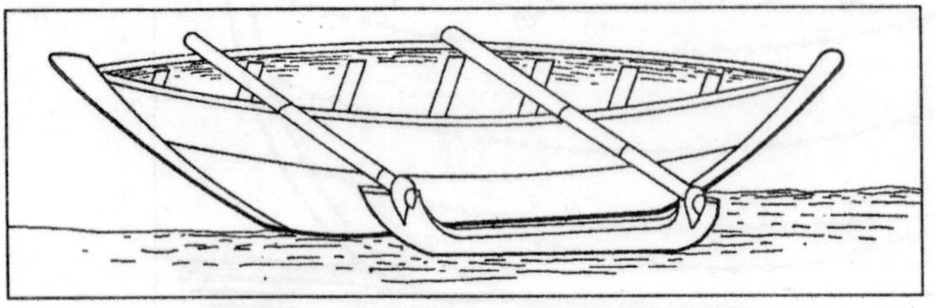

Fig. 32.4 : Rampani boat of Karwar

(4) **Built-up boats** — It is highly specialized indigenous fishing craft. They are commonly used along the Bombay coast and north Ratnagiri. There are small variation according to different places such as :

(a) *Ratnagiri type* boat, which has pointed bow, straight and narrow keel and low gunwale.

(b) *Machwa* — It is provided with broad hull, straight keel and pointed bow. It is popular in Bassein hence called *Bassien type.*

(c) *Satpati* or *Galbati type* — It has straight keel, high gunwale, medium pointed bow and broad beam. Satpati can be mechanized with a motor engine without any modification in design.

(d) *Broach type* — It has flat bottom and is widely used in inshore and estuarine water.

(5) **Coracle** — It is used in rivers, reservoirs and canals for fishing. One or two fishermen can operate this craft. It is made like a round basin and its frame is made with split bamboos. The outer surface is covered with leather.

(6) **Shoe Dhonie** — It is shaped like a shoe. It is used both marine and inland fishery. It is constructed by teak wood with planks grooved with ribs and frames fitted with nails. It is used for fishing with gill nets.

(7) **Kakinada nava**— It is commonly used for inshore fishing. It is made by teak wood. It is keel less but ribs are fitted in frame with nails. It is about 9-10 m in length.

Marine Fishing Gears

Various types of gears are used for fishing in sea. They may be of different size, shape and designs. These gears may be made by fishermen. They are also manufactured in cottage industries. The most commonly and widely used fishing gear are different types of nets. They are used for catching large fishes offshore. The main type of nets being used are boat seine, shore seine, bag nets, fixed or stationary nets, drag nets, drift nets and cast nets.

(i) **Seines** —These are specially designed and large fishing nets. They are generally used in running water. When they are spread in sea; they collect large numbers of fishes. Seines are rectangular in shape mounted on wire. They are spread vertically in the water. Seines are of two types, boat and shore seiries.

(ii) **Boat seine** —These nets are conical in shape provided with wings. The mesh of the nets is smaller in center and increase in size towards outer ends of flanks. This seine is operated in sea by catamarans or boats. The seine traps the fishes. Towing is done with the help of coir.

(iii) **Shore seine** — It is operated from seashore. It is popularly called as *Ber Jal* (Fig. 32.5) in Orissa, Pedda or Alivi vala in Andhra coast, Periya vala or Mada valai in coromandel coast and Kara valai in the Gulf of Mannar. It is a conical bag-like with two wings. One end of net is kept fixed to shore and other end is spread into sea with the help of a boat, in the form of semicircular fashion. When the net is filled up with fishes, the two ends are slowly dragged by group of fisherman.

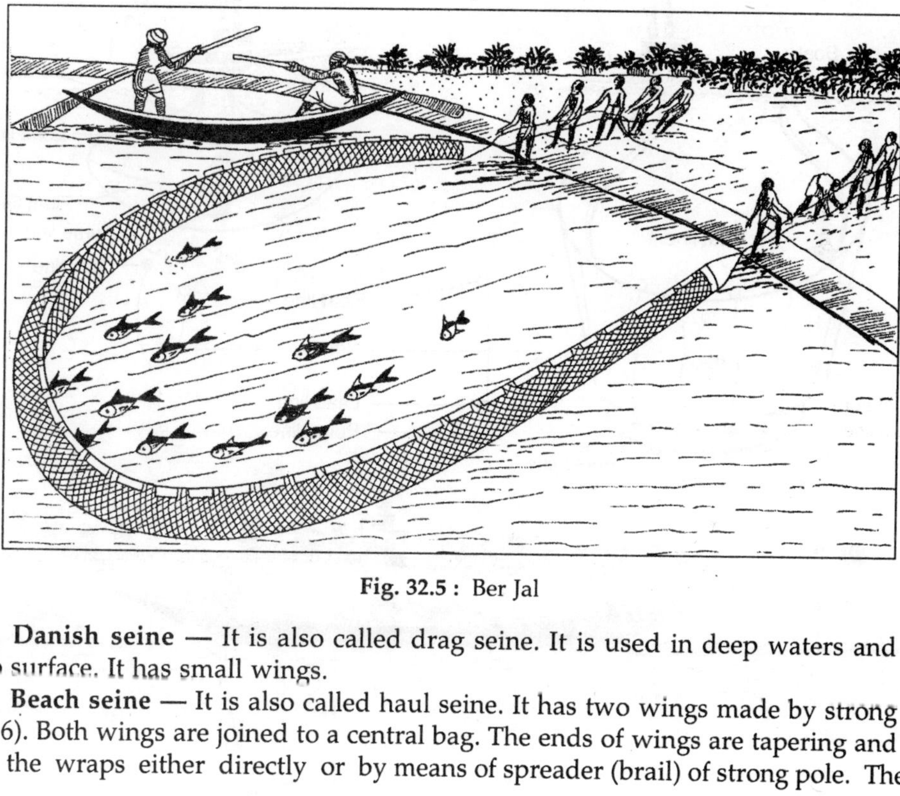

Fig. 32.5 : Ber Jal

(iii) **Danish seine** — It is also called drag seine. It is used in deep waters and do not reach to surface. It has small wings.

(iv) **Beach seine** — It is also called haul seine. It has two wings made by strong twines (Fig. 32.6). Both wings are joined to a central bag. The ends of wings are tapering and connected to the wraps either directly or by means of spreader (brail) of strong pole. The mesh

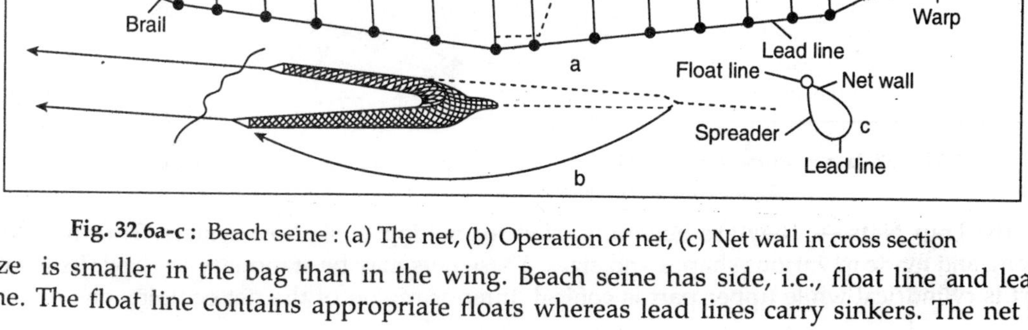

Fig. 32.6a-c : Beach seine : (a) The net, (b) Operation of net, (c) Net wall in cross section

size is smaller in the bag than in the wing. Beach seine has side, i.e., float line and lead line. The float line contains appropriate floats whereas lead lines carry sinkers. The net is used in such a way that its one wing remains on the beach. The other wing is spread on right angle in such a way that, when it is dragged slowly it encircles that part of water section. Both float and lead line do not allow fish to escape.

(v) **Purse seine** — It is used to catch pelagic and migratory fishes. It is purse-like (Fig. 32.7). It has two main lines—float line, which remains on the surface and a lead line, which sinks into the water, but does not touch the bottom. The fishes are trapped and do not escape because net is pursed during the operation. At the time of use, one end of the net is tucked with a boat and another end is laid down with the help of a cruise making a circle and bringing this end back to the boat. The net then takes the shape of purse.

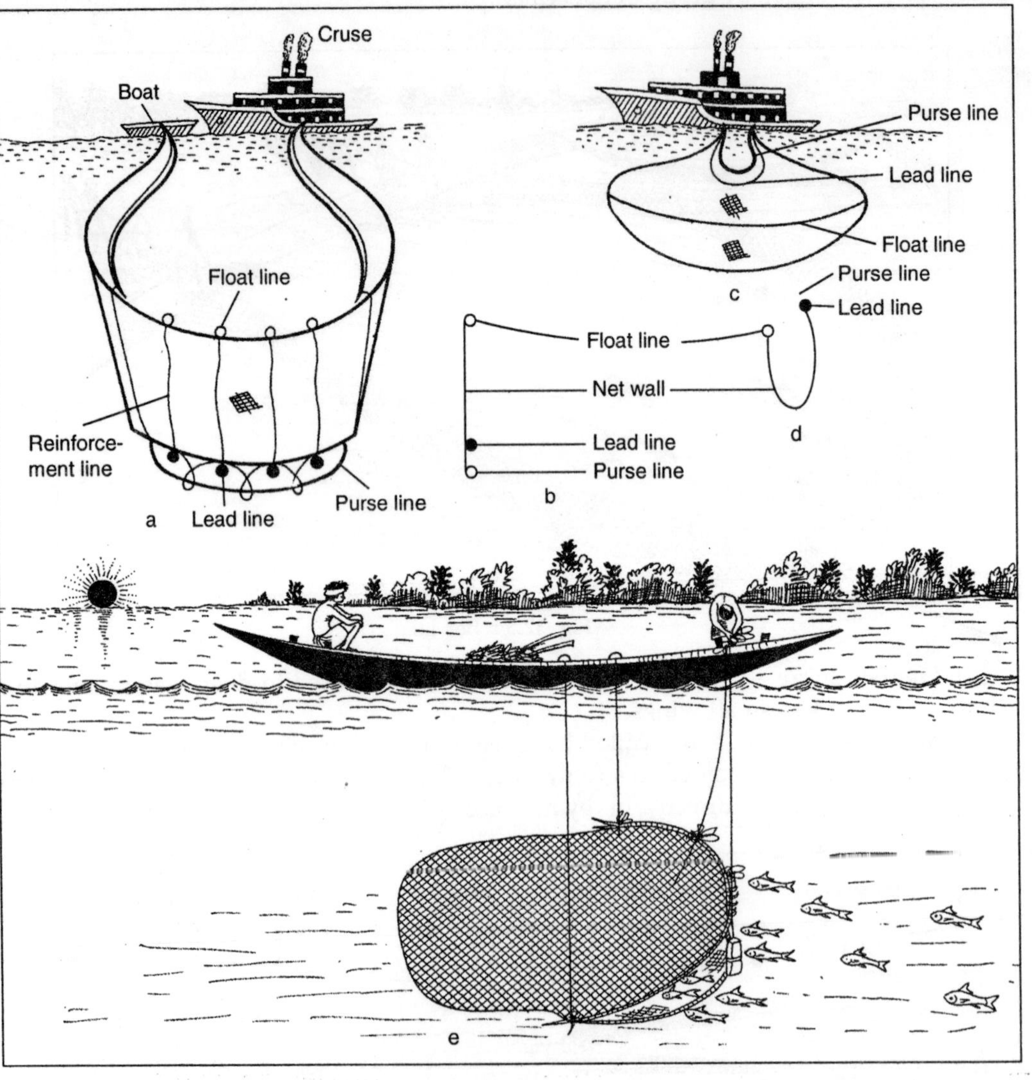

Fig. 32.7a-e : Purse seine : (a) before pursing, (b) net wall at (a), (c) after pursing, (d) net wall at, (e) Shanglo Jal

(II) **Trap Nets** — They are generally used for fishing in shallow waters. Trap nets are strong and made in various shapes and sizes. These nets may be stationary or fixed. Its lower part is cylindrical while upper part is conical. Interior region of the net contains one or two cone-shaped necks to prevent escape of fish. Large trap nets are called a *pound net*, which has a chamber with a wide gate.

(III) **Drop Net** — It is square in shape and mounted with to supple loops at the corners that tied in a cross at the top and is attached to a pole. Drop net is operated with a boat. It is dropped and pulled to catch fishes.

(IV) **Cast Net** — It is a circular and cone shaped net. It is spread from the edges of water. Its circumference is attached to lead line while its centre is attached with a rope. The net assumes shape of umbrella when it is spread on the water. When the net sinks to the bottom it is pulled and fishes are collected.

(V) **Drift Nets and Gill Nets** — These types of nets are made by nylon materials (Fig. 32.8). Gill nets are kept overnight in the water and then dragged. The fishes get entangled in the meshes. There are two types of nets—simple and trammel nets.

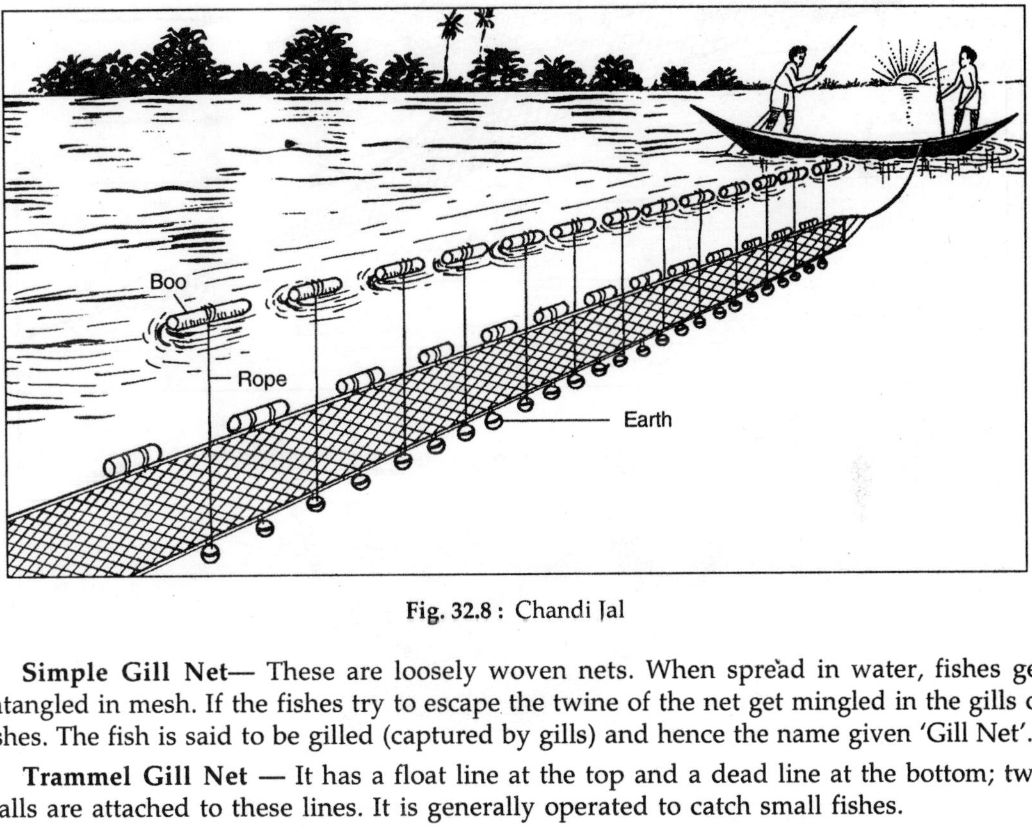

Fig. 32.8 : Chandi Jal

Simple Gill Net— These are loosely woven nets. When spread in water, fishes get entangled in mesh. If the fishes try to escape the twine of the net get mingled in the gills of fishes. The fish is said to be gilled (captured by gills) and hence the name given 'Gill Net'.

Trammel Gill Net — It has a float line at the top and a dead line at the bottom; two walls are attached to these lines. It is generally operated to catch small fishes.

(VI) **Fixed or Stationary Net** — These nets are used to catch fish at inshore water during low tides. These nets are kept fixed with the help of floats, sinkers and stakes. It is rectangular or conical in shape. They are available in various sizes. In West Bengal and Orissa conical fixed nets are used, which are called Ghurni Jal, or behundi, Kathia-kool Jal, Panch-Kathiaber Jal and Panch. However, rectangle nets used are called as Mal Jal in West Bengal, Barnada Jal in Orissa and Kakavalai, Jadi or Mtagh Jal in Tanjore.

(VII) **Bag nets** — It is conical in shape without wings (Fig. 32.9a, b). Some commonly used bag nets are iroga in Andhra, thuriwala in Tamil Nadu and koliwala in Kerela. These nets are used with the help of two catamarans or boats. In the coasts of Mumbai and Gujarat a special type of bag net called 'dol' is used. It is conical with wide mouth. The mouth is fixed on a bamboo.

(VIII) **Scoop Net or Dip Net** — It is round in shape and is used to capture delicate fishes. It is like a finger bowl and can be moved swiftly in a scooping manner, collecting the fish.

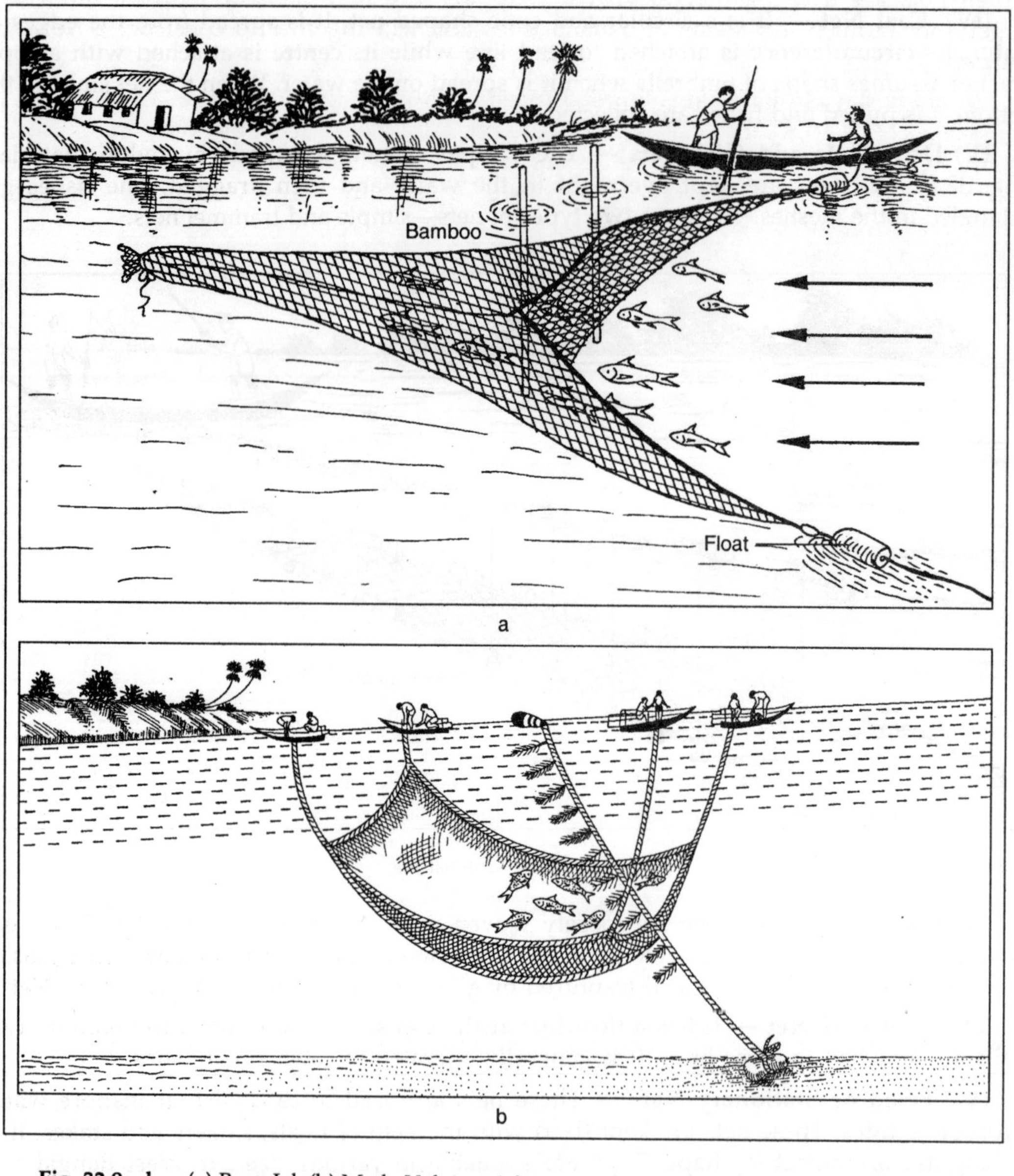

Fig. 32.9a, b : (a) Been Jal, (b) Mada Valai — A bag net of Coromandel coast (after Hernell).

(IX) **Hooks and Lines** — It consists of two types of hand lines and long lines. Various types of hooks are used, such as chain hooks, baited hooks, revolving and non-revolving hooks for capture of larger fishes.

(X) **Trawls** — These are large dragging type nets (Fig. 32.10). There may be two types of trawls with beam called *beam trawls* and *otter trawls*.

Inland Fishing Crafts and Gears

Inland fishing crafts— Rafts and Dongas are age-old inland fishing crafts used in steady waters.

Rafts — Rafts are traditional crafts made by different types of materials. In Bihar earthen pots are tied together to provide support to a high bamboo platform. In West Bengal and Tamil Nadu stems of banana trees are held together to construct a floating platform.

In olden days buffalo skins are tied together to make a crude raft. In West Bengal a simple types of dugout canoe called *Dongas* are used. It is constructed by hollowing the stem of palm tree. It is commonly used for fishing in shallow waters. Vallum is stronger dugout canoe used in brackish waters of Kerala. In Tanjore and Tiruchirapalli chatty rafts made by earthenware are commonly used.

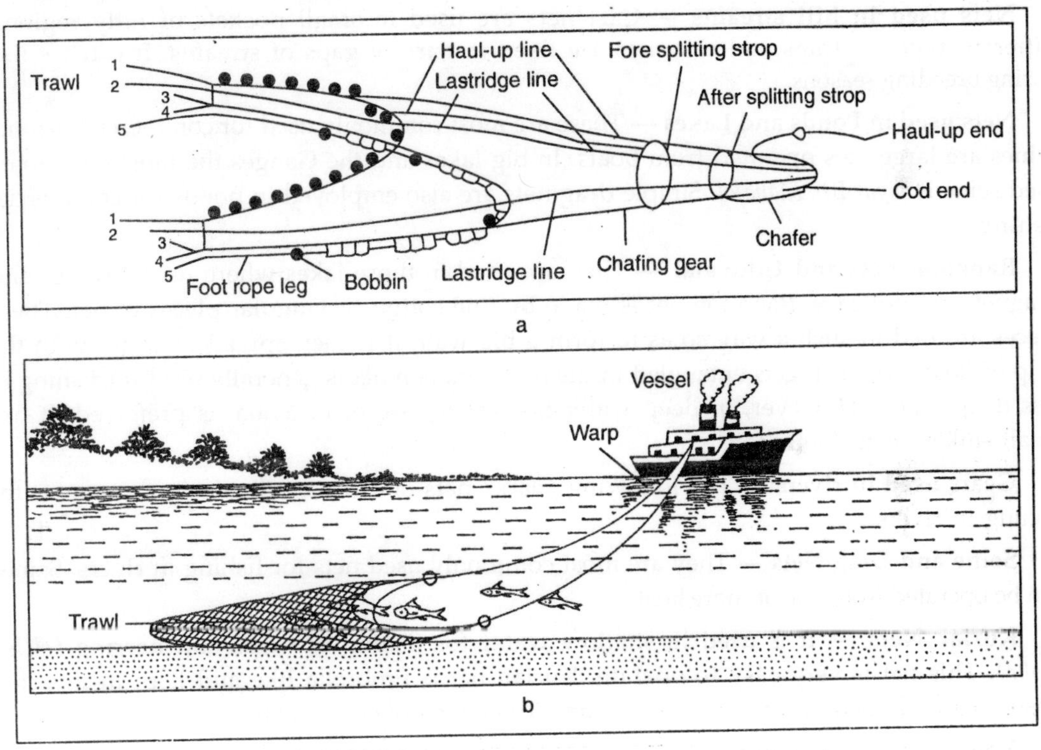

Fig. 32.10a, b : Trwa : (a) The Gear, (b) The net in operation

Boats — Following types of boats are used for fishing in rivers.

Plank-built boat — These boats are sturdy and are used for fishing in rivers with strong tides and currents. In different regions different types of plank boats are operated. One of common types is 'Dinghis' used in west Bengal. Dinghi is narrow and it has a tapering bow and stern. It has no Keel. Dinghi is generally used to operate purse nets and dip nets. Another type is Chandi nauka, which is used to operate drift nets. It is 18 m long and 3 m wide.

In Calcutta (Kolkata) and the other parts of West Bengal a medium sized boat called *Mechho bachari* is used to transport live fishes. However, in Chilka lake (brackish water lagoon) another type of plank boat called *nava* is used.

Kulnawa — It is a specialized boat used in the river Ganges for fishing of minnows. Kulnawa means open wale boat. This boat is used in calm water during night from February to April. It is brought from Tarai parts of Nepal. Kulnawa is made up of 3 parts, viz., a frilled pole, screen platform and boat with one of the walls open. It is made by Sal wood or *Kathal* wood. It is 7 m long and is dinghy type of boat without any keel. The boat is painted by coaltar to keep it dark in colour.

The screen platform is made by 5 m × 0.5 m bamboo splinters, which are woven interlacing plastic cords. Screen is painted with enamel paint. The free margin of screen is always immersed in water. The frilled pole is 4-6 m long bamboo pole kept drooping as frill. 6% dry grass bunches are tied to these poles.

Inland Fishing Gears — India has a wide variety of diversified water bodies, hence the nets used are also diversified.

Nets used in hill streams — Cast nets are used in small pockets of hilly regions. Different types of traps and cast nets are fixed in narrow gaps of streams. It catches fish during breeding seasons.

Nets used in Ponds and Lakes — These are most commonly used for commercial fishing. Seines are large nets operated from boats. In big lakes and the Ganges the most commonly used seine is *Jagat Ber Mahajal*. Simple drag nets are also employed in ponds for commercial fishing.

Rangoon nets and Uduvalai — They are used in those lakes where use of seines and dragnets are not easy. Rangoon net is made by fine cotton rectangular pieces of net. These pieces are tied in such a way so as to form a big wall. It is then spread in water with the help of floats. The fish gets entangled in the net. Rangoon net is generally used for fishing in less deep water. However, in deep water bodies the use of uduvalai is preferred. It has small sinkers with footrope.

Gears used in rivers — The following gears/ nets of different dimensions are used for fishing in rivers.

Seine and Drag Nets — They are most commonly used nets for fishing in rivers. Seines can be operated from one or more boats.

Kuriar Nets — It is used in shallow water for fishing of carps and herring *(Hilsa ilisha)*. This net is very easy to operate. Kurian net is umbrella like and kept in inverted position and dipped in water for some times. Then it is pulled out with capture.

Kona Jal or Bhasa Gulli— It is a special type of large cotton seine net used for fishing of *Hilsa*.

Kona Jal— It is a special type of seine net of about 90 m × 9 m in size. It has conical pockets of small mesh size. These pockets are fixed at a distance of 8-10 m all along the net. It is made by cotton, fish once caught, cannot escape because of a valve-like flap present in the pocket. It is also called Bhasa Gulli.

Moi or Moia Jal — It is a simple net commonly used in shallow waters for fishing.

Jagat Ber or Maha Jal — It is a simple net used all around in rivers.

Chunti Jal — It is a type of drag net used in rivers of Bihar. Two fishermen operate it.

Kharra Jal — Hela jal, bhesal jal and firki jal are also commonly used for fishing of carps and *Catla* in the river Narmada.

Khorsula Jal or Koila Jal — It is a special type of dip net. Bhil tribal fishermen use this net in Narmada river for fishing of migratory *Hilsa*.

Jhanda Net — It is like an open bag operated in shallow waters. It is rectangular in shape with small mesh. This net is stretched on a bamboo frame.

Suti Jal — It is a long-tube like stationary bag net. It has a long wing.

Bada Jal — It is a modified type of Suti jal, which has a wide opening, being kept open with the help of a rod.

Purse Net — Special types of purse nets used for fishing in rivers are kharki jal and shanglo jal or sharki jal. In shanglo jal the mouth of purse can be opened or closed with a vertical cord. However, in kharki jal a vertical bamboo rod is fixed to the lower part of mouth of the purse to open or close it. These nets are operated from a dug out canoe.

Machanized Craft

Selection of Gear

Selection of appropriate gear for catching fish is of utmost importance in order to increase catch. Correct gear may be chosen taking into consideration of the following points :

- For catching the big and strong fish, the gear should also be strong and sturdy with proper mash size.
- To catch fish at various level of the water body, different nets should be used, viz. surface gill nets, column gill nets and bottom gill nets.
- Fishes which swim in shoals, may be caught by using encircling nets like drag nets, purse nets, etc. Similarly using hooks and lines can catch the individual fish.

Maintenance of Gears

Proper care and handling of fishing gears after their use is as important as their use. Proper maintenance increases the durability of the gears. Following care is necessary :

- The gear should be washed thoroughly with the clean water and weeds and mud, etc. should be removed carefully.
- Then dip the net in dilute $KMnO_4$ or $CuSO_4$ or common salt solution to get rid of harmful bacteria.
- Wash again with clean water and then spread in shade for drying.
- To increase durability and strength of the fibre of gear, it may be kept immersed for 10-15 min in hot tar diluted with kerosene.

REFERENCE

Jyoti, M. K. & Sharma, A. 2006. Fishes, Aid to Collection, Preservation and identifications Daya Publishing House, New Delhi.

Chapter 33

AQUARIUM AND ITS MAINTENANCE

All over the world maintaining and keeping an aquarium is gaining popularity among the common people. Installation of a beautiful aquarium becomes the focal point of the room. It is a fascinating view to watch the brilliantly coloured alive fishes, which gives a glimpse of the mysteries of the underwater world and brings nature right to the home. An attractive piece of interior decoration, the aquarium also indicates creativity of a man. It can be set up with diverse ornamental fishes with different green foliar plants, pebbles and rocks in artistic manner, giving an opportunity of creative expression.

The most important aspect of aquarium is its educational value. Maintaining an aquarium is helpful in learning habits of different fishes and their behaviour and breeding patterns. We can observe the spawning, hatching and their developments also. A beautifully set aquarium is a good entertainment. In many hospitals, hotels, offices and schools aquarium is maintained, which prove to be most beneficial as they create interest and help to take their mind off their circumstances. Even in a small room, aquarium of 3-4 feet containing 10-15 beautiful fishes can be kept as a fascinating hobby, and to watch these silently swimming fishes is a great pleasure (Alderson, 2003; Crow *et al.*, 2004).

How should be an aquarium — The aquarium should be of size according to the space and money available. Generally a large size ensures stable water conditions like temperature and pH. A wide fish tank allows the water surface exposed to the air.

Types of aquaria —There are many types of aquaria in common usage.

(1) Glass aquarium — It may be oval, rounded or rectangular in shape. They may be built with metallic frames or may be frameless, in which glasses are joined together with silicon sealer.

(2) Plastic aquaria — They are less expensive than glass aquaria but tend to scratch easily.

How to construct an aquarium—For constructing an aquarium of 24″×12″ × 15″, we require five sheets of $\frac{1}{4}$″ thick plate glass as follows :

(1) Base glass sheet of 24″× 12″

(2) Front glass sheet of 24″× $14\frac{1}{2}$″

(3) Back glass sheet of 24″× $14\frac{1}{2}$″

(4) Side 1 glass sheet of $11\frac{1}{2}$″× $14\frac{3}{4}$″

(5) Side 2 glass sheet of $11\frac{1}{2}$″ × $14\frac{3}{4}$″

The edges of each glass sheet should be carefully smoothen with a dry abrasive paper followed by cleaning with soap water, rinsing with clean water and allowing to dry. Now fix the two appropriate glass sheets with commercial silicon glass sealant. When all glass sheets are jointed, the tank should be checked for any leakage.

Accessories of Aquarium — To maintain the normal conditions of water and also to beautify the aquarium a variety of accessories are used, which are as follows.

(1) **Hood or lid** — A slant roofed lid or hood is placed on the aquarium to prevent entry of any undesirable things or other harmful creatures. It also prevents fish from jumping out of water. The lid can be made of aluminum sheet painted with pleasant colour.

(2) **Aquarium stand** — Aquarium can be placed on shelves or on variety of beautifully designed and suitable iron stands. Nowadays very attractive aquarium stands are available in aquarium shops, which enhances beauty of room.

(3) **Lighting devices** — Proper illumination in the aquarium is the most important factor as it not only notably makes fish visible but also enhances the beauty of tank. Natural light is always beneficial as it is necessary for photosynthesis of small plants of aquarium. Light can also be provided artificially through the bulb or fluorescent tubes of appropriate intensity. Bulb can be installed on the hood or lid of the aquarium. Generally 20 watts bulb is switched on for 10 to 12 hours a day, which is sufficient for the 24" size aquarium.

(4) **Heating devices** — In such areas where the temperature falls below 1°C, it is desirable to maintain the temperature of aquarium to the normal. For this purpose many heating devises are used. These are specially designed electric heater and thermostat. It is recommended that if an aquarium containing 10 gallons of water to be heated to 70 ° F in as unheated room will need 10 watts electrical energy. To monitor and check whether the aquarium water is maintained an required temperature, a reliable thermostat must be fixed in the aquarium.

(5) **Aquarium filters** — A variety of filters are used in aquarium. The main purpose of these filters is filtration of water and removal of waste food and decaying plant leaves and making the water more suitable for healthy fishes and plants. Following three types of filters are used commonly.

Mechanical filters — These are simple pumps containing a nylon filter wood or a foam cartridge. When the aquarium water is passed through it, the suspended particulate matters are trapped and effectively removed from the water. As a result this type of filter is clogged and hence regularly cleaned.

Chemical filter — These filters use absorptive properties of activated carbon. This tiny activated carbon can attract and remove from water certain dissolved organic fish waste products and certain other compounds.

Biological filters — In these filters aquarium gravel, sand, acrylic sheet and porous asbestos used in the form of 2 to 3 inches layer at the tank floor. The gravel bed gently disturbed every 2 to 3 weeks and the accumulated waste is collected with siphon. In biological filtration method the harmless bacteria found on gravels attack the organic waste and convert them in much less toxic compounds.

(6) **Aerators and diffusers** — Proper aeration and oxygen supply throughout in the aquarium water is essential for growth of fishes. Hence, oxygen is supplied through aerators to balance carbon dioxide and oxygen in the aquarium. The air is blown with an electric pump through porous stones, from which a series of fine air bubbles rise above the surface. A

variety of beautiful stones and models are available in shops with adjustable aperture for air supply. It is compulsory to take precaution to clean the porous stone as it often gets chocked with suspended particles. Pumic-stone diffuser of different shapes provided with air tube is also used. Connecting valves, T-pieces, and clamps are very useful for splitting the air supply from an air pump to serve more than one filter or air stone or more than one tank. They also permit fine control of filters and aeration.

(7) **Feeding rings** — These rings are made of plastic or glass tubing in the form of a square or a circle. They float on the water and serve to limit the spread of dried food. Some aquarium specialists do not recommend the use of these rings as the larger species tend to get the greater share then the food be not confined to one area.

(8) **Thermometers** — A good quality thermometer must be placed at vertical position, which can be easily read. Most are dually calibrated for reading in Fahrenheit and centigrade scales.

(9) **Planting sticks** — They may be used occasionally when it is desired to plant bunch of delicate plants in between the rocks where it is difficult to reach with hand. The planting stick can be made of thin wooden or metal strip.

(10) **Ornamental plants** — Fixing of natural plants in aquarium is also necessary for oxygenation of the tank. Some of the important plants used in aquarium are *Vallisnaria, Ceretophyllum, Myriophyllum, Clodia* and *Fontinalis*. Aquarium plants may be of two types, viz., rooted plants and floating plants. Among rooted plants *Amazan sored* is colourful with long leaves, *Eelgrass* is green with narrow leaves and *Crystocdmiyes* is coloured with broad leaves, whereas some important floating plants are *Salvania* with long leaves and *Crystal worth* with small leaves.

Artificial colorful plants are also used to enhance the beauty of the aquarium. But one should take the precaution to use only plants made of soft material, as hard plants can damage the delicate fishes. Nowadays many attractive and beautiful paintings are available in the aquarium shops that can be placed at the back panel of aquarium.

Location of aquarium— Aquarium should be kept in the place of room, which is well lit, safe and clearly visible, so that it may draw attention of the people entering the room. It should not be kept in front of window as direct sunlight will penetrate in the aquarium and tend to increase the water temperature and also enhances photosynthetic activity in the aquarium. Such place is preferred where sufficient diffused light easily falls on it.

Setting of Aquarium — Set the aquarium on the stand, clean all the glass sheets with damp cloth. Spread the sand at the bottom of aquarium. It should be spread slant i.e. 2 inches at the back to 1 inch at the front. This will allow sediments to accumulate in front and can be easily removed by siphoning. Sand should be thoroughly washed with $KMNO_4$ before placing in the aquarium. Care should be taken in the selection of sand. It should not be too coarse, as it will allow food particles to accumulate. Neither it should be too fine such as seashore sand, pack too firmly to allow roots of plants to spread and flourish. The amount of sand required is approximately 12 pound per square feet aquarium base. Then spread coloured natural gravels in the form of small islands.

Now fix the selected plants at the back and sides along the glass. Fill the aquarium with clean water. The best way to fill the water in aquarium is with a rubber tube pouring water slowly in a small jug kept in the aquarium. Water should not be poured with force as it may disturb the arrangement of the aquarium. The tank is filled up to *within 2" of the* top and the water will be clear. Sometimes tap water contains high chlorine contents, which is harmful to fishes. In these circumstances the chlorine contents can be reduced by drawing tap water and allowing it sand for several hours in the open air. Well water may also be used to fill aquarium as it usually contains wealth of minerals. Fresh pond water is considered to be excellent provided it is free from dangerous larvae.

Now place other accessories like bulb, thermostat, thermometer, filters aerators and feeding cups at appropriate places. Keep the aquarium for 3 to 4 days in this state and then introduce fishes.

Selection and Introduction of Aquarium Fishes

It is essential to take the following precautions while selecting the aquarium fishes :

(1) Healthy fishes should be taken from reliable sources.

(2) Behaviour of the fish should be lively with well spread fins and actively swimming.

(3) No sign of sluggishness should be visible in the fishes.

(4) Fishes should be disease-free with no infection of fungus or other parasites.

(5) Selected species should be compatible to each other.

(6) The size of all fishes to be kept in aquarium should be in uniformity; otherwise the larger fishes may harm smaller fishes.

(7) Fishes housed in aquarium should be attractive and brilliantly coloured.

(8) While bringing the fishes from the aquarium shop, they should be properly packed in well-oxygenated polythene water bags at the normal temperature.

Following are some species of aquarium fishes, which can be easily obtained from aquarium shops.

Gold fish *(Carassius carassius)*, Guppy *(Lebistes reticulates)*, Mollies *(Mollienesia latipinna)*, Angel fish *(Pterophyllum eimekei)*, *Gambusia affinis*, *Colisa latia*, *Barbus ticto*, *Betta splendens*, *Tilapia macrocephala*, *Trichogaster trichopterus*, *Danio melabarius.*

Select any 5 to 6 fishes of choice and then introduce them gently in the aquarium. Avoid adding large number of fishes in tank over a short period.

Maintenance of Aquarium

Food and Feeding — It is preferred to supply dry and artificial fish feed, which is a balanced diet for the fishes. Fish should be fed regularly. It can be obtained at the shops. Sometimes live feed is also given. Water fleas forms the most common live feed for aquarium fishes and can be collected by small net from a lake or pond. It is recommended to supply dry food and live food alternatively. For better utilization of feed, it should be given in small amount more frequently. Because large amount of feed given at once tend to settle down quickly and fishes cannot consume the required amount. Overfeeding should be avoided as overfed fish gets lazy and unhealthy. Also unused feed settles at the bottom of tank and decomposes, which lead to disturbance of ecosystem of aquarium (Kasseimann, 2005 & Hunziker, 2005).

(1) **Replenishment of Aquarium Water**— Water should be added when its level falls below the certain limit. Once in a month three fourth of aquarium water is replaced with fresh water by siphoning. This helps in removing excess of waste matter.

(2) **Proper aeration** — Aerators should be switched on frequently in short interval of time to keep the oxygen level high.

(3) **Cleaning of Aquarium** — The most common problem found in the aquarium is appearance of green or brown algae on the front and side glass panels, which must be removed carefully. It should be done at the time of replenishment of water. It can be wiped off with the help of soft mop or scraper. Scraper is a long rod bent to an angle of 45 degrees at the bottom to which is attached a blade, rubber or a piece of felt. The large floating algae can be removed with the help of a small stick. Gravels also can be washed in dilute $KMNO_4$, if algae are deposited over them. Water filter should be efficient and be cleaned regularly.

Other Required Device

Siphon tube — It is a simple 6 to 7 feet long rubber tube with $\frac{1}{4}$ or $\frac{1}{2}$ inch internal diameter. It is most essential for filling and emptying an aquarium. Pinching the tube can regulate the flow of water in the tube. The simple method of removing the water from aquariums is, first put a thumb on one opening of tube to close it and then fill in the water from other opening. When the tube is fully filled with water, immerse either end of rubber tube deep in the aquarium water. Now remove thumb from another end, which is kept outside the aquarium. The aquarium water will start draining out. The siphoning will prevent unnecessary disturbance in setting of aquarium.

Nets — The nets are used for capturing and removing fish or other undesirable matters from the aquarium. Different nets made of nylon with various mesh size, are available in market. A net should be deep enough so that captive may not escape.

Buckets and cleaning brushes are also required at the time of cleaning the aquarium.

Transgenic Aquarium Fishes (Hallerman, 2002)

The tiny zebra fish that lives in aquariums, a popular laboratory animal, was genetically modified to produce a fluorescent red pigment, and is being promoted for sale as a household aquarium pet, the "glofish". The glofish caused a stir in the United States because regulation of such transgenic pets is murky and none of the major regulatory agencies: FDA, USDA or EPA, has been willing to take the lead in regulating the glofish (even though USDA does deal with pet animals).

Breeding Aquarium Fish

Breeding fish is not as dull as it may sound. Here are some hints to get you started.

How do fish breed?

Fish either produce live young or eggs. Guppies, Swordtails and Mollys, are live bearers and are easy to breed. A female Guppy will give birth every four to six weeks and produce up to thirty young at a time. Live bearers are the easiest to start with for the novice.

However, the majority of fish breed from eggs. Even with egg bearers, there is considerable variation in the way in which the eggs are propagated.

Many fish, including Goldfish, Tetras and Zebra fish, are egg scatterers. These fishes scatter their eggs spontaneously around the tank as they swim. In a stream or creek, the eggs would be safely swept away but in an aquarium, the eggs are often eaten by the adult fish. This is a common problem with Goldfish. Placing pebbles or marbles in the bottom of the brooder tank provides places for the eggs to rest and develop, out of the way of the adult fish. Goldfish eggs will also stick to plants in the tank. Once the fish has spawned, the plants can be removed to a nursery tank for development. Goldfish eggs take from three to six days to hatch.

Fish such as Cichlids deposit their eggs on broad-leafed plants, on overhanging rocks or in small plant pots placed in the tank. Splashing Tetras will even lay eggs on the leaves of plants just above water level, to protect the eggs from other fish predators.

Selection of Breeding Pair

Select a pair that are already showing courting behaviour. A breeding pair is often noticeable because of the interest they show to each other.

The fish can be brought into peak condition by feeding them good quality food, especially live food, prior to breeding.

For those that lay eggs, careful observation at the time of spawning (egg laying) is important. Male fish often show too much attention to the female and they need to be separated after spawning has occurred. In other cases, the fish and their eggs will need to be separated to prevent the fish from eating the eggs.

Feeding the Fry

Do not give food to your new brood until they are able to consume it otherwise the food will decay and pollute the tank.

While the young of live-bearing fish will be able to eat immediately, the young of egg-laying fish often remain immobile for several hours while they feed off the remainder of their yolk-sac.

Commercially available fry food comes in liquid or powdered form. The size of the particles must be such that the fish can eat it and for this information, you should consult your aquarium shop.

Brine shrimp are commonly fed to new fry but they can also be fed microworms and infusoria. Microworms are cultured from a mixture of cornflour, water and yeast. Infusoria are a collection of ciliated protozoans that can be produced by pouring boiling water over dried hay or green lettuce leaves in a jar. Keep the mixture in a warm, dark place for a few days and a healthy broth of Infusoria will be produced in the scum floating on the top of the mixture. Small quantities of the Infusoria can be added to the brooding tank several times a day.

Fry do best on several feedings a day. Only small amounts of food should be given as leftover food will spoil and affect the health of the fish.

REFERENCES

Aldeson, D. (2003), Freshwater Aquariums. Pet—Bowtie press pages 92.

Crow, R. Burkhart, Alice & Kedey, D. (2004). Pet — Bowtie. Pets PRC pages 256

Hallerman, Eric M. (2002), "Genetically Modified Fish and Shellfish: Food for Thought." Virginia Issues and Answers 8.1. Virginia Tech Publications & Outreach Communications.

Hunziker, E.R. (2005), Marine Aquariums : Basic Aquarium setup. Bowtie Press pages 144.

Kasseimam, C. (2005), Planted Aquariums : Creation and maintenance. Kroeger Publishing Company pages158.

Chapter 34

FISH DISEASES

Fish Diseases : Prevention and Treatment

Fish disease is a major concern for fish farmers and fish consumers. The health of cultivated fish can be adversely affectd by numerous biotic and abiotic factors. Various diseases not only affect the growth of the fish but also deteriorate the nutritional value of the fish. Large mortality occurs due to various diseases in fish. This decreases the fish production. Disease is visible and noticeable changes from the normal health of an organism. In fish the diseases can be recognized by unhealthy appearance, unusual body movements, swelling or spotting, lesions, blisters and film like coating on the body (Mark, 1983 & Untergasser, 1999).

There are literally hundreds of afflictions that can effect the health of your fish. Only the most common of these afflictions are listed here. There are several precautions, which can be taken to reduce the possibility of disease and keep it from spreading if it does occur.

- Buy only good quality, compatible fish seed.
- Quarantine new fish before adding them to the pond (A hospital tank can be used for this).
- Avoid stressing the fish with rough handling, sudden changes in conditions, or "bully" tankmates.
- Don't overfeed your fish.
- Remove sick fish to a hospital tank for treatment.
- Disinfect nets with 1% $KMNO_4$ used to move sick fish.

Diseases

Bacterial
- Red Pests
- Mouth Fungus
- Tuberculosis
- Dropsy
- Scale Protrusion
- Tail Rot & Fin Rot

Fungal
- Argulus
- Ichthyoporidium

Parasitic
- Velvet or rust
- Anchor Worm
- Ergasilus
- Flukes
- Nematodes
- Leeches

Protozoan
- Costia
- Hexamita
- Ich
- Neon Tetra Disease
- Glugea and Henne guya
- Chiodonella

Non-infectious
- Tumour
- Congenital Abnormalities
- Injuries
- Constipation

Miscellaneous
- Head and Lateral Line Erosion (HELA)
- Eye Disease
- Swim-bladder Disease

Viral
- Lymphocytis

BACTERIAL DISEASES

Bacterial diseases are usually characterized by red streaks or spots and/or swelling of the abdomen or eye (Noga, 1995).

Red Pest

Symptoms : Bloody streaks on body, fins and/or tail appear, so it is called *red pest.* In severe infection these streaks could lead to ulceration and possibly followed by fin and tail rot with the tail and/or fins falling off.

Treatment : External treatments are usually not effective as the disease is internal. At the appearance of disease :

- Treat the tank with a disinfectant and clean the tank as best as possible.
- To disinfect, use acriflavine (trypaflavine) or monacrin (monoaminoacridine) using a 0.2% solution at the rate of 1 ml per litre. Both disinfectants will colour the water, but the colour disappears as the disinfectants dissipate.
- Do not feed a lot while the fish is being treated.
- If the fish do not appear to respond favorably, discontinue disinfecting. Then add an antibiotic to the food. 1% of antibiotic may be carefully mixed with flake food. If you keep the fish hungry they should eagerly eat the mixture before the antibiotic dissipates. Antibiotics usually available in 250 mg capsules. If added to 25 grams of flake food, one capsule should be enough to treat dozens of fishes. A good antibiotic are chloromycetin (chloramphenicol) and use tetracycline.

Mouth Fungus

Symptoms : White cottony patches around the mouth. It looks like a fungus attack of the mouth, so it is called *mouth fungus.* It is actually caused from the bacterium *Chondrococcus columnaris.* In the beginning a grey or white line appear around the lips and later short tufts arise from the mouth like fungus. This disease may be fatal due to production of toxins and the inability to eat. Hence treatment at an early stage is essential.

Treatment : Penicillin at 10000 units per litre is a very effective treatment. Second dose should be given in two days, or use chloromycetin, 10 to 20 mg per litre, with a second dose in two days.

Tuberculosis

Symptoms : Emaciation, hollow belly, possibly sores. Tuberculosis is caused by the Bacterium *Mycobacterium piscium.* Fish infected with tuberculosis may become hollow bellied, pale, show skin ulcers and frayed fins, and loss of appetite. Yellowish or darker nodules may appear on the body or eyes. The main cause for this disease appears to be over crowding in unkept conditions.

Treatment : There is no known and effective treatment for this disease. The best thing to do is to destroy the infected fish and, if unkept conditions or overcrowding is the suspected cause, it is required to take necessary measures.

Dropsy

Symptoms : Bloating of the body, protruding scales. Dropsy is caused from a bacterial infection (*acromonas*) of the kidneys, causing fluid accumulation or renal failure. The fluids in the body build up and cause the fish to bloat up and the scales to protrude.

Treatment : An effective treatment is to add an antibiotic to the food. With flake food, use about 1% of antibiotic and carefully mix it. Antibiotics in 250 mg capsules if added to 25 grams of flake food will be sufficient to treat dozens of fishes. A good antibiotic is chloromycetin (chloramphenicol), or use tetracycline.

Scale Protrusion

Symptoms : Protruding scales without body bloat. Bacterial infection of the scales and/or body causes scale protrusion. An effective treatment is to add an antibiotic to the food. With flake food, use about 1% of antibiotic like chloromycetin (chloramphenicol), or tetracycline.

Tail Rot & Fin Rot

Symptoms : Disintegrating fins that may be reduced to stumps, exposed fin rays, blood on edges of fins, reddened areas at base of fins, skin ulcers with grey or red margins, cloudy eyes. It is caused by the bacteria *Aeromonas.* If tank conditions are not good an infection can be caused from a simple injury to the fins/tail. Tuberculosis can lead to tail and fin rot. Basically, the tail and/or fins become frayed or lose colour.

Treatment : Treat the water or fish with antibiotics. A good antibiotic is chloromycetin (chloramphenicol) or tetracycline. Treatment of 1% $CuSO_4$ is also effective.

Ulcer

Symptoms : Loss of appetite and slow body movements. It is caused by baceria, haemophilus.

Treatment : Give dip treatment in 1% $CuSO_4$ for one minute for a period of 3 to 4 days. Antibiotics oxytetracycline and chloramphenicol may be useful in acute infection.

Precautions during Treatment of Bacterial Diseases

Bacterial diseases are best treated by antibiotics such as penicillin, amoxicillin, or erythromycin. The most common parasitic disease called "Ich" can be treated most effectively with copper or malachite green in the right dosage. Most medications contain copper as an ingredient. Many water treatments like "Aquari-Sol" will also contain copper as an ingredient. The copper may be harmful to most plants and invertebrates, such as snails. Indeed, most snail removers are copper based.

Antibiotic may disturb biological filtration in the tank. Therefore, it is also recommended to monitor either ammonia and nitrite levels of water, or use an ammonia remover such as "AmQuel" to be sure that the level of ammonia does not exceeds the desired limit.

VIRAL DISEASES

Lymphocytis

Symptoms : Nodular white swellings (cauliflower) on fins or body. Lymphocystis is a virus and being a virus, affects the cells of the fish. It usually manifests itself as abnormally large white lumps (cauliflower) on the fins or other parts of the body. It can be infectious but is usually not fatal. Unfortunately, there is no cure. Fortunately, this is a rare disease.

Treatment : There are two suggested treatments. One treatment is to remove and destroy the infected fish as soon as possible. The other treatment is to simply separate the infected fish for several months and to wait for remission, which usually does occur.

PROTOZOAN DISEASES

Velvet or Rust

Symptoms : Yellow to light brown "dust" on body, clamped fins, respiratory distress (breathing hard).

This disease has the appearance of a golden or brownish dust over the fins and body. The fish may show signs of irritation, like glancing off aquarium decor, shortage of breath (fish-wise), and clamping of the fins. The gills are usually the first thing affected. Velvet affects different species in different ways. Danios seem to be the most susceptible, but often show no discomfort. The disease is highly contagious and fatal.

Treatment : The best treatment is with copper at 0.2 mg per litre (0.2 ppm) to be repeated once in a few days, if necessary. Acriflaving (trypaflavine) may be used instead at 0.2% solution (1 ml per liter). As acriflavine can possibly sterilize fish and copper can lead to poisoning, the water should be gradually changed after a cure has been effected.

Costia

Symptoms : Milky cloudiness on skin.

Treatment : This is a rare protozoan disease that causes a cloudiness of the skin. The best treatment is with copper at 0.2 mg per liter (0.2 ppm) to be repeated once in a few days, if necessary. Acriflavine (trypaflavine) may be used instead at 0.2% solution (1 ml per litre). As acriflavine can possibly sterilize fish, and copper can lead to poisoning, the water should be gradually changed after a cure has been effected.

Hexamita

Symptoms : Caused by intestinal flagellated protozoa that attack the lower intestine. As it is a disease of the digestive tract, characterized by the loss of appetite.

Treatment : An effective treatment is the drug metronidazole. A combined treatment in the food (1% in any food the fish will eat) and in the water (12 mg per liter) is recommended. Repeat the water treatment every other day for three treatments.

Myxosporidisis

Symptoms : It is caused by infection of *Myxosorida.* Cysts appear on the body, internal tissues and organs. Fish becomes weak. Scales become weak, perforated and fall off.

Treatment : Give dip treatment in 10% common salt solution.

Ich (Ichthyphthirious)

Symptoms : Salt-like specks on the body/fins. Excessive slime. Problems in breathing (Ich invades the gills), clamped fins, loss of appetite. Ich, **white spot disease,** whatever the name, this is the most common malady experienced in the home aquarium. Luckily, this disease is also easily cured if noticed in time. Ich is actually a protozoa called *Ichthyophthirious multifiliis.* There are three phases to the life cycle of these protozoa. Normally, to the amateur aquarist, the life cycle is of no importance. However, since Ich is susceptible to treatment at only one stage of the life cycle, an awareness of the life cycle is important.

Adult phase—It is embedded in the skin or gills of the fish, causing irritation (with the fish showing signs of irritation) and the appearance of small white nodules. As the parasite grows it feeds on red blood cells and skin cells. After a few days it bores itself out of the fish and falls to the bottom of the aquarium.

Cyst phase—After falling to the bottom, the adult parasite forms into a cyst with rapid cell divisions occurring.

Free swimming phase—After the cyst phase, about 1000 free swimming youngs swim upwards looking for a host. If a host is not found within 2 to 3 days, the parasite dies. Once a host is found the whole cycle begins a new.

Treatment : The drug of choice is quinine hydrochloride at 30 mg per litre (1 in 30000). Quinine sulphate can be used if the hydrochloride is not available. The water may cloud but this will disappear. By reducing the time (with raised temperature) of the phases, you should be able to attack the free swimming phase effectively. Most commercial remedies contain malachite green and/or copper, which are both effective.

NEON TETRA DISEASE

Symptoms : Whitened areas deep into the fishes flesh. Muscle degeneration leading to abnormal swimming movements.

So named for the fish it was fist recognized on. It is caused by the sporozoa *Plistophora hyphessobryconis.* Even though it is named after Neon Tetras, it can appear on other fish. Whitish patches appear as if just below the skin. In Neon Tetras it destroys the bright blue-green neon stripe. The organisms form cysts which burst and release spores. The spores penetrate further and form more cysts. Eventually, the spores migrate to the water and are eaten by other fish in the food. These spores migrate into the digestive tract, then the muscles, and a new infection starts.

Treatment : There is no known cure. It is best to destroy the infected fish and clean the aquarium.

Glugea and Henneguya

Symptoms : Similar to Lymphocytis, the fish will have nodular white swellings on fins or body.

Treatment : *Glugea* and *Hnneguya* and sporozoans that form large cysts on the fish's body and release spores. Luckily, these diseases are very rare. The fish bloat up, with tumour- like protrusions, and eventually die. No cure, as of yet. It is best to destroy the infected fish before the spores spread.

Bio-Disease

Symptom : It is cuased by protozoa, *Myxobolus pfcifferi.* Large boils of varying size of nut appear in several parts of the body.

Treatment : Give bath in 3% common salt solution or in 1% formalin solution for 10 minutes.

Chilodonella

Symptoms : Dulling of the colours due to excessive slime, frying of the fins, weakness, gill damage. This disease causes a blue white cloudiness on the skin and attacks the gills. Later the skin may be broken down and the gills destroyed. The fish may behave as if they have irritations.

Treatment : Acriflaving (trypaflaving) may be used at 1% solution (5 ml per litre). As acriflavine can sterilize fish, the water should be gradually changed after a cure has been effected. It also helps to raise the temperature to about 80 °F.

Whirling Disease

Symptoms : It is also a protozoan disease, caused by *Myxosoma cerebralis.* Blakeening of tail, caudal band and deformity of anal region are common symptoms.

Treatment : Destroy all the diseased fishes by applying quicklime at the rate of 1kg/ha.

Knot Disease

Symptoms : It is caused by protozoa, *Myxobolus exigus.* Salt konts appear on the skin.

Treatment : There is no effective treatment. Therefore, all infected fishes should be removed immediately and killed.

FUNGAL DISEASES

Fungus (Saprolegnia)

Symptoms : Tufts of dirty, cotton-like growth on the skin, can cover large areas of the fish, fish eggs turn white. Fungal attacks always follow some other health problems like parasitic attack, injury, or bacterial infection. The symptoms are a grey or whitish growth in and on the skin and/or fins of the fish. Eventually, if left untreated, these growths will become cottony looking, The fungus, if left untreated, will eventually eat away on the fish until it finally dies.

Treatment : Use a solution of phenoxethol at 1% in distilled water. Add 10 ml of this solution per litre. Repeat after a few days if needed, but only once more as three treatments could be dangerous inhabitants. If the symptoms are severe the fish can be removed and treated with small amount of providone iodine or mercurochrome. For attacks on fish eggs, most breeders will use a solution of methylene blue adding 3 to 5 mg/l as a preventive measure after the eggs are laid.

Ichthyosporidium

Symptoms : Sluggishness, loss of balance, hollow belly, external cysts and sores.

Ichthyosporidium is a fungus, but it manifests itself internally. It primarily attacks the liver and kidneys, but it spread everywhere else. The symptoms vary. The fish may become sluggish, losely balanced, show hollow bellies, and eventually show external cysts or sores. By then it is usually too late for the fish.

Treatment : Phenoxethol added to food as a 1% solution may be effective. Chloromycetin added to the food has also been effective. But both of these treatments, if not watched with caution, could pose a risk to your fish. It is best, if diagnosed soon enough, to destroy the affected fish before the disease spreads.

PARASITIC DISEASES

Argulosis

Symptoms : Caused due to *Argulus* (Fish louse). The fish scrapes itself against objects, clamped fins, visible parasites about 1/4 inch in diameter are visible on the body of the fish. The fish louse is a flattened mite-like crustacean about 5 mm long that attaches itself to the body of fish. They irritate the host fish which may have clamped fins, become restless, and may show inflamed areas where the lice have been.

Treatment : With larger fish and light infestations, the lice can be picked off with a pair of forceps. Other cases can best be done with a 10 to 30 minute bath in 10 mg per litre of potassium permanganate or treat the whole tank with 2 mg per liter, but this method is messy and dyes the water.

Anchor Worm (Lernaea)

Symptoms : The fish scrapes itself against objects, whitish-green threads hang out of the fish's skin with an inflamed area at the point of attachment.

Ahchor worms are actually crustaceans. The youngs are free swimming and borrow into the skin, go into the muscles and develop for several months before showing. They release eggs and die. The holes left behind are ugly and may become infected. The anchor worm is too deeply imbedded to safely remove.

Treatment : A 10 to 30 minute bath in 10 mg per litre of potassium permanganate, or treat the whole tank with 2 mg per litre, but this method is messy and dyes the water.

Ergasilus

Symptoms : The fish scrapes itself against objects, whitish-green threads hang out of the fish's gills. This parasite is like the anchor worm, but is smaller and attacks the gills instead of the skin.

Treatment : Treatment can best be done with a 10 to 30 minute bath in 10 mg per litre of potassium permanganate.

Flukes

Symptoms : The fish scrapes itself against objects, rapid till movement, mucus covering the gills or body, the gills or fins may be eaten away, the skin may become reddened. There are many species of flukes, which are flatworms about 1 mm long, and several symptoms That are visible. They infest gills and skin much like Ich, but the difference can be seen with a hand lens. You should be able to see movement and possibly eyespots, which is not found in Ich. Gill flukes will eventually destroy the gills thus killing the fish. Symptoms of heavy infestations are pale fish with drooping fins, rapid respiration, glancing off aquarium decor, and/or hollow bellies.

Treatment : Treatment can best be done with a 10 to 30 minute bath in 10 mg per litre of potassium permanganate. Or treat the whole tank with 2 mg per litre, but this method is messy and dyes the water.

Nematoda

Symptoms : Worms hanging from the anus. Nematodes (threadworms) infect just about anywhere in the body but only shows itself when they hang out of the anus. A heavy infestation causes hollow bellies.

Treatment : Two treatments have been suggested. First treatment : soak the food in parachlorometaxylenol and give the fish a bath or treat the aquarium with 10 ml per litre. The bath should last for several days. Second treatment : find special food containing thiabendazole as a nematode (threadworm) cure and hope the fish will eat it.

Leeches

Symptoms : Leeches are visible on the fish's skin.

Leeches are external parasites and affix themselves on the body, fins, or gills of the fish. Usually they appear as heart shaped worms (they are just curled up) attached to the fish. Since leeches are sucking and borrowing into the surface of the fish, removal with forceps can cause great damage, if not death, to the fish.

Treatment : If the fish is bathed in a 2.5 percent solution of salt for 15 minutes, most of the leeches should just fall off. Those that do not will be affected enough to remove with forceps with minimal damage. Another treatment is to add Trichlorofon at 0.25 mg/l.

MISCELLANEOUS

Head and Lateral Line Disease (also known as Hole-in-the-Head disease)

Symptoms : Begins as small pits on the head and face, usually just above the eye. If untreated, these turn into large cavities and then the disease progresses along the lateral line.

Head and Lateral Line Disease is attributed to a nutritional deficiency of one or more of vitamin C, vitamin D, calcium, and phosphorous. It is thought to be caused by a poor diet or lack of variety, lack of partial water changes or over filtration with chemical media such as activated carbon.

Treatment : HLLE has been reversed by one or more of the following treatments :

- Increase frequent water changes.
- Add vitamins to frozen foods.
- Add flake foods, as they are enriched with vitamins.
- Add green, either frozen or in leaf form, to the diet.
- Decrease the amount of beef heart as it lacks many critical nutrients.

(This disease is often confused with another disease called Hexamita, because both these diseases are often seen simultaneously in the same fish. Haxamita is a protozoan disease that attacks the lower intestine).

Eye Problems

Symptoms : Cloudy cornea, opaque lens, poor eye, swelling, blindness.

- Cloudy cornea can result from a bacterial invasion. Antibiotics may help.
- Opaqueness can result from poor nutrition or a metacercaria invasion (grubs). Try foods with added vitamins and change the diet to include variety.
- Pop eye (exophthalmia) can result from rough handling, gas embolism, tumours, bacterial infection, or vitamin A deficiency. Gas bubble or bacterial infection can be treated successfully with penicillin or amoxicillin.
- Blindness can be caused by poor nutrition or excessive light. Lowering the light level and a change of diet to include lots of variety may help prevent it.

Swim-bladder Disease

Symptoms : Abnormal swimming pattern, difficulty in maintaining equilibrium.

Swim bladder problems usually indicate another problem listed here. If you suspect swim-bladder problems in a fish, first check and treat it for other diseases as listed below :

- Congenitally deformed bladder.
- Cancer or tuberculosis in organs adjacent to swim bladder.
- Constipation
- Poor nutrition
- Serious parasitic and bacterial infestation.

NON-INFECTIOUS MALADIES

Tumours

Tumours can be caused by a virus or a cancer, but most tumours are genetic. The genetic tumours may be caused from too much hybridization, common amongst professional breeders. Practically all tumours are untreatable. If the fish is in distress, it should be destroyed.

Congenital Abnormalities.

Abnormalities usually occur when professional breeders are trying to acquire certain strains in breeds.

Physical Injuries

As in the human world, accidents happen in the lives of fishes also. If the cause of the injury is obvious, it should be remedied. Then the injury should be treated. Then injury should be touched with 2% mercurochrome, which is supplied commercially. Also, depending on the fish's tolerance to water conditions, keeping the fish in slightly acid-water should speed recovery (pH 6.6). Minor injuries, if the water conditions are good, should just heal themselves.

Constipation

Some fishes are more susceptible to constipation than others. Usually fish with more compressed bodies like angelfish and silver dollars. Symptoms are loss of appetite and swelling of the body. The cause is almost always diet. Usually, with a change of diet, the condition rights itself. But in stubborn cases try dried food that has been soaked in medicinal paraffin oil. Glycerol or castor oil may also be used. If the diet is changed on a regular basis and live foods offered are occasionally, this condition may never occur.

The intensification and further development of freshwater aquaculture in India urgently requires knowledge, research facilities and research and expertise on fish diseases and fish health protection.

REFERENCES

Mark, D.L. (1983), Fish Diseases. T.F.H. Publication Inc. New Jersey.

Noga J.E. (1995), Diagnosis and Treatment. Blackwell Publishing pages 1-339.

Unterasser Dieter (1999), Hand Book of Fish Diseases. La Broke's Cardon Path and pond LLC. 8 Pitt Street Brushton pages 160.

PART III

GENETICS

Chapter 35

THE CHROMOSOMES

During cell divisiôn, cytologist observed densely staining prominent structure in the nucleus of the eukaryote cells. These dense staining structures were designated as chromosome (chromo = colored; soma = bodies). In 1902, Walter Sutton first time observed that chromosomes segregate during the process of gamete formation. Later, it is now accepted that genes, the hereditary material, are present on the chromosomes. The morphology of eukaryotic chromosomes can be studied during mitosis. Each species existing in nature could be distinguished by its specific number of chromosomes and is the recent way of classification. The normal structure of chromosomes, different types of chromosomes, size of the chromosomes, the position of centromere and pattern of light and dark staining when chromosomes are stained by different chemical dyes are collectively called as chromosome morphology.

Eukaryotic chromosomes are composed of chromatin, a combination of nuclear DNA and proteins. At metaphase, which is a phase in the cell cycle after the DNA in the nucleus has been replicated; each chromosome contains two identical strands of DNA (each strand contains two complementary strands of nucleotides). The two strands of DNA, or chromatids are held together at a single point, the centromere, or primary constriction point.

The chromosomes are classified on the basis of centromere. It is difficult to see chromosome but they can be seen at the time of cell division, the best appearance is at mataphase stage. Since beginning, there were two methods for describing the morphology of chromosomes, one old method was on the configuration of chromosome at anaphase and second method, presently in use is the configuration at metaphase stage. At metaphase there are the best condensation of chromosomes. According to anaphase configurations, the chromosomes can be classified as V-shaped, J-shaped or rod shaped, it is now seldom used. However, now classification of chromosomes is based on metaphase stage and on the bases of position of centromere. The centromere is visible as the region where two sister chromatids are attached to each other and is the site for the attachment of spindle fibres during chromosomal movements. The centromere may be median, submedian, subterminal or terminal. So there are four morphological classes as observed on metaphase configuration namely, *metacentric* (V-shaped), *submetacentric* (J-shaped), *acrocentric*, or *telocentric* (rod shaped). If the centromere is subterminal, the chromosomes are known as subtelocentric (Rod shaped). Often chromosome with terminal and subterminal centromeres are described as acentric. A diagrammatic sketch is given for the understanding of various parts of chromosome (Fig. 35.1).

The metacentric chromosomes (m) have the centromere in the middle (medial) and the two arms are of equal size. In submetacentric chromosomes (sm), the centromere is submedial hence both the arms are similar but are not equal. In acentric chromosomes the centromere is very near to one end so one arm is very short and another is very long. The distinction between submetacentric and acrocentric is subjective and vary from author to author. Telocentric chromosomes (t) have the centromere adjacent to telomere, and these chromosomes have only one visible arm. Some authors have mentioned that if the centromere is subterminal, the chromosome type is subtelocentric (st). Fishes possess all the above four types of chromosomes.

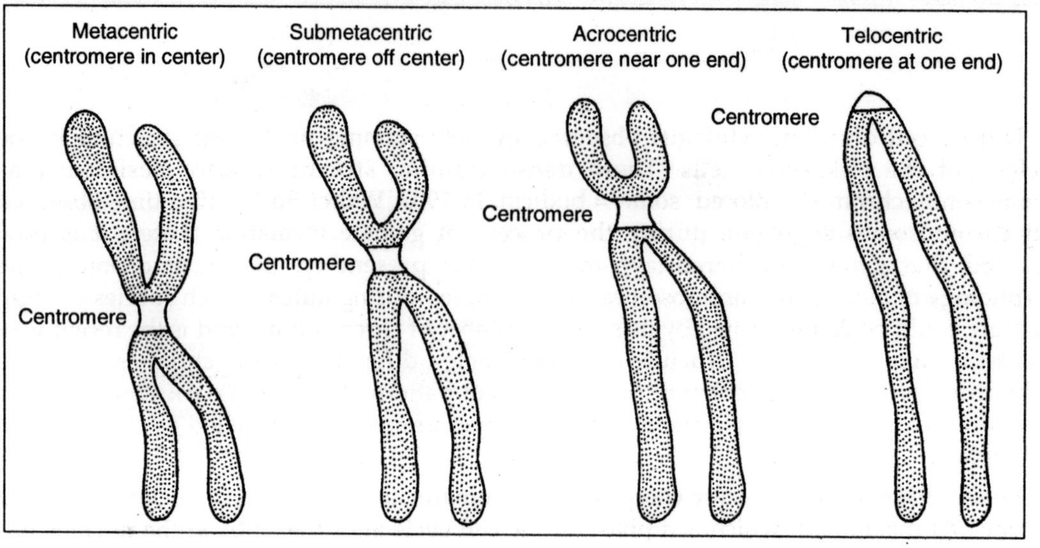

Fig. 35.1 : Diagrammatic representation of Chromosomes on metaphase stage

The size of fish chromosomes is small and the number is very high and overlapping each other. There is confusion in centromere positions too. This confusion may be eliminated by assigning definite numerical values for each category of chromosome type based on arm ratio as suggested by Laven *et at.* (1964). He suggested ratio of the length of the long arm divided by the length of the short arm. He has recommended the following measurement for designating chromosome. The nomenclature is followed in fishes very extensively (Fig. 35.2).

Centromeric Position	Arm Ratio (L/S)	Chromosome Type
Median	1.00-1.70	Metacentric
Subterminal	1.71-3.00	Submetacentric
Subterminal	3.01-7.00	Subtelocentric
Terminal	7.01-	Acentric

Fig. 35.2 : Type of Chromosomes on arm ratio basis (Laven *et al.* 1964)

There are great variation in somatic chromosomes of fishes. The diploid number in fishes ranges from 12 or 16 to 239 + 7. The highest number of 239 has been recorded in *Acipencer naccarii* as reported by Fontana and Colombo (1974). The DNA value ranges between 0.4 P.G. in Odonoptformes and 142 P.G. in dipnoi (Kirpichnikov, 1981). The review on the chromosomes both somatic and sex chromosomes were made by Ojima (1985), Rishi (1979) and Manna (1989) (Sharma *et al.,* 2002). It is unanimously accepted that a majority of families of fishes have chromosome number in the range of 2N = 44 – 52 (Gold *et al.*, 1980) with predominantly acrocentric or subtelocentric chromosomes. However, in magarily (modal number) of diploid chromosome is 2N = 48. In about 138 species the peak is 2N = 44 – 52 (Gold *et al.*, 1980) with predominantly acrocentric or subtelocentric chromosomes. However, in magarily (modal number) of diploid chromosome is 2N = 48. About 138 species the peak is 2N = 46. In about 238 species it was noticed that the peak is 2N = 50. These numbers are counted on well spread metaphase stage and based upon colchicina citrate dry method. The karyotype arrangement is based on homomorphic pairing. The metacentric, submetacentric, subtelocentric, and telocentric chromosomes arrangements are observed in many species. The chromosomes of *Psilorhynchus sucatio* measures 3.46 μm and 1.34 μm in length. The maximum difference was between the chromosome number one and two and was 0.46 m. Khuda-Bukhsh and Chanda (1989) reported 25 homomorphic pairs containing eleven pairs of metacentric (m) – (numbers 5, 10, 13–16, 18–19, 21–23), nine pairs of submetacentric (s.m.)(numbers 1–4, 7–9, 11 and 17) 5 pairs of telocentric (t) chromosomes (number 6, 12, 20, 24, and 25) leading to the chromosomal formula of N = 11 m + 9 s.m. + 5t. They were unable to distinguish sex chromosomes in their observations. In about 8 species of *Labeo* 2 N = 50 chromosomes but in *Labeo cerulaeus* 2N = 48 chromosomes. The three species of hybrid also follow the same pattern (Manna and Khuda-bukhsh, 1977, a, b and Khuda-bukhsh, 1980, 1982). Nayyar (1962) reported 2N = 54 rod-like chromosomes in *Labeo dero* but Khuda-buksh and Chanda (1989) found that 2N = 50 with a large number of biarmed chromosomes in these species.

For average induced breeding the chromosomes structure of Indian major carps are important. In *Labeo rohita*, the 2n number of chromosomes is 50, out of which metacentric chromosomes are 18,the submetacentric chromosomes are 8, and subtelocentric are 24. Similarly, another important major carp is *Catla catla*, total number of chromosomes are 50 pairs which are morphological distinguished into metacentric (m) as 8 pairs, submetacentric are 16 pairs while subtelocentric are 26 pairs. The number of fundamental arms is 78 as reported by Manna (1988). The third Indian major carp is *Cirrhinus mrigala*. It possesses 50 pairs of chromosomes. They are classified as 6 pairs as metacentric, 26 pairs as submetacentric while 18 pairs of chromosomes are subtelocentric,

In fish, *Astyanax scabripinnis*, de Marco Ferro *et al.*, (2003) reported metacentric macrochromosome (Bm),similar in size to the first chromosome of the karyotype, with two variant forms, a large submetacentric(Bsm) and a small metacentric (Bm), both showing reduced frequencies in the population.

The karyotype of rohu-catla hybrids consist of 4 (m), 24 (sm) and 22 acrocentric pairs. They are different with the parent types in morphological features(Khuda-Buksh,1982).

Apart from the above normal karyotype, a few investigators also noticed B chromosomes in fishes. The B chromosome is a chromosome also known as supernumerary chromosome; it is not essential for life of a species. Thus the population would consist of individuals with 0,1,2,3 (etc.) **B chromosomes** (White, 1973). In *Theringicththys labrosus*, the B microchromosome appeared totally heterochromatic (de Carvalho & Dias, 2005). The **B chromosomes** have the following effects on the structure of A chromosome.

1. Increase asymmetry chaisma distribution
2. Increase crossing over and recombination frequencies: increase variation
3. Cause increased unpaired chromosome: infertility.

Sex Chromosomes (Heterogamy)

It is difficult to distinguish between male fish from female fish on the basis of sex chromosomes but sexes do occur in fishes. Male and female fishes in major carps could be distinguished during breeding season by pectoral fin and by pressing the belly to see whether milt or eggs are coming out. The distinct sex chromosomes, like that of human, are not visible in the cytological preparations in fishes. Campos and Hubbs (1971) suggested that sex chromosomes may be present in some fishes while absent in majority of cases. Sharma and Tripathi (1988) reported in *Lepidocephalichthys guntea* diploid number of 52 chromosomes in male and 51 in the female. They reported that one telocentric chromosome being less in females. They suggested that female is heterogametic as ZO, while the male is homogametic having ZZ. Similar results were described earlier in *Trichogester fasciatus* by Rishi (1975, 1979). In female, he has observed that 2n = 16m + 16sm + 15t while in male they found 2n = 16 m + 16sm + 16t. Rishi (1976) described female heterogamety of ZZ : ZO type (ZZ = male; ZO = female), where as Z represent the smallest of the acrocentric chromosomes. The 2nd chromosome number and heterogamety in some fishes are given in the table.

The list of some important species are given as follows :

Sl. No.	Species	2N NO and Karyotype	Sex	Sex Clnonosous	Reference
1.	*Anguilla anguilla*	38 = 22M, SM + 16A	?	ZW	Ojima, 1985
2.	*Mystus tengara*	54 = 10M + 38 SM, st = 64	?	XY	Manna, 1984
3.	*Colisa fasciatus*	48 = 17M + 16 SM + 15A	?	ZW	Ojima, 1985
4.	*Lepomis cyanellus*	48 = 48a	?	XY?	Manna, 1984
5.	*Cynoglossus puncticeps*	39 = IM + 38T	?	ZW?	Manna, 1984
6.	*Carassius auratus*	100 = 12M + 36SM + 52ST, T	?	XY	Kirpichnikov, 1981
7.	*Cyprinus carpio*	101 = 12M + 355M + 52 ST, Λ + 2M	?	XY	Ojima, 1985
8.	*Hoplias lacerdae*	50 = 36 M + 14 SM	?	XY	Ojima, 1985
9.	*Gambusia farmosa*	48 = IM + 2 SM+ 2SM + 45A	?	ZW	Ojima, 1985
10.	*Poceilia farmosa*	46 = IST + 45 A	?	ZW	Manna,1984
11.	*Oncorhyncus nerka*	57 = 45M, SM + 2ST + 10A	?	XO	Manna, 1984

Contd.

Sl. No.	Species	2N NO and Karyotype	Sex	Sex Clnonosous	Reference
12.	*Sambo gairdneri*	89 = 67M, SM + 3ST + 19A	?	X_1X_2Y	Ojima, 1985
13.	*Tetrodon rigropunctatus*	37 = 15N + 14 SM + 8 ST	?	ZW	Manna, 1984
14.	*Aspidoparia mavav*	2n = 50 (8m + 12sm + 12st + 18t)			
15.	*Clossocheilus latius*	2n = 50 (6m + 12sm + 16st + 16t)			
16.	*Labeo bangusia*	2n = 70 (16m + 6sm + 16st + 32t)			
17.	*Perilampus asper*	2n = 48 (4m + 65T + 38t)			
18.	*Puntius chrysopterus*	2n = 50 (2m + 2sm + 4st + 42t)			

REFERENCES

Campos, H H. and Hubbs, C. 1971. Cytology of six species of gambusiine fishes. Copea 3 : 566-569.

Cytologia 53 : 81-86.

de Carvalho, R.A. & Dias, A. L. (2005). Karyotypic Characterization of *Iheringichthys labrosus* (Pisces, Pinelodidae) : C, G and restriction endonuclease bonding. Genet. Mol. Res. 4(2) : 663-667.

de Marco Ferro,D.A. Moreira-Filho,O & Bertollo, L.A.C. 2003. B chromosome polymorphism in Fish, *Astyanax scabripinnis*. Genetica. 119 : 147-157.

Fontana, F. and Colombo, G. (1974). Experientia, 30 : 739.

Gold, J. R., Karel, W. J. and Strand, M. R. (1980). Chromosomal change and rectangular evolution in North American cyprinid fishes. Genet. Res. 35 : 157–164.

Khuda-Bukhsh, A. R. (1982). Karyomorphology of two species of *Tor* (Pisces, Cyprinidae) with a high number of chromosomes. Experientia, 38 : 82–83.

Khuda-Bukhsh, A. R. and Chanda, T. (1989). Somatic chromosomes of three species of hillstream fishes from Assam. In : Fish Genetics in India, (Eds. Das, P. & Jhingran, A. G.). Today & Tommorow's Printers & Publishers, New Delhi. pp. 69-73.

Khuda-Bukhsh, A.R. (1980). A high number of chromosomes in the hillstream cyprinid, *Tor putitora* (Pisces). Experientia, 36 : 173–174.

Kirpichnikov, V. S. (1981). Genetic Basis of Fish Selection. Springer–Verlag, Berlin.

Laven, A, Fredga, K. & Sandberga, A. 1964. Nomenclature for centromeric position in chromosomes. Heteditas (Lund) 52(2) : 201-220.

Manna G. K. and Khuda-Bukhsh, A. R. (1977a). Karyomorphology of cyprinid fishes and cytological evaluation of the family. Nucleus 20 : 119-127.

Manna, G. K. (1983). Cytogenetics studies on fishes and amphibia. Proc. XV Intnatl Congr. Genet, New Delhi. In : Genetical research in India. ICAR, Pub. NND. New Delhi. pp. 244-273.

Manna, G. K. (1984). Progress in fish cytogenetics. Nucleus, 27 : 203-231.

Manna, G. K. (1989). Fish cytogenetics related to toxonomy, evolution and monitoring aquatic genetoxic agents. In : Fish Genetics in India, (Eds. Das, P. & Jhingran, A.G.) Today's and Tomorrow's Printers and Pub. New Delhi. pp. 21–26.

Manna, G. K. and Khuda-Bukhsh, A. R. (1977b). A checklist of chromosomes in cyprinid fishes. J. Zool. Res., 1 : 34-43.

Nayyar, R. P. (1962). Karyotype studies in two cyprinids. Cytologia 27 : 229–231.

Ojima, Y. (1985). Chromosome data retrieval system.

Rishi K.K. (1976). Karyotypic Studio on four species of fishes. Nucleus 19 : 95-98

Rishi, K.K. 1975. Somatic and meiotic chromosomes of *Trichogester fasiatus* (Bl. & Sch.) Teleostei; Perciformes; Osphronemidae. Genen. Phaenen 18(2-3) : 49-98.

Rishi, K.K. 1979. Somatic G banded chromosomes of *Colisa fasciatus* (Perciformes : Belentidae) and confirmation of female heterogamety. Copea. 1 : 146-149.

Sharma, O.P, Tripathi, N.K. Sharma K.K. 2002. A review on chromosome banding in Fishes. In : Some aspects of Chromosome Structure and Functions. Ed. by R.C. Sobti *et al.* Narosa Publishing House, New Delhi, India.

Sharma, O.P. and Tripathi, N.K. (1988)—Female heterogamy in two teleostean fishes.

White M.J.D. (1973)—Animal Cytology and evolution 3rd Edition. Cambridge University Press, Cambridge.

Chapter 36

CHROMOSOME BANDING

As described in the earlier chapter, centromere position and arm ratios can assist in identifying specific pairs of chromosomes, but inevitably several or many pairs of chromosomes appear identical by these criteria. The ability to identify specific chromosomes with certainty was revolutionized by the discovery that chromosomes at metaphase when stain with certain specific dyes would produce reproducible patterns of bands. **Chromosome banding** has since become a standard and indispensable tool for cytogenetic analysis. A unique banding pattern is now commonly used to identify each chromosome and to diagnose chromosomal aberrations, including chromosome breakage, loss, duplication or inverted segments.

What is **Chromosomal Band?**

A band is defined as that part of a chromosome, which clearly distinguished from its adjacent segment by appearing light (bright) and dark stripes or bands, which appear along its length after being stained with specific dyes. The stained chromosomes when visualized under microscope show a continuous series of bright and dark bands (or fluorescent versus non-fluorescent). Importantly, each chromosome displays a unique banding pattern, analogous to "bar code". which allows it to be reliably differentiated from other chromosomes of the same size and centromeric position (Fig. 36.1).

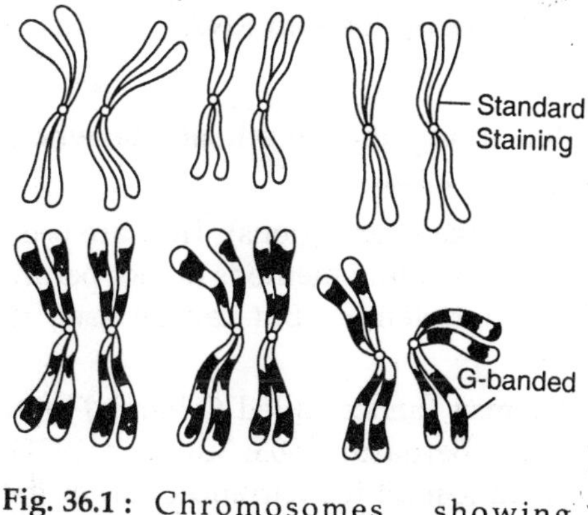

Fig. 36.1 : Chromosomes showing standard staining & G-bands.

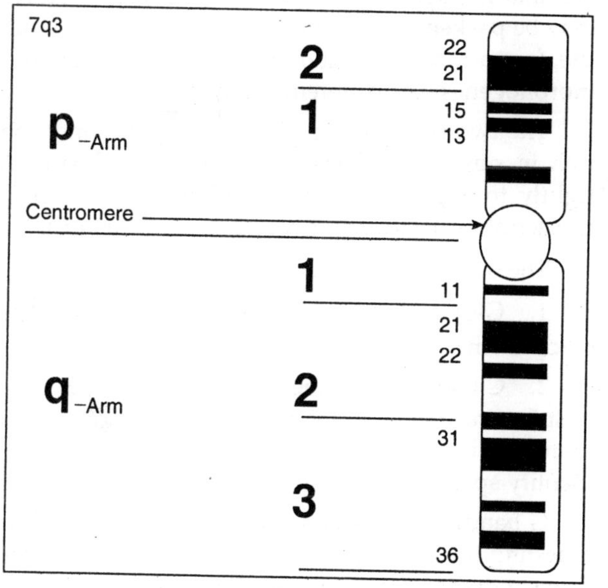

Fig. 36.2 : Chromosomes 7 and q3 bands showing Wolf Parkinson Syndrome.

The causes of many diseases in human beings can be identified now on the basis of molecular genetics. Wolf Parkinson Syndrome is caused due to mutation in 7q3, which means that the defect is due to chromosome 7 and in the q arm at band three (Fig. 36.2).

Structure of Chromosome

To understand what chromosomal bands represent, it is essential to know structure of chromosome. Eukaryotic chromosomes are composed of chromatin, a combination of nuclear DNA and proteins. There are two varieties of chromatin, one is known as *heterochromatin* and another is called as *euchromatin*. They can be distinguished cytological into segments, the heterocromatin, it takes dark stain while other is euchromatin, which takes lighter stain. Heterochromatin is localized in the periphery of the nucleus (Fig. 36.3).

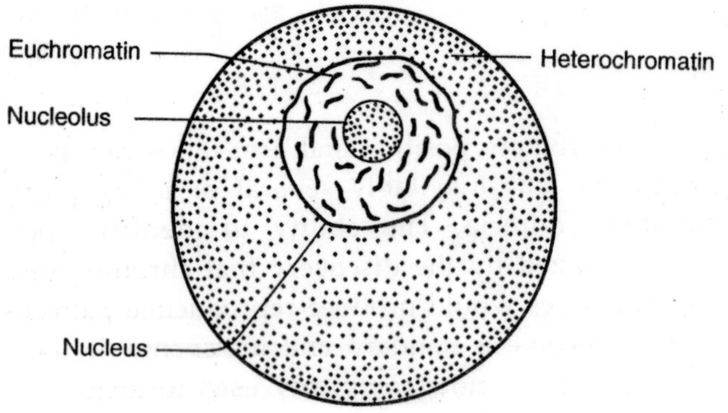

Fig. 36.3 : Diagrammatic representative of cell showing the location of heterochromatin

Heterochromatin is believed to serve several functions, from gene regulation to the protection of the integrity of chromosome. The constitutive heterochromatin occurs around the chromosome centromere and near telomeres. All cells of a given species will package the same region of DNA in constitutive heterochromatin, and in all cells any genes contained within the constitutive heterochromatin will be poorly expressed. The facultative heterochromatin will not be consistent within the cells of a species, and thus sequences in one cell that is packed in facultative heterochromatin (and the genes within poorly expressed) may be packed in euchromatin in another cell (and genes within no longer silenced). Hence, the formation of facultative heterochromatin is regulated, and is often associated with morphogenesis or differentiation.

The two types of proteins are histones and non-histones proteins. Histones are proteins rich in positively charged amino acids, lysine and arginine. For this reason they bind tightly to negatively charged phosphates in DNA. The non-histone proteins are mostly transcription factors that regulate that part of DNA which transcribe into RNA.

The banding techniques falls in two groups.

1. G-Q and R-bands, these bands are distributed along the length of the whole chromosome.

2. C-bands (centromeric bands) and NOR's (nucleolus organizer regions). They are used to stain a restriction number of specific bands or structures. C-banding methods do not permit identification of every chromosome in the somatic cell complement but can be used to identify specific chromosomes.

G banding : G bands are obtained by stain with Giemsa (hence called G-bands) after digesting the chromosomes with trypsin or by acetic-saline (Bernardi, 1988; Summer, 1971). Trypsin demonstrated a conspicuous banding pattern in almost all the chromosomes of the complement. In the G-band, dark region contains heterochromatin, which is late replicating and is rich in adenine and thymine (AT). The centromeres for the most part stained weakly. That is, they were negative for G-banding, showing that these regions are sensitive to the proteolytic action of trypsin, while most of the telomeres, which are heterochromatic, displayed strong staining, and were therefore not digested by trypsin. The B' microchromosome could not be visualized in the G-banding preparations.

In fish, *I. labrosus*, G-banding was prominent in almost all the diploid number of 56 chromosomes as noticed by de Carvalhoc and Dias (2005). However, centromeres showing negative G-banding. In a study of the family Pimelodiade Swarca *et al* (2005) utilized G-banding in chromosome preparations of *Steindachneridion scripta* and *Pseudoplatystoms corruscans* and found a pattern of longitudinal chromosomal differentiation in the three species.

When the restriction endonuclease, *BamHI* was used, showed the presence of a supernumerary microchromosome (B chromosome), with both inter and intra-individual variation. de Carvalho and Dias (2005) reported that telomeres remained intact, while some centromeres were weakly digested. The B chromosome was also not digested by this enzyme. The first pair of chromosomes showed a pattern of longitudinal bands, both with G-banding and by *BamHI* this was more evident with G-banding. This banding pattern can be considered a chromosomal marker for this population of *I. labrosus.* These authors also reported the occurrence of C-banding in the heterochromatin of the telomeric regions in most of the chromosomes of *I. labrosus* from the Capivara Reservoir at the two localities, few chromosomes showed C-banding positive centromeres. When present, the supernumerary or B microchromosome appeared totally heterochromatic.

G-bands were also observed in Indian fishes, such as *Channa punctatus, Colisa* fascieatus, *Mystus tengara, Puntius sophore* and *Labeo rohita* (Rishi, 1979). Lakra and Krishna (1994) reported G-banded karyotypes in Indian major carps. Sharma and Sharma (1998) also reported G-bands in chromosomes of large numbers of Indian fishes.

Regarding heterochromatin distribution pattern in other populations of *I. labrosus* have shown that each population has a characteristic pattern Summer (1977) found a large quantity of heterochromatin in interstitial and terminal regions in population of *I. labrosus* from the Jurumirim reservoir. On the other hand, Swarco *et al.*, (2005) found centromeric and telomeric heterochromatin in a population from the Mogi-Guacu River (SP), and practically Abe & Muramoto (1975) observed that heterochromatin is distributed mainly in telomeric regions from the Tibagi River (TR). They suggested that these chromosomes are used as marker for this population.

R-bands are approximately the reverse of G-bands (the R stands for "reverse"). The dark region is euchromatic and where as bright regions are heterochromatin. R-banding is obtained by heat and use of Giemsa or fluorescence. Florescence G- R-bands are the photographic negatives of the bright field versions i.e. the reverse of the bright field G-band and R-bands.

C banding : C-banding is done by depurification with acid, denaturing is carried out by base and extraction of nonheterochromatic DNA in hot salt solutions as stated by Coming (1978) then stained with Giesma stain. It stains areas of constitutive heterochromatin, which is tightly packed and contains repetitive DNA. C-banding regions were evidenced in telomeric regions of most of the chromosomes of catfish, *lherigichthys labrosus*, taken from the Capivara Reservoir. C-banding pattern in some chromosomes were obtained by Alul (endonuclease), which identifies and cleaves the DNA specific sequence AG/CT (Bernard, 1998). Swarca *et al.*, (2005) also obtained banding patterns similar to C-banding with *Alul* in *Pinirampus pirinampu* and *Pimelodus maculatus*, as did Swarca *et al.*, (2005) in *Steindochneridion* sp and S. *scripta*. In Indian fishes, Rishi and Rishi (1992) obtained

C-bands in *Labeo rohita,* while Sharma and Sharma (1998) recorded C-bands in *Mastacembelus pancalus, Ompak bimaculata, Channa gachua, Schziothorax richardsoni* etc.

Q banding : In this technique the chromosomes are stained by fluorescent quinacrine dye and the chromosomes show intense flourescent (quinacrine) Q-bands. These bands are rich in adenine and thymine (AT). When exposed to ultraviolet light they show intense fluorescence.

NOR Silver-Staining

This procedure help in identifying genes for ribosomal RNA in a previous cell cycle. NOR banding is carried out with the help of silver staining with fluorochrome chromomycin A3 (CMA) distinguishing chromosomal sites of 18s ribosomal RNA. The region is rich in Guanine and Cytosine (CG). If stain with mithraycin, it apparently stain DNA. It has been studied in about 200 species of fishes. In *Cyprinus carpio* one pair of NOR was noticed while interstitial NOR was reported in *Channa punctatus* by Rishi (1972). Recently, Swarca *et al.,* (2005) observed secondary constrictions on the short arm of the first pair of acrocentrics, shown to be associated with molecular organizer region in *Zungaro zungaro* (Pisces, Pimelodidae).

Restriction Endonuclease Banding (Zenze & Voiculecu, 1975) and Fluorescence in Situ Hybridization is molecular cytogenetic technique to be adopted extensively in the study of fish chromosomes.

REFERENCES

Abe, S. and Muramoto,J. (1974). Differential staining of chromosomes of two salmonids, Salkvelinuus leucomaenis (Pallas) and *S. malma* (Walbaum). Proc. Jap. Acad. 50(7): 507-511.

Bernardi, G. (1998). The isochore organization of human genome. Annu. Rev. Genetics, 23: 637-661.

Coming, D. E. (1978). Mechanism of chromosome banding and implications for chromosome structure. Annu. Rev. of Genetics, 12 : 25-46.

de Carvalho, R.A and Dias, A. (2005) Karyotic characterization of *Iheringichthys labrosus* (Pisces, Siluriformes, Pimeleodidae) : C, G and restriction endonuclease banding. Genet, Mol. Res. 4 : 663-667.

Lakra, W.S. & Krishna, G. (1994). G-banding karyotypes of three species of Indian major carps. CI. S. 57: 23-25.

Rishi, K.K. & Rishi, S. (1992). C-heterochromatin in fish chromosomes and its evolutionary implications. Persp. in Cytol. and Genet. VII: 307-332.

Rishi, K.K. (1979). Somatic G-banding chromosomes of *Colisa fasciatus* (Perciformes: Belontidae) and confirmation of female heterogametry. Copia (1): 146-149.

Sharma, O.P. & Sharma, K.K. (1998). Aryotype analysis of a Schizothracinefish, Ptycho-barbus corirostris (Cyprinidae: Cyprinidae).

Summer, A.T. (1977). Banding as a level of chromosome organization. In current chromosome Research. Ed. Jones K. and Brandham, P.E., Amsterdam : Elseviewr/ North Holland Biomedical Press: 17-22. Perspective Cyt. Genet. 9 : 307-311.

Swarco, A.C. Fenocchio, A.S. Cestari, M.m and Dias, A (2005). First chromosome data on *Steindachneridion scripta* (Pisces, Siluriformes, Pimeleodidae) from Brazilian rivers : Giemsa, CBG, G, and RE banding. Genet, Mol. Res. 4 : 734-741.

Zenzes, M.T., Voiculescu., I. (1975). C-banding pattern in *Salmo trutta,* a species of tetraploid origin. Genetica 45: 531-536.

Chapter 37

GENETIC VARIATIONS OR MUTATIONS

Every living organism including human beings *(Homo sapiens)* is different. Neither two human beings nor homozygous twins are similar. The homozygous twins are similar in many respects but they are not identical. Further, you have just your mother's colour and daddy's eyes. It is clear that characters are transmitted from parents to offspring. According to Mendel, genes (factors) constitute the hereditary material which is passed on from generation to generation. Dissimilarities amongst members of same species are due to genetic variations.

Mendel was the first to recognize the genetic basis of variation. He was the first to point out that during crosses if two true breeding parents are crossed, new genetic recombination or characters appear in the offspring which are genetically different from the parents. The transmissions of characters from one generation to next generation (parent to offspring) follow the Mendel's Laws of inheritance. These laws are law of segregation and law of independent assortment.

New Concept of Genetics

With the advent of researches in last 56 years after the discovery of double helix model of DNA(Fig. 37.1). Genetics is divided into following branches, which are interlinked and overlapping areas of research.

(a) Transmission Genetics (some times referred to as Mendelian Genetics).

(b) Molecular genetics and

(c) Population/Evolutionary Genetics.

All these genetics together are responsible for understanding the process and transmission of genetic variations from generation to generation.

Finally, it is settled that DNA is the genetic material. The appearance of character or phenotype in the organism is due to genetic variation, i.e., the changes in sequence of coding region of the gene and in the formation of new protein. The changes also occur in the non-coding part of the DNA/RNA also. Now it is clear that genetic variations are the sole cause even for evolution. Genetic variations play an important role in population genetic too.

Genetic Variations and their Causes

Mutations are the original sources of all genetic diversity. It is now proved beyond any doubt that the genetic materials are DNA or RNA. So the changes in DNA (small or large) in an organism are the reasons for genetic variations. These changes may be produced either by internal or by external mechanism or by certain agents and are termed as mutations. The sharp difference between true mutation and other changes in an organism is its heritability. The germ line mutations are important as they are heritable and passed on to the next generation (White, 1976; Watson & Crick, 1953).

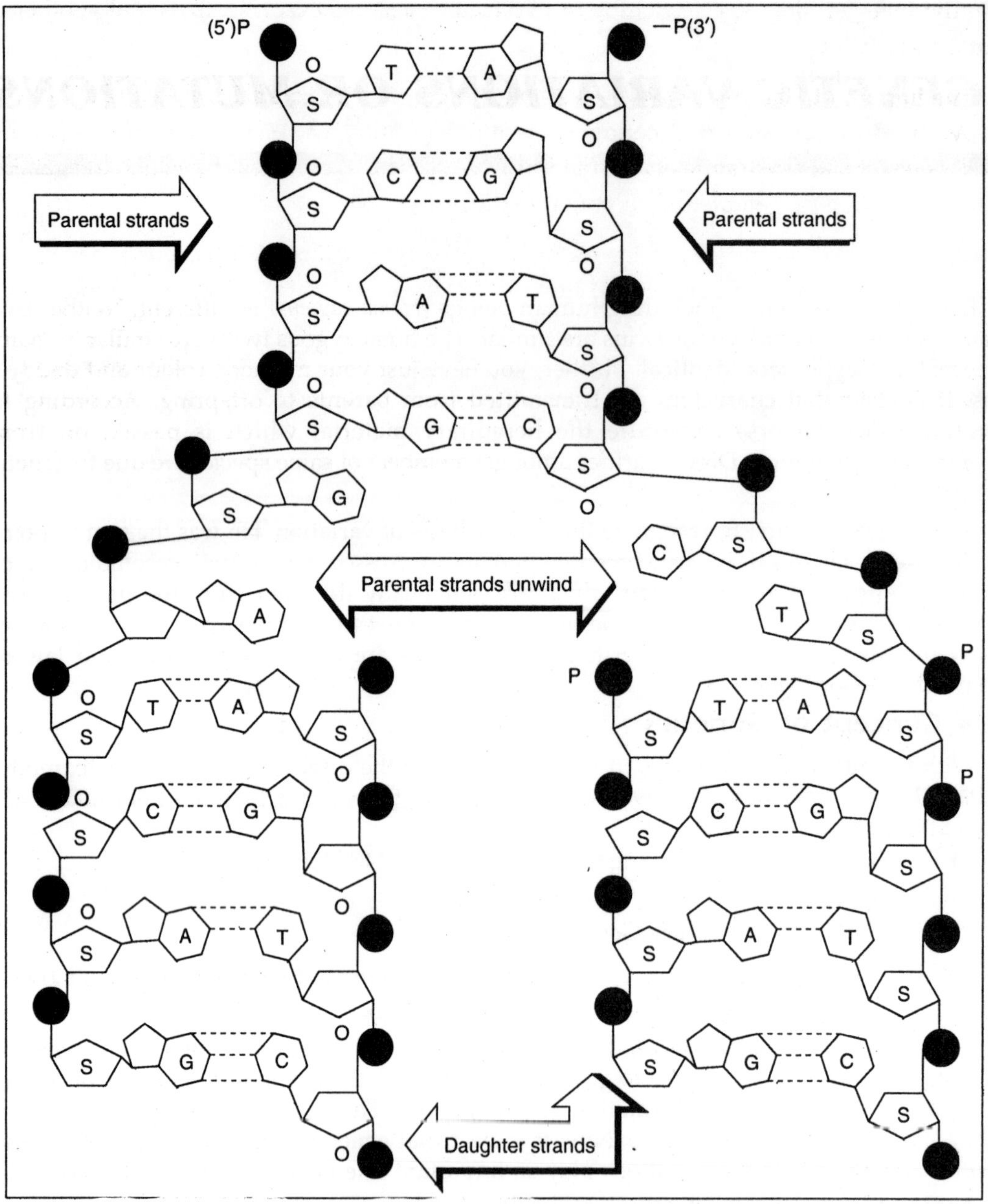

Fig. 37.1 : Double helix model of DNA showing faithful replication of DNA

Mutations are rare and occur when a gene alters itself for no apparent reason. Mutations can be *harmful, neutral* or helpful. Harmful mutations hinder the survival of the organism or cause death. In this case, the individual usually die before they can reproduce and thus the mutant gene is eliminated. Some mutants are neutral, meaning that they neither help nor hinder the survival individual. In this case the organism may survive to reproduce and pass on the neutral mutated gene to the next generation. Sometimes the mutation turns out to be helpful, meaning the mutation helps the individual to survive in the environment.

Mutations are classified as **gene mutations** and **chromosomal mutations**. The uniqueness of individuals within a species is due to two factors; one is DNA (Fig. 37.1) and another is sexual reproduction.

The important feature of DNA is that one strand of DNA could serve as a template for the synthesis of new strand. Secondly, formations of mRNA, which codes protein (amino acids) is generated from anti-sense strand of DNA. This is the process by which the genetic material could be perpetuated from parent to offspring.

The genetic code consists of a long series of consecutive codons. Each codon is a triplet of three nucleotides, which code for one amino acid (20 amino acids which form protein). The names of these amino acids with their abbreviations are given in Fig. 37.2. The protein is formed by coding region of the DNA. The primary structure of protein is determined by sequences of nucleotides or bases which codes sequences of amino acids. It is also important to note that different combination of three nucleotide often code similar amino acid (Fig. 37.3).

L-Amino acids	Abbreviation	Properties of sides chains
Alanine	Ala (A)	Nonpolar
Valine	Val (V)	
Isoleucine	Ile (I)	
Leucine	Leu (L)	
Phenylalanine	Phe (F)	
Proline	Pro (P)	
Methionine	Met (M)	Contains sulfur
Glycine	Gly (G)	Polar
Serine	Ser (S)	
Threonine	Thr (T)	
Tyrosine	Tyr (Y)	
Tryptophan	Trp (W)	
Asparagine	Asn (N)	
Glutamine	Gln (Q)	
Cysteine	Cys (C)	Contains sulfur
Aspartic acid	Asp (D)	Acidic
Glutamic acid	Glu (E)	
Lysine	Lys (K)	Basic
Arginine	Arg (R)	
Histidine	His (H)	

Fig. 37.2 : Groups of amino acids in protein

1st letter (5' end)	2nd letter U	2nd letter C	2nd letter A	2nd letter G	3rd letter (3' end)
U	Phe UUU	Ser UCU	Tyr UAU	Cys UGU	U
U	Phe UUC	Ser UCC	Tyr UAC	Cys UGC	C
U	**Leu UUA**	Ser UCA	STOP UAA	STOP UGA	A
U	**Leu UUG**	Ser UCG	STOP UAG	Trp UGG	G
C	**Leu CUU**	Pro CCU	His CAU	Arg CGU	U
C	**Leu CUC**	Pro CCC	His CAC	Arg CGC	C
C	**Leu CUA**	Pro CCA	Gin CAA	Arg CGA	A
C	**Leu CUG**	Pro CCG	Gin CAG	Arg CGG	G
A	lie AUU	Thr ACU	Asn AAU	Ser AGU	U
A	lie AUC	Thr ACC	Asn AAC	Ser AGC	C
A	lie AUA	Thr ACA	Lys AAA	Arg AGA	A
A	Met AUG	Thr ACG	Lys AAG	Arg AGG	G
G	Val GUU	Ala GCU	Asp GAU	Gly GGU	U
G	Val GUC	Ala GCC	Asp GAC	Gly GGC	C
G	Val GUA	Ala GCA	Glu GAA	Gly GGA	A
G	Val GUG	Ala GCG	Glu GAG	Gly GGG	G

Fig. 37.3 : The genetic code showing that combination if three nucleotides often code similar amino acid

The "central dogma of molecular biology" states that genetic information flows from DNA to RNA to Protein (Fig. 37.4).

DNA

↓ Transcription

RNA

↓

mRNA

↓ Translation

Protein (out side nucleus)

Fig. 37.4 : Central dogma of Biology.

GENE MUTATIONS

The gene mutations are further classified as follows (Dobzhansky, 1972; Atherly *et al.*, 1999).

(A) Spontaneous mutations.

(B) Insertion & deletion mutations or Frame shift mutations

(C) Transposons

(D) Induced mutations.

(A) **Spontaneous mutations :** The spontaneous mutations or background mutations result due to internal factors, such as a replication error of DNA, mistake in the recombination, mispairing of DNA damage, depurination, deamination of the bases, and movement of transposons. They occur not by chance but due to definite biochemical changes. These are further classified as follows :

(1) Base pair substitution

(2) Neutral mutations

(3) Missense mutations

(4) Nonsense mutations (Amber mutations).

1. Substitution of Base Pairs

The most common DNA mutations (gene mutations) are caused due to base pair (purine to purine, pyrimidine to pyrimidine and pyrimidine to purine or vice versa) substitution in the coding region of DNA. As a rule, if in one strand of DNA, G (nucleotide) is present then in another strand automatically C (nucleotide) will be present as they are complimentary. If in one strand of DNA one base pair for example G is replaced A, then the former combination of the GC is to be replaced by AT. This can be further classified as transition mutations or transversion mutations. In transition mutation, purine is replaced by another purine in the same strand of DNA or a pyrimidine is replaced by pyrimidine in the same strand of DNA i.e. GC is replaced by AT and AT is replaced by GC. In transversion, the purine is replaced by pyrimidine on the same strand of DNA or a pyrimidine is replaced by purine in the same strand of DNA i.e. GC to CG or TA and AT to AT to TA or GC.

Silent Mutations

It is interesting to note that sequences replacement or gene mutation will not produce visible phenotypic changes always as stated earlier. Such types of mutations are known as **silent mutations.** For example, if in a codon CUU due to mutation now becomes CUA or CUG or CUC will code amino acid, leucine. From the chart it is clear that different codon code the same amino acid (Fig. 37.3). For example there are six combinations of codons which codes leucine. The reason is that though a base pair change has occurred in an allele's codon due to mutation, but due to the formation of same amino acid as end product, there is no change in amino acid sequences in the protein. The genetic code is degenerates and secondly because many codons are responsible for coding same amino acids. Analine has four codon (GCU, GCC, GCA,GCG), while histidine has two codons (CAU,CAC).

Neutral Mutation

Neutral mutations are also base pair substitution in the allele's codon. Though codon produces a different amino acid, the change of a few amino acids in the primary structure does not change the function of protein. For example if in the original allele's codon is CUU, the codon CUU will code leucine. But if CUU is replaced due to mutation and is changed to AUU, the amino acid isoleucine will be coded. The two amino acid, leucine and isoleucine, are chemically similar, hence the change in amino acid would not alter the function of the protein thus there will be no phenotypic change. The another example is insulin hormone.

The human insulin is heterodimeric protein, composed of an α– chain having 21 amino acids and a β-chain with 30 amino acids (Fig. 37.5). The insulin of other animals is also a dimmer similar to human insulin. However, the insulin of pig is different from human insulin only in one amino acid at position 30 of β chain, instead of Thr it is Ala. Otherwise there is no change in the sequences of amino acid in α and β-chains. The insulin of cow is different from human in three amino acids at positions α8 (Ala instead of Thr), α10 (Val instead of IIe) and β-30 (Ala instead of Thr). Though some amino acids are changed but the change in these amino acids are not critical in the function of insulins. These insulins are available in the market for human use. They are manufactured by rDNA technology.

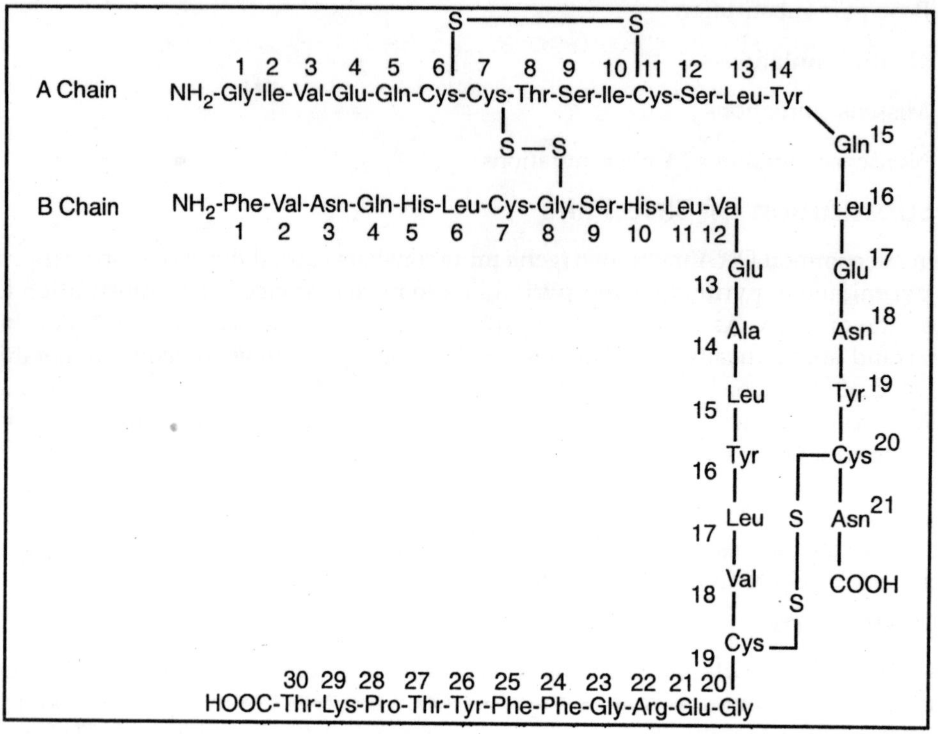

Fig. 37.5 : Molecular structure and Amino acid sequence of human insulin, showing α-2β chain

Missense mutation. Another class of mutation is known as missense mutation, where there is a substitution in only one base pair resulting in the formation of a new amino acid. Some times it causes some diseases. Hypertrophic cardiomyopathy in human beings is caused due to missense mutations in exon 13 of MHC (Myosine heavy chain) β chain resulting in the change of Adenine for guanine and resulting in the formation of gluatamine instead of arginine (Fig. 37.6). This missense mutation causes enlargement of heart (left ventricle) Mac Rac, *et al.*, (1995).

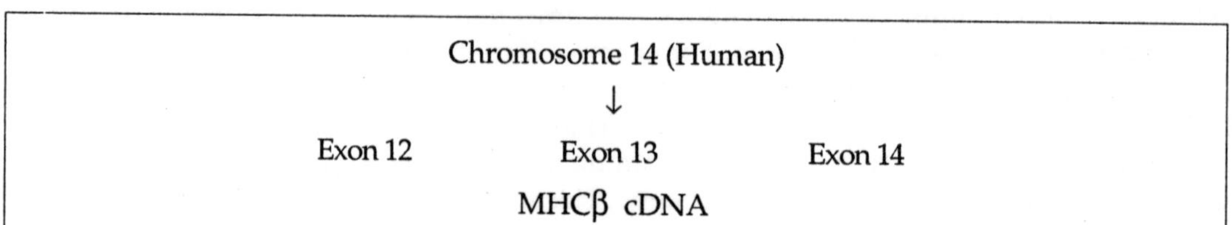

Fig. 37.6 : Missense mutation. Change of one base pair, in MHCβ cDNA Adenine to Guanine results in change of amino acid Glutamine for Arginine.

Nonsense Mutation (Amber Mutations)

It is a form of mutation in which substitution of base pairs results in the codon UGA, UAA or UAG. These codons are nonsense codon. In such mutation no other amino acid is formed except the production of original protein. Unlike missense mutation, nonsense mutations rarely display partial activity because the protein product of the alleles is changed so radically.

Frame Shift Mutations/Insertion and Deletion Mutations

In these mutations, there is insertion or deletion of one or two base pairs (not multiple of three) in DNA. This results in altered reading frame of mRNA. For example, if the DNA coding strand CAT CAT CAT CAT CAT has a single base pair deletion at base pair 6, the mRNA will read CAU CAC AUC AUC AUC, and so forth. The frame shift mutation usually has radical effect on the protein product.

DNA replication errors may cause mutations (Tautomerism)

All the bases (A,G,T,C) can exist in nature in two tautomeric forms either the keto or enol form if it has a hydroxyl group, or the imino and amino forms it has an amino group. Tautomeric shift causes mutation because the uncommom forms of the bases do not always pair properly during DNA replication. Such mutations exist in nature in one in 10,000 bases or 10 × 10. These alternative structures do not pair properly with its complementary bases (Fig. 37.7a & b).

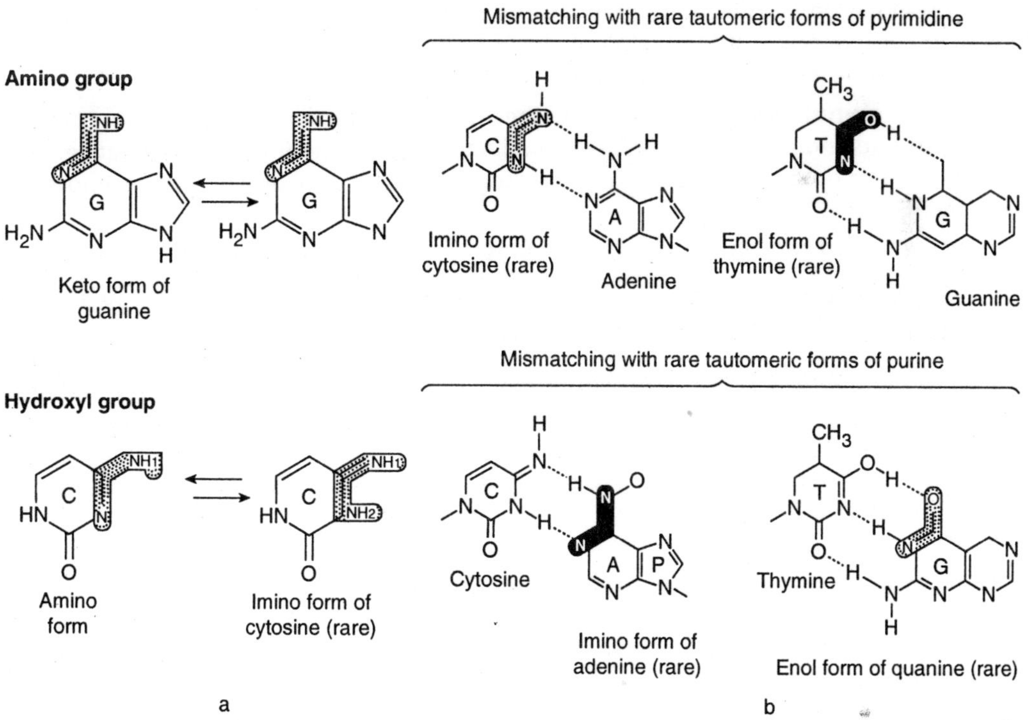

Fig. 37.7(a), (b) : (a) The rare tautomeric structure of the bases, sharing rare movement of a hydrogen atom results in an alternative form of the bases resulting this alternative structure does not pair properly with complementary form. (b) Mismatch bases. Charge of position of hydrogen mismatch pairing. The mismatching result in mutation if it happens during DNA replication.

Transposon Insertion

These are mobile elements present in the genome and can jump and insert in the DNA. It is stated that 1-10 kb DNA are capable of movement within the genome. It is also known that 50 to 80% of the mutations caused by the disruption of gene. These are also responsible for genetic variation.

Chromosomal Aberrations are Responsible for Origin of Species

The difference between the chromosomal and gene mutations is that the rearrangement involves long segments of DNA, rather then single bases. It generally occurs at the time of DNA replication. They can be seen in microscopic picture at prophase1 at the time of chiasma formation. Further recombination involves in non homologous sister chromatids (single DNA molecule from non homologous chromatids) instead of sister chromatids. Chromosomal theory of inheritance suggests that genes (DNA) are physically located on chromosomes and that Mendelian inheritance can be explained in terms of chromosome behavior during cell division. The chances of mutations are more and can be explained by the following example.

If the chromosome number in diploid organism is 10 pairs, then 10 are coming from male (sperm) and 10 are coming from female-ovum. Then the possible combinations would be $(2)^{10}$ =1024 (Beaumont & Hoare, 2003). Such random combinations are possible according to principal of independence assortment of Mendel. This means that such large numbers of genetic variations are possible. Although chromosomes variations are no longer used as markers in population studies, they play an important part in evolution and formation of new species. The examples of the fusion of chromosomes resulting in the formation of new species are available in *Drosophila* genus.

Chromosomal mutation is a visible change in chromosome structure. Chromosomes themselves mutate and evolve and before the advent of allozyme markers some geneticists spent much of their time squinting down microscopes following inheritance of chromosomal rearrangements.

Chromosomal aberrations are classified as under :

(a) Translocation (b) Inversion

(c) Deletion (d) Duplication

Chromosome number for each species is fixed if the number of chromosome changes, normally; in the broader sense, it would be a new species. The sexual reproduction plays a prime role in the creation of genetic variations. Most of chromosomal rearrangements arise as a result of mistake during meiosis. Chromosomal theory of inheritance suggests that genes (DNA) are physically located on chromosomes and that Mendelian inheritance can be explained in terms of chromosome behavior during cell division.

For human beings, the number of chromosomes are 46 (23 pairs; 22 autosomes and one pair of XX or XY), but in egg or in sperm the number is only 23 (haploid). In *Drosophila melanogaster*, the number of chromosomes is 8 (4 pairs; 3 pair of autosomes and one pair either XX or XY).

Inversion : In inversion there is no deletion or addition of hereditary material. A fragment of one chromosome breaks off and re-attaches to its original position in reversed orientation. The original chromosome may contain the centromere (pericentric inversion) or it may not (paracentric). Chromosome heterozygote inversions can be recognized by the presence of loops in the cytological preparations of cell at pachytene stage of meiosis.

Deletion : Chromosome deletions occur when the DNA strand breaks but fails to mend. The fragments or pieces of chromosome (DNA) of that do not contain centromere (acentric fragments) will be lost during subsequent cell division. A disease known as Cri due Chat syndrome in which metal retardation, growth restriction and cat like cry occur in human being, is due to **deletion in chromosome.**

Duplication : Chromosomal duplication provides an additional copy of a block of DNA (pieces of chromosome) having a complete gene sequence. When duplication contains a complete gene sequence natural selection can operate independently on both the new and the old sequence to produce divergent variants.

Translocation Role and the Formation of New Species

The examples of the fusion of chromosomes resulting in the formation of new species are available in *Drosophila* genus. There are five species of *Drosophila*, namely, *subobscura*, *psuedoobscura*, *melanogaster*, *ananassae* and *willistoni*. They are derived by the fusion of chromosomes and translocation between non-homologous chromosomes. The fusion of chromosome occurs when two non-homologous chromosomes fuse into one. The ancestral condition exists in *Drosophila subobscura*, which possesses five pairs of acrocentrics (rod shape) and one pair of dot like chromosomes (Fig. 37.8). *Drosophila pseudoobscura* contains 4 pairs of autosomes and one pair of dot like chromosomes. It is said that 4 pairs instead of five originated due to fusion of one pair of autosome with X chromosomes of *subobscura*. The 4 pairs of acentric autosomes are fused into two pairs of metacentrics in *Drosophila malanogaster* and *D. ananassae*, but in the latter species a pericentric inversion has transformed the acentric X chromosome into a small metacentric. In *Drosphila willistoni* there are only three pairs of chromosomes, the ancestral dot-like chromosome being incorporated into the X chromosome. The evolution of karyotype in many other groups has been worked out.

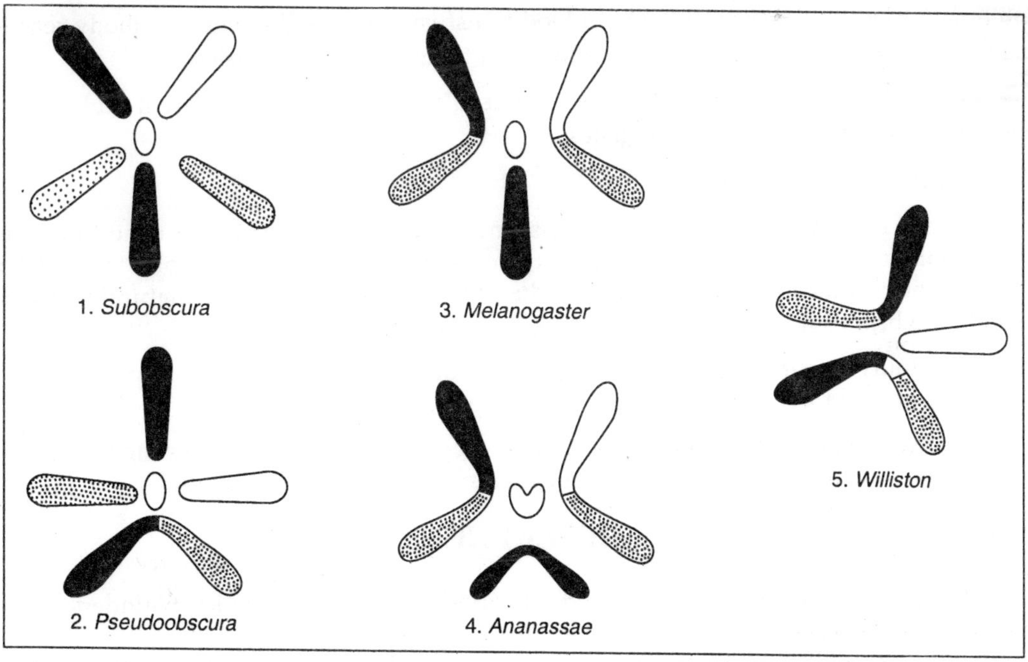

Fig. 37.8 : Evolution of various species of *Drosophila* due to translocation

Highly Repetitive DNA Sequences

The DNA which is capable for coding protein in human being is very small. Only 3% of DNA is functional and the rest is junk DNA. Some of this junk DNA contains **pseudogenes**, gene due to unknown reason are nonfunctional. Yet other parts of non-coding DNA composed of dispersed or clustered repeated sequences of varying length, from one base pair(bp) to thousands of bases (kilobases,kb) in length. They are spread over the genome region called variable numbers of tandem repeat (VNTR). These are classified as follows :

(1) Simple tandem Repeat (STR)

(2) Simple sequence length polymorphism (SSLP), which contains tandem (i.e., linked chains). These sequences may be short (1 to 10 base pairs) or much longer. The main feature of these tandem repeats is that number of repeat can vary between individuals. It is reported that increase and decrease in the number of repeats occur during copying by the recombination or replication slippage. They are not point mutation but occur at much faster rate. Variations in the number of repeats at these satellite (repeats 100 to 5000 bp), minisatellite (5 to100 bp) or microsatellite (2 to 5bp).

Many human diseases could now be recognized or diagnosed on the basis of triple nucleotide (DNA) repeats.

Name of Disease	Sequences	Chromosome	Abnormal Number
X Syndrome	CGG	Xp27	200
Hungtington Syndrome	CAG	4	31-121

It is now demonstrated that ABO blood types in humans is controlled by one gene with multiple alleles. At the time in human blood transfusion to avoid antigen antibody reaction, blood-grouping test is done which is nothing but to know multiple alleles.

Segregation and complementary tests are used to know whether different mutations are allels of the same gene or different genes.

Polyploidy

Increase in number of chromosomes is known as polyploidy. It is a condition where individuals have more than two copies of each chromosome. For example, triploid have three set of chromosomes and tetraploid have four. Polyploidy occur naturally in some plants. The best example is wheat which is hexaploid.

Tetraploidy occurred in the recent history of salmonoid fishes. Polyploidy can be artificially induced in normally diploid species for aquaculture processes.

Organisms change through time and can develop into new organisms through the process of evolution. The most important cause of evolution is genetic variations.

REFERENCES

Atherly, A.G.; Girton, J.R.; McDonald, J.F. (1999). Science of Genetics. Saunders College Publishing. New York .

Beaumont, A.R. & Hoare, K. (2003). Biotechnology and Genetics in Fisheries and Aquaculture. Blackwell Publishing. 1-158.

Dobzhansky, Th. (1972). Evolution, Genetic and Man. Wiley Eastern, New Delhi.

MacRae, C.A., Ghaises, N. Kass, S. (1995). Familial hypertrophic cardiomyopathy with Wolff-Parkinson-White syndrome maps to a locus on chromosome 7q3. JClin Invest. 1216-1220.

Watson, J.D. & Crick, F.H.C. (1953). Molecular structure of nucleic acids : A structure for deoxyribose nucleic acid. Nature. 171 : 737-738.

White, M.J.D. (1976). Animal Cytology and evolution, Cambridge university Press. London.

Chapter 38

THE RECOMBINANT DNA TECHNOLOGY AND MOLECULAR BIOLOGY IN FISH BREEDING

The discovery of recombinant DNA technology has made a revolution in the modern molecular biology. Through this technique, large quantities of proteins present in trace amount, as well as other biologically active substances, can be generated through biotechnology and these genetically engineered macromolecules have very little side effects. It can also be used for performing diagnostic *in situ* hybridization. This technique also helps to isolate and identify genes, responsible for hereditary diseases.

In 1983, recombinantly made plasminogen activator (rt PA) made a paradigmatic shift in the therapy of myocardial infarction (heart attack). Now with the use of recombinant DNA technology, many other product such as hirudin, superoxide dismutase, urokinase, prokinase, etc. are in the market (Roberts, 1993).

In aquaculture particularly in induced breeding, the use of crude piscian and mammalian pituitaries are used (Weil, *et al.*, 1986). The pituitary extract contains gonadotropin growth hormones and their releasing factors. They are required in excess quantities for the fish breeding. These polypeptides are too large to be synthesized chemically and these polypeptides can be generated biotechnologically through manipulation of DNA, will have least side effects and will be better instead of synthetically prepared analogs already in use for mass breeding. Recently success has been obtained in fish breeding in improving many quantitative characters such as egg diameter, body weight growth rate, body length and in hyper-immunity in fishes such as *Tilapia zillii* and *Oreochromis niloticus* simply by injecting Shark DNA in muscle by syringe. The random amplified polymorphic DNA (RAPD) analysis showed that polymorphic fragments varied between 54.55% for primer 1 (5'--ACC GGG AACG – 3') and 14.29% for primer 2 (5'–ATGACCTTGA–3') as reported by Sudha *et al.*, 2001, El– Zaeem & Assem, 2004 and Assem 2 EL-Zaeem, 2005).

Fish proteins are nowadays used in medicine also. Protamine is used to reverse anticoagulation during cardiac surgery and catheterization (Harrow, 1985). Salmon calcitonin is better than mammalian calcitonin for treatment in Paget's disease and certain form of osteoporosis (Wisneski, 1990). Fish antifreeze protein is used in cryopreservation of human organs, as well as in preserving foodstuff (Knight, *et al.*, 1984).

The cell culturing and recombinant DNA technology in fishes are slow and need more extensive study (Bols, 1991). However, fish genes for some potential products have been expressed in bacteria and yeast (Sekine *et al.*, 1985) (Peters *et al.* 1989; Hayami *et al.*, 1989). Production of salmon calcitonin (Kampe *et al.*, 1988) and antifreeze proteins (Peters *et al.*, 1989) has been explored in microbial cells.

There is a great need to obtain gene or mRNA for the polypeptide such as fish gonadotropin growth hormones, cytokinin, protamine, calcitonin and antifreeze protein, etc. Once the gene/mRNA is synthesized, it can be converted into cDNA through the enzyme reverse transcriptase. Once the cDNA is obtained for a particular protein, then through DNA cloning (a technique used to produce large quantities of a specific DNA fragment to produce million copies of a DNA fragment by routine bacterial cloning techniques), large quantity of DNA could be obtained.

The DNA fragment of our interest can be introduced or inserted into the plasmid (present as extrachromosomal self-replicating bodies) in bacteria or virus (replicate in bacteria, bacteriophages) or artificially developed vectors called as cosmids.

For insertion, the DNA of the vector is cut by endonucleases (Smith and Wilcox, 1970 and Kelly and Smith, 1970) and the cDNA is then inserted and finally joined by enzymes known as DNA ligases (Olivera *et al.*, 1968). The product now contains a DNA piece in the vector DNA, hence the technique is known as recombinant DNA. This vector is then amplified in an appropriate host either bacteria, mammalian cell or piscine cell culture.

The size of cloning is limited. Plasmid can accommodate 15000 bp, phages up to 25000 bp and cosmids up to 45000 bp. The size has been overcome by the technique known as yeast artificial chromosomes (YAC) (Schwarz and Cantro, 1984).

Watson and Crick (1953) described DNA as a double strand helix structure (Fig. 38.1). The DNA is a polynucleotide molecule. The nucleotides are joined together by a phosphodiester linkage. It has a massive size and each chromosome is a continuous DNA molecule. The molecule consists of a base, a deoxyribose sugar and a phosphate. The genetic information inherited by individual is encoded in DNA.

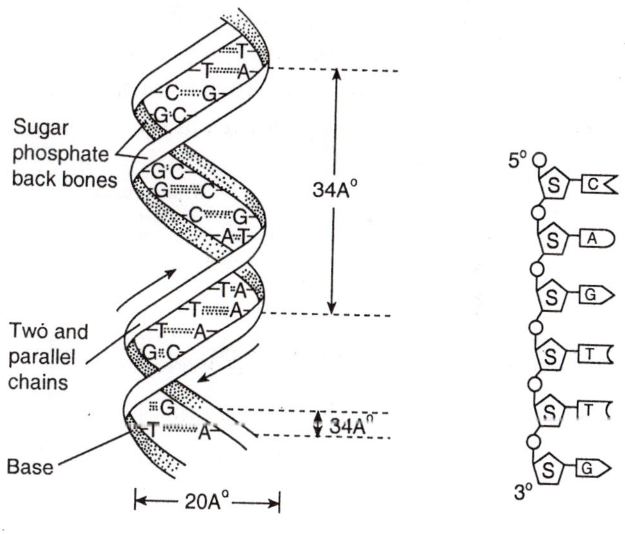

Fig. 38.1 : Watson and Crick Model of DNA double helix

The two chains of the DNA, one is 5′–3′ and another is 3′– 5′ in directions, i.e., the two chains are opposing each other. The direction is important because replication of DNA starts from 5′ to 3′ and also the sequence essential for gene regulation is located on 5′ to 3′ end of the gene. The hydrogen pairing between bases are specific, adenine (A) always pairs with thymine (T) and guanine (G) always pairs with cytosine in DNA.

DNA has one more characteristic feature that the two strands could be separated if the temperature is raised to 95 °C, this is known as denaturing. These two separated strands can be joined to form original structure if the temperature is lowered to 55 °C, the phenomenon is known as hybridization or annealing (Marmor and Lane, 1960). This characteristic feature is more important for recombinant DNA technology.

The central dogma of molecular biology is that, DNA gives rise to mRNA using DNA as a template. The proces is known as *transcription*. The mRNA is responsible for the formation of protein, the phenomenon known as translation, and protein synthesis is the cell function (Fig. 38.2).

Nucleus

DNA
Transcription
mRNA (1° transcript)
Processing
mRNA (Mature)

Cytoplasm

mRNA
Translation
Protein
Post-translation processing
Subcellular localization

Fig. 38.2 : Central dogma of molecular biology, showing process of transcription and translation

The formation of mRNA is complicated, during maturation of mRNA, the introns are spliced out and exons are rearranged. The mRNA comes out of the nucleus for coding specific protein. Before coming into the cytoplasm the mRNA forms methylated cap to the 5′ and a poly (A) tail to 3′ ends. The cap is essential in the initiation of translation and poly A tail at 3′ region is essential for messages in the cytoplasm (Fig. 38.3).

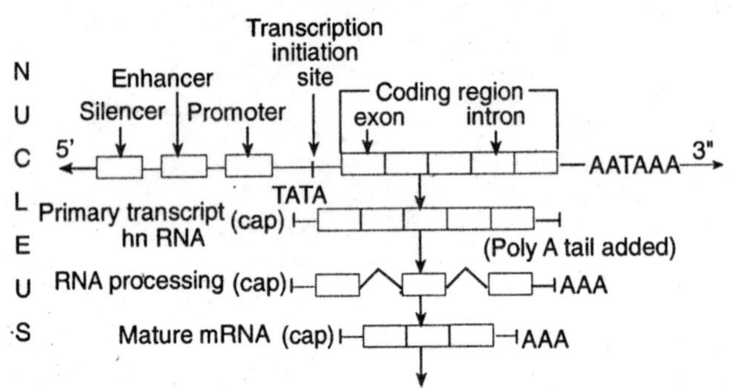

Fig. 38.3 : Structure of a gene. These small functional units within the nucleus contain the coding information for the synthesis of a polypeptide and on its 5′ end, having regulatory sequence that include silencers, enhancers and promotor. The TATA is the initiation site for polymerase and is present in most eukaryotes at about 10 to 30 bp 5′ from start codon TAC of the coding regions.

A break-through in molecular biology established when it was discovered (Termin and Mizutani, 1970) in retrovirus the RNA can be converted to DNA by the enzyme reverse transcriptase (Baltimore, 1970; Termin and Mizutani, 1970). The discovery is the backbone of recombinant DNA technology. The mRNA can be converted to cDNA (complementary DNA) with the help of enzyme reverse transcriptase (Fig. 38.4).

mRNA
| Reverse Transcription
cDNA

Fig. 38.4 : The enzyme reverse transcriptase converts mRNA to DNA referred to as complementary DNA (cDNA). DNA is single-strand and complementary to the sequence of RNA except that thymine is replaced by uracil. The DNA Polymerase helps in converting this single strand-DNA into double strand cDNA.

So in other words, it is possible to translate mRNA into complementary DNA (cDNA) by using mRNA as a template. The cDNA so formed contains appropriate complementary bases for mRNA except, of course, with thymine replacing uracil. The cDNA derived from mRNA is nowadays used in the diagnosis of many diseases by labelling it radioactively.

The reverse transcriptase also helps in cloning a gene. The first gene was cloned at Stanford (Cohen *et al.*, 1973) and first molecule of recombinant DNA technology then was formed, a revolution in molecular biology.

The discovery of endonucleases or restriction enzymes help in cutting the DNA up to 3 to 8 nucleotides. The endonucleases have been obtained from about 400 strains of bacteria and these enzymes can recognize about 100 different sites in DNA. Some of the endonucleases as well as their recognizing sites are (Fig. 38.5).

Name	**Procedure for naming**			**Sequence recognized by the enzyme**	
EcoR1	E	*Escherichia*	Genus	GAATTC	5′ – 3′
	co	*coli*	species	CTTAAG	3′ – 5′
	R	Ry 13	strain		
Bam H1	B	*Bacillus*	genus	GGATCC	5′ – 3′
	am	*amyloliquifaciens*	species	CCTAGG	3′ – 5′
	H	H	strain		
Hind III	H	*Haemophilus*	genus	AAGCTT	5′ – 3′
	m	*influenza*	species	TTCGAA	3′ – 5′
	d	Rd	strain		
Hpal	H	*Haemophilus*	genus	GTTAAC	5′ – 3′
	Pa	*parainfluenzae*	species	CAATIG	3′ – 5′

Fig. 38.5 The substrate that are attacked by restriction endonucleases are palandromic sequence. The palandromic reads the same from forward and reverse directions, e.g., MADAM. The best example of a restricted enzyme is EcoR1 made from *E.coli* strain Ry 13. The EcoR1 attacks the nucleotide sequence GAATTC, CTTAAG .

The DNA ligases help in joining the insert in the DNA of the vector and vector with original and insert DNA can be amplified in the appropriate host. Sanger and Coulson

(1975), and Maxim and Gilbert (1977) developed techniques for the rapid sequencing of DNA and RNA. Now polymerase chain reaction (PCR) is available, which can provide 1,000,000 copies of a DNA fragment in 3 to 4 hours and billion copies in 24 hours (Roberts *et al.*, 1992, Roberts, 1993).

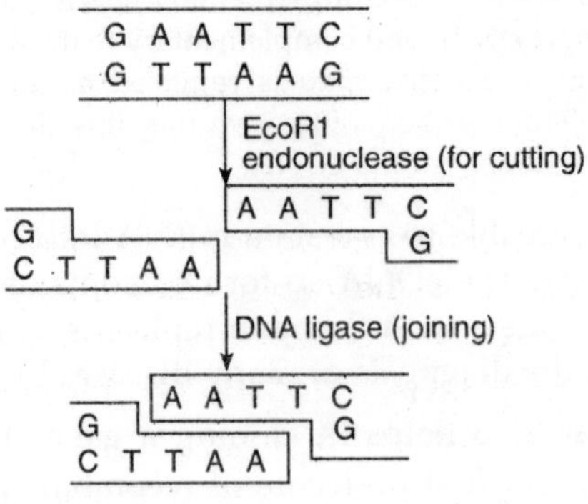

Fig. 38.5 : Restriction endonuclease E coR1 recognizes specific sequence and cut 3 to 8 bp. DNA ligase helps to ligate together. Important steps in recombinant DNA.

With the use of these techniques, aquaculture can be improved and fish protein can be made biotechnologically. Therefore, extensive research in these lines are essential for gene expression, either in bacteria, virus or fish cell culture.

REFERENCES

Assem SS. & El-Zaeem Sy. (2005). Application of biotechnology in fish breeding. 11 : production of highly immune genetically modified Sedbelly trlapia, *Tilapia zillii*. African Journal of Biotechnology. 4(5) 449-459.

Baltimore, D. (1970). Viral RNA dependent DNA polymerase. Nature 226 : 1209-1211.

Bols, N. C. (1991). Biotechnology and Aquaculture, the role of cell cultures. Biotech. Adv. Vol. 9 : pp. 31-49.

Cohen, S.; Change, A.; Boyer, H.; Helling, R. (1973). Construction of biological functional bacterial plasmids in vitro. Proc. Natl. Acad. Sci. USA 70 : 3240-3244.

El-Zaeem Sy. Assem SS.; (2004). Application of biotechnology in fish breeding 1. Production of highly immune genetically modified Nile tilapia, *Oreochromis niloticus* with accelerated growth by direct injection of shark DNA into skeletal muscle. Egypt. J. aquat. Biol. Fish 8(3) : 67-92.

Hayami, T.; Seto, N.; Ichirya, T.; Inoue, Y.; Marata, K.; Kimura, S.; Nonaka, M. and Kimura, A. (1989). Production of recombinant tuna growth hormone by a yeast, *Saccharomyces cerevisi*. Agrio. Biol. Chem. 53, 2917-2922.

Harrow, J. C. (1985). Protamin—a review of its toxicity. Anesth. Analog. 64 : 348-361.

Kampe, F.; Chow, F. Fasa, D.N.; & Heath, H. (1988). (Homoserine 31)-salmon calcitonn. 1 : Fully active analoguie of calcitonin synthesized by recombinant DNA techniques. Bio/Technology. 6 : 190-192.

Kelly, T.J. Jr. & Smith, H. O. (1970). A restriction enzyme from *Hemophilias Influenza-II* : Base sequence of the recognition site. Mol. Biol. 51 : 393-409.

Knight, C.A.; Devries, A. L.; and Oolman, L.D. (1984). Fish anti-freeze protein and the freezing and recrystallization of ice. Nature. 38 : 295-296.

Marmor, J. and Lane, L. (1960). Strand separation and specific recombination in deoxyribonucleic acids. Biological Studies. Proc. Natl. Acad. Sci. USA. 46 : 453-461.

Maxim, A.M. and Gilbert, W. (1977). A new method of sequencing DNA. Proc. Natl Acad. Sci, USA. 74 : 560-564.

Olivera, B.M.; Hall, Z.W. & Lehman, I.R. (1968). Enzymatic joining of polynucleotides. V.A. DNA adenylate intermediate in the polynucleotide —joining reaction. Proc. Natl. Acad. Sci. USA. 61: 237-244.

Peters, I.D.; Hew, C.L. and Davis, P. L. (1989). Biosynthesis of winter flounder antifreeze protein in E.coli, Protein Eng. 3 : 145-151.

Roberts, R. (1993). Introduction to the techniques of molecular biology and the molecular basis for cardiac growth pp. 3-30. In Basic foundation of cardiology vol. 1. In : Hurst. The Heart (Eds. Rourke, R.A.O., Roberts, R., and Soumenblick, E.H.) Mc.Graw Hill. Inc. N. Y. pp. 3-20.

Roberts, R.; Towbin, J.; Parkar, T. G. & Beles (Eds) (1992) : A Primer of Molecular Biology, New York. Elsevier Science.

Sanger, F. and Coulson, A. R. (1975). A rapid method for determining sequences in DNA by primed synthesis and DNA polymerase. J. Mol. Biol. 94 : 444-448.

Sekine, S.; Mizukami, T.; Nishi, T.; Kuwana, Y.; Saito, M.; Itoh, S.; and Kawauchi, H. (1985). Cloning and expression of cDNA for S growth hormone in *Escherichia coli*, Proc. Natl. Acad. Sci. 82 : 4306-4310.

Schwarz, D.C., & Cantro, C. R., (1984). Separation of Yeast chromosome-Sized DNAs pulsed field gradient gel electrophoresis. Nucleic acids, Res. 37-36.

Smith, H. O. & Wilcox, K.W. (1970). A restriction enzyme from *Hemophilias Influenza*. I : Purification and general properties. J. Mol. Biol. 51 : 379-391.

Sudha PM, Low S. Kwang J, Gang., Z (2001). Multiple tissue transformation in adult Zebrafish by gene gun Bombardment and muscular injection of naked DNA. Mar. Biotechnol. 3 : 119-125.

Termin, H.M. and Mizutani, S. (1970). RNA dependent DNA polymerase in virions of Rous sarcoma virus. Nature. 226 : 1211-1213.

Watson, J. D., & Crick, F.H.C. (1953). Molecular structure of nucleic acid Nature. 171 : 737-738.

Weil, C.; Hansen, P.; Hyam, D.; LeGac. F. Breton, B. and Crim, L.W. (1986). Use of pituitary cells in primary culture to study the regulation of Gonadotropin Hormone (GtH) secretion in rainbow trout : setting up and validating the system as assessed by its responsiveness to mammalian and salman gonadotropin releasing hormone, Gen. Comp. Endo. 62 : 202-209.

Wisneski, L.A. (1990). Salmon calcitonin in the acute management of hypercalcemia, calcif. Tissue Res. 46 : 526-530.

Chapter 39

FISH MITOCHONDRIAL DNA

In Eukaryotes, DNA is present in the chromosomes (inside nucleus) and in addition to chromosome, DNA is also present in two organelles, mitochondria (animals and plants) and chloroplasts (in green plants). These organelles are essential components of the cytoplasm of the eukaryotic cells. The mitochondrial DNA is also organized in a chromosome, which is circular in shape and is closely identical to the chromosome of the bacteria. In human beings, the disease known as myoclonic epilepsy and ragged-red fibres (MERRF) are due to inheritance of mitochondrial mutations in heteroplasmic individual. An individual whose organelles have more than one allele is termed as heteroplasmic. In this disease, cell is unable to produce sufficient amount of ATP hence cell death is caused due to lack of energy (Atherly *et al.*, 1999).

The mitochondria of animal and plant cells are the energy generating organelles while the chloroplasts are the photosynthetic organelles and occur only in plants. One unique feature that distinguishes mitochondria and chloroplasts from other organelles is that they contain their own genomes. The DNA present in the organelles is often known as extra chromosomal genes and is the examples of extra nuclear inheritance. There are three types of extra nuclear inheritance : maternal effect, genes residing in cytoplasmic organelles (maternal inheritance), and infectious inheritance. The inheritance of the genome/genes of mitochondrion behaves positively in non-Mendelian fashion.

Mitochondria are sausage-shaped structure in the living cell (Langley, 1968). They are surrounded by two membranes. The inner membrane forms elongated sacs called cristae which extend into the interior of the mitochondrion (Fig. 39.1). The insides of the inner

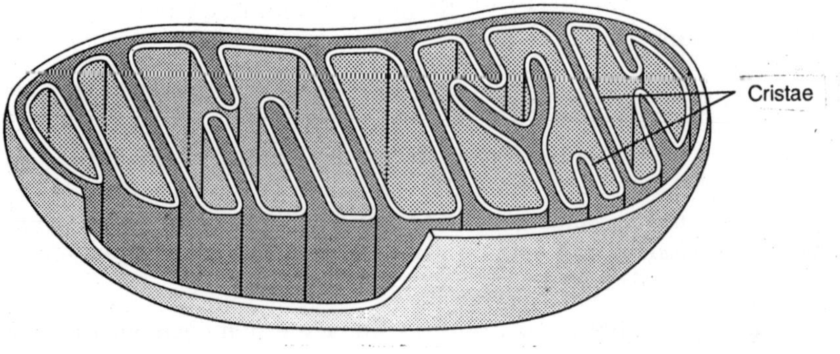

Fig. 39.1 : Structure of Mitochondrion showing cristae

membrane and outside of the outer membrane are covered with small particles. The inner membrane particles are composed of base, stalk and head while the particles of the outer membrane are devoid of base and stalk. In cristae, sugar and phosphates are broken down into CO_2 and water in series of chemical reactions known as oxidative phosphorylation.

The function of the mitochondrion is to produce high energy ATP. It does this by complex procedure in which pairs of electrons are passed along from the outer membrane to the inner membrane particles and through a series of **four complexes** . DPNH hand over the electrons to **Complex I.** Succinate transfers electrons to **Complex II.** In both cases the electrons pass from the base of the particles into the stalk where they are then handed over to **Complex III.** Next the electrons move on to the **Complex IV in the head of the particle** (Fig. 39.2, 39.3). Ultimately, as is typical of the respiratory chain, two electrons are accepted and carried off by a molecule of oxygen.

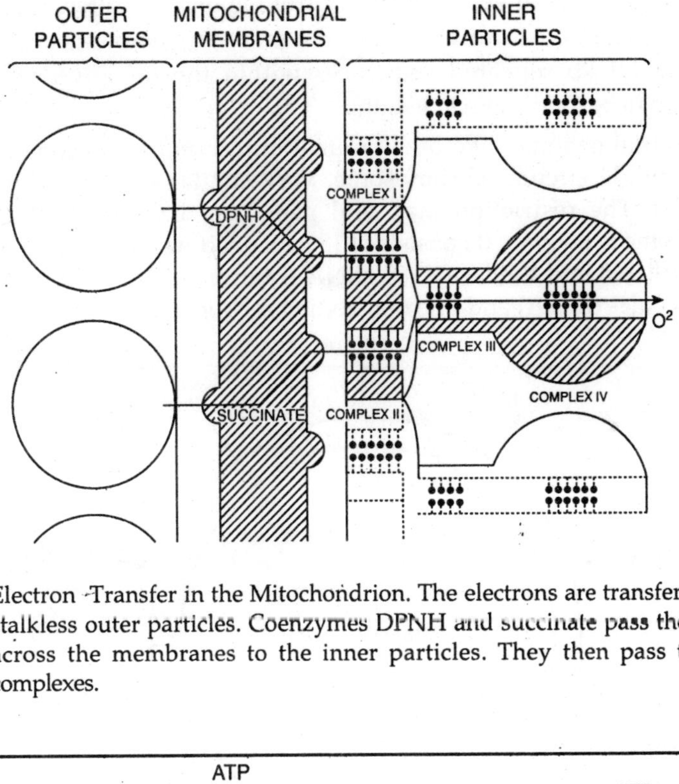

Fig. 39.2 : Electron Transfer in the Mitochondrion. The electrons are transferred by the stalkless outer particles. Coenzymes DPNH and succinate pass the electrons across the membranes to the inner particles. They then pass to various complexes.

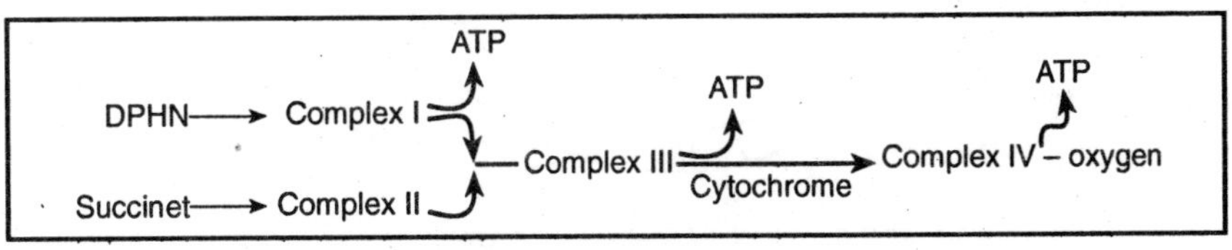

Fig. 39.3 : Diagrammatic sketch showing electron transmission through various complexes. Redrawn, Scientific American, 1964.

The production of ATP is carried out by large enzyme complexes embedded in the mitochondrial inner membrane. Some of the components of these **complexes** are determined by **genes in the mitochondrial genome,** others, by genes in the nucleus. Assembling these complexes requires a close interaction between these two genomes.

The mutations of the mitochondrial and chloroplast genes show maternal effects different than the nuclear inheritance. These cytoplasm organelles are usually inherited with the egg cytoplasm from the maternal parent. This is known as maternal inheritance. Genes in the cytoplasm organelle are widely studied today to determine how they function and how they interact with the nuclear genes. They are useful tools in the study of evolution. The fish mitochondrial DNA has been extensively used in phylogenetic studies.

Structure of Mitochondrial DNA or Mitochondrial Chromosome

In contrast of nuclear genome, the mitochondrial genome (genes) of animals is very efficient. The mt DNA has genes and these mitochondrial genes do not have introns so there is no junk DNA and there is no need for splicing of intron before coding protein (Atherly, *et al*, 1999; Bequmont & Hoare, 2003; Benzie 2000 & Doyle *et al.*, 1996).

There is another difference from the nuclear genome that there are no repetitively sequences in the mitochondrial genome; the control region does often vary in length due to **tandem repeats**. Exception to this rule are the scallops, many species of which exhibit several large upto 1.4 kb repeated sequences within the mt DNA genome which can consequently extend to beyond 30 kb in length.

Fish mitochondrial genome, like other animal mitochondrial genome, is almost always organized into a single circular chromosome, very similar to the chromosome found in bacteria (Fig. 39.4). The restriction map and gene organization of *Protopterus dolloi* of mitochondrial genome is known. It consists of 13 genes coding for proteins. Two genes are for coding for the small 12S and the larger16S rRNA (ribosomal RNA), 22 genes are for coding transfer RNA molecules (tRNA) and one non-coding section of the DNA which acts as the

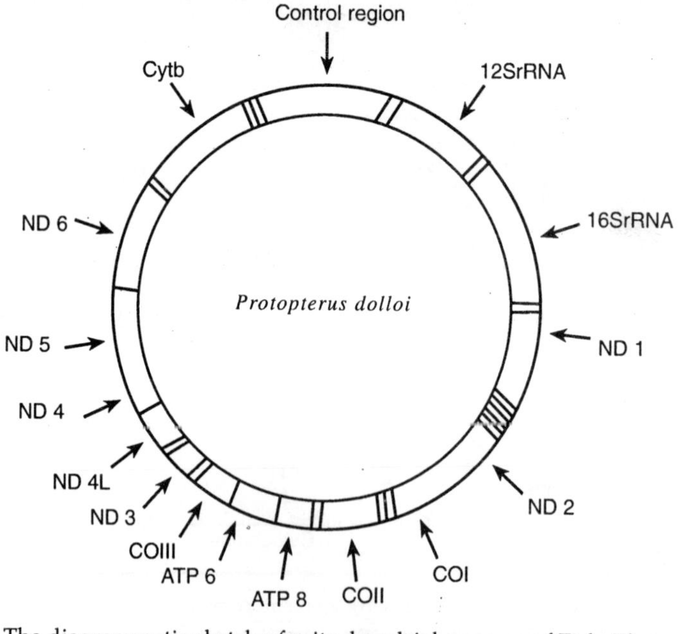

Fig. 39.4 : The diagrammatic sketch of mitochondrial genome of Fish. The position of the 13 protein coding genes are indicated on the molecule. They are : seven subunits of the enzyme NADH dehydrogenase (ND 1, 2, 3, 4, 4L, 5, 6), cytochrome b (Cyt b), three subunits of cytochrome c (COI, II, III) and two subunits of the enzyme adenosine triphosphate synthetase (ATP 6 and ATP 8). 12SrRNA = 12S ribosomal RNA 16SrRNA = 16S ribosomal RNA.

initiation site for mt DNA replication and RNA transcription. This is known as the control region.

The size of the mitochondrial chromosome varies from species to species, but is constant within species (King *et al.*, 1993). Significant genetic differentiation of six different halotypes has been found in the mitochondrial DNA amongst Atalantic salmon (*Salmo salar* L) in Denmark and seven other European salmon population by amplification analysis by RPLF using four restriction endonucleases assay (Nielson, *et al.*, 1996). The fish mitochondrial DNA has been extensively used in phylogenetic studies.

Mitochondrial genomes contain only a small number of genes but most cells contain many mitochondria so mitochondrial DNA may comprise a significant fraction of the total DNA in a cell. The rate of mutation in animal mtDNA is higher than in the nuclear DNA (about 5 to10 times higher). This means that the evolution is greater in mtDNA than in the nuclear DNA, and this feature is of importance to us when we are looking for genetic markers which will reflect changes in the more recent past (Nielson *et al.*, 1996).

REFERENCES

Atherly, A.G., Girton, J.R.; McDonald,J.F. (1999). Science of Genetics. Saunders College Publishing.

Beaumont, A.R. & Hoare, K. (2003). Biotechnology and Genetics in Fisheries and Aquaculture. Blackwell Publishing, 1-158.

Benzie, J.A.H. (eds) (2002). Genetics in Aquaculture VII, Elsevier, Amsterdam.

Doyle. R.W.; Herbinger, Ball, C.M. Gall, G.A.E.; (Ed). (1996). Genetics in Aquaculture V. Elsevier, Amsterdam.

King, D.P.F., Hovey, S.J, Thomopson, D and Scott. A. 1993. Mitochondrial DNA variation in Atlantic salmon, *Salmo salar* L., population, J. Fish Biol. 42, 25-33.

Langley, L.L. (1968). Cell Function. Van Nostrand Reinhold Company, affiliated East-West Press Pvt. Ltd. New Delhi.

Nielson, E.E., Hansen, M. M. and Loeschcke, V. 1996. Genetic structure of European population of *Salmo salar* L. (Atlantic salmon) inferred from mitochondrial DNA. Heredity. 77, 351-358.

Chapter 40

QUALITATIVE TRAITS AND THEIR PATTERN OF INHERITANCE

Distinction between qualitative and quantitative traits in Fishes.

There are two main phenotypic categories of traits (characters), one is *qualitative* and another is *quantitative*. The inheritance pattern of quantitative traits can be used now widely in aquaculture to obtain fishes with desired phenotypic character according to the need of the marked.

The qualitative character could be distinguished phenotypically (externally) from quantitative character. The qualitative trait is defined in qualitative terms for instance the colour of the flowers is either red or white. Another example is that the colour of eye of *Drosophila* is either red (wild) or white (mutant). These colours are discrete categories or distinct colour variant. The colour of the wild variety of *Cyprinus carpio* is black but in population some fishes show golden colour (variant), the golden colour fishes have great demand in the European market. These examples are the examples of *qualitative* or *discontinuous variations.*

The quantitative variations are the variations in the body weight at different age, larval growth rate and percentage of fat in the meat, weight of fish, time required for the fish to acquire maturity, fin size or shape of fins are the example of quantitative variations. The early maturity and quantity of meat in the fish are important for aqua farmers.

The qualitative and quantitative trait could also be distinguished on the basis of inheritance depending upon their specific genotype combination. If the offspring in F_1 generation is phenotypically resemble one of the parents while in F_2 the phenotype ratio of the offspring show 3 : 1 $\left(\frac{3}{4} \text{ or } \frac{1}{4}\right)$ dominant/recessive respectively. If the white roses are crossed with red roses and the colour of the roses in F_1 is either red or white (resemble one of the parents) then this is the qualitative character. Another example of qualitative inheritance is red and white colour coats in cattle. If in F_1 generation, the phenotypic colour would be either red or white and in the F_2 phenotypic colour would be 3 : 1 in monohybrid cross, clearly shows that the colour trait is qualitative and not the quantitative character. In *Cyprinus carpio,* if black colour fishes (wild & dominant) are crossed with golden colour fishes (variant, recessive) and in the F_1 generation, all fishes are Black and if in F_2 generation the ratio is 3 : 1 (3 black dominant and one golden recessive is a clear example of qualitative phenotype).

On the other hand, if the phenotypic colour in the F_1 generation would be intermediate between the colours of the parent and in the F_2 generation the colour would be in the form of continuous distribution between the colours of the parents, this is called quantitative inheritance (Fig. 40.1).

Cyprinus carpio

	Black Colour (wild)	X	Golden Colour (variant)
	(Black dominant W W)		(Golden recessive, GG)
F_1 Generation	All Black (wild); (similar to parent) (WW, WG)		If, intermediate colour (not similar to parent)
F_2	similar to one parent colour (3 : 1) **qualitative**		different shades colour of parents **quantitative**

Fig. 40.1 : Showing phenotypic and genotypic differences between inheritance pattern of qualitative and quantitative inheritance

With the help of both phenotypic and genotypic combinations, it is possible to take the advantage to produce the required characters in the fish in the aqua farms by the fish farmers.

Since the rediscovery of Mendel's law in 1900, the law of segregation and law of independent assortment, controversy raised that there is incompliance of Mendel's law. However, it is resolved that the quantitative as well as qualitative traits follow Mendelian inheritance. Incompliance is due to the influenced by variation of alleles at one or more loci and also as quantitative traits are controlled by multiple genes. Finally, it is settled that the inheritance of quantitative trait depends upon allele's combination of gene, which has occurred due to mutations, and in totality these traits appear due to gene behavior and **pattern of segregation during meiosis.** So it is important to know the pattern of inheritance.

Gene is a length of DNA that contains thousands or hundred of thousands of nucleotides pair, and sequences of nucleotide pairs in this DNA contain the information that makes it gene.

According to Mendel, a single phenotypic character of an organism is controlled by two alleles or allelomorphs. It was not known at that time that alleles could have many different allelic relationships.

It is known that genes not only have two alleles but also have several alleles. Today we know that the products of different genes interact, and that of alleles of one gene generate different phenotype depending upon whether the individual also has particular alleles of another gene. It is now suggested that Mendelian genes control both the quantitative and qualitative traits too (Atherly *et al.*, 1999; Doyle *et al.*, 1996).

The alleles or allelomorphs of a single gene interact in various ways. It is accepted that mutation has also occurred. These mutant alleles can give large range of different phenotypes. For example, more than 350 alleles of human cystic fibrosis gene have been discovered.

The relationships in the biological world are as follows: These relationships are also noticed in fishes and it is now possible to obtain marked friendly phenotype variants,

(1) Complete dominance over the other alleles when the alleles are situated at the same loci,

(2) Incomplete or partial dominance.

(3) Over-dominance.

(4) Co-dominance (equal expression).

A. Alleles of the same geneloci can show complete dominance

The common examples given by Mendel are the tall and dwarf of pea plants. The genetic combination would be homozygous dominant (TT), heterozygous (Tt) and homozygous recessive (tt). According to Mendel's Law of inheritance, genes have only two alleles: one dominant and another recessive. These two genes are located at the same loci, however, we know now that a gene may have many alleles at the same locus. If the organism inherited the same version of the gene from both parents, it is said to be homozygous dominant/ recessive (TT, tt). If the version is different then it is heterozygous (Tt). Each version of a particular gene is called as allele. In the F_1 generation on cross between tall homozygous and homozygous dwarf plants, the offspring will be having tall plants, showing complete dominance over the dwarf plants, but in F_2 the offspring will be both tall and dwarf in the ratio of 3 : 1 (monohybrid cross).

The example of complete dominance in fishes for the benefit of aqua farmers is the body colour of *Cyprinus carpio*.

In wild type, body colour of fishes are grey but in population fishes of golden body colour are also found. These golden colour fishes are called as *variant*. Variant is an allele or phenotype that differs from standard or wild type but is not deleterious or abnormal, hence survived. The phenotype golden colour that differs from standard wild type grey colour is due to genotype mutations in the alleles which the fish gets through inheritance. The gold variant is of high market value in European countries. The phenotypic golden colour is a qualitative trait. The genetic combinations are known.

In *Cyprinus carpio*, two alleles for wild type colour are due to allele, W and variant gold colour is due to presence of allele, G. The phenotype gold colour will appear in the fish if two homozygous alleles (GG) exist in the offspring at the same locus. The offsprings having the genotype combination of homozygous dominant (WW) or heterozygous (WG) will show wild colouration if they exist at the same locus. As in this case, wild (W) allele is dominated while the gold (G) allele is recessive and these alleles are present at the same loci. This is a straight forward and perfect example of dominance recessive relationship, because the phenotype (golden colour) has not appeared in heterozygous condition. Alleles of the same gene can show complete dominance in homozygous dominant or heterozygous condition. Gold colour phenotype will appear if there is homozygous recessive. Two alleles (allelomorphs) of a single gene show complete dominance over the action of other which is recessive. This gives the classical ratios of monohybrid, 3 : 1 (Fig. 40.2) or dihybrid 9 : 3 3 : 1. It is intra-allelic or allelic genetic interaction. The following would be the result of monohybrid cross and is represened as follows (Fig. 40.2).

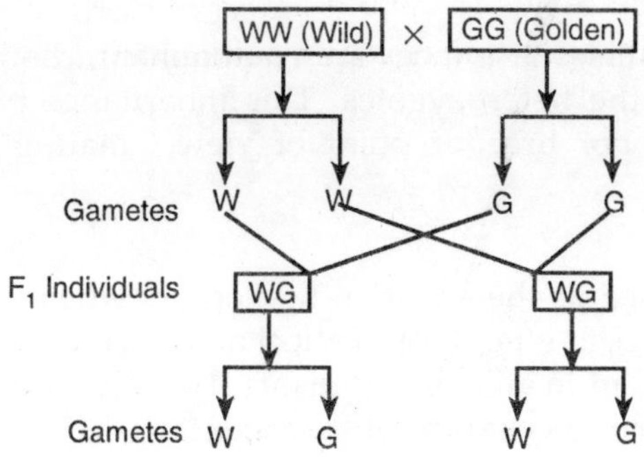

Punnet square

Gamets	W	G
W	W W wild	W G wild
G	W G wild	G G Gold

Ratio : Phenotypic 3 : 1

Genotypic, 1 : 2 : 1 (1 Homozygous dominant : 2 heterozygous : 1 Homozygous recessive)

Fig. 40.2 : Punnet square to show the genotypic constitution for the breeder to obtain combination of genotype to get market friendly fish

Taken the advantage of this fact, the breeder has to select both male and female of golden colour (homozygous recessive) and breed them in hatchery resulting only golden colour fishes. Golden colour is rare in nature.

B. Alleles of the same gene can show incomplete dominance (Semi-dominance)

Dominance may be incomplete, semi-dominant alleles may produce the same product but in a lesser quantity. Semi-dominance pattern of inheritance is also seen in fishes. Though the golden colour in *Cyprinus carpio* is the example of complete dominance, but in Tilapia species, *Oreochromis mossambicus* the colour variant is the clear case of incomplete dominance. The wild type fish has normal black colour having homozygous combination as WW. The variant has gold colouration as GG and the gold colouration will appear in the fish, if the fish is having homozygous alleles as GG at the same locus. If these fishes are crossed, the F_1 generation will have a heterozygous fish with the genotype combination WG. The heterozygous fish having WG combination has a **bronze** colour instead of either **black** or **golden** similar to parents. The occurrence of bronze phenotype colour in F_1 clearly indicates that there is no dominance and recessive situation but the dominance is incomplete (semi-dominance). In semi-dominance, the heterozygous shows **intermediate phenotype** between the dominant and recessive (Benzie, 2000; Beaumont & Hoare, 2003).

The Taiwan Fisheries Research Institute has developed red variants of *O. mossambicus* that are controlled by recessive alleles. Hybrid of red and white strain of *O. niloticus* has been developed after a great artificial selection. Another very common example is noticed in snapdragons, heterozygous for colour alleles has pink flower in contrast to red and white colour phenotype will appear if the allele's combination is homozygous. The semi-dominant pink phenotype colour will appear if it is homozygous dominant or homozygous recessive are breed. It is also certain that one allele is dominant and the other is recessive will not apply if large number of alleles are present.

C. Co-dominant

Sometimes the two alleles at a locus are co-dominant, both contribute equally to the phenotype character of the heterozygotes. This inheritance pattern of a single allele is difficult to distinguish. For breeder point of view, mating must occur between two homozygous.

D. Epistasis in Fish

The epistasis is defined as the situation that occurs when an allele of one gene affects the expression of the alleles at other location/loci on the same genome. The phenomenon of epistasis is commonly found in animal and plants. There is a functional interaction between different genes when allele or genotype at the locus "masks" or inhibits the expression of a non-allele or genotype at a distinct locus. The example of this phenomenon is found in fishes, *O. niloticus* and *Cyprinus carpio.*

In *O. niloticus,* scales have **opalescent-white** designated as colour and is controlled epistatically at **two loci**. At one locus the wild type allele is designated as W while another is Z and pearl-type allele as P and at another locus it is L. The colour pearl will appear in those fishes which during independent assortment get **P** and **L** alleles together at different loci. This does not involve the standard type of dominance because pearl-type allele can be expressed in the heterozygous condition. It is simply the presence of the pearl-type alleles at both loci that confers the pearl phenotype on its carrier. At the **second locus** the wild-type allele is designated as **Z** and the pearl-type allele as **L**. If the combination is **WPZL**, it will be pearl. If the combination is **WPLL**, it will also be pearl. If the combination is **PPZL,** it will also be pearl and if the combination is PPLL it will also be pearl. The ratio of phenotype of wild and pearl will be 7 (wild): 9 (pearl). It is simply the presence of the pearl phenotype type alleles **P** and **L** at both loci that express the pearl phenotype in fishes. WWZZ (wild type) & PPLL (pearl). The pearl type if the following double heterozygote F_1 individuals are crossed (WPZL crossed to WPZL) (Fig. 40.3).

Double	WP		WP
Pearl............	Z L	Cross	Z L
Meiosis, Gametes	WZ, WL, PZ, PL		WZ, WL, PZ, PL.

Punnet square

Gametes	WZ	WL	PZ	PL
WZ	WW ZZ Wild-type	WW ZL Wild-type	WP ZZ Wild-type	WP* ZL* Pearl
WL	W W ZI Wild-type	WW LL Wild-type	WP* ZL* Peart	WP* LL* Pearl
PZ	WP ZZ Wild-type	WP* ZI* Pearl	PP ZZ Wild-type	PP* ZI* Pearl
PL	WP* ZL* Pearl	WP* ZL* Pearl	PP* ZL* Pearl	PP* LL* Pearl

Results : 7(wild) : 9 (Pearl) PL at different loci so epistasis

Fig. 40.3 : Pearl type will express if two alleles P & L will come together either as two alleles or even single otherwise they will mask the appearance of pearl colour

Another example of epistatic interaction is also noticed in the pattern of scales in common carp, *Cyprinus carpio*. The mating is between fishes, which are double heterozygote. Advantage of these techniques has been used in aquaculture also. Some desired commercial importance quantitative traits have been produced in some fishes such as *Cyprinus carpio*, which is cultured throughout world. *Cyprinus carpio* is the most important cultured fish throughout the world. It has great commercial value. *Cyprinus carpio*, an exotic major carp is well adopted in Indian water too. It is now successfully bred by induced breeding throughout the world. It has three main varieties of commercial importance as under.

(1) *Cyprinus carpio* (Communis) : The common name is scale carp. The body is covered with regularly arranged small scales. The pattern is wild type. The wild type allele is represented by W while the variant allele is denoted by the word S. At another locus the wild type allele is Z and the variant allele is N. The occurrence of Z allele modifies the pattern of scales in the heterozygotes, but is lethal in the homozygous state (ZZ). This locus is epistatic to theW/S and it modifies the pattern of scales, heterozygous (ZN) survive and NN homozygous have wild type of scales.

(2) *Cyprinus carpio* (Specularis) : It is commonly known as mirror carp. The body is covered with large unevenly bright yellow scales normally restricted to lateral line hence a large area of the body remains as scaleless.

(3) *Cyprinus carpio* (Nudus) : The common name is leather carp. The body is provided with a single row of somewhat degenerate scales along the base of dorsal fin. Rest of the body is devoid of scale. So *Cyprinus carpio* (Nudus) is reduced scale pattern. These fishes have high value in European market while *Cyprinus carpio* (Communis) having wild type pattern of scale is preferred in Asian countries. The fish breeders according to the commercial need can now produce them by taking the use of genotype combination, its pattern of inheritance and, applying the techniques described above. If the cross is made between double heterozygous, then the result is described as follows (Fig. 40.4) :

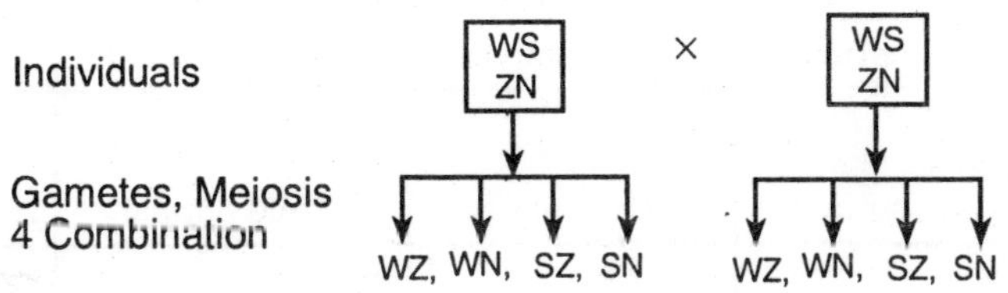

Punnet square

Gametes	WZ	WN	SZ	SN
WZ	WW ZZ lethal	WW ZN line	WS ZZ lethal	WS ZN line
WN	WW ZN line	WW NN wild-type	WS ZN line	WS NN wild
SZ	WS ZZ lethal	WS ZN line	SS ZZ lethal	SS NS leather
SN	WS ZN line	WS NN wild-type	SS ZN leather	SS NN mirror

Results : ratio 6 line : 4 lethal : 3 wild : 2 leather : 1 mirror

Fig. 40.4 : Punnet square showing the result of cross to obtain required fish with specific combination

E. Another situation will appear if alleles of two character present at two loci but no interlinking or influence of each other

In some fishes there are two important characters of commercial value but they are controlled by two loci and these loci are not interlinked. The best example of this is *Lebistes reticulates*, a guppy. The phenotypic characters are gold colour and spinal curvature. The two alleles for colour are W dominant for wild (black colour) and G for gold colour, which is recessive in character. For curvature of the fish, the recessive is represented by C and dominant wild is represented by Z. If two heterozygote are crossed. The result would as follows (Fig. 40.5) :

Genetic combination of

Adult Fish genotype — WG ZC × WG ZC

Meiosis Gametes formation — WZ, WC, GZ, GC   WZ, WC, GZ, GC

one gene of character goes in one gamete

Punnet Square

Gametes	WZ	WC	GZ	GC
WZ	WW ZZ Gray normal	WW ZC Gray normal	WG ZZ Gray normal	WG ZC Gray normal
WC	WW ZC Gray normal	WW CC Gray curved	WG ZC Gray normal	WG CC Gray curved
GZ	WG ZZ Gray normal	WG ZC Gray normal	GG ZZ Gold normal	GG ZC Gold normal
GC	WG ZC Gray normal	WG CC Gray curved	GG ZC Gold normal	GG CC Gold curved

Fig. 40.5 : Ratio of phenotype : 9 (grey normal) : 3 (grey curved) : 3 (gold normal) : 1 (gold curved)

This clearly indicates that two gene allele present at the two loci have no influence on each other. The breeder can take the advantage of this character and breed the fishes so that only fishes with gold body colour with normal spine would appear in their hatcheries (McAndrew & Penman, 1999).

F. Unlinked Loci

In guppies *(Lebistes reticulates)* there are two characters, one is gold colour and another is spinal curvature. They are situated and controlled independently at separate loci. At first locus, dominant wild type allele (W) produces grey colouration, while the recessive (G) allele produces gold colouration in homozygous (GG) fish.

Because of the complete dominance of the W allele, heterozygous (WG) fish is grey. At the second locus a recessive variant (C) causes curvature of the spine in the (CC) homozygotes. The dominant fish has a normal spine fish.

The quantitative inheritance has been noticed in fishes also. The advantage of quantitative traits has been used in aquaculture. Generally the colour of the wild variety of *Cyprinus carpio* is grey but in population some fishes show golden colour. So the phenotypic golden colour is a qualitative trait. These golden colour fishes are called as variant. Variant is an allele or phenotype that differs from standard or wild-type but is not deleterious or abnormal. The phenotype golden colour that differs from standard wild type grey colour is due to genotype mutations in the alleles which the fish gets through inheritance but the mutant is not deleterious or abnormal, hence they survive. The gold variant is of high market value.

REFERENCES

Atherly, A., Girton, J.R. & McDonald. (1999). The Science of Genetics. Saunders College Publishing. Harcourt Brace College Publishers. London, Tokyo.

Benzie, J. A. H(eds). (2000). Genetics in Aquaculture. VII. Amsterdam.

Beaumont, A.R. and Hoare, K. (2003). Biotechnology and Genetics in Fisheries and Aquaculture. Blackwell Publishing.

Doyle, R.W.; Herbinger, C.M.; Gall, G.A.E. (1996). (eds.) Genetics in Aquaculture V. Elsevier, Amsterdam.

McAndrew, B. & Penman, D. (eds.). (1999). Genetics in Aquaculture VI, Elsevier, Amsterdam.

Chapter **41**

MEASUREMENT OF QUANTITATIVE INHERITANCE

Quantitative Inheritance: The Techniques for the Measurement

After the rediscovery of **Mendel's law,** many scientists conducted similar experiments and found that there was fluctuation in Mendel's law of inheritance. However, they were unanimous that genes inherit characters. Gene is a stable entity. It controls the phenotype of the individual. An organism carrying the normal gene is called **wild type,** the organism carrying an altered gene is called a **mutant**. For example, wild type *Drosophila* has **bright red-eye** colour while mutant has white-eye colour. The body colour of *Cyprinis carpio* is **grey colour** but the mutant has body colour **golden**. These are the instances of qualitative characters. On the other hand, the body weight at age or the body weight of Fish/organisms and milk production of cow and length of fishes of similar age group are not discrete entity (colour never be equal) but is continuous variable and, therefore, the examples of quantitative characters. The quantitative trait phenotype is determined by alleles of many genes acting together. The phenomena are called as **polygenic**.

The quantitative characters are not only under polygenic (multiple alleles) control but the environment influences the quantitative trait too. Two individuals with similar genotype if raised in **different environmental** have a different quantitative phenotype. For instance, fishes having the same genotype but, if one group is raised in environment having rich fed while other group is reared in environment having less rich fed, the former group will naturally grow faster showing quantitative phenotype. It is clear that quantitative traits would depend upon genotype as well as environment.

There are three characteristic features of quantitative inheritance :

(1) They are continuous variable. There is **no distinct** phenotype in the offspring. For example, if the cross is made between white and red, the offspring will be having continuous distribution of colour intermediate between the colours of the parents with different intensities.

(2) It is polygenic in nature. The single character is controlled by alleles of many different genes. The population is having large numbers of different genotype and different genotype may have the same phenotype. The examples are been cited in the previous chapter.

(3) The environment influences the quantitative traits as described above. When we use the term environment, it covers all aspects in which organisms interacted with physical and biological environment around it during its lifetime. Here, it is clear that the genotype environment also plays important role in the appearance of phenotye of quantitative character. The example is again cited that a well-fed fish, *Cyprinus carpio* will grow faster than the poorly fed carp irrespective of genotype.

Hence it is established that environment and genotype both play role in the determination of quantitative character. Based upon the researches done between 1903-1918, it is accepted that inheritance of quantitative traits follow Mendelian fashion.

R.A. Fisher, a British statistician, described the biometrical approach and suggested that Analysis of variance and Heritability are needed for measurment for quantitative inheritance (Mc Andrew & Penman, 1999; Anthery *et al.*, 1999).

(1) **Analysis of variance**

Before understanding the variance it is essential to know that there is relationship between the mean ($\bar{X}$) and standard deviation (S) in normal distribution curve. It is also necessary to know that there is a relationship between the standard deviation and variance. This relationship is given as under.

Standard deviation square $(S)^2$ = Variance

Standard deviation S $= \sqrt{V}$

The quantitative traits are continuous variable and distributed normally.

The normal distribution graph is bell shaped. The following are the characteristics of the normal distribution (Benzie, 2000; Kothari, 2004).

(1) It is a symmetrical distribution.

(2) In the normal distribution the mean of the population ($\bar{X}$) occurs at the peak of the curve or the ordinate at the mean is the highest ordinate. Similarly the height of the mean ordinate and the height of other ordinates at various standard deviations from the mean happen to be a fixed relationship with the height of the mean ordinate. One standard deviation from the mean on each side is always 34.13 of the total area (68.26) of the curve (Figs. 41.1, 2 & 3).

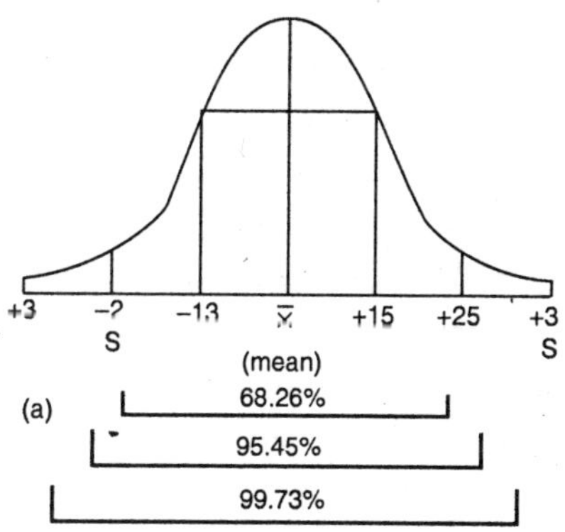

Fig. 41.1 : Properties of the normal distribution

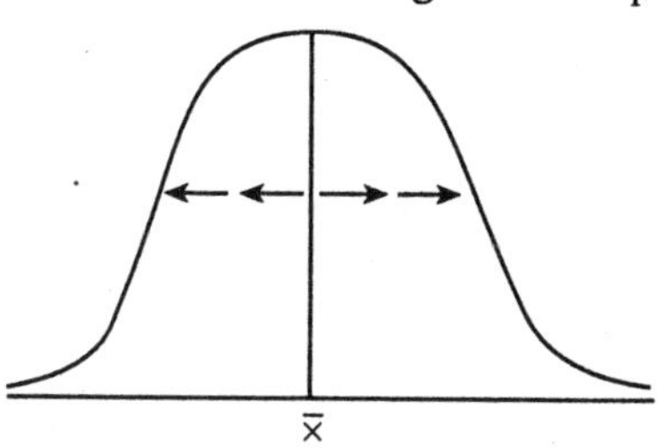

Fig. 41.2 : Showing broader distribution of population with large variance

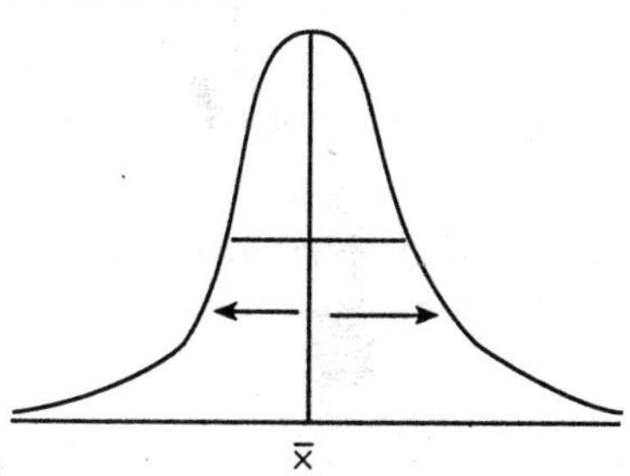

Fig. 41.3 : Graph showing less variance of population

(3) The curve is asymptotic to the base line, which indicates to continute to approach but it never touches the horizontal axis.

(4) The standard deviation or variance or $(S)^2$ defines the spread of the curve. So if we examine two graphs having the same mean $(\overline{X})$ but showing the difference of variance we can conclude whether the variance is more or less.

(a) **Formula for the calculation of Variance**

In quantitative character, the variance is known as phenotypic variance and is represented as Vp. It is well accepted tool used in quantitative genetics, which includes the analysis of the components of the phenotypic variance.

Vp = Vg + Ve.

The formula is based on the following concept. The **phenotypic variance** (Vp) has three additive components: the genetic variance (Vg), the environmental variance (Ve) and the interaction variance (Vi).

The formula actually is as follows :

Vp = Vg + Ve +Vi. Vp = phenotypic variance

Vg = genetic variance/variance due to genes (different alleles and loci, QTL)

Ve = environmental variance; we have no way of measuring interaction variance Vi, it is usually assumed to be zero, So Vp = Vg + Ve.

(a) **How to calculate genetic variance (Vg)?**

The variations caused by genes may come from three different sources. Firstly, the variation may be related to straightforward presence or absence of particular allele at quantitative trait loci (QTL). This is additive genetic variance and denoted by (Va). It is most important as the presence or absence of a particular allele which is passed unchanged to the next generation.

Secondly, in some cases there is presence or absence of particular genotype at QTL. For instance, a particular heterozygous combination of alleles at a locus may confer an advantage on an individual with respect to a particular trait. This is called as dominant genetic variance and denoted as Vd. It is less manageable to simple artificial selection. Because during meiosis segregation and independent assortment of alleles will take place and in the next generation instead of the same heterozygous combination (which is advantageous), different combination may appear hence there is no guarantee of inheritance.

Thirdly, genetic variation is produced by interactions between loci called epistatic or non-allelic interaction genetic variance, denoted by Vi. Hence the genetic variance Vg can be calculated by the following formula:

Vg = Va+Vd+Vi

Va = additive genetic variance

Vd = dominant genetic variance

Vi = interaction genetic variance

Explanation: The genetic variance in population is caused mostly due to the

(i) allelic difference in the genotype. They are additive genetic variance (Va),

(ii) the dominance genetic variance (Vd), genes are dominant or recessive

(iii) and genetic, interaction variance (Vi).

The genetic variance (Vg) can be calculated by substracting the environmental variance from the variance of the F_2 individuals (Vg= VF_2-Ve).

(i) **Va is calculated as follows** : The phenotypic variation that is caused by individuals is due to different alleles of gene at QTL that effect the phenotype. It is most important element of genetic variation for the aquaculture breeder as the presence or absence of particular alleles is a character. The additive variance (Va) can be calculated by subtracting the backcross variance (VB_1 and VB_2) from the F_2 variance using the following formula :

$$Va = 2\ (VF_2 - (VB_1 + VB_2)/2\ )$$

The dominant variance (Vd) is that portion of the phenotypic variance caused by individual having different alleles of genes that effect the phenotype. The dominance variance (Vd) is much less amenable to simple artificial selection because genotype are broken down during meiosis and put together again in different combinations in the next generation. Hence crossing between inbred, highly homozygous lines does not guarantee predictable heterozygosity in the offspring and such F_1 hybrid are commonly used in plants and animal breeding. The development of inbred lines in aquatic organisms is still in infancy but ploidy manipulation techniques are quite successful.

The third component of genetic variation is produced by interactions between loci and is called epistatic or non alleleic genetic interaction variance (Vi). It includes epistasis, enhancement, suppression, etc. It includes an individual may be highly ranked for a trait as it possesses particular combinations of genotypes across two or more QTL. The additive genetic variance and dominance variance can be estimated by measuring the variances of groups of individuals that have known genetic relationship. The values of these variances can be then used to make deductions about the alleles present in the population. The two types of F_1 crosses (F_1 individuals crossed with both sets of parents) are especially valuable for this type of analysis.

(ii) **The dominance variance (Vd)** can be calculated by subtracting the addititive variance from genetic variance. In these calculations, the interactive component is ignored.

(b) **The environment variance (Ve)** is the measure of phenotypic differences that are produced by the different environment such as quality of water, food quality and quantity, temperature and stocking density in which the individuals in the population live. For example a well-fed fish will grow faster than a poorly fed fish. The growth of the plants will be more in the soil where the nutrients in the soil are more in comparison to the plant where the nutrients are low in the same field. So, there are physical and biological environment interacted in a close fashion. It also includes the cellular environment, which is responsible for coding the protein. So the environmental variance should be measured using a genetically uniform population. This could be obtained by inbreeding. Such population would be genetically same and therefore Vg = 0. And all phenotypic variation must be due to environment and therefore Vp = Ve. If two crosses namely P_1 and P_1 are crossed having the same genotype then the F_1 individuals will be genetically uniform and the total phenotypic variance is an estimate of the environmental variance. Assuming that all the populations are raised in the same environment, environmental variance is the average of the variance of parents and the F_1.

$$Ve = (Vp^1 + V_p^2 + V_F^1)/3.$$

Heritability

The proportion of the variance of a trait which is under genetic control is termed as the heritability. **Heritability** is a measure of the genetic component of the variance, and through this technique it is used in predicting the phenotypes of offspring. So first the mean and variance of a quantitative trait is obtained from the parental population then this information can be used to make prediction about the mean of the phenotypic distribution in the offspring (generation). And we can determine how strongly the phenotype traits of the offspring will resemble the phenotype of the parents (Beaumont & Hoare, 2003).

There are two widely used numerical measures of heritability. One is the coefficient of **heritability** (h^2), also called heritability in the **narrow sense.** The other is **the degree of genetic determination (H^2),** also called heritability in the **broader sense**. Both values depend upon the ratio of the genetic variance to phenotype variance. h^2 is the ratio of the additive variance to the total phenotypic variance, and H^2 is the ratio of total genetic variance to total phenotypic variance. The following are the formulae of measuring the heritability:

$$h^2 = Va/Vp$$

$$H^2 = Vg/Vp$$

Both these values indicate that what portion of the variation in a population is the result of genetic variation? Both these values theoretically vary from 1 to 0, if the value is high it shows that large portion of the phenotypic variation is the result of genetic variations. h^2 and H^2 have important limitations. Their values are calculated for one population in one environment, so they cannot be used for other generations of the same population raised in different environments or for other populations. Each population has a different set of genotypes and will have a different proportion of the phenotypic variance caused by genetic variance and different values of h^2 and H^2. An example is cited to explain how h^2 can be used for predicting the phenotype of the offspring.

REFERENCES

Mc Andrew, B. & Penman, D. (eds.) (1999). Genetics in Aquaculture VI, Elsevier, Amsterdam.

Atherly, A., Girton, J.R. & McDonald. (1999). The Science of Genetics. Saunders College Publishing. Harcourt Brace College Publishers. London, Tokyo.

Benzie, J.A.H. (eds.), (2000). Genetics in Aquaculture. VII. Elsevier Amsterdam.

Beaumont, A.R., and Hoare, K. (2003). Biotechnology and Genetics in Fisheries and Aquaculture. Blackwell Publishing.

Kothari, C.R. (2004). Quantitative Techniques. Vikas Publishing House, New Delhi.

Chapter 42

BINOMIAL EXPANSION

Use of Binomial Expansion or Jacob Bernoulli's Theorem to Answer to Some Genetic Questions

Binomial distribution is associated with the name J. Bernoulli (1654-1705), but it was published eight years after his death. Binomial means two 'names'; hence frequency distribution falls into two categories—a dichotomous process. This distribution is a probability distribution expressing the probability of two mutually exclusive events, called p (success) and q (failure), whose combined probabilities add up to one (i.e., $p + q = 1$). Using the multiplication and additive rules and using the Binomial expansion it is possible to answer genetic questions and predicting the probabilities that what would be the particular combination of genotype and phenotype will result. Let us take the example of Mendel's monohybrid cross. He has selected pea and in one of the experiment he has made a cross between two true-breeding strains, one with wrinkle seed and another with round seed, the round and wrinkle phenomena are usually exclusive events. The second character he has selected was seed colour, yellow versus green and according to him this is also exclusive event. He has actually taken 7 contrasting characters for framing laws of inheritance. **Exclusive means** that the colour of the seed would be either yellow or green but it cannot be both. According to Mendel, the result of F_2 was 3 : 1, i.e., three dominant and one recessive. If round were dominant, then in F_2, generation phenotype would be three rounds and one wrinkle. It means that the probability (p; success) of round would be $p = 3/4$ and wrinkle (q; failure) would be $q=1/4$. The binomial theorem can be used to determine the probability that any group of F_2, individuals will have a particular combination of phenotype by calculating the probabilities of all possible combinations of individuals that could make up group, and then summing these probabilities, if the event will happen in n traits, then it will be $(q + p)^n$. For example, for a group of two F_2 seeds ($n = 2$), all possible combinations of phenotype are given by expanding the binomial raised to power 2 or $(p + q)^2 = p^2 + 2pq + q^2 = 1$. To solve our problem of the group of 6 seeds, we need to determine the number of possible combinations in a group of 6 seeds ($n = 6$), which is done by expanding the binomial raised to the power 6, $(p + q)^6$, the coefficients of the terms are 1, 6, 15, 20, 15, 6, 1.

Terms of the Binomial Expansion is as follows (Agarwal, 2004):

1. $(p+q)^1 = p + q$
2. $(p+q)^2 = p^2 + 2pq + q^2$
3. $(p+q)^3 = p^3 + 3p^2q + 3pq^2 + q^3$
4. $(p+q)^4 = p^4 + 4p^3q + 6p^2q^2 + 4pq^3 + q^4$
5. $(p+q)^5 = p^5 + 5p^4q + 10q^3q^2 + 10p^2q^3 + 5pq^4 + q^5$
6. $(p+q)^6 = p^6 + 6p^5q + 15p^4q^2 + 20p^3q^3 + 15p^2q^4 + 6pq^2 + q^6$
7. $(p+q)^7 = p^7 + 7p^6q + 21p^5q^2 + 35p^4q^3 + 35p^3q^4 + 21p^2q^5 + 7pq^3 + q^7$
8. $(p+q)^8 = p^8 + 8p^7q + 28p^6q^2 + 56p^5q^3 + 70p^4q^4 + 56p^3q^5 + 28p^2q^3 + 8pq^7 + q^8$

Pascal's Triangle

1	1	2
2	1 2 1	4
3	1 3 3 1	8
4	1 4 6 4 1	16
5	1 5 10 10 5 1	32
6	1 6 15 20 15 6 1	64
7	1 7 21 35 35 21 7 1	128
8	1 8 28 56 70 56 28 8 1	256
9	1 9 36 84 126 126 84 36 9 1	512

Some properties of the Binomial Distribution are listed as follows:

The mean and the standard deviation of the binomial distribution can be obtained by using the formula as given below :

Mean of population is μ, $\mu = Np$

Standard deviation of population, $\sigma^2 = Npq$

Moment coefficient of skewness, $a^3 = \dfrac{q-p}{\sqrt{Npq}}$

Another easy formula/method to calculate the probability is as follows:

$$\text{probability} = \frac{n!}{w!\,x!}\,p^w q^x$$

w stands for number of individual of one type

x stands for individuals of other types, n stands for total number of individuals in group

(i.e, n = w + x), p for probability of one type and q is the probability of other type. The symbol ! is the symbol of factorial, meaning the multiplication of a number by all of the integers between it and one. For example, $4! = 4 \times 3 \times 2 \times 1 = 24$.

The Chi-Square Test

The chi-square test was used to test that alleles segregate on Mendelian principles. It is required a comparison of expected and observed numbers. It is used in statistics for judging the significance of the sampling data. Prof. Fisher developed chi-square test. Symbolically written as χ^2 (pronounced as Ki-square). It is a statistical measure with the help of which it is possible to assess the significance of the difference between the observed genotypic numbers (frequencies) and the expected numbers (frequencies) obtained from some hypothetical universe. Whether the experimental and predicted ratio is in good agreement or not. This could be tested by Chi square Test. This test determines that in course of any experiment procedure dealing with quantitative data, some variation, called **"experimental error"**, can be attributed to chance error alone (Spiegel, 1981; Atherly, Girton, & McDonald 1999).

This can be calculated by the following formula:

$$\chi^2 = \Sigma = \frac{(\text{observed value} - \text{expected value})^2}{\text{expected value}}$$

Let us take the example of ratio of phenotype and genotype of experiments conducted by Mendel. By applying χ^2, results show that the observed frequencies are in agreement with the predicted ratios. The data of Mendel's actual experiments are given in the following table. The variations in the expected and predicted ratio are due to experimental error alone.

Observed, O s	Expected, E	$(O-E)^2$ Z	$\frac{(O-E)^2}{E}$
315	312.75	5.06	0.016
101	104.25	10.56	0.101
1.08	104.25	14.00	0.135
32	34.75	7.56	0.218
Total 556	556		$0.470 = \chi^2$

Mendel observed in his experiment, the ratio as 9 : 3 : 3 : 1 in dihybrid cross in the F_2 generation while the ratio in monohybrid cross in F_2 generation was 1: 2 : 1. He found 315 rounds, yellow seeds, 101 rounds, green seeds, 108 wrinkle, yellow seeds and 32 wrinkle, green seeds. So the expected numbers of each phenotype are 556 (9/16) = 312.75 round, yellow seeds; 556(3/16) = 104.25 round. Green seeds; 556(3/16) = 104.25 wrinkle, yellow seeds and 556(1/16) = 34.75 wrinkle, green seeds (Atherly *et al.*, 1990). Chi-square will show whether the difference between the actual and predicted ratio is due to experimental error or not. The Chi-square value calculated is 0.470, which is got by applying by the following formula:

$$\chi^2 = \Sigma = \frac{(\text{observed value} - \text{expected value})^2}{\text{expected value}}$$

Degree of freedom is required for the calculation of χ^2, the number of independent constraints determined the number of degree of freedom (or d.f.). The degree of freedom is a measure of the number of independent variables present in a given experiment. It is stated that the chances of error affect only one independent variable. In Mendels experiments stated above the variables are only 4 so the degree of freedom is 4 –1 = 3. The number of any three phenotypic classes is determined, the number of the fourth class is fixed.

To find out the probability we have to consult Chi-square table. The table is given as under (Agarwal, 2004 & Kothari, 2004):

Fisher and Yates (1963); probability associated with values of the Chi-square Statistics

d.f.	0.90	0.50	0.10	0.05	0.01
1	0.016	0.46	2.71	3.84	6.64
2	0.21	1.39	4.61	5.99	9.25
3	0.58	2.37	6.25	7.82	11.35
4	1.06	3.36	7.78	9.49	13.28
5	1.61	4.35	9.24	11.07	15.09
6	2.2	5.35	10.65	12.59	16.81
7	2.83	6.35	12.02	14.07	18.48
8	3.49	7.34	13.36	15.51	20.09
9	4.17	8.34	14.68	16.92	21.67
10	4.87	9.34	15.99	18.31	23.21
11	5.58	10.34	17.28	19.68	24.73
12	6.3	11.34	18.35	21.03	26.22
13	7.04	12.34	19.81	22.36	27.69
14	7.79	13.34	21.06	23.69	29.14
15	8.35	14.34	22.31	25.00	30.58
20	12.44	19.34	28.41	31.41	37.57
25	16.47	24.34	34.38	37.67	44.31

The hypothesis is never agreed or disagreed by the P value. The results of the investigator which are acceptable or unacceptable with respect to hypothesis, evaluate, the results of the Chi-square observations. **The 5 percentage points (0·05) on the table are usually chosen as an arbitrary standard for determining the significance or goodness of fit.**

Table value of χ^2 for 3 degree of freedom at 5% level is 7.82, a chi-square value is 0.47 which is lower than the table value hence it is correct. In other words, it may be said that a probability at 5% level of significance is **7.82** which is more/greater hence the hypothesis is correct. If it is less than 5%, then rejected.

Let us take another example to understand the application of **chi-square tests**.

Genetic theory states that children having one parent of blood type A and the other of blood type B will always be of one of the three types **A, AB, B** and that the proportion of the three types will on an average be as **1 : 2 : 1**. A sample of 300 children was collected—30% were found to be **type A**, 45% — type **AB** and the remainder —type **B**.

Type	Observed Frequency (O)	Expected Frequency (E)	$(O-E)^2$	$\frac{(O-E)^2}{E}$
A	90	75	225	225/75 = 3
AB	135	150	225	225/150 = 1.5
B	75	75	0	0/75 = 0

$$\chi^2 = 4.5$$

The **table value** χ^2 for (3 – 1) **2d.f. at 5%** (0.05) level of significance is **5.99**. The **calculated value of chi-square** is **4.5** which is less than the table value and can be ascribed to have taken place because of the chance error of the experiment. This supports the theoretical hypothesis of the theory of inheritance (segregation of alleles) that **A, AB** and **B** stand in the proportion of **1 : 2 : 1**.

Another example in this connection is as follows :

The proportion of the bean in four groups namely **A, B, C** and **D** should be **9 : 3 : 3 : 1**. A farmer sowed 1600 beans. The results of his experiment show that in four groups the data is 882,313,287 and 118. Does the experimental result support the theory that they are in the ratio of **9 : 3 : 3 : 1**?

Taking the null hypothesis that the experiment results support the theory, the expected frequencies are :

$9/10 \times 1600, 3/16 \times 1600, 3/16 \times 1600, 1/16 \times 1600 = 900, 300, 300, 100$

Thus:

O	882	313	287	118
E	900	300	300	100

χ^2 = (882-900)2 /900 + (313-300)2 /300 + (287-330)2 /300 + (118-100)2/100

= 0.3600 + 0.5633 + 0.5633 + 3.23 = 4.7266

d.f. = 4 –1 = 3.

The tabular value of χ^2 for 3 d.f. at 5% level of significance = 7.815.

Since the calculated value of χ^2 is less than the table value, the hypothesis may be accepted and the results are in line with the theory.

The Chi-square Test is also used to determine whether the population is in Hardy-Weinberg Equilibrium.

One use of the Hardy-Weinberg method is to predict the frequency of individual homozygous for a deleterious recessive allele. These authors have discovered that a random-mating population (panmictic) with **no mutations, no selection pressure, no genetic drift and no migration,** the relative **frequencies** of each gene (allele) tend to remain **constant** from **generation to generation**. The law tells genotype frequencies to gene frequencies in randomly mating populations, the alleles segregating at any given locus can easily compute the expected genotype frequencies in that population. This is known as Hardy-Weinberg Law or often called as law of population equilibrium. Once we know the **frequencies** of the alleles in the population it is possible to calculate the **frequencies** with which the genotype will be present in the offspring from that generation. The population of sexually reproducing diploid organisms at a locus with two alleles at frequencies p and q, after one generation of random mating, the frequencies of the genotype would be AA, Aa and aa will be p^2, 2pq, q^2.

Let us take an example that two alleles of the gene AA dominate while aa i.e., the alleles are recessive gene. According to Mendel's Laws of inheritance, the genotype ratio would be 25% AA (homozygous dominant), 50 would be Aa (heterozygous) and 25% would be aa (homozygous recessive). The statistical formula for this law is $p^2 + 2pq + q^2$. According to Mendel's Laws, the following should be genetic combination.

Individuals
Gamete segregation
Individuals
F_2, gametes segregation

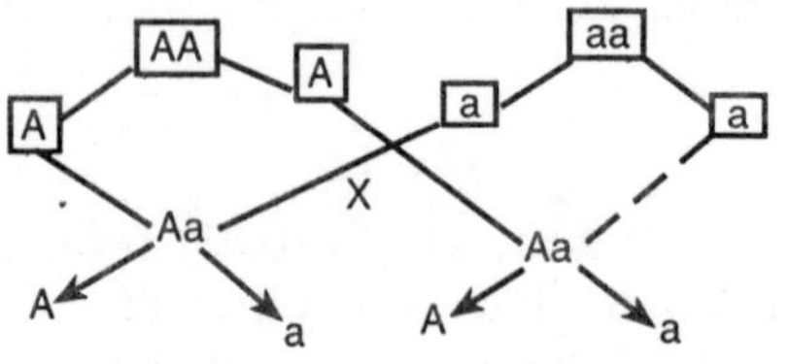

Genetic		A		a
A.	AA	pp or p^2	Aa	Pq
a.	Aa	pq	Aa	qq or q^2

AA	Aa	aa
p^2	2pq	q^2

This shows that alleles remain constant from generation to generation.

The application of chi-square test is slightly different while applying for Hardy–Weinberg law because it deals with frequencies of expected genotype rather than numbers. The degree of freedom is taken **n-2 instead of n-1**.

REFERENCES

Agarwal, D.R. (2004). Quantitative Methods. Vrinda Publications (P) Ltd, Delhi.

Atherly, A. Girton, J.R.; & McDonald. (1999). The Science of Genetics. Saunders College Publishing. Harcourt Brace College Publishers. London, Tokyo.

Kothari, C.R. (2004). Quantitative Techniques. Vikas Publishing House, New Delhi.

Spicgel, M.R. 1981. Schaum's outline Series, Theory and Problems of Statistics. Metric Editions. McGraw — Hill Book company, New York.

Chapter 43

TRANSGENIC FISHES

The application of genetic engineering to animals, like potatoes with built-in insecticide, could provide numerous benefits, including the possibility of a safer, cheaper food supply and the creation of new sources for inadequate pharmaceutical resources. With the advancement in the field of genetic engineering, the application of its commercial use has also increased. Aquatic animals are being engineered to increase aquaculture production. The use of genetic engineering and rDNA technology has done miracles in medical and industrial research. The transgenic fish are being promoted as the first marketable transgenic animals for human consumption. One of the most important aspects between fish and other terrestrial animals for cultivation and genetic improvement is that, usually, fishes have higher levels of genetic variation and hence more scopes for selection than most mammals or birds.

Using the gene transfer technology, scientists now have created a genetically engineered variety of **Atlantic salmon** that grow to market size in about **18 months,** otherwise the fish takes about **24-30 months** for becoming market size fish (Du *et al.*, 1992). It is also hoped that we can now modify a large number of fishes with fast growing characteristics and bring Blue Revolution.

The following are the important points needed for genetic engineering (gene transfer) to produce transgenic fish.

(1) A gene sequence is to isolate for the particular characteristics ; for example, growth hormone gene.

(2) These genes (gene sequence) are then inserted into a circular DNA known as plasmid Vector (enzymes endonucleases and ligases are used).

(3) Plasmids are harvested in the bacteria to produced billions of copies.

(4) Plasmids are introduced into linear DNA. The linear DNA is sometimes called a *gene cassette* because it contains several sets of genetic material in addition to new inserted gene; for example, growth hormone gene. The technology is available to integrate genes in germ line of developing individual (fish) and finally transmitted into further generations.

(5) Making the cassette a permanent part of fish's genetic makeup.

What is Transgenic Fish?

A transgenic fish is one that contains genes from another species. A transgenic fish is an improved variety of fish provided with one or more desirable foreign gene for the purpose of enhancing fish quality, growth, resistance and productivity.

Typically, genes of one or more donor-species are isolated, and spliced into artificially constructed infectious agents, which act as vectors to carry the genes into the cells of recipient species. Once inside a cell, the vector carrying the genes will insert into the cell's genome. A transgenic organism is regenerated from each transformed cell (or egg, in the case of animals), which has taken up the foreign genes. And from that organism, **a transgenic variety** can be bred. In this way, genes can be transferred between distant species, which would never interbreed in nature.

Development of Transgenic Fishes

Development of transgenic fish has focussed on a few species including salmon, trout, carp, tilapia and a few others. Salmon and trout are cash crops while the others primarily provide sources of protein. Currently, about 40 or 50 labs around the world are working on the development of transgenic fish. About a dozen of them are in the U.S., another dozen in China, and the rest in Canada, Australia, New Zealand, Israel, Brazil, Cuba, Japan, Singapore, Malaysia, and several other countries. Some of these labs are associated with companies that expect to commercialize their fish in a few more years (Dunham, 1999). Many of the fish under development are being modified to grow faster than their wild or traditionally bred aquaculture siblings.

Faster growth is usually accomplished by transferring a **fish growth hormone gene** from one species of fish into another. The faster growing fish not only reach market size in a shorter time, they also feed more efficiently. Trout **growth hormone** (GH) was used to produce transgenic carp with improved dressing properties. Such transgenic carp are recommended for production in **earthen ponds**.

Transgenic Salmon : The Atlantic salmon is engineered with a pacific salmon, growth hormone driven by the arctic antifreeze promoter gene. The rapid growth of that transgenic salmon is achieved, not so much by the transgenic growth hormone as by the antifreeze gene promoter that functions in the cool water desirable for salmon flavour. Devlin *et al.*, (1994) research scientists with Fisheries & Oceans, Canada, in West Vancouver, British Columbia has modified the growth hormone gene in coho salmon by developing a gene construct in which all the genetic elements are derived from sockeye salmon. The transgenic coho grew on average 11 times faster than unmodified fish and the largest fish grew 37 times faster. The growth hormone levels in the transgenic fish are high year-round, rather than falling off in the winter as occurs in ordinary salmon. Devlin *et al.*, (2001). The modified salmon are large enough to be marketed after one year, in contrast to standard farmed salmon that do not reach market size for at least three years.

Transgenic Tilapia : Tilapia fish, native to Africa, are cultured world wide as "poor mans food", second only to carp as warm water food fish, and exceeding the production of Atlantic salmon (whose market value is twice that of tilapia). Tilapia has been extensively genetically modified and promoted as a transgenic fish exclusive for isolated or contained production (Hallerman, 2002). Transgenic tilapia, which is modified with **pig growth-hormone**, has three times larger than their non-transgenic siblings (Rahman *et al.*, 1980). Tilapia genetically modified with human insulin grew faster than non-transgenic siblings, and could also serve as a source of islet cells for transplantation to human subjects.

Transgenic Medaka fish : Purdue animal scientist Muir and Howard (1999) used tiny Japanese fish, *Oryzias latipes* called medaka to examine what would happen if male medakas genetically modified with growth hormone from Atlantic salmon. Inserting a gene construct consisting of the human growth hormone driven by the salmon growth promoter into medaka produced the transgenic medaka. The viability of groups of modified and conventional fish was measured at three days of age, and 30 percent fewer transgenic fish survived to that age. The researchers calculated that large males had a four-fold mating advantage, based on observations of wild-type medaka.

In another experiment Silk moth genes were introduced into Medaka fish to create resistance to bacterial pathogens.

Transgenic Zebra fish : The tiny zèbra fish (*Brachydanio rerio*) that lives in aquariums, was genetically modified to produce a fluorescent red pigment, and is being promoted for sale as a household aquarium pet, the "glofish". The glofish caused a stir in the United States because regulation of such transgenic pets is murky and none of the major regulatory agencies : Food and Drug Administration (FDA), United States Department of Agriculture (USDA) or Environment Protection Agency (EPA), has been willing to take the lead in regulating the glofish (even though USDA does deal with pet animals). The glofish is available for sale from January 5, 2004 without regulatory approval in United States (Fig. 43.1).

Gong *et al.* (2003) developed novel varieties of the Zebra fish. Three "living colour" fluorescent proteins, green fluorescent protein (GFP), Yellow fluorescent protein (YFP), and red fluorescent protein (RFP or dsRed), were expressed under a strong muscle-specific mylz2 promoter in stable lines of transgenic zebra fish. These transgenic zebra fish with vivid fluorescent colours (green, yellow, red or orange) fluorescent proteins can be seen with naked eyes under both daylight and ultraviolet light in dark. The green fluorescent protein (GFP) is originally isolated from the jellyfish *(Aequorea tictoria).*

Transgenic common carp: Thomas T. Chen, director of the Biotechnology Centre at the University of Connecticut, Storrs, transferred into common carp the growth hormone DNA from rainbow trout fused to a sequence from an avian sarcoma virus. The genetic material was injected into fertile carp eggs with microinjection. The offspring of the first generation of transgenic fish grew 20 to 40% faster than their unmodified siblings. Chen is also developing transgenic catfish, tilapia, striped bass, trout, and flounder.

Fig. 43.1 : Zebra fish Glofish genetically modified to produced fluorescent red pigment

Research associate Amy J. Nichols and Professor Rex Dunham (1999) in the department of fisheries and allied aquaculture at Auburn University, Auburn, Ala., have developed transgenic carp and catfish that grow 20 to 60% faster than standard farmed varieties. They use **microinjection** and **electroporation** to inject another copy of a fish growth hormone gene into fertile fish eggs. The growth of the resulting modified carp and catfish is stimulated by extra fish growth hormone.

Development of Transgenic Fishes in India

In India, research in transgenic fish was initiated in Madurai Kamaraj University (MKU), Centre for Cellular and Molecular Biology (CCMB), Hyderabad and National Matha College, Kollam with borrowed constructs from foreign scientists. The first Indian transgenic fish was generated in MKU in 1991 using borrowed constructs. Scientist in India has developed experimental transgenic of rohu fish, zebra fish, catfish and singhi fish.

Genes, promoters and vectors of indigenous origin are now available for only two species, namely rohu and singhi for engineering growth.

Transgenic rohu recently produced from indigenous construct at Madurai Kamaraj University has proved to be eight times larger than the control siblings. This transgenic rohu attains 46 to 49 grams body weight within 36 weeks of its birth.

Autotransgenesis : Indian scientists are concentrating on developing transgenic fish through autotransgenesis which involves just increasing the copies of growth hormone genes present in a fish as opposed to allotransgenesis which amounts to transfer of genes from different species.

The increase in growth homone genes leads to an increase in flesh content. Indian scientists feel that autotransgenesis is safer and less controversial. According to T.J. Pandian of the school of biological sciences in the Madurai Kamaraj University, the generation time of most fish species is shorter and breeding frequency is relatively higher. A single female can produce several hundred or thousand eggs and thus provide a larger number of genetically identical eggs. Besides, the most important advantage is that the fertilization is external and can be readily controlled by experimental manipulation.

According to Pandian, "the limited availability of transgenes of piscine origin had been the major hurdle in production of transgenic fish. However, with advancements in molecular biology, more than 8500 genes and cDNA sequences of piscine origin have been isolated, characterized and cloned in the world."

Controlled Culture of Transgenic Fish and Feed

Pond commercial culture is effective for carp and tilapia, but more difficult with salmon and trout. Currently, pond culture is suitable for carp and tilapia because the fish are vegetarians, carnivorous salmon and trout depend on a diet of fish and fishmeal but the worldwide stock of feed fish has diminished and suitable vegetable meat substitutes must be found. Atlantic salmon (as typical cold water carnivores) cannot thrive on a diet of rapeseed oils but the fish can achieve maturity if finished with fish oils at least 20 weeks near the end of their maturity cycle. GM oil rapeseed with enhanced production of long chain fatty acids are proposed to serve as feed for pond cultured fish. And glyphosate-tolerant GM canola meal has been pronounced substantially equivalent to non-GM canola as feed for rainbow trout.

Gene Transfer Technology for Development of Transgenic Fishes

The most commonly used methods in fish biotechnology are chromosome manipulation and hormone treatments, which can be produced triploid, tetraploid, haploid, gynogenetic and androgenetic fish. Other popular methods of gene transfer in fish are microinjection, electroporation of sperms, electroporation of eggs and incubation of sperms. Following are the main steps in gene transfer for development of transgenic fish.

Preparation of DNA Construct

Desired transgene should be a recombinant gene or DNA construct, which is constructed in plasmid that contains an appropriate promoter-enhancer element and a structural DNA sequence. The foreign genes are typically introduced with strong genetic signals, **promoters and/or enhancers**, which enable the foreign genes to be expressed at very high levels continuously (or constitutively), effectively placing those genes outside the normal metabolic regulation of the cell, and of the transgenic organism resulting from the transformed cell. There are three main types of transgenes:

(1) **Gain-of-function**—These transgenes are able to increase particular function in transgenic individual after their expression. For example growth hormone genes from mammal and fish linked to appropriate promoter-enhancer element and a structural DNA sequence to produce GH transgene. This GH transgene when express in transgenic individuals increases production of growth hormone leading to enhanced growth of transgenic animal.

(2) **Reporter function** —These transgenes are able to identify and measure the strength of promoter-enhancer element.

(3) **Loss of function**—This transgene is not yet used for modification of transgenic fish. Such transgenes are used for interfering with the expression of host genes.

The promoter-enhancer elements of transgenes are linked to a growth hormone gene of fish. Hence transgenic fish contain extra DNA sequences that are originally derived from same species. Gene construct is then introduced into fertilized egg or embryo, so that transgene be linked to genome of each cell of egg or embryo.

Gene Transfer by Microinjection

Microinjection is most successfully and widely used technique for gene transfer in fish. One method of microinjection technique involves the use of fine injection needle for introducing DNA into cut site in the cell. While doing so it destroys those cells that are in direct contact with the injected DNA. To ensure the integration of the DNA it should be injected to intact cells close to the cut site. The injection apparatus consists of a dissecting stereomicroscope and two micromanipulators, one with a glass micro-needle for delivering transgene and other with a micropipette for holding fish embryo in place (Fig. 43.2).

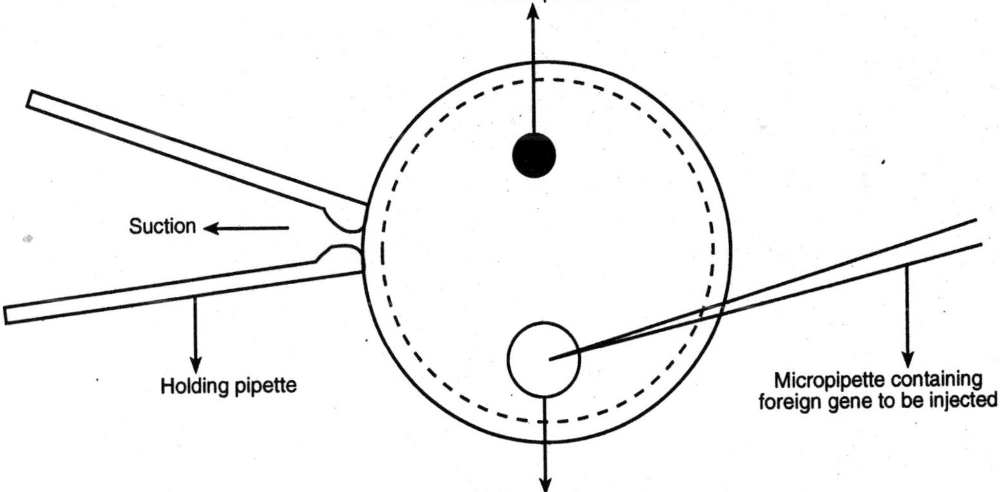

Fig. 43.2 : Micro injection appartus to introduce DNA in intact Cell

The success of microinjection technique depends on the nature of egg chorion. The soft chorion facilitates the microinjection while the thick chorion limits the ability to visualize the target for injection of DNA. In many fishes (Atlantic salmon and rainbow trout) the egg chorion gets tough and hard just after the fertilization or to contact with the water and provides a difficulty in injecting the DNA. But using the following methods can solve this problem :

(1) By using the micropyle (an opening on the egg surface for sperm entry during the fertilization) for inserting the injecting needle.

(2) By using microsurgery for making an opening on the chorion.

(3) By digesting the chorion with enzymes.

(4) By using 1mM glutathione for initiating fertilization and reducing hardness of chorion.

(5) By direct injection to the unfertilized eggs.

Another technique of gene transfer is intra-nuclear microinjection, which involves direct physical approach using a fine needle to deliver DNA into cell or even nuclei.

To facilitate rate of microinjection protoplast with partially reformed cell wall may be attached to a solid support with artificially bound substrate without damaging the cells. Solid support may be of either glass cover slips or slides.

Steps of Microinjection Technique

(1) Desired eggs and sperms are stored separately at the optimum conditions.

(2) Add water and sperms and initiate the fertilization.

(3) Ten minutes after the fertilization, eggs are dechorionated by trypsinization.

(4) Fertilized eggs are microinjected with desired DNA just within a few hours of fertilization. DNA is released into the centre of the germinal disc to the first cleavage in dechorionated eggs. The time available for microinjection is first 25 minutes and that too between fertilization and first cleavage.

(5) After microinjection the embryos are incubated in water until hatching takes place.

Survival rates of microinjected fish embryos is seem to be about 30-80% depending the fish species.

Advantages of Microinjection Technique

This technique has the following merits.

(1) Optimum quantity of DNA can be delivered per cell, increasing chances for integrative transformation.

(2) The delivery of DNA is precise, even into nuclei of target cell again improving chances for integrative transformation.

(3) The small structure can be injected.

(4) It is a direct physical approach, hence it is a host range independent.

Disadvantages of Microinjection Technique

(1) A single cell can be injected at a time, hence it is time consuming process.

(2) It requires sophisticated instruments and specialized skills.

(3) Limited embryonic time restricts injection to more eggs and a low transformation rate.

Gene Transfer by Electroporation

It is a simple, fast, efficient and convenient method for transferring gene. This method involves an electrical pulse to deliver DNA into cells (Fig. 43.3). The cells are exposed to a short electrical shock, which make the cell membrane temporarily permeable to DNA. The desired DNA fragment is placed in direct contact of protoplast membrane, which enters into the cell upon electric shock. Hole may be created as a result and stabilized by a favourable dipole interaction with electric field. Electroporation involves a chain of electrical pulses for permeation of cell membrane, thereby allowing the entry of DNA into fertilized eggs. The rate of DNA integration in electroporated embryo is more than 25% is the surviving rate, which is slightly higher in comparison of microinjected ones.

Advantages of Electroporation Technique

(1) It allows simultaneous entry of DNA constructs.

(2) It is more suitable method for those species, which has very small eggs for microinjection.

(3) This method does not require specialized skill.

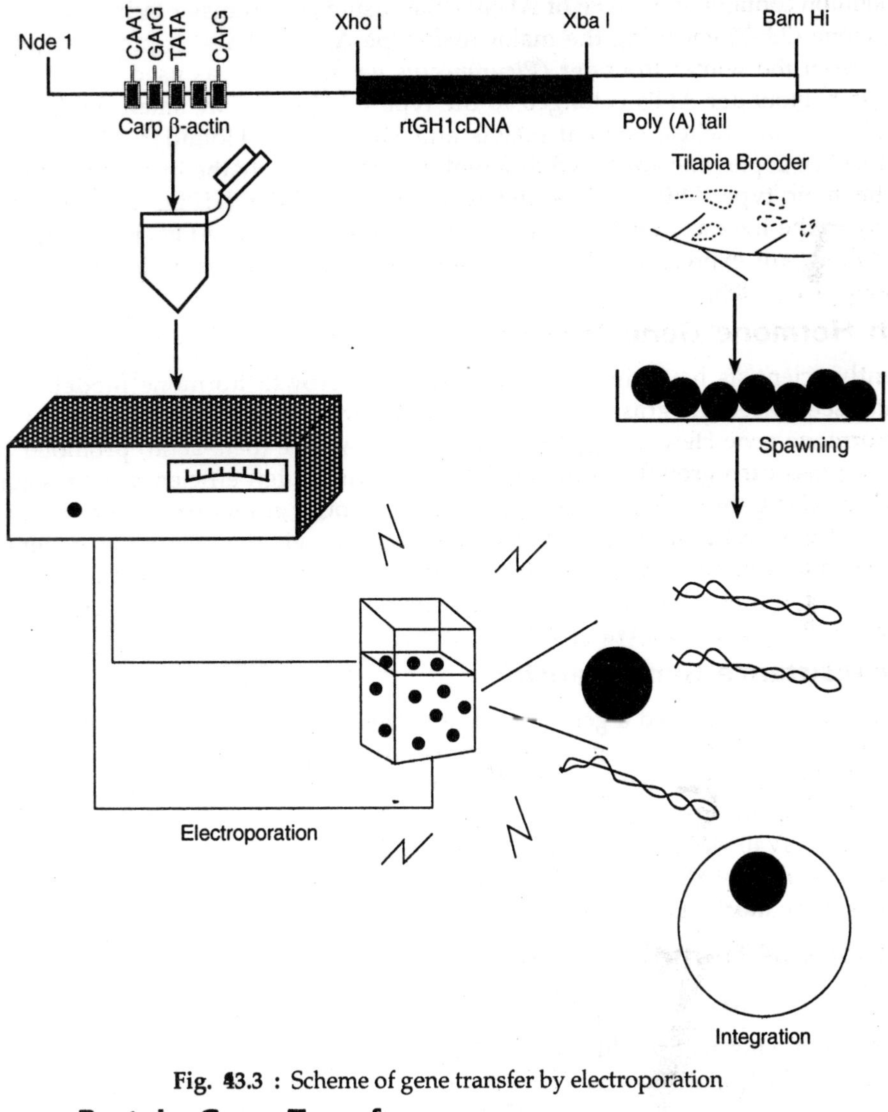

Fig. 43.3 : Scheme of gene transfer by electroporation

Antifreeze Protein Gene Transfer

Many teleost inhabiting icy marine water in the Polar Regions produce antifreeze glycoproteins (AFGPs) or antifreeze proteins (AFPs) in their sera to protect them from freezing. This protein lowers the freezing temperature of solution without altering its melting temperature. Thermal hysteresis, the difference between the freezing and melting temperature, is a unique property of these proteins. AFPs and AFGP have been demonstrated to bind to ice crystals and inhibit ice crystal growth. Despite their similar antifreeze properties, these proteins are quite different in their protein structures. There are one type of AFGP and three types of AFP (Davies and Hew, 1990). Recently fourth type of AFP has also been reported in longhorn sculpin.

The Atlantic salmon *Salmo salar*, lacks any of these AGFPs or AFPs gene(s) and are unable to survive in sub-zero sea water temperature (Hew *et al.*, 1999). An inability to tolerate temperature below – 0.6 °C to – 0.80 °C is one of the major problems of sea cage farming in Northern Atlantic coast. Hew and his co-workers developed antifreeze-resistant Atlantic salmon containing the AFP or AFGP genes using gene transfer technology. They used genomic clone (2A-7) encoding the major liver-type AFP (wflAFP-6, previously known as (HPLC-6) from the winter flounder *(Pleuronectus amaricanus)* was used as a candidate for gene transfer. Flounder AFPs belonged to the type I AFPs that are small polypeptides and high in alanine and helical content (Davis and Hew, 1990). Flounder AFPs is multigene family of 80-100 copies encoding two different isoforms, namely the liver type and skin type AFPs. The liver type AFPs such as wflAFP-6 or wflAFP-8 (HPLC-8), are synthesized exclusively in the liver as prepro AFPs. In contrast the skin-type AFPs, including wfsAFP-2 and wfsAFP-3, are expressed widely in many peripheral tissues as intracellular mature AFPs (Gong *et al.*, 2003).

Growth Hormone Gene Transfer

Recently scientists have developed an "all fish" growth hormone model. They have cloned and sequenced the grass carp and common carp carbonic anhydrase (CA) gene and growth hormone gene Hew *et al.*, (1992). The grass CA gene (beta-actin) promoter has been linked to a grass carp growth hormone cDNA to form a high efficiency expression vector called pCAZ. Using the CAT gene as receptor gene, a pCA grass carp growth hormone was microinjected into fertilized, non-activated common carp via the micropyle, generating "all fish" transgenic carp. The presence of transgene was detected by reverse transcriptase PCR and Northern blotting. These transgenic fish showed about 137% high growth rate of the control (Zhu, 1992, Zhu 1996, Wang 2001).

Disease-resistance Gene Transfer

In China scientists piloted a gene contributing resistance to the grass carp haemorrhagic virus (GCHV). Eleven different gene fragments encoding protein was cloned and isolated from translation in vitro using GCHV genomic single gene fragments. Based on the information of capsid protein SP6 and SP7 gene cDNA, 3 oligonucliotides were synthesized and fused with SV40 MT promoter and transferred into grass carp cytokine-induced killer (CIK) cells via a constructed expression vector and transfected with GCHV. The result indicated that the mortalities were reduced by one order after challenge with the virus.

Applications of Transgenic Fish

Transgenic Fish may be better used for the following purposes.

(1) For increasing fish production to meet the growing due to demand of food due to increase in world population.

(2) For production of pharmaceutical and other industrial products from piscine origin.

(3) For development of transgenic native glow fish varieties for aquarium.

(4) As fish biosensors for monitoring aquatic pollution.

(5) For isolation of genes, promoters and synthesis of effective gene constructs.

(6) For researches in embryonic stem cells and in-vitro embryo production.

(7) For production of anti-freeze protein.

Environmental Concerns about Transgenic Fish

The primary environmental concerns about releases of transgenic fish, for example, include competition with wild populations, movement of the transgene into the wild gene pool, and ecological disruptions due to changes in prey and other niche requirements in the transgenic variety versus the wild populations (Hedrick, 2001 and Stockstad, 2002).

Transgenic Fish could Threaten Wild Populations

West Lafayette, Ind. — Purdue University researchers have found that releasing a transgenic fish to the wild could damage native populations even to the point of extinction. Transgenic fish could present a significant threat to native wildlife (Gross, 2002). "Transgenic fish are typically larger than the native stock, and that can confer an advantage in attracting mates", Muir says. "If, as in our experiments, the genetic change also reduces the offspring's ability to survive, a transgenic animal could bring a wild population to extinction in 40 generations" (Muir and Howard, 2002).

Although at Canadian research facilities, elaborate precautions are being taken to prevent the release of transgenic fish into the environment. The fish are often raised in ponds covered with nets to keep birds out; enclosed by electric fences to keep muskrats, raccoons, and humans out; and the outlets are fitted with screened drains to prevent the loss of small fishes or eggs.

Gene Flow

One of the larger environmental concerns raised by transgenic fish is the possibility that a transgenic species raised in open water pens will escape and spread novel traits into the ecosystem by breeding with wild relatives, a biological process known as "gene flow." Gene flow between transgenic or conventionally bred fish and wild populations is an environmental concern, because it may present a threat to natural biodiversity. Some researchers believe that the genetic differences introduced to a transgenic fish may impact its net fitness, a scientific term meaning an organism's ability to survive and pass its genes to future generations. The concept, which factors in characteristics such as the juvenile and adult viability of a fish, the number of eggs produced by a female, and the age at which a fish reaches sexual maturity, provides a useful barometer for discussing some gene flow scenarios (Anderson, 1986).

According to one scientific model, if a transgenic fish escapes and mates with a wild fish, gene flow could follow one of three scenarios :

Purge Scenario — When the net fitness of a transgenic fish is lower than that of its wild relatives, natural selection will quickly purge from the wild population any novel gene(s) introduced by the transgenic fish. In theory, evidence of the novel trait will disappear from subsequent generations.

Spread Scenario — When the net fitness of a transgenic fish is equal to or higher than the net fitness of a wild mate, gene flow is likely to occur and the genes of the transgenic fish will spread through the wild population. This means evidence of the transgenic genome would persist in subsequent generations.

Trojan Gene Scenario —When the net fitness of a transgenic fish is altered such that the fish has enhanced mating success but reduced adult viability (i.e., chances of surviving long enough to mate), introduction of that fish into the wild population could result in a rapid decline of the wild population (Muir and Howard, 1999). Essentially mating success would ensure the spread of the novel gene throughout the population, but the inability to survive would reduce the population size of subsequent generations and potentially lead to extinction.

A declining fish population would also have secondary impacts on other aquatic species that feed on, or otherwise depend on it. Populations unable to successfully switch over to another food source, or those whose survival or reproduction depends directly on the declining population, would also suffer.

Transgenic Fish Invasive Species

Even if they do not breed with wild relatives, transgenic fishes that escape into natural ecosystems could be an environmental nuisance by becoming an invasive species. This danger mainly arises for those transgenic fishes endowed with new genes that improve such fitness traits as breeding capabilities and the ability to withstand harsh conditions. The establishment of a thriving transgenic fish population in an ecosystem where it has never existed could crowd out native fish populations.

Risk Mitigation

It is important to note that developers of transgenic fish are attempting to reduce or eliminate both gene flow and invasive species risks by sterilizing transgenic fish. Sterilization is relatively easy and inexpensive but success rates are highly variable. In addition, sterilization does not necessarily neutralize environmental risks (Bright, 1996). Academic scientists note that an escaped, sterile fish might still engage in courtship and spawning behavior, disrupting breeding in wild populations. Waves of escaped sterile fish could also create ecological disruptions as each group is replaced by another equally strong group of transgenic sterile fish (Kapuscinski and Hallerman, 1991).

Food Safety Issues

One important food safety issue involves the extent to which fish absorbs and stores environmental toxins, such as mercury, high levels of which could pose a danger to humans who eat the contaminated fish (Maclean and Laight, 2000). Some scientists worry that discrete biological changes induced by the genetic engineering process might enable transgenic fish to absorb a toxin that conventional fish cannot absorb or to better tolerate higher levels of a toxin already known to cause concern. Some scientists have expressed concern that the genetic engineering process could increase the allergic potential of fish, particularly through the introduction of novel proteins that never before existed in the food chain. However, it is equally possible that genetic engineering will form their diet.

Genetically engineered plant crop had faced protest in various countries regarding safety for food and environment. There is a need for the regulating transgenic animals for debate.

REFERENCES

Anderson, M. (1986). Evolution of condition dependent sex ornaments and mating preferences: sexual selection based on viability differences. Evolution, 40: 804-816.

Bright, C. (1996). Understanding the threat of biological invasions. In: Streke L. (ed.),

Devlin, R.H.; Baiagi, C.A.; Yesaki, T.Y.; Smailus, J. C.; and Byatt, J.C. (2001). Growth of domesticated transgenic fish. Nature, 409: 781-782.

Devlin, R.H.; Yesaki, T.Y.; Baiagi, C.A.; Donaldson, E.M.; Swanson, P.; Chan, W.K.; (1994). Extraordinary salmon growth. Nature, 371, 209.

Du, S. J.; Gong, Z.; Fletcher, F.L.; Shears, M.A.; King, M.J.; Idler, .D.R.; Hew, C.L. (1992).

Dunham, R.A. (1999). Utilization of transgenic fish in developing countries: Potential benefits and risks. J. World aquacult. Soc. 30 : 1-11.

Gong, Z.; Wan, H.; Tay, T.L.; Wang, H.; Chen, and Yan, T. (2003). Development of transgenic fish for ornamental and bioreactor by strong expression of fluorescence protein in the skeletal muscle. Biochemical and Biophysics Research Communications 58-63.

Gross, M. R. (2005). Will farmed Atlantic salmon invade the ecological niches of wild Atlantic salmon? In : Gallaugher P and Orr C. (eds.) Aquaculture and protection of wild salmon. 25-28, Simon Frasier University Cont. Stud. Sci. Burnaby, British Coloumbia.

Growth enhancement in Atlantic salmon by the use of an "all fish" chimeric growth hormone gene construct. Biotechnology 10: 176-180.

Hallerman, Eric M. (2002). "Genetically Modified Fish and Shellfish: Food for Thought." Virginia Issues and Answers 8.1, Winter Virginia Tech Publications & Outreach Communications.

Hedrick, P.W. (2001). Invation of transgenes from salmon or other genetically modified organism into natural populations. Can. J. Fish Aquat. Sci. 58: 841-8440.

Hew, C.; Poon, R.; Xiong, F.; Gauthier, S.; Shears, M.; King, M.; Davis, P.; and Fletcher, G. (1999). Liver specific and seasonal expression of transgenic Atlantic salmon harboring the winter flounder antifreeze protein gene. Transgenic Research 8 : 405-414. Academic Publishers, Printed in Netherlands.

Hew, C.L.; Fletcher, G.L. and Davis, P.L. (1992). Transgenic Salmon: tailoring the genome for food production. J. Fish Biology 47: 1-19.

Kapuscinski, A.R. and Hallerman, E.M. (1991). Implications of introduction of transgenic fish into natural ecosystems. Can J. Fish Aquat. Sci. 48 : 99-107.

Maclean, N. and Laight, R.J. (2000). Transgenic fish: an evaluation for benefit and risks. Fish Fish 1: 146-172.

Muir, W. M. and Howard, R.D. (1999). Possible ecological risks of transgenic organism release when transgenes affect mating success : sexual selection and the Trojan gene hypothesis. Proc. Natl. Acad. Sc. 96 : 13853-13856.

Muir, W.M., and Howard, R.D. (2002). Assessment of possible ecological risks and hazards of transgenic fish with implications for other sexually reproducing organism. Transgenic research 11: 101-114.

Rahman, M.A.; Mak, R.; Ayed, H.; Smith, A., and Maclean. (1980). Expressions of a noval piscine growth hormone gene result in growth enhancement in transgenic Tilapia *(Oreochromis mosambicus)*. Transgenic Res. 7: 357-369.

State of World. A World watch Institute Report on Progress of a Sustainable Society. pp. 95-113.

Stockstad, E. (2002). Engineered fish: friend or foe of the environment? Science 1797-1798.

Wang (2001). Construction of all-fish gcACGH carp and genetic analysis of its F_1 generation. Chinese Science bulletin 46 (3), 226-229.

Zhu, Z. (1992). Growth hormone gene and transgenic fish. Agricultural biotechnology. (Edts. C.B. You and Z.Z. Chen.). China Science and Technology Press, Beijing. pp. 106-116.

Zhu, Z. (1996). Foreign gene introduction via fish germ mediated technique. Chinese J. of Hydrobiology. 20 (3): 242-247.

Chapter 44

GYNOGENESIS

The gynogenesis is defined as the process of development of ovum after penetration of sperm, but in the process the sperm nucleus is unable to form syngamy with the nucleus of ovum. So gynogenetic fish inherit genes only from mother. The penetration of sperm is responsible for initiation of development alone. Gynogenetic haploid embryos contain a haploid chromosome complement representing a single product of a meiotic division, one quarter of a tetrad embryo do not develop entirely normally.

Gynogenetic diploid embryo contains a diploid chromosome complement, also derived from mother only and are obtained by treatments to prevent either the second division in an egg fertilized with UV-irradiated sperm.

In fish, the gynogenesis is achieved both naturally and artificially. In some fishes, gynogenesis is a natural mode of reproduction, these unisex fish may be often triploid. In a few fishes such as silver crucian carp (Prussian crucian carp of European goldfish) and *Poeciliopsis*, unisex form present in natural population.

Population, which are produced by natural gynogenesis (Cherfas, 1969 and Schultz, 1967). Artificial gynogenesis can also be produced under suitable experimental conditions. During artificial gynogenesis, manipulations in event of fertilization are made by destroying DNA of the spermatozoa and to restore diploidy to the ovum with either meiosis or cell cleavage. In this process the male inheritance is excluded, the offspring obtained in this way possesses inheritance of female characters and they are of only one sex, i.e., female. The unisexual population obtained by artificial gynogenesis are used for fisheries management.

Gynogenesis can also be used to produce clonal lines of tilapia (Sarder *et al.*, 1999). Hybrid among such inbred lines are expected to exhibit significant heterosis and uniform performance with respect to growth rate and body conformation. The most important breeding goal of tilapia is to improve growth rate and feed conversion efficiency, which are prime factor regulating profitability of the industry. Sex determination mechanisms, genetically male tilapia have higher growth rates than even hormonally reverse female (Mair *et al.*, 1995).

Ioannise *et al.*, (2004) observed that gible carp, *Carassius gibelio* exhibits gynogenetic reproduction by utilising the sperm of other species which resulted in the increase population in the lake as gible carp females have little opportunity for mating with gibel carp male. According to them Ylikiensis roach is the best available sperm donar option for gynogenetically gibel carp of the lake. These authors made artificial fertilization of gibel carp eggs with sperms of other species such as ylikiensis roach (CrXr), gold fish (Cr.XG) adb common carp (CrXC), hatching occurred after three days at 14 °C in all groups.

METHOD FOR PRODUCING GYNOGENETIC FISHES

There are large numbers of ways available for the initiation of gynogenesis in the development of embryo. The classical method is to prick each ovum with a needle dipped in serum of blood. According to Lestage (1933), the weak electrical currents if passed through ovum may result in parthenogenesis. In artificial gynogenesis sperm with denatured DNA is successfully used. The denaturing of DNA in spermatozoan is carried out by the following methods :

1. The spermatozoa are first subjected to high dose of radiation which destroy DNA without seriously altering cytoplasmic components. Such spermatozoa after penetration into ovum initiate the development process and result in gynogenesis. Doses of 100 kiloroentgens of X-rays are recommended (Golovinskaya *et al.*, 1963). Purdon (1969) and Purdon and Lincoln (1974) recommended gamma radiation from cobalt.
2. Ultraviolet radiation of 15 W sterilization has been used to inactivate DNA of frog spermatozoa (Beathy, 1964) and can also be used for fishes.
3. In addition to these, dyes such as trypaflavine toluidine blue and thiazine are also used to denature DNA of fish spermatozoa.

As male inheritance part has been removed, so the gynogenetic individual must be haploid, but haploid individuals fail to develop beyond the larval stage (Stanley and Sneed, 1974). In artificial gynogenesis many haploid but only a few diploid individuals are obtained. In natural gynogenesis, on the basis of principle that they receive only maternal genetic material hence must be haploid. But it is surprising to note that in gynogenetic fishes, if polyploid chromosomes number are formed it is constant. Why these fishes have constant triploid chromosome? Chromosome is summarised by Stanley and Sneed (1974). They suggested four mechanism (Fig. 44.1A-D).

In *Poeciliopsis*, the gynogenesis is a natural one. The number of chromosome is triploid. In this method the chromosome replication occurs without cleavage. In Fig. 44.1A the oogonium contains triploid number. Then the chromosomes become double without undergoing cleavage, the number of chromosome becomes 6n they are reduced to half *i.e.* 3n in meiosis during cleavage.

In *Carassius*, the silver crucian carp is gynogenetic triploid (Fig. 44.1B), the chromosome number replicate by the process of endomitosis. In such fish the first mitotic division does not occur, hence number is not reduced, hence at cleavage each daughter cell retained triploid (3n) number.

In third case the chromosome number is diploid. This is present in *Misgurnus.* Here chromosomes undergo replication by endomitosis. They become 4n. In this Case for restituting chromosome number is the combination of the 2nd polar body with the female pronucleus, which is equivalent to failure of the second meiotic division. Such mechanism has been reported in loach by Ramashov and Belyaeva (1964) and Purdon (1969) (Fig. 44.1C).

In the fourth mechanism, meiosis occurs but replication of chromosomes without cleavage during the first mitosis restores diploidy, as noticed in silver crucian carp (Lieder 1955).

Artificial gynogenesis in Indian carp has been successfully carried out in Central Institute of Freshwater Aquaculture of Bhubaneshwar. Gynogenetic *rohu* and *catla* were produced (John *et al.*, 1988, 1989).

Success has been achieved in obtaining gynogenetic offsprings in *Cyprinus carpio* (Golovinskya, *et al.*,1969) and grass carp (Stanley and Sneed, 1974).

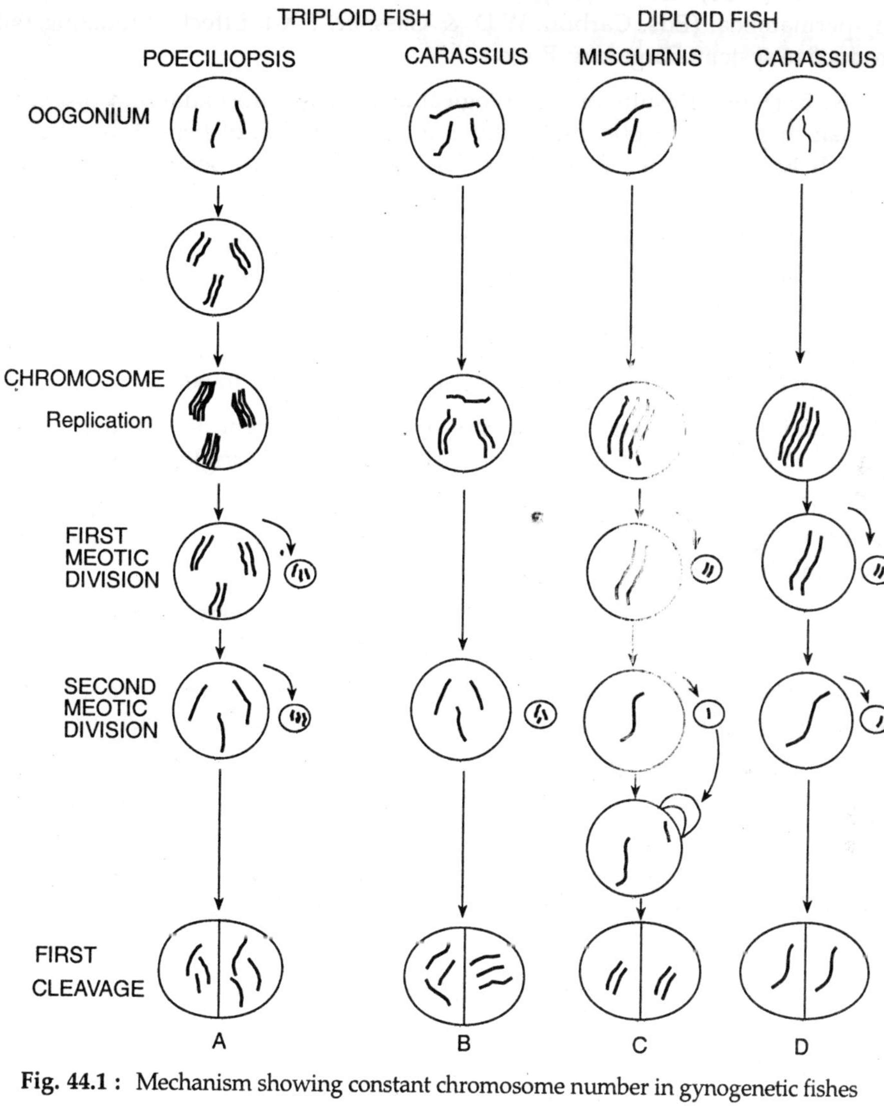

Fig. 44.1 : Mechanism showing constant chromosome number in gynogenetic fishes

Uses : It is useful in genetic studies and selection breeding Gynogenetic offsprings are useful for selecting uncommon recessive traits or new mutants.

Artificial gynogenesis is useful in producing inbred lines which can subsequently be crossed to produce hybrid vigour.

It is a useful tool to obtain homozygous females, and different lines of gynogenetic fish might be crossed to produce heterosis in the offspring.

It will help to control over reproduction in natural populations. Overpopulation due to excessive spawning leads to stunted fish and gynogenesis will help in complete elimination of reproduction and thus helps in regulation of population size.

factors, individual male and female may have very different genotype with respect to sex determination. According to Kallman (1984), the theory was never critically tested, nevertheless it has been accepted by most investigators in this field.

Kallman (1968) pointed out that autosomal factors in southern platyfish (*Xiphophorus maculatus*) cannot be sex gene because they interact with specific independently segregatting sex chromosome to bring about their effect. Kallman (1968) designated them as transfer gene of sex or regulatory gene.

Similarly, Avtalion and Hammerman (1978) have proposed a model of sex determination for *Sarotheroden* (Tilapia) that involves a pair of gonosomes and single autosomal locus. Each species is homozygous for autosomal factor, but different alleles have become fixed in species with male and female heterogamety.

2. XX and XY mechanism : A very clear pair of sex chromosomes has been reported in male deep sea fish, *Bathylegus wesethi,* the largest in Karyotype is presumably X and the smallest Y. In *Dasyatis sabina* is the only cartilaginous fish, which has differentiation of sex chromosomes and autosomes.

In about 48 species the XX and XY mechanism is present. The female is homogamatic (XX) while heterogamety is normally present in male. In about 15 forms, XX and XO types of mechanism is present in karyotype. The females have XX and males have XO. Female heterogamety of ZZ and ZW has been maintained for 25 species and ZZ and ZO for 2 forms (Rishi, 1989, Ojima, 1985).

In platyfish (*Xiphophorys maculatus*) three sex chromosomes are distinguished W, X and Y, having Wyy, WX and XX which cause female. This is due to gene mutation or chromosomal rearrangement. Due to the presence of certain regulatory genes XX, WY, WX and WW platyfish can develop into functional male.

Kallman (1984) and Thompson (1978) found that male determining gene is present on all sex chromosomes. In species with XX-XY mechanism, it is a constitutive locus on Y, but a facultative locus on X (and W), while it has to be induced by appropriate signals. It follows that operator like control elements, O^1, O^2, O^3 adjecent to the male determining gene on the X, W and Y chromosomes cannot be identical. In this model the control loci play the key role in sex determination (Fig. 45.1).

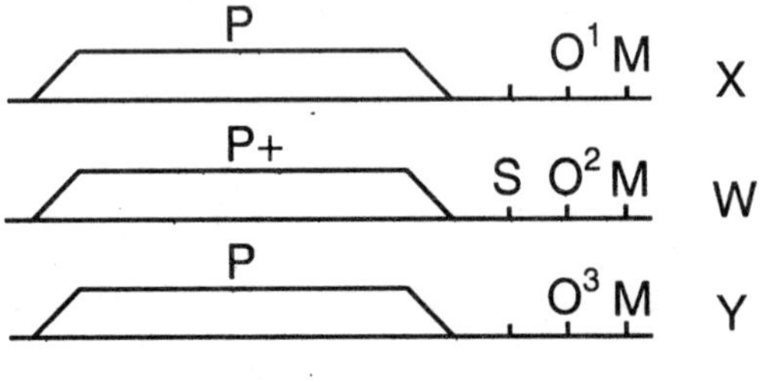

Fig. 45.1 : Male determining gene M. is present on X, W, Y chromosomes. M, male determining gene; O, operator like gene; X chromosome; W & Y chromosomes.

The male determining gene on the sex chromosome could be the H-Y locus. Even if H-Y locus is autosomal, Kallman theory will not change much in any greater detail. The male determining gene on the sex chromosome would then become an activator for autosomal H-Y locus. The differentiation of XX, WX, WW platyfish into functional males to effect the mutant autosomal regulatory gene that activates the male determining gene or X and W chromosome.

In hermaphrodite fishes, Peter (1964) and Kosswig (1964) stated that in sword tail, *Xiphophorus helleri* has polyfactorial basis and that in late maturing males the sum of the valencies fairly predominates over sum of female valencies, whereas in early maturing males the ratio of male to female valency is much larger.

MULTIPLE MECHANISMS

Multiple sex elements related to male heterogamety have been found in 11 forms and those of female heterogamety in one species. They are XI X1 X2 X2 : X1 X2 Y (Rishi, 1989; Ojima, 1985) X1 X2 Y sex chromosomes have also been found in male ZO, ZW and ZZ 1 females were also present (Manna, 1989).

The above phenomenon is due to varied processes of reproductive system. Fishes are unisexual, bisexual, hermaphrodite and gonochorian.

Table 1 : Identified sex chromosomes XO, XY, X1 X2 Y in male heterogamety and ZO, ZW in female heterogamety (Manna, 1989).

Anguilla australis	?	ZO/XO
Argentina silus	?	XY
Mystus tengara	?	XY
Trichogaster fasciatus	?	ZW
Callionymus beneteguri	?	X1 X2 Y
C.omatipinnus	?	X1 X2 Y
C. sagitta	?	XY
Carassius auratus	?	XY
Cyprinus carpio	?	XY
Monocantris japonica	?	XY
Poecilia formosa	?	ZW
Poecilia formosa	?	ZO
Xiphophorus maculatus	?	XY
Oncorhynchus nerka	?	XO
Epinephalus tauvina	?	ZW

REFERENCES

Avtalion, R.R. and Hammerman, I. S. (1978). Sex determination in *Sarotherodon* (Tilapia) 1. Introduction to a theory of autosomal influence. Bamidgeh 30 : 110-115.

Kallman, K. D. (1968). Evidence for the existence of transformer genes for sex in teleost, *Xiphophorus maculatus*. Genetics. 60 : 811-828.

Kallman, K. D. (1984). A new look at sex determination in Poecitiid Fishes. In : Evolutionary Genetics of fishes (Ed. Turner, B.J.). P. enum Press. New York & London. pp. 95-171.

Kosswig, C. (1935). Die Kreuzung Zweier XX-bZW-XY-Geschlechter miteinander under Ersastzeines Y-Chromosome einer Art Durch das X-Chromosome einer anderew. Zuchter 7 : 40-48.

Kosswig, C. (1964). Polygenic sex determination. Experientia 20 : 190-199.

Manna, G. K. (1989)–Fish Cytogenetics related to taxonomy, evolution and monitoring aquatic genotoxic agents. In : Fish Genetics in India. (Eds. Das, P. & Jhingran, A.G.) Today & Tomorrow's Printers and Publishers, New Delhi, pp. 21-51.

Ojima, Y. (1985). Chromosome data retrieval system.

Peter, G. (1964). Vergbichende Untersychungen an drei Subspecies von *Xiphophoru-shelleri* Heckel (Pisces) Z. Zool. Syst. Evolutionsforsch 2 : 185-271.

Rishi, K. K. (1989). Current status of Fish cytogenetics. In : Fish Genetics in India. (Eds. Das, P. & Jhingrin, A.G.) Today & Tomorrow's Printers and Publishers, New Delhi, pp. 1-20.

Thompson, F.H. (1978). H-Y antigen loci. Science 201-842.

Wing, O. (1934). The experimental occurrences of sex chromosomes into autosomes and vice versa, as illustrated by *Lebistis.* C. R. Trav. Lab. Carlsberg Ser. Physiol. 21 : 1-49.

Zaborski, P. (1982), Expression of H-Y antigen in non-mammalian vertebrates and its relation to sex differentiation. In : Reproductive Physiology of Fish (Eds. Richter, CJ. J & Goos, H. J. Th). Centre for Agriculture Publishing and Documentation, Wageningen. pp. 64-68.

Chapter 46

MONOSEX CULTURE AND SEX REVERSAL

The sex in fishes are genetically determined by sex chromosomes. The chromosomes are distinguished as autosome and sex chromosomes. However, this distinction on cytological basis is difficult as the chromosomes in fishes are tiny and usually numerous. The clear distinction has been noticed in deep sea fish, *Bathylegus wesenthi* where the X chromosomes in its karyotype are larger than the Y chromosomes, which is short. In mammals the female is homogametic (XX) and the male is heterogametic. In birds opposite system prevails, female (WY) and (XY) male (YY). In fishes both systems are present but the system is not differentiated at the time of hatching till the gonads are not differentiated.

An interesting phenomenon has been observed that the sex in fishes can be reversed, male can be converted to female functionally and female can also be converted into functional male, this is known **as sex reversal** or **monosex culture**.

This can be done artifically with the treatment by male hormones, androgens (include testosterone an androgenic C 19 steroid) and female hormone, estrogens. Estrogens are C 18 steroids. Although, sex reversal has many advantages in fish farming but the great disadvantages is that such hormone treated fish cannot be used for human consumption and due to government regulation on the use of chemical on food fish. The advantage is that in male fishes typically grow faster than female. Male catfish has an average 15%. larger than female. However, this difficulty could overcome by means of induced androgenesis which will be discussed later.

In many fishes sex is reversed in the following way :

In rainbow trout, Atlantic Salmon, *Coho salmon, Tilapia,* gold fish and grass carp, sex can be reversed by administration of sex hormone at early stages of development (Yamamota, 1969; Shreck, 1974; Lam, 1982; Donaldson and Hunter, 1982; Refstie, 1982; Shelton, 1982; David & Pandian, 2006). The female grass carp or the fishes which are homogamatic are treated with male hormone, methyltestosterone MT. MT causing sex reversal of female fish to phenotypic male with sex genotype as XX. This sex reversed fish is phenotypic male but genotypically female (XX) and when such fishes are crossed with normal females (XX) should produce only XX progeny (Shelton, 1982) (Fig. 46.1).

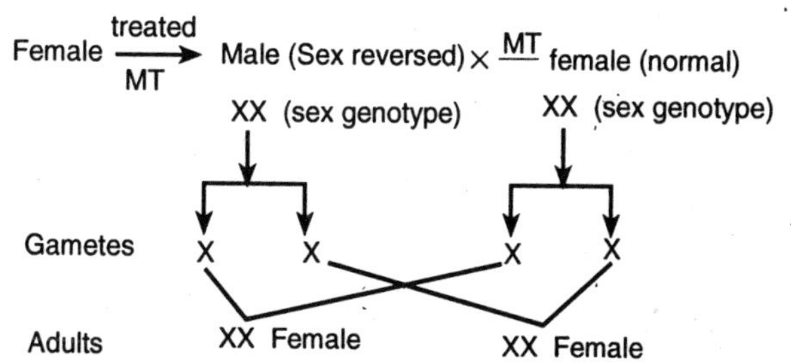

Fig. 46.1 : Mating of sex reversed male fish (genotype sex XX) with normal female showing all fishes female

Shelton (1982) made experiment in 1977, 1978, 1979, according to his experiments he obtained mature sex reversal male (XX – male) and the successfully grown 8000 to 10,000 presumed monosex fish. Yamamoto (1969) also successfully obtained monosex in *Lebestes* and *Oryzias*. He used androsterone for transfer female (XX) into functional male (XX).

He also reported that the sex of male can be reversed as functional female phenotypic with XY genotype by treatment with female hormones.

Pandian and Kiran Kumar (2003) reproted that **neuroendocrine** peptide can induced sex reversal in protogynous labrid teleost, similar to androgenic and antiestrogenic steroids. He noticed that instead of administration of dietary supplemented, it would more potent to induced sex-reversal if metabolities of 17 a-meythyltestosterone (MT) are circulated in the culture water. According to him, it would be 200-1,000 times cheaper and 90-95% steriod would be eliminated within 1-2 weeks. However, sex reversal in chinook salmon, *O. tshawytscha* require dietary supplimentation.

Among fishes hermaphroditism occur in many species. It can either be protoandrous, protogynous or synchronous. Sex reversal in certain coral reef fishes has been reported in *Labroides dimidiatus*. This fish is an ectoparasite and is known as cleaner fish because it removes ectoparasite from the skin.

The administration of hormones may be either at the eye eggs, alevins or fry. They are immersed in water containing steroid at 200 ug/l. Donaldson and Hunter (1982) gave treatment of 2 hours duration of 7 days. Fry were fed for first 90 days, a diet containing either 5 mg/kg estradiol or 10 mg/kg methyltestosterone. The estrogen and androgen treated group contained 97% success and rest as sterile fish. Both groups have similar rate of maturation. The sterile fishes grow at a slower rate.

Monosex can also be achieved by gynogenesis. *Carassius* and viviparous teleosts such as *Poecilia* can produce exclusive female forms by gynogenesis. They also occur naturally in population (Kallman, 1962; Manna 1989).

The problem of hormone treatment can be solved in homogametic female (XX) when they can be reversed to functional male and if breed with normal female (XX) resulting in obtaining all female population without the treatment with hormones. This is time consuming and expensive.

Recently, induced androgenesis i.e., progeny with only a paternal genome has been used for the production of all male progeny without the use of sex reversing steroids. At Auburn Univ., in collaboration with University of Oklahoma. YY-male (supermales) in tilapia and channel catfish could be obtained by estrogen treatment but androgenesis offers direct induction for conversion of super male YY without chemical treatment and without the need of progeny testing. In course of experiments with sex steriod hormones, peculiar results were obtained. Some cat-fish when treated with sex steroid hormones in their diet. When these fishes start taking feed, gonods of all fishes developed ovaries resulting all female populations. Shelton *et al.*, (1997). Mayer *et al.*, (1995) Kocher *et al.*, 1997; Kocher, 2001; McConnell *et al.*, 2000. However, the paradoxial feminization response to androgens was unexpeced, half of the fish population developed testes and developed into male as well as reversed females (with sex genotype as XY). The mating of females with normal males (sex genotype XY) found super male (sex genotype as YY) (Fig. 46.2).

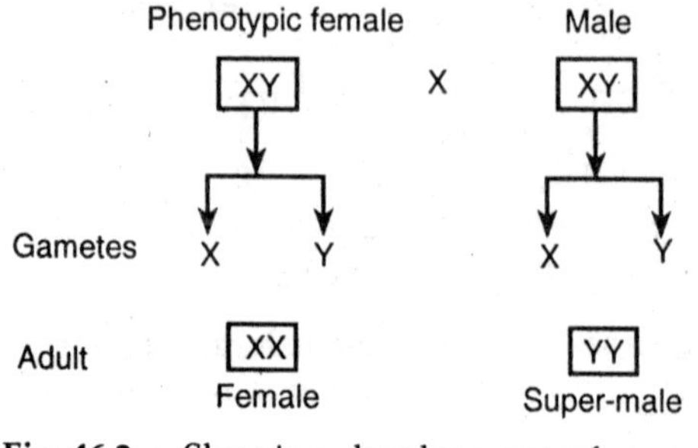

Fig. 46.2 : Showing development of super male, during the cross between phenotypically reverse female with normal male

Advantages

1. Monosex culture is a solution to overpopulation caused by high fecundity.
2. Monosex culture has advantage when one of the sex have superior production traits, better growth rate or higher marketing value.

It the super-males (YY) are mated with normal females (XX), the all offspring from these YY males will produce nearly 100% male (Fig. 46.3).

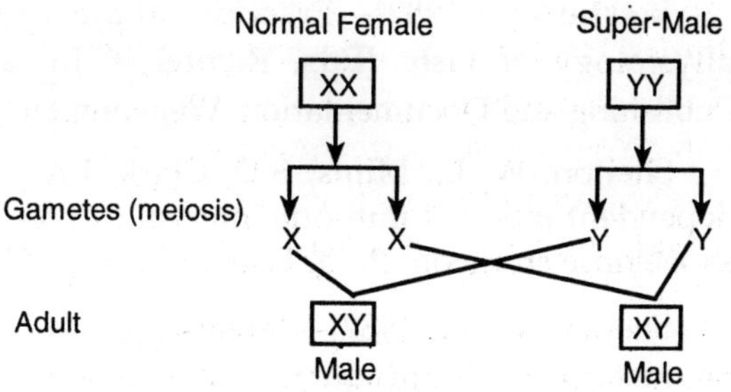

Fig. 46.3 : Showing mating of super male with normal female showing all offspring only normal male by the process of androgenesis

REFERENCES

David, C. J., & Pandian T. J. (2006). Cadaveric Sperm induces intergencise androgeneses in the Buenos Avies tetra, Hemigrammus Caudovittatus. Theriogenology, 65 : 1048-1070.

Donaldson, E. M. and Hunter, G. A. (1982) – Sex control in fish with particular reference to salmonids. Can. J. Fish Aquat. Sci. 39 : 99–110.

Kallman, K. D. (1962). Gynogenesis in teleoest, *Molliensia formosa* (Giravd), with a discussion on detection of parthenogenesis in vertebrates by tissue transplantation J. Genet. 58 : 7–24.

Kocher Thomas D. Lee, Sobolewska, W.J., ; Penman, D. & McAndrew. Maya *et al.* 1997. A genetic linkage map of a chchlid fish, the tilapia (*Oreochromis niloticus*). Genetics-148 : 1225-1232.

Kocher. Thomas D. (2001) : Biotechnology Interface : The site of impact workshop. Tilapia Genetics and Application.

Lam, T. J. (1982)—Application of endocrinology to fish culture. Can J. Fish Aquat. Sci. 39 : 111–137.

Manna, G. K. (1989)—Fish cytogenetics related to taxonomy evolution and monitoring aquatic genotoxic agents. In : Fish Genetics of India. (Ed. Das, P. and Gingran, A. G.) Today's and Tomorrow's Printers and Publishers. New Delhi. pp. 21– 46.

Mayer, J.M. *et al., (*1995). Inducton of diploid andrigenetic and mitotic gynogntic Nile tilapia (Oreochromis niloticius) Theor. Appl. Genet. 90 : 205-210.

Mayer, J. M. & Hershberger, W.K. (1991). Artifical Spawning of tilapia egg. J. World Aquacult. Soc. 22 : 72-82.

McConnell, S.K. J. Beynon, C. Leamon, J. & Skibinski D.O.F. (2000). Microsattalite market based genetic linkage map of *Oreochromis* aureus and *O. nilolicus* (Cichllans) : extensive linkage group segment homologes revealed. Animal Genetics. 31 : 214-218.

Pandian, J.P. & Kiran Kumar, S. (2003). Recent advances in hormonal induction of Sex reversal in fish. Journal of Applied Aquaculture 13; 205-230.

Refstie, T. (1982)—Practical application of sex manipulation. In : 'Reproductive physiology of Fish' (Eds. Richter, C.J.J. and Goos, H.J.T.H.) Centre for Agricultural Publishing and Documentation, Wegeningen, pp. 73-77.

Shelton, W. L., Mims, S.D. Clark, J.A. Hiott, A.E. and Wan, C. 1997. A temperature-dependent index of mitpotic interval (to) for chromosome manipulation in paddle fish and shovelnose sturgeon. Prog. Fish-Guilt. 59 : 152-168.

Shelton, W. L. (1982)—Monosex grass crop. production through breeding sex reversed broodstock In : 'Reproductive physiology of Fish' (Eds. Richter, C.J.J. and Goos, H.J.T.Th) centre for Agriculture Publishing Documentation Wegeningen pp. 82.

Shreck, C. B. (1974)—Hormonal treatment and sex manipulation in fish. In : Control of sex in fishes. C. B. Schreek (Ed.) Extension Division. Virginia. Polytechnique Inst. Virginia Polytechnique Institute and State Univ. Blackburg. V-A. pp. 84-106.

Wohlfath, G.W. 1991., The heredity of sex determination in *Tilapia*. Aquaculture 92 : 143-156.

Yamamoto, T. (1969)—Sex differentiation. In : Fish Physiology, Vol. 3 (Eds. Hoar, W. S.; Randall, D. J.) Academic Press. New York. pp. 117-175.

Chapter 47

FLUORESCENT IN SITU HYBRIDIZATION

It is a laboratory technique used to determine how many copies of the specific segment of DNA are present in a cell. It identifies specific chromosomal region in cytological preparations. In the lab, a segment of DNA is chemically modified or stained by fluorescent dye and when examined under fluorescent microscope looks very bright coloured fluorescent. Fluorescence in situ hybridization has greatly increased the sensitivity, specificity and resolution of chromosome analysis.

The advantages of this technique are numerous but some are mentioned as follows :

(1) The resolution of this method is much better than that of traditional chromosome banding and approximately 2 megabases (Mb) in length.

(2) The protocol of this technique is easy and is applicable for both dividing cells as well as non-dividing cells. This technique can identify a range of mutations including deletion, duplication, aneuploidy and presence of derivative chromosomes. This technique is also commonly used to monitor recurrent or residual disease in bone marrow transplant patients.

FISH (Fluorescent In Situ Hybridization) is generally performed on stimulated peripheral blood for both chromosomal analysis and the identification of specific translocation and deletion. For example, it is used to demonstrate Philadelphia chromosome (Ph^1) formed by the translocation between chromosome 9 and 22 and causes a chronic myelogenous leukemia (CML). It is also used in the identification of the translocation between chromosome 8 and 14 that causes Burkitt's lymphoma and between chromosome 18 and 14 that results B-cell leukemia. It is also used for the diagnosis of Down Syndrome (Strachean *et al.*, 1999; Genetic link, 2005 a & b).

In FISH procedure, cell cultures can be prepared. The mataphase/prometaphase chromosomes are fixed onto a glass slide. In addition, FISH can be modified for analysis of interphase nuclei. Then, the chromosomes are denatured; a labelled DNA probe is then introduced and then probe hybridized to the chromosome on the slide. Therefore, the term 'in situ' hybridization is given.

The DNA probes are labelled with fluoro-chrome, such probe gives fluorescent signal and could be seen with fluorescence microscopy. The fluoresence could be deducted with the help of filters attached in the flourescent microscope. FISH probes are approximately 40 kilo bases (kb) in length, and are detected as a double fluorescent signal on each member of a chromosome pair. The double-dot corresponds to the signals from each of the two aligned sister chromatids.

In its simple application, a normal FISH pattern for any such probe would be two sets of double-dots, one set for each member of the normally paired chromosomes. These double-dots often fuse to form one signal. Cells monosomic for the chromosomal region in question would show only a single set of double-dot per nucleus, while trisomic cells would show three sets of double-dots.

REFERENCES

Genetical Link, (2005). (a) Diagnostic Laboratory Services, INC, Chromosome Analysis-Cytogenetic Tests. Professional Staff, Genetic Link. pp. 1-2.

Genetic Link, (2005) (b) Fluorescence In Situ Hybridization (FISH) on Human Chromosomes and Interphase Nuclei. Department of Medical Genetics, University Hospital Gent. Belgium. Genetic Link. pp. 1-6.

Strachean, Tom and Read, Andrew, P. (1999). Human Molecular Genetics 2, 2nd ed. Bios Scientific Publishers, Oxford.

Chapter 48

POLYMERASE CHAIN REACTION (PCR)

Polymerase chain reaction (PCR) is a technique that can be used to amplify the small amount of DNA (or RNA) several many fold using DNA polymerase and a primer for both strands. This results in an exponential increase in DNA copy number with cycles of synthesis. Kary B. Mullis invented this technique in 1983 and he was awarded Nobel Prize for this technique in 1993. The automated polymerase chain reaction (PCR) is commonly the first step in the vast majority of DNA analysis. This procedure, conveniently and without tedium (tiresome) of cloning, can provide 1000000 copies of a DNA segment in 3 to 4 hours and a billions copies within 24 hours. Actually it is a chemical reaction involving enzymes (Newton and Graham, 1997; Cline, Braman and Hogrefe. 1996, Schrijver *et al.*, 2005).

The principles on which the PCR function are as follows. There are three sequential steps.

(1) **Denaturing :** The first step is to denature the DNA (both Primer DNA and Native DNA). Separation of two strands of DNA is known as denaturing while bringing two stands together is called as reannealing or hybridization (Fig. 48.1). The discovery of denaturing and reannealing of DNA was made artificially by Marmor and Lane (1960). According to them, the double stranded DNA could be separated into two separate single strands (denaturing) by heating at a high temperature (90-95 °C) and hybridization or joining of two strand of DNA (double stranded DNA) which is possible by decreasing the temperature (50-56 °C). This is the main principle on which the PCR functions. The second principle is extension of the primer with the help of DNA polymerase enzyme.

The native DNA, primer and DNA polymerase enzyme (Tag polymerase) are added in a PCR tube and the temperature is raised and lowered in the PCR machine to denature and hybridize.

(2) **Annealing :** After denaturing the DNA, next step is to allow the synthesize primer to anneal to the opposite DNA strands of the desired target sequence. Annealing (hybridization) can occur if the temperature from 95 °C is lowered to 50/ 56 °C. The temperature is then brought down in the machine. As soon as the temperature is lowered, the native DNA strands will bind not only each other but also bind the primers. This specific hybridization, or "recombination" of complementary nucleotides, provides both rationale and the practical basis for much of recombinant DNA (Fig. 48.2).

(3) **Extension of DNA** (Taq Polymerase Enzyme is needed)

After the primers have annealed they should be extended using the oligonucleotide primer as the point of initiation. (Schekman, Weiner, & Kornberg, 1956) described that enzyme DNA polymerase is necessary for the synthesis of DNA. Therefore, for extension of primer we need a thermo stable DNA polymerase so that enzyme is not destroyed in high temperature. The thermo stable polymerase (Taq polymerase) is isolated from *Thermus aquaticus*. The Taq polymerase has a tremendous advantage in performing the PCR reaction. It has an optimal activity temperature of around 70 °C and fresh enzyme is not to be added to each test tube after the heating and cooling process. Thermo stable DNA polymerase is now manufactured by different companies with different sources (other than *Thermus aquaticus*) under different trade names (TM).

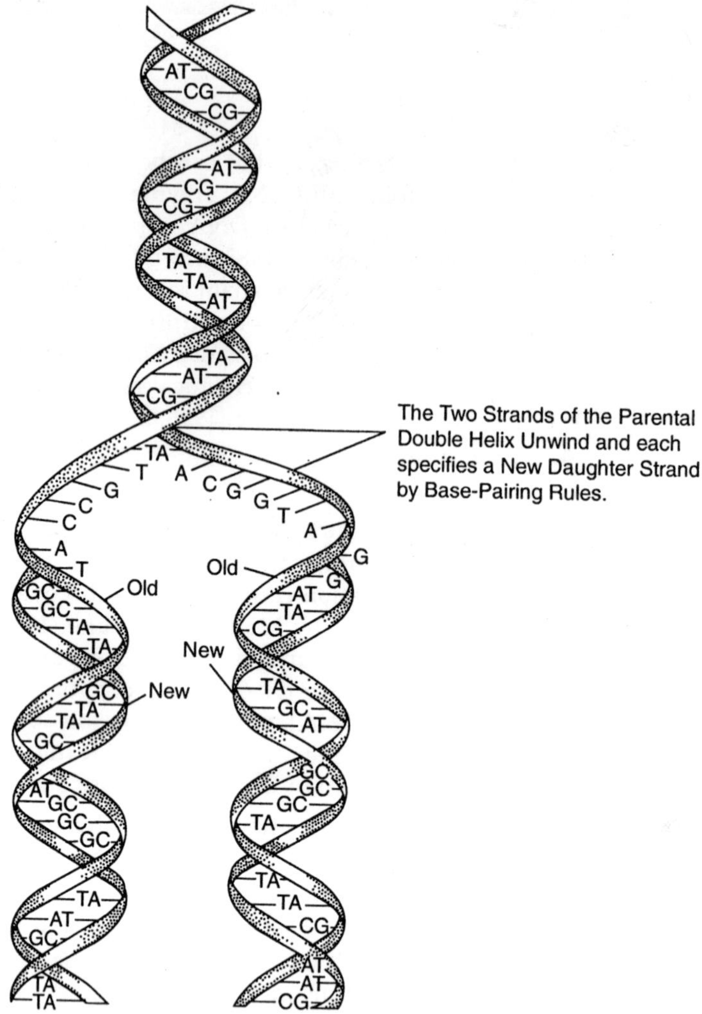

Fig. 48.1 : DNA replication conserves the nucleotide sequence. DNA is double-stranded helical molecule bound together by the nucleotide bases contained on each individual strand. During cell division two identical copies of the original parental strand are made by unwinding the DNA, then synthesizing a complementary second strand to make two identical new daughter strands.

For the extension of primer, the thermo stable DNA polymerase is added in the tube and the temperature is then raised to 65-70 °C.

Subsequent cycles of strand separation (denaturing), primer hybridization (annealing), and strand synthesis (extension) ensure exponential amplification of the target sequences by using the amplified sequence if the previous cycle as a new template, there are three cycles (Fig. 48.2).

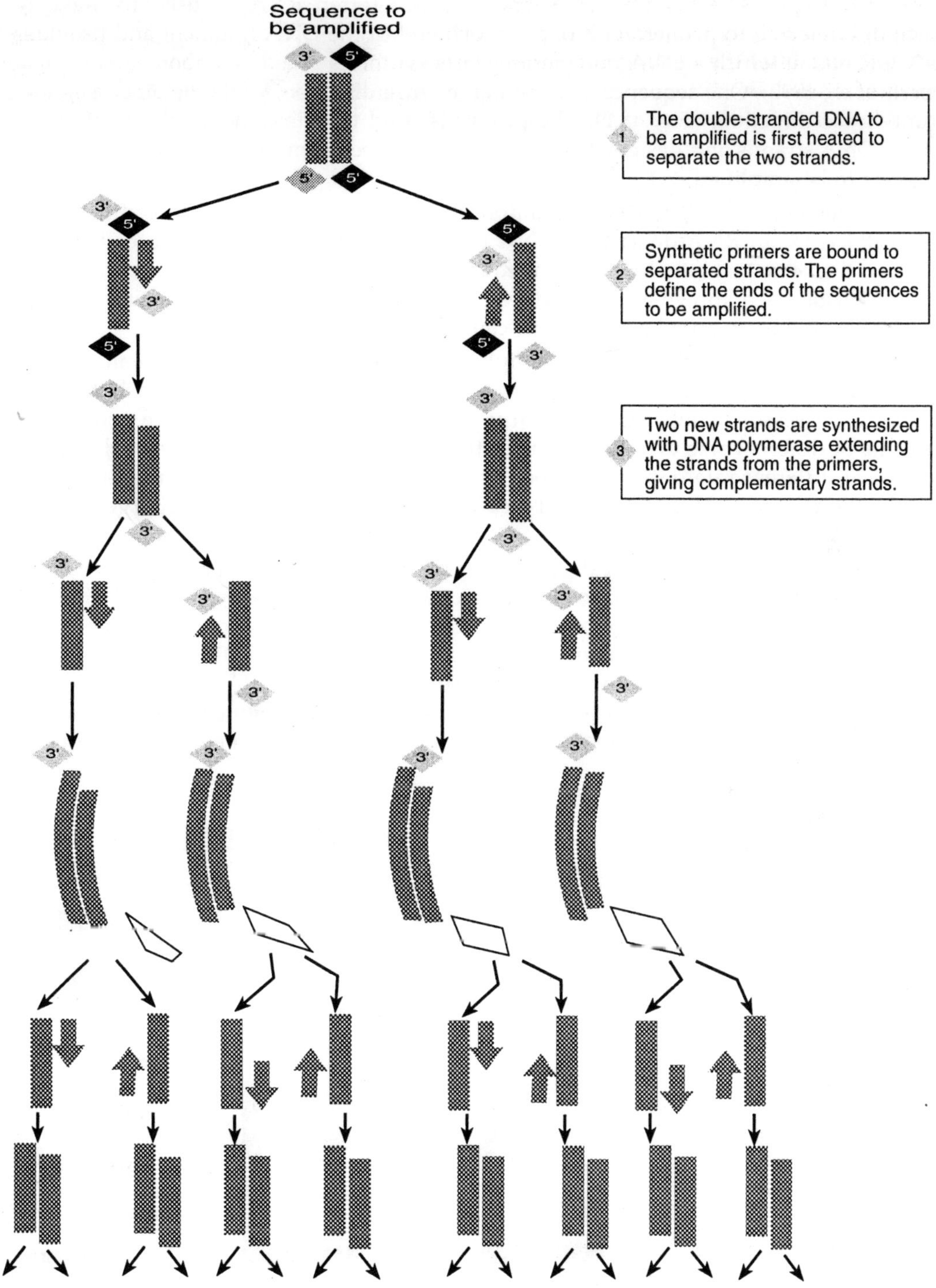

Fig. 48.2 : DNA replication conserves the nucleotide sequence. DNA is a double-stranded helical molecule bound together by the nucleotide bases contained on each individual strand. During cell division two identical copies of the original parental strands are made by unwinding the DNA, then synthesizing a complementary second strand to make two identical new daughter strands.

Primer making. The primers are short sequence (often RNA) is used to make two nucleotides, referred to primer, with one for each end of the DNA fragment and provides a free 3 -OH end at which a DNA polymerase starts synthesis of a deoxyribonucleotide chain (Roberts *et al.*, 1992). One sequence is made in the sense direction, while the other is made in the anti-sense direction. (Fig. 48.2). The primer is used to prime the synthesis of complimentary DNA strands and designed such that the DNA between the primers is the fragment of interest to be amplified.

If the messenger RNA (mRNA) is amplified it must be converted to complimentary DNA (cDNA) by an RNA-dependent DNA polymerase, reverse transcriptase. The cDNA complex is then changed into double stranded DNA, becoming template for a subsequent PCR reaction. This is often known as reversed transcriptase polymerase chain reaction (RT PCR). The primers can be generated artificially or can be purchased from the firms.

After the reaction is completed, the amplified products (DNA) can be visualized by gel electrophoresis. The important physical property of DNA molecule is that each individual nucleotide possesses a net negative charge resulting from the phosphate group. Thus fragments of different size exposed to an electric field tend to migrate towards the positive electrode at different rate depending upon the size, with small fragments migrating faster than the larger one. This process of separation based up electric charge is called *electrophoresis*.

The DNA samples, after being digested into fragments of different sizes by a restriction enzyme (endonuclease), are added to support medium such as agrose gel or acrylamide. The pores of the gel depend upon its percentage and can be compared to a (molecular) sieve. Thus, migration speed of the DNA fragments through gel depends on both DNA fragment size and applied voltage. The procedure to separate and deduct specific DNA fragment is Southern blotting. The importance of this technique is that single strand fragment can be transferred by capillary action to a solid support medium (Nylon or cellulose membrane) and permanently fixed by heating (Fig. 48.3).

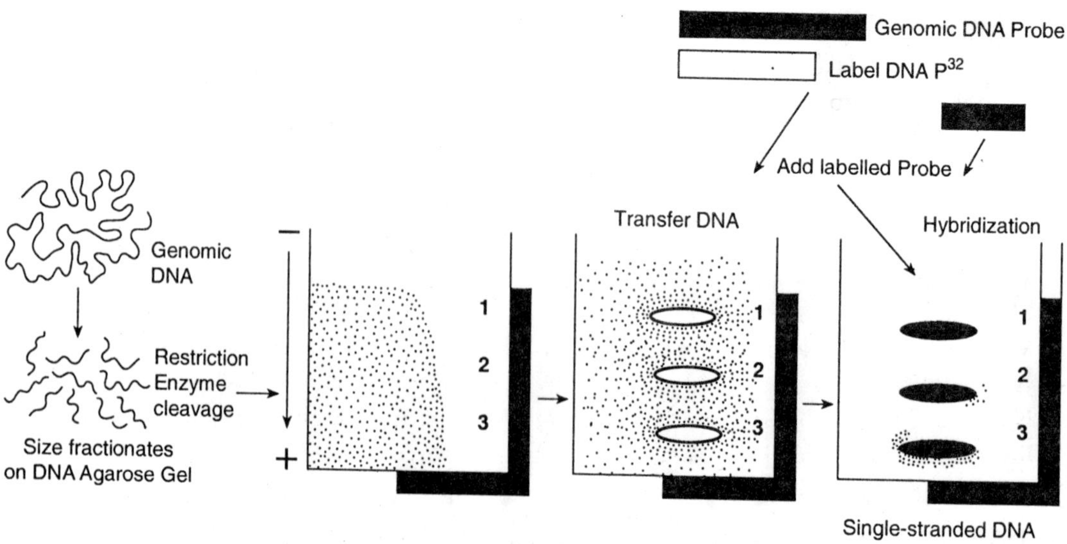

Fig. 48.3 : Southern blotting technique. The DNA is cleaved by appropriate restriction enzyme; digested fragments are separated by electrophoresis on gel. Then denatured and transferred to a membrane and treated with labelled DNA prove through autoradiography.

After electrophoresis, the pattern of DNA can be visualized under UV lamp after treating with ethidium bromide stain; the ethidium bromide is a flourescent dye. The agrose gel electrophoresis distinguished fragments from 100 base pairs to 60000 base pairs (60 kb) whereas, polyacrylamide gels effectively separate fragments than 1000 base pairs (1 kb).

Pulse gel electrophoresis (PFGE) has made possible the separation of DNA fragments even up to 2kb. In this procedure the electric field is altered in different directions forcing the molecules of DNA to reorient between each pulse, or electric current. This is the best technique to identify a known gene.

The endonucleases can recognize 3,4,5,6 or 8 base pairs and are available in the market.

PCR buffers/salts.

For Taq DNA polymerase, the buffer is made as below :

50 mM KCI

10 mM Tris-HCI at room temperature

Dimethyle sulphoxide(DMSO) and glycol are used as co-solvent in PCR buffers when the target has very high denaturation temperature.

Deoxynucleotide triphosphate (*dNTPs*) is ion cofactors.

Now biotechnology companies are providing prepared solutions. The basic principle is described as follows :

There are four dNTPs namely, *dATP* (adenine triphosphate), *dCTP* (cytosine), *dGTP* (gunine), *dUTP* (uracil) which are used according to the base pair rules to extend the primer and ultimately to copy the target sequence. The dNTP binds with Mg^{2+}. The amount of dNTPs present in a reaction will determine the amount of free Mg^{2+} available for optimal enzyme activity. Stock *dNTP* solutions should be neutralized with Tris base to pH 7.0 and their concentration should be determined spectrophotometrically.

Substrate and Substrate Analogs

Taq polymerase works (*dNTP*) very effectively. However, there are some modified substrates available in place of Tag DNA polymerase substrate. They are dogoxigenin-dNTP, biotin-11-*dUTP*, C7deaza-*dGTP*, and fluorescently labelled *dNTPS*.

The following types of PCR are in general use:

(1) Nested PCR, (2) RT-PCR; (3) Anchor PCR; (4) Asymmetrical PCR; (5) Inverse PCR; (6) DOP-PCR.

REFERENCES

Cline, J. Braman and Hogrefe, H.H. (1996). PCR fidelity of Pfu DNA polymerase and other thermostable DNA polymerase nucleic Acids Res. 24. 3546-3551. (Pub Med) (full text in PMC).

Marmor, J. and Lane, L. (1960). Strand separation and specific recombination in deoxyribonucleic acids. Biological studies. Proc. Natl Acad Sci USA. 46, 453-461.

Newton, C.R. and Graham, A. (1997). PCR, 2nd edn. Bios. Scientific Publishers, Oxford.

Roberts, R.; Towbin, J.; Parker, T. G.; Blue, R. (Eds) (1992). A Primer of Molecular Biology, New York, Elsevier Science.

Schekman. R.; Weiner, A.; Kornberg, A. (1956). Multienzyme systems of DNA replication. Science 186 : 987-993.

Schrijver, I.; Zehnder, J.; Cherry, A.M. (2005). Cytogenetic and molecular genetic diagnostic tools. web.

Chapter 49

ADVANCE FLOW CYTOMETRY

Flow cytometry is the recent technique developed during the last two decades to study the cell contents particularly nuclear packing efficiency. In addition to *NPE*, the physical and chemical characteristics of the contents of the various cells and pathological conditions could be analyzed quickly with correct conclusion. The flow cytometers are available in the market in the form of a machine. The machine consists of fluid system, an optical system to collect light signals generated as cells passing through the laser. The machine is also provided with electronic system which converts the light signals to proportional electronic signals that are ultimately digitized for computer analysis. About 5000 cells per second, measuring individually one after the other, from the suspension of heterogeneous population could be scanned.

The sophisticated designed **Fludics System** is provided with lasers, optics, electronics and sorting. The Fludics system has Fluorescence Activated Cell Sorter (FACS Calibur fluids system), during sorting it can sort single cell count accuracy. The result is high purity with recovery that falls between Single cell and Recovery. It has electronically controlled with three buttons to select PRIME, STAND BY, and RUN operational mode. The rate are controls as High; (60ul/min.), medium (35 ul /min) and Low (12 ul/min).

The excitation and emission, fluorochromes, filters and colour compensation are the important part of the equipment. 530/30 BP filter measures signals emitted by Fluorescence Isothiocyanate (Stain). If Phycoerythrin and Popidium Iodide is used then 585/42 BP filter is used. 650 LP filters are used if Peridinin Chlorophyll and 7-Amino-Actinomycin D Signal are generated. The laser signals are read by Becton Dickinson cell quest soft ware. Florescent detector and band filters are to be adjusted as per the instruction and requirement of the instrument. 488 nm Argon laser light sources and 623 nm band pass filter are commonly present in equipment.

Fig. 49.1 : Flowcytometer Machine

It is also having system to display data and analysis. The important parameters are histograms, multiparameter data, and list mode data and gating. The machine analysis of the data with the help of software provided and give histograms of the DNA content versus count which are displayed on the computer automatically (Figs. 49.1, 49.2) (Loken, 1990; B.D., India, 2002).

The applications of flow cytometer are numerous and are a boon for biological research and for the clinical investigations in many human diseases. Flow cytometer is used for the study of blood lineages in the hemopoitic tissue in the head kidney cells of air breathing fishes. It can be easy to distinguish between the pluripotent hematopoietic stem cell and phenotype of the mature functional, peripheral lymphocyte population. The significant use of this equipment is to study the apoptosis or cell death and also to study the necrosis of cell due to pesticide toxicity.

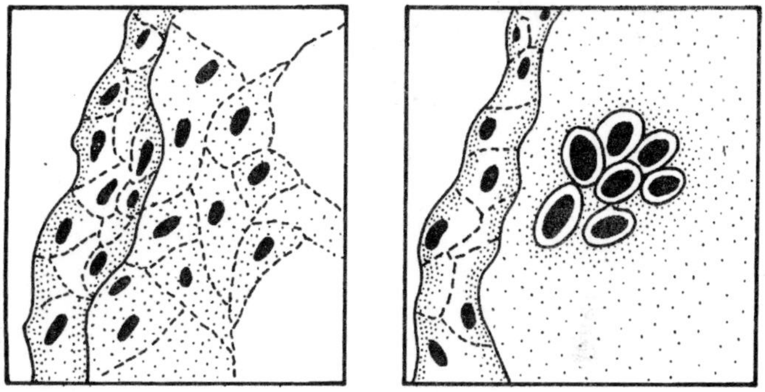

Fig. 49.2 : Showing normal and malignant nuclei

Due to wide-angle scatter signals it is possible to resolve lymphocytes, monocytes and granulocytes from lysed whole blood specimen. It is also possible to separate platelets from the background noise. It is used for immune system and in immunophenotyping. It is used to study the leukemia, and lymphoma, transplantation research, reticulocyte enumeration and analysis of cellular DNA contents. Due to the pulse processing mechanism, it is advantageous in DNA cell-cycle analysis. NPE analysis is used to differentiate primary breast cancer and metastasis, primary colon cancer and colon metastasis as well as gastric and ovarian cancer (Fig. 49.2).

The protocol used is that the first cells were removed and fixed in 3% paraformaldehyde. They are permeabilized with 0.5% Triton x 100. The nuclear DNA was then labelled with popidium iodide after RNase treatment using cycle Test Plus DNA kit (Loken, 1990).

REFERENCES

B.D. India, 2002. Flow cytometry : Some basic Aspects. In INDO-US Workshop on Advance Cytometry. Panjab University, Chandigarh. 24-31.

Loken, M.R. (1990). In Immunofluorescent technique in flowcytometry and sorting. 2nd Edition. Willy.

PART IV

LEGISLATURE

Chapter 50

FISHERIES LAWS, 1897

Introduction

Laws are essential for the regulation and proper functioning of the society in its desirable and benign form—especially to eliminate unhealthy delinquencies and deviations. Thus, law plays an important role in developing a civilised society. Every nation, therefore, has a constitution based on its social, economic, and political ideologies and notions. The constitution lays down the provisions to frame the laws.

After independence, India is a federal republic subdivided in State and Union Territories and is governed by its new constitution known as 'Constitution of India' which came into force on 26 January 1950. The Constitution empowered the Parliament and State legislatures to frame laws and regulations on a number of subjects including air, water, forest, wildlife and fisheries for the preservation, protection and prevention of animal diseases. The Constitution of India was amended, adopted and repealed from time to time according to the changing needs.

At the Central level there are several key laws and regulations relevant to aquaculture.

Indian Fisheries Act, a century old Act, was introduced in 1897, called as The Fisheries Act, 1897. It has been amended time to time according to the needs before and after independence. Before Independence in 1947, under British rule, the Indian part was known as "British India" which later was replaced by the term "British Possession". The definitions of both are defined in General Clauses Act, 1897, as per A.O.1950. Some parts of present India were also under France and Portugal.

As per Clause (5) of General Clauses Act, 1897, "British India" shall mean, as respect to the period before the commencement of Part 111 of Government of India Act, 1935, all territories and places within His Majesty dominions which were for the time being governed by His Majesty's through the Governor General of India or through any Governor or officer subordinate to the Governor General of India, and as respect to any period after that date and before the date of establishment of the Dominion of India means all territories for the time being comprised within the Governors' Province and the "Chief Commissioner" except that a reference to British India in an Indian law passed or made before the commencement of Part 111 of the Government of India Act, 1935, shall not include a reference to Berar . Due to change in the British Constitution, British Possession replaced the word British India as per the power bestowed by General Clause Act, 1897. The definition is prescribed in clause and the "British Possession" shall mean any part of Her Majesty's dominions under both Central and local legislature. All parts under Central legislature shall, for the purpose of this definition, be deemed to be one "British Possession".

In 1897, the Act was passed for the necessity of the conservation and protection of Fish and their natural habitat, the inland, brackish and marine water—as fish was used as human food. People consumes animal protein—fish and prawns are the major source of animal protein. Fish contain 16 to 22% of protein, and therefor, fish is the best food for human consumption. Hence there is necessity to repeal some provisions, to add fresh provisions and amend The Fisheries Act, 1897, to include laws for the fish food security and for the sustainable development to captures fisheries, culture-fisheries and culture based capture of fisheries. This Act was amended in 1903 during the British rule and later, in independent India too, it was amended time to time, but is known as The Fisheries Act, 1897.

The definition of the Acts is given in General Clauses Act, 1897. They are framed with reference to an offence or civil wrong, which shall include a series of acts, and words which refer to Acts done extend also to illegal omission. Hence, "Central Act" shall mean an Act of Parliament while the State Act shall mean an Act by the state legislature.

Fisheries Act gives definition of fish and aquatic life, water (inland, brackish and marine), explosion and poisoning put in the water causing damage to aquatic life, use of net and gears for capturing fish. It also prescribed punishment that violates the law. Now this Fisheries Act, 1897, is under the umbrella of Environment (Protection) Act, 1986.

The Fisheries Act, 1897, is reproduced : (Bare Act)

1. Title and extent

1. The Act may be called the Indian Fisheries Act, 1897 (Act No. 4 of 1897).

2. It extends to whole of India except 3* (the territories which, immediately before 1st November 1956, were comprised in Part B States) 4* 5*

2. Act to be read as supplemental to other Fisheries Laws

Subject to the provisions of Sections 8 and 9 of the 6* General Clauses Act, 1887(1 of 1887), this Act shall be read Supplemental to any other enactment 7* for the time being in force relating to fisheries in 8* (the territories to which this Act extends) 4***

(Notes on Sections 1 & 2): Territories of India, acquisition of foreign territories, and General Clause Act):

At the time of independence of India from British rule, some part of present India was in the possession of French Government and some part was in the possession of Portugal Government. The Constitution of India has the power to include foreign territories in Indian Union as per the General Clauses Act. There is always need to reorganise existing states to create new states and therefore redistribution of territories is a must. This could be done as per the law prescribed in the constitution of India to create Act or through Amendment in the Constitution of India. Reorganisation of territories and creation of new states are been done by the "Central Act".

"Territories of India" is defined in Articles 1 2 3 of Part 1 of the Constitution of India. It includes some decisions of court also. The Constitution provides the power for admission, formation of new States and alteration of areas, boundaries or names of existing States as well as acquisition of foreign territories.

According to the Reorganization Act, 1987, the territories specified in Section 4 of the Goa, Daman and Diu are included in Union Territories of India.

Laccadive, Minicoy and Amindivi islands (name altered) were included in Union Territories of India by the Act of Parliament, 1973, w.e.f. 1.11.1973.

Pondicherry, which was in French establishment in the British India as well at the time of independence, was included in Union Territories of India before 16th day of August 1962, by the Twelfth Amendment in the Constitution by Act, 1962.

The territories and formation of Madhya Pradesh was done by the sub-section (1) of Section 9 of the States Reorganization Act, 1956, excluding the territories specified in Section 3 of the Madhya Pradesh Reorganization Act, 2000, due to the formation of Chattisgarh State.

This Act is subject to the provisions of General Clauses Act. Originally this Act was promulgated in 1868 and 1887. Whereas it expedient to consolidate and extend the General Clauses Acts, 1868 (1of 1868) and (1 of 1887).

3. Definitions

In this Act, unless there is anything repugnant in the subject or context:-

(1) Fish include shellfish;

(2) Fixed engine means any net, cage, trap or other contrivance for taking fish, fixed in the soil or made stationary in any other way; and

(3) "Private water" means water, which is the exclusive property of any person or in which any person has for the time being an exclusive right of fishery whether as owner, lessee or any other capacity.

Explanation. Water shall not cease to be "private water" within the meaning of this definition by reason only that other persons may have, custom, a right of fisheries therein.

4. Destruction of fish by explosives in inland waters and on coasts

(1) If any person uses any dynamite or other explosive substance in any water with intent thereby to catch or destroy any of the fish that may be therein, he shall be punishable with imprisonment for a term which may extend to two months, or with fine which may extend to two hundred rupees.

(2) In sub-section (1) the word "water" includes the sea within a distance of one marine league (= 3 nautical miles = 3.456 miles) of the sea coast, and an offence committed under that sub-section in such sea may be tried, punished and in all respects dealt with as if it had been committed on the land abutting on such coast.

(Notes on Sections 3 & 4).

Section 3 deals with shellfish which include fishes and any aquatic animal with shell, especially edible one. Aquaculture amongst animals principally deals with Fishes, Crustaceans and Molluscs because their culture is commercially viable.

Engines are equipments used for catching shellfish, fish and molluscs from the water resources for commercial use. Engines are described in detail in chapter on Fish and Craft.

Section 4 deals with explosives, which create noise in the water-bodies causing death of aquatic organism. According to this provision the use of dynamite or gunpowder or any other explosive substance in inland or coastal waters are banned. The explosion of dynamite or gunpowder will produce explosion, tending to burst forth noisily causing the damage to the life of fishes or shellfish. This act shall be punishable and the punishment is the imprisonment for a term, which may extend to two months, or with fine, which may extend to two hundred rupees.

"Water" in this section means/includes the coastal region, estuaries and inland water. The law is also prescribed as coastal laws. They are both National and International laws.

5. Destruction of fish by poisoning water

(1) If any person put any poison, lime or noxious material into any water with intent thereby to catch or destroy any fish, he shall be punishable with imprisonment for

a term, which may extend to two months, or with fine which may extend to two hundred rupees.

(2) The State Government may, by notification in the official Gazette, suspend the operation of this section in any specified area, and may in like manner modify or cancel any such notification.

(Notes of Section 5)

If the damage is made to the fish and shellfish by applying poison, the persons will be liable for the punishment as per the provisions given in this section. The punishment is even the imprisonment for a term of two months or a fine of Rs. 200 depending on the gravity of the crime.

6. Protection of fish in selected waters by rules of the state Government

1. The State Government may make rules 1* for the purpose hereinafter in this section mentioned, and may, by notification in the official Gazette, apply all or any of such rules to such water, not being private water, as the State Government may specify in the said notification.

2. The State Government may also, by like notification, apply such rule or any of them to any private water with the consent in writing of the owner thereof and of all persons having for the time being any exclusive right of fisheries therein.

3. Such rules may prohibit or regulate all or any of the following matter, that is to say:-

For rules under 6, see different local Rules and Orders :

(a) The power to make rules under this section is subject to the condition that they shall be made after previous publication;

(b) the construction of weirs; and

(c) the dimension and kind of the nets to be used and the modes of using them.

4. Such rules may also prohibit all fishing in any specified water for a period not exceeding two years.

5. In making any rule under this section the State Government may :

(a) Direct that a breach of it shall be punishable with fine which extend to one hundred rupee, and, when the breach is a continuing breach, with a further fine which may extend to Rs. 10 for every day after the date of the first conviction during which the breach is proved to have been persisted in, and

(b) provide for —

(i) the seizure, forfeiture and removal of fixed engine, erected or used, or nets used, in contravention of the rule, and

(ii) the forfeiture of any fish taken by means of any such fixed engine or net.

6. Arrest without warrant for offences under this Act.

Arrest without warrant for offences under this Act : (1) Any police-officer, or other person 1*specially empowered by the State Government in this behalf, either by name or as holding any office, for the time being, may, without an order from a Magistrate and without warrant, arrest any person committing in his view any offence punishable under Section 4 or 5 or under any rule under Section 6—

1. If the name and address of the person is unknown to him, and if the person declines to give his name and address, or if there is reason to doubt the accuracy of the name and address if given.
2. A person arrested under this section may be detained until his name and address have been correctly ascertained.

Provided that no person so arrested shall be detained longer than may be necessary for bringing him before a Magistrate, except under the order of a Magistrate for his detention.

The power to make rules under this section is subject to the condition that they shall be made after previous publication.

BIBLIOGRAPHY

Bakshi, P. M., 2005, Constitution of India, Universal Law Publishing Co., New Delhi.

Khera, R. C., 2005, Khera's Central Civil Act, Allied Book Company, New Delhi, India.

Footnotes

1*. For rules under 6, see different local Rules and Orders :

(a) the construction of weirs; and

(b) the dimension and kind of the nets to be used and the modes of using them.

7. Such rules may also prohibit all fishing in any specified water for a period not exceeding two years.

8. In making any rule under this section the State Government may

(a) direct that a breach of it shall be punishable with fine which may extend to one hundred rupees, and, when the breach is a continuing breach, with a further fine which may extend to Rs. 10, may for every day after the date of the first conviction during which the breach is proved to have been persisted in, and

(b) provide for—

(i) the seizure, forfeiture and removal of fixed engine, erected or used, or nets used, in contravention of the rule, and

Chapter 51

MADHYA PRADESH LAWS, 1948 (NO. VIII OF 1948)

Introduction

Just after Independence in 1947, the State of Madhya Pradesh formulated law for the protection, conservation and development of fisheries in 1948 and this Act of 1948 was again amended in 1981. Both are presented herewith. Madhya Pradesh is a state which is not connected directly with sea; it contains only inland water. Before the carving out of Madhya Pradesh, some part of present Madhya Pradesh was known as Central Province and Berar when India was 'British India'. The capital of C.P. & Berar was Nagpur. The Saugor and Nerbudda Territories were also marged in it. Till the time many princely States were kept separate—either as "B" and "C". Bhopal and Rewa were placed under "C" categories whereas Madhya Bharat was placed in "B" states.

The present Madhya Pradesh after reorganization in 1956 was the largest State situated in the heart of Republic of India till November 2000 when the new State of Chattisgarh was created. Madhya Predesh is now bordered by Uttar Pradesh, Chattisgarh, Gujarat and Rajasthan. The territories of Chattisgarh were specified as per the provision of the Sections of Madhya Pradesh Reorganization Act.

Madhya Pradesh was having about 4.13 lakh hectares of water, out of which 1.19 lakh hactares is available in village ponds while 2.71 lakh hactares under irrigation reservoir. As per State Government, 2.18 lakh hectares of water is used for capture Fisheries and some is used for culture Fisheries.

For the substainable development of Fisheries, the protection, management and conservation of Fish, Shell-fish and aquatic fauna is essential. It is also essential that the water be protected in such a way that biological material could survive and reproduce properly. Aquaculture provides huge employment to both rural and urban areas. Hence, Madhya Presesh has made Fisheries Act, as per the provisions of the Constitution of India. The following changes are made in the Principal Act, known as the Fisheries Act of 1897.

The words "in the provinces" omitted by A.O. 1950.

Act to be read as supplemental to other Fisheries Laws.

The Act is known as The Madhya Pradesh Fisheries Laws, 1948 (No. VIII of 1948).

The State of Madhya Pradesh has formulated laws for the protection, conservation and development of fisheries in 1948 and this Act was amended in 1981. Both are presented herewith. Madhya Pradesh is a state which is not connected directly with sea; it contains only inland water. The act is known as The (Madhya Pradesh) Fisheries Act, 1948 **(No. VIII of 1948).**

(Received the assent of the Governor on the 26th March 1948 and the assent was first published in the Central Provinces and Berar Gazette on the 2nd April 1948).

An Act to provide for the protection, conservation and development of fisheries in [3] (Madhya Pradesh).

Whereas it is expedient to provide for the protection, conservation and development of fisheries in [3] (Madhya Pradesh).

It is hereby enacted as follows :-

Preamble 1. (1) This act may be cited as the [4] (Madhya Pradesh) Fisheries Act, 1948;

[5](2) It extends to and shall be in force in the whole of (Madhya Pradesh).

[6](3) Omitted

2. In this Act, unless there is anything repugnant in the subject or context.

Definition

(i) "Fish" includes shellfish, and fish in all stages in its life history.

(ii) "Fishery Officer" means such officer as the State Government may appoint for the purpose of this Act and includes any officer or class of officers empowered by the State Government to act as Fishery Officer.

(iii) "Fixed engine" means any net, cage, trap or other contrivance for taking fish, fixed in the soil or made stationery in any other way.

(iv) "Private Water" means water :

(a) which is the exclusive property of any person;

(b) in which any person has for the time being an exclusive right of fishery capacity.

It does not include any river, canal, stream, jhil, or any piece of water which ordinarily has direct communication with any river, canal, stream or jhil.

Explanation : Water shall not ceases to be "Private water" within the meaning of this definition by reason only of the fact that persons other than the owners may have by custom a right of fishery therein;

(v) "Prescribed" means prescribed by rules made under this act.

Protection of Fisheries by Making Rules

3. (i) The State Government, after previous publication, may make rules for the purpose hereinafter in this section mentioned and shall by making in such rules declare the waters, not being private waters, to which all or any of them shall apply.

Footnotes :

For statement of objects and reasons, due Central Provinces and Gazette, dated the 26th Sept. 1947. Part II, page 277. For Report of the Committee, see Central Provinces and Berar Gazette, dated 16th Jan. 1948, Part II, page 18, for proceedings in Assembly Proceedings, 19 Vol. IV, pages 14-27 of No. 14, dated the 30th October 1947 and Vol. V, pages 62-64 of 4, dated the 4th March 1948.

The Act came into force in the former state of Madhya Pradesh on 2nd July 1948—see Agriculture Department Notification No. 2 1583-X, dated the 29th June 1948, published in Central Province and Berar Gazette, dated the 2nd July 1948, Part I, page 356.

(ii) The State Government may, by notification, apply such rules or any of them to any private waters with the consent in writing of the owner thereof and of all persons having for the time being any exclusive right of fishery therein.

(iii) Such rules may :

(a) prohibit or regulate all or any of the following matters, that is to say—

(i) the erection and due to fixed engines;

(ii) the construction, temporary or permanent of weirs, dams and bunds;

(iii) the dimension, size of mesh and kind of nets to be used and the mode of using them;

(iv) the use of more than one method of catching fish at one time.

(b) prohibit the destruction of, or any attempt to destroy fishes by explosives, gun, bow, arrow or the like in inland waters;

(c) prohibit the destruction of, or any attempt to destroy fishes by using a chemical or any other substance likely to cause pollution of water;

(d) prohibit the throwing into any water of any solid or liquid which may be harmful to fishes in such water;

(e) prohibit fishing except under a license and regulate the grant of such licenses, the fees payable thereof and the conditions to be inserted therein;

(f) prescribe the seasons during which the killing, catching or sale of fishes of any prescribed species shall be prohibited; and

(g) prescribe a minimum size or weight below which no fish of any prescribed species shall be sold.

Power to Prohibit of Sale of Fish

4. The State Government may, by notification, prohibit in any Power to any specified areas, the offering or expensing for sale prohibit or barter of any fish capture of

Footnotes :

1. The Act was extended to the whole of Madhya Pradesh by 5.3, the M.P. Extension of Laws Act, 1958 (23 of 1958); vide item Part A of the Schedule to the said Act, with effect from 1st January 1959 vide Govt. of M.P. Law Deptt. Notification No. 4177 XXI, date, 31-12-58 published on P-17 of the M.P. Gazette dated 01.01.1959.
2. Substituted for "the Central Provinces and Berar", by Adopt Order, 1950.
3. Substituted by 5.3 (4) of the M.P. Extentenion of Laws Act, 1958, for the words, "Central Provinces and Berar".
4. Substituted by 5.3 (3) of the M.P. Extentsion of Laws Act, 1958, for the words, "it extends to the whole of Mahakoshal".
5. Omitted by 5.3(3) of the M.P. Extension of Laws Act, 1958.

(iv) Such rules may also prohibit all fishing in any specified waters for a specified period.

(v) In making any rule under this Section the State Government may provide for -

(a) the seizure, removal and forfeiture of any fixed engine or apparatus erected or used for fishing in contravention of the rules, and

(b) the forfeiture of any fishes taken by means of any such fixed engine or apparatus.

which has been made unlawful sale of fish by any rule made under Section 3 of this Act, although caught outside (Madhya Pradesh) and although their capture might be legal at the place where it was caught.

5. If any person contravenes any of the provisions of this Act or rules made thereunder, he shall, on conviction, be punishable with imprisonment for a term which may extend to three months or may extend to five hundred rupees or with both.

Arrest Warrant

6. (1) Any sub-Inspector of Police, or other person specially. Arrest empowered by the State Govt. in this behalf may, without/with warrant, arrest any person committing of attempting to commit under this Act a breach of any provisions of this Act or rules made thereunder if the person decline to give his name and address or if there is reason to doubt the accuracy of the name & address, if given.

(2) A person arrested under this section may be detained until his name and address have been correctly ascertained.

V 1898

Provided that no person so arrested shall be detained longer than may be necessary for bringing him before a Magistrate except under an order of the Magistrate for his detention according to the provisions of the Code of Criminal Procedure, 1898.

(3) All Fishery Officers shall have the same powers of search and investigation in respect of an offence under this Act as Police Officers of the rank of a Sub-Inspector has under the provisions of the Code of Criminal Procedure, 1898.

Jurisdiction inferior to that of Magistrates of the second class excluded.

7. (1) No court inferior to that of a Magistrate of the second class shall try any offence under this Act.

(2) No court shall take cognizance of any offence under this Act, except on a report in writing of the facts constituting such offence made by a Fishery Officer or a Police Officer not below the rank of Sub-Inspector or any other persons or class of persons authorized by the State Government in this behalf.

Power to compound offences

8. (1) Any offence specified in the Schedule may be compounded by the [1]Collector by acceptance of a sum not exceeding one hundred rupees.

(2) On composition of an offence, the accused shall be discharged and property seized from his possession shall be released.

Officers to be deemed public servants

9. All persons empowered to exercise powers and perform duties under this Act shall be deemed to be within the public meaning of Section 21 of the Indian Penal Code, 1860.

Footnote:

1. Substituted by 5.3 (3) of the M.P. Extension of Laws Act, 1958 (23 of 1958) for the words "Mahakoshal region".

Indemnity

10. No suit, prosecution or other legal proceedings shall lie against any persons for anything which is in good faith done or intended to be done under this Act.

Repeal

11. The Indian Fisheries Act, 1897, in its application to the Central Provinces and Berar, is hereby repealed.

THE SCHEDULE

Offences compoundable under Section 8

Description

1. Fishing with a net having a smaller mesh than that prescribed under the rules made under this Act.
2. Fishing without and licence.
3. Selling or attempting to sell fish of a size or weight less than the standard prescribed under this Act.
4. Killing or catching or selling or attemption to kill, to catch or to sell any fish of a prohibited species during a close season.
5. Fishing or attempting to catch fish with any gear or method other than the prescribed gear or method.
6. Using at any one time more than one method of catching fish when prohibited under the rules made under this Act.
7. Licence-holder employing or engaging non-licensees to help him with their nets while fishing.
8. Fishing or attempting to fish in prohibited waters.
9. Offering or exposing for sale or barter any fish, the sale of which is prohibited in any specified area by a notification issued under Section 4.

BIBLIOGRAPHY

Bare Act of Madhya Pradesh

Chapter 52

MADHYA PRADESH LAWS, 1981

The Madhya Pradesh Fisheries (Amendment) Act, 1981 (No. 34 of 1981)

(Received the assent of the President on the 5th October 1981; assent first published in the "Madhya Pradesh Gazette (Extra Ordinary)", dated the 7th October 1981.

An act further to amend the Madhya Pradesh Fisheries Act, 1948.

Be it enacted by the Madhya Pradesh Legislature in the Thirty Second Year of the Republic of India as follows :-

1. **Amendment in Section I :** This Act may be called the Madhya Pradesh Fisheries . (Amend – short titlement) Act, 1981.

2. **Amendment in Section 3 :** In Section 3 of the Madhya Pradesh Fisheries Act, 1948 Amendment (No. 8 of 1948)

(Hereinafter referred to as the Principal Act) Section 3.

(a) In sub-section (3), of Section 3 for clause (f), the following clauses shall be substituted, namely;

(f) prohibited sale or purchase of fish or transport or movement of fish within or outside the State in excess of the prescribed quantity except under a licence;

(ff) Prescribe the seasons during which

(i) the killing or catching of fish of any prescribed species shall be prohibited;

(ii) sale, movement or transport of any prescribed species shall be prohibited except under a licence;

(fff) regulate the grant of licences under clause (f) and sub-clause (ii) of clause (ff), and/or prescribe the fees payable thereof and the conditions to be attached thereto; and

(b) In sub-section (5), of Section 3 for—clauses (a), (b) and (c) the following clause shall be substituted, namely :

(a) the seizure, removal and forfeiture of any fixed engine, apparatus or equipment erected or used for fishing in contravention of the rules;

(b) the forfeiture of any fish procured by means of any such fixed engine, apparatus or equipment or transporation in contravention of the rules; and

(c) the seizure, removal and forfeiture of any animal carts, vessels, rafts, boats or vehicles used for transporting or catching fish in contravention of the rules."

3. **Amendment of Section 5**

For Section 5 of the Principal Act, the following sections shall be substituted, namely :

Penalties :

"5. If any person contravenes any of the provisions, of this Act or Rules made thereunder, he shall on conviction, be punishable with imprisonment for a term which may extend to one year or with fine which may extend to five thousand rupees, or with both.

Cognizance of offence.

5-A. Every offence under this Act shall be cognizable".

4. **Amendment of Section 6**

In section 6 of the Principal Act,

(a) for the proviso to sub-section (2) of Section 6 the following proviso shall be substituted, namely :

"Provided that no person so arrested shall be detained longer than may be necessary for bringing him before a Judicial Magistrate except under an order of the Magistrate for his detention according to the provisions of the Code of Criminal Procedure, 1973 (No. 2 of 1974)";

(b) for sub-section (3), the following sub-section shall be substituted, namely :

"(3) All Fishery Officers shall have the same powers of search, seizure and investigations in respect of an offence under this Act as police officer of the rank of a Sub-Inspector has under the provisions the Code of Criminal Procedure, 1973 (No. 2 of 1974)

5. **Amendment of Section 8**

For sub-section (1) of Section 8 of the Principal Act, the following sub-section shall be substituted, namely :

"(1) Any offence specified in the Schedule may be compounded by the Collector by acceptance of a sum not exceeding one thousand rupees."

6. **Amendment of Schedule**

In the Schedule to the Principal Act, after entry 4, the following entries shall be inserted, namely :

"4-A. Transporting or attempting to transport any fish of a prohibited species during a closed season except in accordance with the conditions of the licence granted under the rules.

4-B. Selling or transporting or attempting to sell or transport fish in excess of the prescribed quantity except under and in accordance with the conditions of the licence granted under the rules."

7. **Repeal. The Madhya Pradesh Fisheries (Amendment) Ordinance, 1981;**

Some changes are been made in the Schedule.

Bhopal, 10th November 1972

RULES

Rules there notified on No. 5042-3862-14-vety. In exercising of the powers conferred by Section 3 of the Madhya Pradesh Fisheries Act, 1948 (VIII of 1948) and in supersession of all rules made or order issued previously on the subject, the State Government hereby makes the following rules, the same having been previously published as required by sub-section (1) of the said section, namely :

1. **Short title** — These rules may be called the Madhya Pradesh Riverine Fisheries Rules, 1972.

2. **Definitions** — In these rules, unless the context otherwise requires :

   (a) "Act", means the Madhya Pradesh Fisheries Act, 1948 (VIII of 1948);

   (b) "Director", means the Director of Fisheries, Madhya Pradesh ;

   (c) "Form", means a form appended to these rules;

   (d) "Licence", means a licence granted under these rules;

   (e) "Licensing Authority", means the Assistant Director of Fisheries of the District or any other officer specially empowered by the Director in this behalf in respect of rivers and streams falling within his respective jurisdiction;

   (f) "Month", means the period commencing from the date of the month on which the licence is issued and ending on the day immediately before such date of the next month;

   (g) "Nalla", means a depression anywhere in the land area, whether private or Government, from where water flows during rainy season and connects with any perennial water source like river, tank or reservoirs;

   (h) "Schedule", means a schedule appended to these rules;

   (i) "Specified Waters", means water specified in Schedule I.

   (j) "Society", means Registered Fishermen Co-operative Society.

3. **Prohibition of fishing in specified waters —**

   (1) Subject to these rules fishing in specified waters except under a licence granted under these rules shall be prohibited.

(2) Notwithstanding anything contained in sub-rule (1) fishing in specified waters shall be closed —

(i) in areas shown in column (4) of Schedule II throughout the year, and;

(ii) in other areas during the period from 16th June to 15th August, each year.

BIBLIOGRAPHY

Bare Act published in Gazette of M.P. Govt.

Chapter 53

MARINE PRODUCT EXPORT DEVELOPMENT AUTHORITY ACT, 1972 (NO. 13 OF 1972)

Marine Product Export Development Authority (MPEDA) was constituted in 1972 under the Marine Product Export development Authority Act, 1972 (No. 13 of 1972). The Authority functions under the Ministry of Commerce, Government of India, and acts as coordinating agency with different Central and State Government establishments engaged in fisheries production and allied activities.

The main purpose is to control and to improve to trade in Seafood export to foreign countries in accordance to Hazard Analysis and Critical Control Points (HACCP). The Seafood HACCP Regulation (Appendix 7) is now accepted in EU, US and throughout the World. HACCP is a process or control system designed to identify hazards from raw material procurement, manufacturing, distribution, usage of food product, including measures for hazard control and prevent microbial and other hazards in food production.

Having great potential both in the marine capture fisheries, (India's roughly 8,129 km coastline which encompasses as Exclusive Economic Zone (EEZ) of 2.02 million km square; as well as 2.86 million hectares (ha) of aquaculture production and capture fisheries from rivers and lakes and aquaculture) Government of India has provided special budget of Rs. 450 crore to promote Research and Development and marketing so that our seafood products should be consumed in foreign markets—specially in European and US markets.

It is important to note that Seafood exports fell from 612,641 tonnes in 2006-2007 to 541,701 tonnes in 2007-2008 causing loss of at Rs 7,620 crores. This has happened because our Seafood products exported were refused as they were not safe in accordance with HACCP Regulations. They were rejected because Fish and fisheries products contain filth and *Salmonella*.

Authority (MPEDA) is undertaking many precautions for expanding production as well as necessary sanitary conditions from capture site, processing, packing and delivery to the export market. It covers under four major heads

1. Export Production—Capture Fisheries;
2. Export Production—Culture Fisheries;
3. Introduction of new Technology and Modernization of Processing Facilities;
4. Marketing Promotion.

The Fin fish, Crustacean (mainly shrimps) and Cephalopods (Squid, Cuttlefish and Octopus) are the main constituents in the supply chain in India for fish and fisheries product.

The precautions needed for safe sanitary are :

(a) water quality,

(b) free from microbial pathogens (*Vibrio parahaemolyticus*), chemical contaminants by heavy metal (Mercury, Arsenic, Cadmium, Zinc etc.), antibiotic (unsustainable use of these antibiotics—mainly chloramphenical and nitrofuran—in shrimp farms) and radio-activity in the products.

MPEDA Act became more significant in view of the US Public Health Security and Bioterronism Preparedness and Response Act, 2002 (so called Bioterrorism Act).

The raw materials originate from either marine capture or aquaculture products from India has faced a number of challenges meeting hygiene requirement for fish and fisheries products in the major export markets, especially European Unions (EU) and United States (US). Hence Act was made. The Act is reproduced as under:

THE MARINE PRODUCTS EXPORT DEVELOPMENT AUTHORITY ACT, 1972

ACT NO. 13 OF 1972 [20th April 1972]

An Act to provide for the establishment of an Authority for the development of the marine products industry under the control of the Union and for matters connected therewith.

BE it enacted by Parliament in the Twenty-third Year of the Republic of India as follows:

(1) Short title, extent and commencement

1. This Act may be called the Marine Products Export Development Authority Act, 1972.
2. It extends to the whole of India.
3. It shall come into force on such date 1* as the Central Government may, by notification in the Official Gazette, appoint: Provided that different dates may be appointed for different provisions of this Act.

(2) Declaration as to expediency of control by the Union. It is hereby declared that it is expedient in the public interest that the Union should take under its control the marine products industry.

(3) Definitions. In this Act, unless the context otherwise requires,— (a) "Authority" means the Marine Products Export Development Authority established under Section 4; (b) "Chairman" means the Chairman of the Authority; (c) "Conveyance" includes a carrier vessel or a vehicle; (d) "Dealer" means a dealer in any of the marine products.

(e) "Director" means the Director of Marine Products Export Development appointed under Section 7;

(f) "Export" and "Import" mean Respectively, taking out of, or bringing into, India by land, sea or air;

(g) "Fishing Vessel" means a ship or boat fitted with mechanical means of propulsion which is exclusively engaged in sea-fishing for profit;

Footnote :

1. The provisions of Sections 11 to 19 (both inclusive) shall come into force w.e.f. 15-1-1973; vide Notifin. No. G.S.R. 8(E), dated 10- 1-1973, Gazette of India, Extraordinary, Pt. II, Sec. 3(i), p. 15. The provisions of sections other than Sections 11 to 19 (both inclusive) shall come into force w.e.f. 12-7-1972; vide Notifin. No. G.S.R. 338(E), dated 12-7-1972, Gazette of India, Extraordinary, Pt. II, Sec. 3(i), p. 377/378.

(h) "Marine Products" includes all varieties of fishery products known commercially as shrimp, prawn, lobster, crab, fish, shell-fish, other aquatic animals or plants or part thereof and any other products which the Authority may, by notification in the Gazette of India, declare to be marine products for the purposes of this Act;

(i) "Member" means a member of the Authority;

(j) "Owner", in relation to any fishing vessel or in relation to any processing plant or storage premises for marine products or in relation to any conveyance used for the transport of marine products, includes— (i) any agent of the owner; and (ii) a mortgagee, lessee or other person in actual possession of the fishing vessel, processing plant, storage premises or conveyance;

(k) "Prescribed" means prescribed by rules made under this Act;

(l) "Processing", in relation to marine products, includes the preservation of such products such as canning, freezing, drying, salting, smoking, peeling or filleting and any other method of processing which the Authority may, by notification in the Gazette of India, specify in this behalf.

MARINE PRODUCTS EXPORT DEVELOPMENT AUTHORITY

4. Establishment and constitution of the Authority :

(1) With effect from such date as the Central Government may, by notification in the Official Gazette, specify in this behalf, there shall be established for the purposes of this Act, an Authority to be called the Marine Products Export Development Authority.

369 (2) The Authority shall be a body corporate by the name aforesaid, having perpetual succession and a common seal, with power to acquire, hold and dispose of property, both movable and immovable, and to contract, and shall by the said name sue and be sued.

(3) The Authority shall consist of the following members, namely:— (a) a Chairman to be appointed by the Central Government; (b) the Director of Marine Products Export Development, ex-officio; (c) three members of Parliament of whom two shall be elected by the House of the People and one by the Council of States; (d) five members to represent respectively the Ministries of the Central Government dealing with— (i) agriculture, (ii) finance, (iii) foreign trade, (iv) industry, and (v) shipping and transport; (e) such number of other members not exceeding twenty as the Central Government may think expedient, to be appointed by that Government by notification in the Official Gazette from among persons who are in its opinion capable of representing— (i) the Governments of the States or Union Territories having a sea-coast; (ii) the interests of owners of fishing vessels, processing plants or storage premises for marine products and conveyances used for the transport of marine products; (iii) the interests of dealers; (iv) the interests of persons employed in the marine products industry; (v) the interests of persons employed in research institutions engaged in the researches connected with the said industry; and (vi) such other persons or class of persons who, in the opinion of the Central Government, ought to be represented on the Authority.

370 (4) The number of persons to be appointed as members from each of the categories specified in clause (e) of sub-section (3), the term of office of the members other than the member referred to in clause (b) of that sub-section, and the manner of filling vacancies among, and the procedure to be followed in the discharge of their functions by the members shall be such as may be prescribed. (5) Any officer of the Central Government, not being a member of the Authority, when deputed by that Government in this behalf, shall have the

right to attend meetings of the Authority and take part in the proceedings thereof but shall not be entitled to vote. (6) The Authority shall elect from among its members a Vice-Chairman who shall exercise such of the powers and perform such of the functions of the Chairman as may be prescribed or as may be delegated to him by the Chairman.

5. Acts or proceedings of Authority or its Committees not to be invalidated. No act or proceeding of the Authority or any Committee appointed by it under Section 8, shall be invalidated merely by reason of— (a) any vacancy in, or any defect in the constitution of, the Authority or such Committee; or (b) any defect in the appointment of a person acting as a member of the Authority or such Committee; or (c) any irregularity in the procedure of the Authority or such Committee not affecting the merits of the case.

6. Salary and allowances of Chairman. The Chairman shall be entitled to such salary and allowances and such conditions of service in respect of leave, pension, provident fund and other matters as may, from time to time, be fixed by the Central Government.

7. Executive officers of the Authority and other staff. (1) The Central Government shall appoint a Director of Marine Products Export Development to exercise such powers and perform such duties under the Chairman as may be prescribed or as may be delegated to him by the Chairman. (2) The Central Government shall appoint a Secretary to the Authority to exercise such powers and perform such duties under the Chairman as may be prescribed or as may be delegated to him by the Chairman. (3) The Director and the Secretary to the Authority shall be entitled to such salaries and allowances and such conditions of service in respect of leave, pension, provident fund and other matters as may be fixed by the Central Government.

371 (4) Subject to such control and restrictions as may be prescribed, the Authority may appoint such other officers and employees as may be necessary for the efficient performance of its functions and pay them such salaries and allowances as it may determine from time to time. (5) The Chairman, the Director, the Secretary and other employees of the Authority shall not undertake any work unconnected with their duties under this Act except with the permission of the Central Government.

8. Committees of the Authority. (1) The Authority may appoint such Committees as may be necessary for the efficient discharge of its duties and performance of its functions under this Act. (2) The Authority shall have the power to co-opt as members of any Committee appointed under sub-section (1) such other number of persons who are not members of the Authority, as it may think fit.

9. Functions of the Authority. (1) It shall be the duty of the Authority to promote, by such measures as it thinks fit, the development under the control of the Central Government of the marine products industry with special reference to exports. (2) Without prejudice to the generality of the provisions of sub-section (1), the measures referred to therein may provide for— (a) developing and regulating off-shore and deep-sea fishing and undertaking measures for the conservation and management of off-shore and deep-sea fisheries; (b) registering fishing vessel, processing plants or storage premises for marine products and conveyances used for the transport of marine products; (c) fixing of standards and specifications for marine products for purposes of export; (d) rendering of financial or other assistance to owners of fishing vessels engaged in off-shore and deep-sea fishing and owners of processing plants or storage premises for marine products and conveyances used for the transport of marine products, and acting as an agency for such relief and subsidy schemes as may be entrusted to the Authority; (e) carrying out inspection of marine products in any fishing vessel, processing plant, storage premises, conveyance or other place where such products are

kept or handled, for the purpose of ensuring the quality of such products; (f) regulating the export of marine products;

372 (g) improving the marketing of marine products outside India; (h) registering of exporters of marine products on payment of such fees as may be prescribed; (i) collecting statistics from persons engaged in the catching of fish or other marine products, owners of processing plants or storage premises for marine products or conveyances used for the transport of marine products, exporters of such products and such other persons as may be prescribed on any matter relating to the marine products industry and the publishing of statistics so collected, or portions thereof or, extracts therefrom; (j) training in various aspects of the marine products industry; and (k) such other matters as may be prescribed. (3) The Authority shall perform its functions under this section in accordance with and subject to such rules as may be made by the Central Government.

10. Dissolution of the Authority. (1) The Central Government may, by notification in the Official Gazette and for reasons to be specified therein, direct that the Authority shall be dissolved from such date and for such period as may be specified in the notification. Provided that before issuing any such notification, the Central Government shall give a reasonable opportunity to the Authority to make representations against the proposed dissolution and shall consider the representations, if any, of the Authority. (2) When the Authority is dissolved under the provisions of sub-section (1), — (a) all members notwithstanding that their term of office has not expired, shall, from the date of dissolution, vacate their offices as such members; (b) all powers and duties of the Authority shall, during the period of dissolution, be exercised and performed by such person or persons as the Central Government may appoint in this behalf; (c) all funds and other property vested in the Authority shall, during the period of dissolution, vest in the Central Government; and (d) as soon as the period of dissolution expires, the Authority shall be reconstituted in accordance with the provisions of this Act.

REGISTRATION

11. Registration of fishing vessel, processing plant, etc. (1) Every owner of a fishing vessel, processing plant or storage premises for marine products or conveyance used for the transport of marine products shall, before the expiration of one month from the date on which he first became owner of such fishing vessel, processing plant, storage premises or conveyance, or before the expiration of three months from the date of coming into force of this section, whichever is later, apply to the Authority for registration under this Act of every such fishing vessel, processing plant, storage premises, or conveyance owned by him: Provided that the Authority may, for sufficient reason, extend the time-limit for registration by such period as it thinks fit. (2) Registration once made shall continue to be in force until it is cancelled by the Authority.

12. Application, cancellation, fee payable and other matters relating to registration. The form of application for registration under Section 11 and for the cancellation of such registration, the fee payable on such applications, the particulars to be included in such applications, the procedure to be followed in granting and cancelling registration and the registers to be kept by the Authority shall be such as may be prescribed.

13. Returns to be made by owners. (1) Every owner referred to in sub-section (1) of Section 11 shall furnish to the Authority at the prescribed time and in the prescribed manner such returns as may be prescribed. (2) The Authority may authorise a member or any of its officers to inspect any fishing vessel, processing plant, storage premises or conveyance at any time to verify the accuracy of any return made under this section.

FINANCE, ACCOUNTS AND AUDIT

14. Imposition of a cess on marine products exported. (1) There shall be levied on all marine products which are exported, a cess for the purposes of this Act at such rate not exceeding three per cent, ad valorem, as the Central Government may, by notification in the Official Gazette, fix.

374 (2) The cess levied under sub-section (1) shall be in addition to any cess or duty leviable on marine products under any other law for the time being in force. (3) The provisions of the Customs Act, 1962 (52 of 1962), and the rules and regulations made thereunder, including those relating to refunds and exemptions from duty, shall, as far as may be, apply in relation to the levy and collection of the cess leviable under sub-section (1) as they apply in relation to the levy and collection of a duty of customs under that Act or those rules and regulations.

15. Payment of proceeds of cess to the Authority. The proceeds of the cess levied under Section 14 shall first be credited to the Consolidated Fund of India and the Central Government may, if Parliament by appropriation made by law in this behalf so provides, pay to the Authority, from time to time from out of such proceeds, after deducting the expenses of collection, such sums of money as it may think fit for being utilised for the purposes of this Act.

16. Grants and loans by the Central Government. The Central Government may, after due appropriation made by Parliament by law in this behalf, pay to the Authority by way of grants or loans such sums of money as the Central Government may consider necessary.

17. Constitution of the Fund. (1) There shall be a Fund to be called the Marine Products Export Development Fund and there shall be credited thereto— (a) the proceeds of the cess made over to the Authority by the Central Government; (b) all fees levied and collected in respect of registration made under this Act; (c) any other fee that may be levied and collected by the Authority under this Act or the rules made thereunder; (d) any grants or loans that may be made by the Central Government for the purposes of this Act; (e) any grants or loans that may be made by any institution for the purposes of this Act; and (f) all sums realised by the Authority in carrying out the measures referred to in Section 9. (2) The Fund shall be applied— (a) for meeting the salaries, allowances and other remuneration of the officers and other employees of the Authority; (b) for meeting the other administrative expenses of the Authority.

375 (c) for meeting the cost of the measures referred to in section 9; and (d) for repayment of any loans from the Central Government or from any institution.

18. Borrowing powers of the Authority. Subject to such rules as may be made in this behalf, the Authority shall have power to borrow on the security of the Marine Products Export Development Fund or any other asset for carrying out the purposes of this Act. 19

19. Accounts and audit. (1) The Authority shall maintain proper accounts and other relevant records and prepare an annual statement of accounts, including the profit and loss account and the balance-sheet in such form as may be prescribed by the Central Government in consultation with the Comptroller and Auditor-General of India. (2) The accounts of the Authority shall be audited by the Comptroller and Auditor-General of India at such intervals as may be specified by him and any expenditure incurred in connection with such audit shall be payable by the Authority to the Comptroller and Auditor-General. (3) The Comptroller and Auditor-General of India and any person appointed by him in connection with the audit of the accounts of the Authority shall have the same rights and privileges

and authority in connection with such audit as the Comptroller and Auditor-General has in connection with the audit of Government accounts and, in particular, shall have the right to demand the production of books, accounts, connected vouchers and other documents and papers and to inspect any of the offices of the Authority. (4) The accounts of the Authority as certified by the Comptroller and Auditor-General of India or any other person appointed by him in this behalf together with the audit report thereon shall be forwarded annually to the Central Government and that Government shall cause the same to be laid before each House of Parliament.

CONTROL BY CENTRAL GOVERNMENT

20. Power to prohibit or control imports and exports of marine products. (1) The Central Government may, by order published in the Official Gazette, make provision for prohibiting, restricting or otherwise controlling the import or export of marine products, either generally or in specified classes of cases. (2) All marine products to which any order under sub-section (1) applies, shall be deemed to be goods of which the import or export has been prohibited under Section 11 of the Customs Act, 1962 (52 of 1962), and all the provisions of that Act shall have effect accordingly.

376 (3) If any person contravenes any order made under sub-section (1) he shall, without prejudice to any confiscation or penalty to which he may be liable under the provisions of the Customs Act, 1962 (52 of 1962), as applied by sub-section (2), be punishable with imprisonment for a term which may extend to one year, or with fine, or with both.

21. Directions by Central Government. The Authority shall carry out such directions as may be issued to it from time to time by the Central Government for the efficient administration of this Act.

22. Returns and Reports. (1) The Authority shall furnish to the Central Government, at such time and in such form and manner as may be prescribed or as the Central Government may direct, such returns and statements and such particulars in regard to any proposed or existing programme for the promotion and development of the marine products industry, as the Central Government may, from time to time, require. (2) Without prejudice to the provisions of sub-section (1), the Authority shall, as soon as possible after the end of each financial year, submit to the Central Government a report in such form and before such date, as may be prescribed, giving a true and full account of its activities, policy and programmes during the previous financial year. (3) A copy of the report received under sub-section (2) shall be laid before each House of Parliament.

MISCELLANEOUS

23. Penalty for making false returns. Any person who being required by or under this Act to furnish any return fails to furnish such return or furnishes a return containing any particular which is false and which he knows to be false or does not believe to be true shall be punishable with fine which may extend to five hundred rupees.

24. Penalties for obstructing a member or officer of the Authority in the discharge of his duties and for failure to produce books and records. Any person who— (a) obstructs any member authorised by the Chairman in writing or any officer or other employee of the Authority authorised by it in this behalf or any person authorised in this behalf by the Central Government or by the Authority, in the exercise of any power conferred, or in the discharge of any duty imposed, on him by or under this Act; or (b) having control over or custody of any account book or other record, fails to produce such book or record when required to do so by or under this Act.

377 shall be punishable with imprisonment for a term which may extend to six months, or with fine which may extend to one thousand rupees, or with both.

25. Other penalties. Whoever contravenes or attempts to contravene or abets the contravention of the provisions of this Act or of any rules made thereunder other than the provisions, punishment for the contravention whereof has been provided for in Sections 20, 23 and 24, shall be punishable with imprisonment for a term which may extend to six months, or with fine which may extend to one thousand rupees, or with both, and in the case of a continuing contravention with an additional fine which may extend to fifty rupees for every day during which such contravention continues after conviction for the first such contravention.

26. Offences by companies. (1) Where an offence under this Act has been committed by a company, every person who at the time the offence was committed was in charge of and was responsible to, the company for the conduct of the business of the company as well as the company shall be deemed to be guilty of the offence and shall be liable to be proceeded against and punished accordingly: Provided that nothing contained in this sub-section shall render any such person liable to any punishment, if he proves that the offence was committed without his knowledge or that he exercised all due diligence to prevent the commission of such offence. (2) Notwithstanding anything contained in sub-section (1), where an offence under this Act has been committed by a company and it is proved that the offence has been committed with the consent or connivance of, or is attributable to any neglect on the part of, any director, manager, secretary or other officer of the company, such director, manager, secretary or other officer shall also be deemed to be guilty of that offence and shall be liable to be proceeded against and punished accordingly. Explanation for the purposes of this section— (a) "company" means any body corporate and includes a firm or other association of individuals; and (b) "director", in relation to a firm, means a partner in the firm.

27. Jurisdiction of court. No court inferior to that of a Presidency Magistrate or a Magistrate of the first class shall try any offence punishable under this Act.

28. Previous sanction of Central Government. No prosecution for any offence punishable under this Act shall be instituted except with the previous sanction of the Central Government.

29. Protection of action taken in good faith. No suit, prosecution or other legal proceedings shall lie against the Government, or the Authority or any Committee appointed by it, or any member of the Authority or such Committee, or any officer or other employee of the Government or of the Authority or any other person authorised by the Government or the Authority, for anything which is in good faith done or intended to be done under this Act or the rules made thereunder.

30. Power to delegate. The Central Government may, by order notified in the Official Gazette, direct that any power exercisable by it under this Act (not being the power to make rules under Section 33) may also be exercised, in such cases and subject to such conditions, if any, as may be specified in the order, by such officer or authority as may be specified therein.

31. Suspension of operation of Act. (1) If the Central Government is satisfied that circumstances have arisen rendering it necessary that certain of the restrictions imposed by this Act should cease to be imposed or if it considers necessary or expedient so to do in the public interest, the Central Government may, by notification in the Official Gazette, suspend or relax to a specified extent, either indefinitely or for such period as may be

specified in the notification, the operation of all or any of the provisions of this Act. (2) Where the operation of any provision of this Act has under sub-section (1) been suspended or relaxed indefinitely, such suspension or relaxation may at any time while this Act remains in force be removed by the Central Government by notification in the Official Gazette.

32. Application of other laws not barred. The provisions of this Act shall be in addition to and not in derogation of the provisions of any other law for the time being in force.

33. Power of Central Government to make rules. (1) The Central Government may, by notification in the Official Gazette, make rules for carrying out the purposes of this Act. (2) In particular, and without prejudice to the generality of the foregoing power, such rules may provide for all or any of the following matters, namely:— (a) the number of persons to be appointed as members from each of the categories specified in clause (e) of sub-section (3) of Section 4, the term of office and other conditions of service of members, the manner of filling vacancies among, and the procedure to be followed in the discharge of their functions by, such members.

379 (b) the circumstances in which and the authority by which a member may be removed; (c) the holding of a minimum number of meetings of the Authority every year; (d) the procedure to be followed at meetings of the Authority for the conduct of business and the number of members which shall form a quorum at a meeting; (e) the maintenance by the Authority of records of business transacted by the Authority and the submission of copies thereof to the Central Government; (f) the powers of the Authority, its Chairman, the Director and Committees of the Authority with respect to the incurring of expenditure; (g) the conditions subject to which the Authority may incur expenditure outside India; (h) the preparation of budget estimates of receipts and expenditure of the Authority and the authority by which the estimates are to be sanctioned; (i) the form and manner in which the accounts should be kept by the Authority; (j) the deposit of the funds of the Authority in banks and the investment of such funds; (k) the conditions subject to which the Authority may borrow; (l) the conditions subject to which and the manner in which contracts may be entered into by or on behalf of the Authority; (m) the additional matters in respect of which the Authority may undertake measures in the discharge of its functions; (n) the remuneration and other allowances payable to the person or persons referred to in clause (b) of sub-section (2) of Section 10; (o) the form of, and the particulars to be contained in, any returns or reports to be made to the Authority under this Act; (p) the form of, and the manner of making applications for registration and for its cancellation by the Authority, the fee payable on such applications and the procedure to be followed in granting and cancelling registration and the conditions governing such registration.

380 (q) the collection of any information or statistics in respect of marine products; (r) any other matter which is to be or may be prescribed by, or provided for by rules under this Act. (3) Every rule made under this section shall be laid, as soon as may be after it is made, before each House of Parliament, while it is in session, for a total period of thirty days which may be comprised in one session or in two or more successive sessions, and if, before the expiry of the session immediately following the session or the successive sessions aforesaid, both Houses agree in making any modification in the rule or both Houses agree that the rule should not be made, the rule shall thereafter have effect only in such modified form or be of no effect, as the case may be; so, however, that any such modification or annulment shall be without prejudice to the validity of anything previously done under that rule.

34. Power to make regulations. (1) The Authority may make regulations not inconsistent with this Act and the rules made thereunder, for enabling it to discharge its functions under this Act. (2) Without prejudice to the generality of the foregoing power, such regulations may provide for all or any of the following matters, namely:— (a) the procedure to be followed at meetings of the Committees appointed by the Authority and the number of members which shall form a quorum at a meeting; (b) the delegation to the Chairman, members, Director, Secretary or other officers of the Authority of any of the powers and duties of the Authority under this Act; (c) the travelling and other allowances of members of the Authority and of Committees thereof; (d) the pay and allowances and leave and other conditions of service of officers (other than those appointed by the Central Government) and other employees of the Authority; (e) the maintenance of its accounts; (f) the maintenance of the registers and other records of the Authority and its various Committees; (g) the appointment by the Authority of agents to discharge on its behalf any of its functions; (h) the persons by whom, and the manner in which, payments, deposits and investments may be made on behalf of the Authority.

381 (3) No regulation made by the Authority shall have effect until it has been approved by the Central Government and published in the Official Gazette, and the Central Government, in confirming a regulation, may make any change therein which appears to it to be necessary. (4) The Central Government may, by notification in the Official Gazette, cancel any regulation which it has confirmed and thereupon the regulation shall cease to have effect. 1* "(5) Every regulation made under this section shall be laid, as soon as may be after it is made, before each House of Parliament, while it is in session, for a total period of thirty days which may be comprised in one session or in two or more successive sessions, and if, before the expiry of the session immediately following the session or the successive sessions aforesaid, both Houses agree in making any modification in the regulation or both Houses agree that the regulation should not be made, the regulation shall thereafter have effect only in such modified form or be of no effect, as the case may be; so, however, that any such modification or annulment shall be without prejudice to the validity of anything previously done under that regulation. "1. Ins. by Act 4 of 1986, s. 2 and Sch. (w.e.f. 15-5-1986).

HACCP has been approved by the food standard setting organization known as Codex Alimentarius Commission with full endorsement by the National Academy of Sciences USA and National Advisory Committee on Microbiological Criteria for foods. The sea food HACCP Regulations Fish & fisheries are dealt in section 123 and important provisions are reproduced as follows.

Some of the important Regulations in connection with Fish and Fisheries Products are reproduced as follows:

The HACCP Regulations are reproduced as under

APPENDIX 7

SEAFOOD HACCP REGULATION

(U.S. Food & Drug Administration Center for Food Safety & Applied Nutrition)

PART 123–FISH AND FISHERY PRODUCTS

General Provisions

Subpart A—General Provisions

Sec. 123.3 Definitions.

The definitions and interpretations of terms in Section 201 of the Federal Food, Drug, and Cosmetic Act (the Act) and in part 110 of this chapter are applicable to such terms when used in this part, except where they are herein redefined.

The following definitions shall also apply:

(a) Certification number means a unique combination of letters and numbers assigned by a shellfish control authority to a molluscan shellfish processor.

(b) Critical control point means a point, step, or procedure in a food process at which control can be applied, and a food safety hazard can as a result be prevented, eliminated, or reduced to acceptable levels.

(c) Critical limit means the maximum or minimum value to which a physical, biological, or chemical parameter must be controlled at a critical control point to prevent, eliminate, or reduce to an acceptable level the occurrence of the identified food safety hazard.

(d) Fish means fresh or saltwater finfish, crustaceans, other forms of aquatic animal life (including, but not limited to, alligator, frog, aquatic turtle, jellyfish, sea cucumber, and sea urchin and the roe of such animals) other than birds or mammals, and all molluscs, where such animal life is intended for human consumption.

(e) Fishery product means any human food product in which fish is a characterizing ingredient.

(f) Food safety hazard means any biological, chemical, or physical property that may cause a food to be unsafe for human consumption.

(g) Importer means either the U.S. owner or consignee at the time of entry into the United States, or the U.S. agent or representative of the foreign owner or consignee at the time of entry into the United States, who is responsible for ensuring that goods being offered for entry into the United States are in compliance with all laws affecting the importation. For the purposes of this definition, ordinarily the importer is not the custom house broker, the freight forwarder, the carrier, or the steamship representative.

(h) Molluscan shellfish means any edible species of fresh or frozen oysters, clams, mussels, or scallops, or edible portions of such species, except when the product consists entirely of the shucked adductor muscle.

(i) Preventive measure means physical, chemical, or other factors that can be used to control an identified food safety hazard.

(j) Process-monitoring instrument means an instrument or device used to indicate conditions during processing at a critical control point.

(k) (1) Processing means, with respect to fish or fishery products: handling, storing, preparing, heading, eviscerating, shucking, freezing, changing into different market forms, manufacturing, preserving, packing, labeling, dockside unloading, or holding.

(2) The regulations in this part do not apply to:

(i) Harvesting or transporting fish or fishery products, without otherwise engaging in processing.

(ii) Practices such as heading, eviscerating, or freezing intended solely to prepare a fish for holding on board a harvest vessel.

(iii) The operation of a retail establishment.

(l) Processor means any person engaged in commercial, custom, or institutional processing of fish or fishery products, either in the United States or in a foreign country. A processor includes any person engaged in the production of foods that are to be used in market or consumer tests.

(m) Scombroid toxin-forming species means tuna, bluefish, mahi and other species, whether or not in the family Scombridae, in which significant levels of histamine may be produced in the fish flesh by decarboxylation of free histidine as a result of exposure of the fish after capture to temperatures that permit the growth of mesophilic bacteria.

(n) "Shall" is used to state mandatory requirements.

(o) Shellfish control authority means a Federal, State, or foreign agency, or sovereign tribal government, legally responsible for the administration of a program that includes activities such as classification of molluscan shellfish growing areas, enforcement of molluscan shellfish harvesting controls, and certification of molluscan shellfish processors.

(p) Shellstock means raw, in-shell molluscan shellfish.

(q) "Should" is used to state recommended or advisory procedures or to identify recommended equipment.

(r) Shucked shellfish means molluscan shellfish that have one or both shells removed.

(s) Smoked or smoke-flavoured fishery products means the finished food prepared by:

1. Treating fish with salt (sodium chloride), and
2. Subjecting it to the direct action of smoke from burning wood, sawdust, or similar material and/or imparting to it the flavour of smoke by a means such as immersing it in a solution of wood smoke.

(t) Tag means a record of harvesting information attached to a container of shellstock by the harvester or processor.

Sec. 123.5. Current good manufacturing practice.

(a) Part 110 of this chapter applies in determining whether the facilities, methods, practices, and controls used to process fish and fishery products are safe, and whether these products have been processed under sanitary conditions.

(b) The purpose of this part is to set forth requirements specific to the processing of fish and fishery products.

Sec. 123.6 Hazard Analysis and Hazard Analysis Critical Control Point (HACCP) Plan

(a) Hazard analysis. Every processor shall conduct, or have conducted for it, a hazard analysis to determine whether there are food safety hazards that are reasonably likely to occur for each kind of fish and fishery product processed by that processor and to identify the preventive measures that the processor can apply to control those hazards. Such food safety hazards can be introduced both within and outside the processing plant environment, including food safety hazards that can occur before, during, and after harvest. A food safety hazard that is reasonably likely to occur is one for which a prudent processor would establish controls because experience, illness data, scientific reports, or other information provide a basis to conclude that there is a reasonable possibility that it will occur in the particular type of fish or fishery product being processed in the absence of those controls.

(b) The HACCP plan. Every processor shall have and implement a written HACCP plan whenever a hazard analysis reveals one or more food safety hazards that are reasonably likely to occur, as described in paragraph (a) of this section.

A HACCP plan shall be specific to:

1. Each location where fish and fishery products are processed by that processor; and

2. Each kind of fish and fishery product processed by the processor. The plan may group kinds of fish and fishery products together, or group kinds of production methods together, if the food safety hazards, critical control points, critical limits, and procedures required to be identified and performed in paragraph (c) of this section are identical for all fish and fishery products so grouped or for all production methods so grouped.

(c) The contents of the HACCP plan.

The HACCP plan shall, at a minimum:

1. List the food safety hazards that are reasonably likely to occur, as identified in accordance with paragraph (a) of this section, and thus must be controlled for each fish and fishery product. Consideration should be given to whether any food safety hazards are reasonably likely to occur as a result of the following :

(i) Natural toxins;

(ii) Microbiological contamination;

(iii) Chemical contamination;

(iv) Pesticides;

(v) Drug residues;

(vi) Decomposition in scombroid toxin-forming species or in any other species where a food safety hazard has been associated with decomposition;

(vii) Parasites, where the processor has knowledge or has reason to know that the parasite-containing fish or fishery product will be consumed without a process sufficient to kill the parasites, or where the processor represents, labels, or intends for the product to be so consumed;

(viii) Unapproved use of direct or indirect food or color additives; and

(ix) Physical hazards.

2. List the critical control points for each of the identified food safety hazards, including as appropriate:

(i) Critical control points designed to control food safety hazards that could be introduced in the processing plant environment; and

(ii) Critical control points designed to control food safety hazards introduced outside the processing plant environment, including food safety hazards that occur before, during, and after harvest.

3. List the critical limits that must be met at each of the critical control points.

4. List the procedures, and frequency thereof, that will be used to monitor each of the critical control points to ensure compliance with the critical limits.

5. Include any corrective action plans that have been developed in accordance with Sec. 123.7(b), to be followed in response to deviations from critical limits at critical control points.

6. List the verification procedures, and frequency thereof, that the processor will use in accordance with Sec. 123.8(a).

(ii) The taking of corrective actions. The purpose of this review shall be, at a minimum, to ensure that the records are complete and to verify that appropriate corrective actions were taken in accordance with Sec. 123.7. This review shall occur within 1 week of the day the records are made.

(iii) The calibrating of any process control instruments used at critical control points and the performing of any periodic end-product or in-process testing that is part of the processor's verification activities. The purpose of these reviews shall be, at a minimum, to ensure that the records are complete, and that these activities occurred in accordance with the processor's written procedures. These reviews shall occur within a reasonable time after the records are made.

(b) Corrective actions. Processors shall immediately follow the procedures in Sec. 123.7 whenever any verification procedure, including the review of a consumer complaint, reveals the need to take a corrective action.

(c) Reassessment of the hazard analysis. Whenever a processor does not have a HACCP plan because a hazard analysis has revealed no food safety hazards that are reasonably likely to occur, the processor shall reassess the adequacy of that hazard analysis whenever there are any changes that could reasonably affect whether a food safety hazard now exists. Such changes may include, but are not limited to changes in: Raw materials or source of raw materials, product formulation, processing methods or systems, finished product distribution systems, or the intended use or consumers of the finished product. The reassessment shall be performed by an individual or individuals who have been trained in accordance with Sec. 123.10.

(d) Record keeping. The calibration of process-monitoring instruments, and the performing of any periodic end-product and in-process testing, in accordance with paragraphs (a)(2)(ii) through (iii) of this section shall be documented in records that are subject to the recordkeeping requirements of Sec. 123.9.

Sec. 123.9 Records.

(a) General requirements. All records required by this part shall include:

(1) The name and location of the processor or importer.

(2) The date and time of the activity that the record reflects.

(3) The signature or initials of the person performing the operation, and

(4) Where appropriate, the identity of the product and the production code, if any. Processing and other information shall be entered on records at the time that it is observed.

(b) Record retention.

1. All records required by this part shall be retained at the processing facility or importer's place of business in the United States for at least 1 year after the date they were prepared in the case of refrigerated products and for at least 2 years after the date they were prepared in the case of frozen, preserved, or shelf-stable products.
2. Records that relate to the general adequacy of equipment or processes being used by a processor, including the results of scientific studies and evaluations, shall be retained at the processing facility or the importer's place of business in the United States for at least 2 years after their applicability to the product being produced at the facility.

3. If the processing facility is closed for a prolonged period between seasonal packs, or if record storage capacity is limited on a processing vessel or at a remote processing site, the records may be transferred to some other reasonably accessible location at the end of the seasonal pack but shall be immediately returned for official review upon demand.

(c) Official review. All records required by this part and all plans and procedures required by this part shall be available for official review and copying at reasonable times.

(d) Public disclosure.

1. Subject to the limitations in paragraph (d)(2) of this section, all plans and records required by this part are not available for public disclosure unless they have been previously disclosed to the public as defined in Sec. 20.81 of this chapter or they relate to a product or ingredient that has been abandoned and they no longer represent a trade secret or confidential commercial or financial information as defined in Sec. 20.61 of this chapter.
2. However, these records and plans may be subject to disclosure to the extent that they are otherwise publicly available, or that disclosure could not reasonably be expected to cause a competitive hardship, such as generic-type HACCP plans that reflect standard industry practices.

(e) Tags. Tags as defined in Sec. 123.3(t) are not subject to the requirements of this section unless they are used to fulfill the requirements of Sec. 123.28(c).

(f) Records maintained on computers. The maintenance of records on computers is acceptable, provided that appropriate controls are implemented to ensure the integrity of the electronic data and signatures.

Sec. 123.10 Training.

At a minimum, the following functions shall be performed by an individual who has successfully completed training in the application of HACCP principles to fish and fishery product processing at least equivalent to that received under standardized curriculum recognized as adequate by the U.S. Food and Drug Administration or who is otherwise qualified through job experience to perform these functions. Job experience will qualify an individual to perform these functions if it has provided knowledge at least equivalent to that provided through the standardized curriculum.

(a) Developing a HACCP plan, which could include adapting a model or generic-type HACCP plan, that is appropriate for a specific processor, in order to meet the requirements of Sec. 123.6(b);

(b) Reassessing and modifying the HACCP plan in accordance with the corrective action procedures specified in Sec. 123.7(c)(5), the HACCP plan in accordance with the verification activities specified in Sec. 123.8(a)(1), and the hazard analysis in accordance with the verification activities specified in Sec. 123.8(c); and

(c) Performing the record review required by Sec. 123.8(a)(3). The trained individual need not be an employee of the processor.

Sec. 123.11 Sanitation control procedures.

(a) Sanitation SOP. Each processor should have and implement a written sanitation standard operating procedure (herein referred to as SSOP) or similar document that is specific to each location where fish and fishery products are produced. The SSOP should specify how the processor will meet those sanitation conditions and practices that are to be monitored in accordance with paragraph (b) of this section.

(b) Sanitation monitoring. Each processor shall monitor the conditions and practices during processing with sufficient frequency to ensure, at a minimum, conformance with those conditions and practices specified in part 110 of this chapter that are both appropriate to the plant and the food being processed and related to the following :

1. Safety of the water that comes into contact with food or food contact surfaces, or is used in the manufacture of ice.
2. Condition and cleanliness of food contact surfaces, including utensils, gloves, and outer garments.
3. Prevention of cross-contamination from insanitary objects to food, food packaging material, and other food contact surfaces, including utensils, gloves, and outer garments, and from raw product to cooked product.
4. Maintenance of hand washing, hand sanitizing, and toilet facilities.
5. Protection of food, food packaging material, and food contact surfaces from adulteration with lubricants, fuel, pesticides, cleaning compounds, sanitizing agents, condensate, and other chemical, physical, and biological contaminants.
6. Proper labeling, storage, and use of toxic compounds.
7. Control of employee health conditions that could result in the microbiological contamination of food, food packaging materials, and food contact surfaces.
8. Exclusion of pests from the food plant.

The processor shall correct, in a timely manner, those conditions and practices that are not met.

(c) Sanitation control records. Each processor shall maintain sanitation control records that, at a minimum, document the monitoring and corrections prescribed by paragraph (b) of this section. These records are subject to the requirements of Sec. 123.9.

(d) Relationship to HACCP plan. Sanitation controls may be included in the HACCP plan, required by Sec. 123.6(b). However, to the extent that they are monitored in accordance with paragraph (b) of this section they need not be included in the HACCP plan, and vice versa.

Sec. 123.12 Special requirements for imported products.

This section sets forth specific requirements for imported fish and fishery products.

(a) Importer verification. Every importer of fish or fishery products shall either:

1. Obtain the fish or fishery product from a country that has an active memorandum of understanding (MOU) or similar agreement with the Food and Drug Administration, that covers the fish or fishery product and documents the equivalency or compliance of the inspection system of the foreign country with the U.S. system, accurately reflects the current situation between the signing parties, and is functioning and enforceable in its entirety; or
2. Have and implement written verification procedures for ensuring that the fish and fishery products that they offer for import into the United States were processed in accordance with the requirements of this part. The procedures shall list at a minimum:

(i) Product specifications that are designed to ensure that the product is not adulterated under Section 402 of the Federal Food, Drug, and Cosmetic Act because it may be injurious to health or have been processed under insanitary conditions, and,

(ii) Affirmative steps that may include any of the following:

(A) Obtaining from the foreign processor, the HACCP and sanitation monitoring records required by this part that relate to the specific lot of fish or fishery products being offered for import.

(B) Obtaining either a continuing or lot-by-lot certificate from an appropriate foreign government inspection authority or competent third party certifying that the imported fish or fishery product is or was processed in accordance with the requirements of this part.

(C) Regularly inspecting the foreign processor's facilities to ensure that the imported fish or fishery product is being processed in accordance with the requirements of this part.

(D) Maintaining file copy, in English, of the foreign processor's HACCP plan, and a written guarantee from the foreign processor that the imported fish or fishery product is processed in accordance with the requirements of the part.

(E) Periodically testing the imported fish or fishery product, and maintaining file copy, in English, of a written guarantee from the foreign processor that the imported fish or fishery product is processed in accordance with the requirements of this part.

(F) Other such verification measures as appropriate that provide an equivalent level of assurance of compliance with the requirements of this part.

(b) Competent third party. An importer may hire a competent third party to assist with or perform any or all of the verification activities specified in paragraph (a)(2) of this section, including writing the importer's verification procedures on the importer's behalf.

(c) Records. The importer shall maintain records, in English, that document the performance and results of the affirmative steps specified in paragraph (a)(2)(ii) of this section. These records shall be subject to the applicable provisions of Sec. 123.9.

(d) Determination of compliance. There must be evidence that all fish and fishery products offered for entry into the United States have been processed under conditions that comply with this part. If assurances do not exist that the imported fish or fishery product has been processed under conditions that are equivalent to those required of domestic processors under this part, the product will appear to be adulterated and will be denied entry.

Subpart B—Smoked and Smoke-flavored Fishery Products

Sec. 123.15 General.

This subpart augments subpart A of this part by setting forth specific requirements for processing smoked and smoke-flavored fishery products.

Sec. 123.16 Process controls.

In order to meet the requirements of subpart A of this part, processors of smoked and smoke-flavored fishery products, except those subject to the requirements of part 113 or 114 of this chapter, shall include in their HACCP plans how they are controlling the food safety hazard associated with the formation of toxin by *Clostridium botulinum* for at least as long as the **shelf life** of the product under normal and moderate abuse conditions.

Subpart C—Raw Molluscan Shellfish

Sec. 123.20 General.

This subpart augments subpart A of this part by setting forth specific requirements for processing fresh or frozen molluscan shellfish, where such processing does not include a treatment that ensures the destruction of vegetative cells of microorganisms of public health concern.

Sec. 123.28 Source controls.

(a) In order to meet the requirements of subpart A of this part as they apply to microbiological contamination, chemical contamination, natural toxins, and related food safety hazards—processors shall include in their HACCP plans how they are controlling the origin of the molluscan shellfish they process to ensure that the conditions of paragraphs (b), (c), and (d) of this section are met.

(b) Processors shall only process molluscan shellfish harvested from growing waters approved for harvesting by a shellfish control authority. In the case of molluscan shellfish harvested from U.S. Federal waters, the requirements of this paragraph will be met so long as the shellfish have not been harvested from waters that have been closed to harvesting by an agency of the Federal government.

(c) To meet the requirements of paragraph (b) of this section, processors who receive shellstock shall accept only shellstock from a harvester that is in compliance with such licensure requirements as may apply to the harvesting of molluscan shellfish or from a processor that is certified by a shellfish control authority, and that has a tag affixed to each container of shellstock. The tag shall bear, at a minimum, the information required in Sec. 1240.60(b) of this chapter. In place of the tag, bulk shellstock shipments may be accompanied by a bill of lading or similar shipping document that contains the information required in Sec. 1240.60(b) of this chapter. Processors shall maintain records that document have met the requirements of this section.

These records shall document:

1. The date of harvest.
2. The location of harvest by State and site.
3. The quantity and type of shellfish.
4. The date of receipt by the processor.
5. The name of the harvester, the name or registration number of the harvester's vessel, or an identification number issued to the harvester by the shellfish control authority.

(d) To meet the requirements of paragraph (b) of this section, processors who receive shucked molluscan shellfish shall accept only containers of shucked molluscan shellfish that bear a label that complies with Sec. 1240.60(c) of this chapter. Processors shall maintain records that all shucked molluscan shellfish have met the requirements of this section.

These records shall document:

1. The date of receipt.
2. The quantity and type of shellfish.
3. The name and certification number of the packer or repacker of the product.

PART 1240—CONTROL OF COMMUNICABLE DISEASES

2. The authority citation for 21 CFR part 1240 continues to read as follows:

Authority: Secs. 215, 311, 361, 368 of the Public Health Service Act (42 U.S.C. 216, 243, 264, 271).

3. Section 1240.3 is amended by revising paragraph (r), and by adding new paragraphs (s), (t), and (u) to read as follows:

Sec. 1240.3 General definitions.

(r) Molluscan shellfish. Any edible species of fresh or frozen oysters, clams, mussels, and scallops or edible portions thereof, except when the product consists entirely of the shucked adductor muscle.

(s) Certification number means a unique combination of letters and numbers assigned by a shellfish control authority to a molluscan shellfish processor.

(t) Shellfish control authority means a Federal, State, or foreign agency, or sovereign tribal government, legally responsible for the administration of a program that includes activities such as classification of molluscan shellfish growing areas, enforcement of molluscan shellfish harvesting controls, and certification of molluscan shellfish processors.

(u) Tag means a record of harvesting information attached to a container of shellstock by the harvester or processor.

4. Section 1240.60 is amended by revising the section heading, by redesignating the existing text as paragraph (a) and adding the word "molluscan" before the word "shellfish" the two times that it appears, and by adding new paragraphs (b), (c), and (d) to read as follows:

Sec. 1240.60 Molluscan shellfish.

(b) All shellstock shall bear a tag that discloses the date and place they were harvested (by State and site), type and quantity of shellfish, and by whom they were harvested (i.e., the identification number assigned to the harvester by the shellfish control authority, where applicable or, if such identification numbers are not assigned, the name of the harvester or the name or registration number of the harvester's vessel). In place of the tag, bulk shellstock shipments may be accompanied by a bill of lading or similar shipping document that contains the same information.

(c) All containers of shucked molluscan shellfish shall bear a label that identifies the name, address, and certification number of the packer or repacker of the molluscan shellfish.

(d) Any molluscan shellfish without such a tag, shipping document, or label, or with a tag, shipping document, or label that does not bear all the information required by paragraphs (b) and (c) of this section, shall be subject to seizure or refusal of entry, and destruction.

4. Limits of national and international water of the world of oceans. Polymetallic nodules are important minerals present in the international seabed.

Fig. 54.2 : Polymetallic nodules of Magnesium present in the seabed (Oceans)

The seabed (or floor of the sea) contains non-living organic or inorganic solid, liquid or gaseous substances which are potentially useful to man. The seabed is also rich in mineral resources known as polymetallic nodules in the form of reserves of metal present scattered loosely over the sea floor (Fig. 54.2).

They are called polymetallic nodules because they possess nickel, copper, cobalt and manganese in grades. There are number of factors that contribute to these polymetallic nodules which possess rich deposits of nickel, copper, cobalt-rich ferromanganese and fossil animals (Fig. 54.3).

In addition to these elements, geologists and biologists have also stated about the occurrence of polymetallic sulphides in the marine environment. Sulphides occur around hydrothermal vents sprouting from volcanic areas of the seabed. These mineral nodules have commercial viability. They appeared very competitive with the land-based sources of such metal. It is felt that recovering them from seabed could yield substantial return to investors. This resource is also useful for future development of nations.

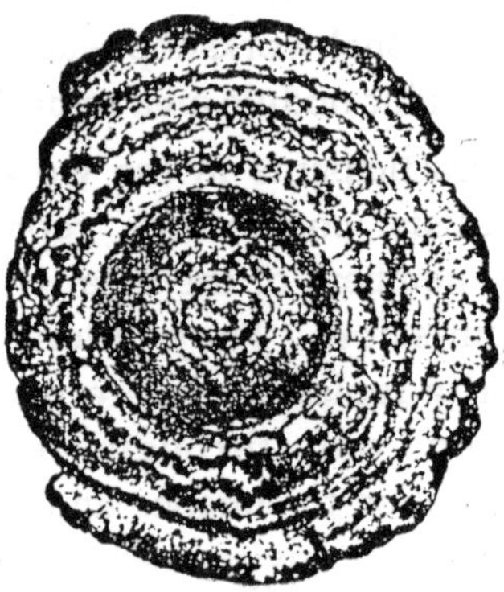

Fig. 54.3 : Gastropod shells in the iron stone deposits in seabed (Ion West & Tonya West (c), 2007)

The origin of marine metal and nonmetal, non-fuel minerals deposits present in ocean basin were considered to be primarily derived from erosion of the continental rock and carried by the rivers to the sea as solid minerals in the oceans, where they are converted as polymetallic nodules by the dissolved chemicals. These polymetallic nodules are located in the reserved area of the Clarion-Clipperton Fracture Zone (CCZ).

The survey has been made by six pioneer investors (UNO) and they concluded that billions of tons of copper, manganese, cobalt and nickel are present in the zone. The present knowledge of these deposits indicate that deposits are available at 3,000/4,000km below the ocean surface and extend to significant depths in the sea. The submerged mountains range about 60,000 km in the location of significant hydrothermal activity which lead to the formation of mineral resources. It is suggested that recovering or mining from the depth of up to 6 km from the ocean surface would not only be feasible but could yield substantial return to the investor.

Sagar Nidhi, a Rs 232 crore vessel, is being used by Indian scientists to explore the mysteries that lie below the sea. Using hydrosweep data the Sagar Nidhi has completed the survey of 150,000 sq. kms. The chemical and biological parameters have been collected under The Board of Management of Polymetallic Nodules programme and the work is in progress.

BIBLIOGRAPHY

A pocketbook on the Ocean with special reference to the water around India, 1999, National Institute of Oceanography (Council of Scientific and Industrial Reseach), Dona Paula, Goa.

Amendment Notification of Coastal Regulation Zone, Ministry of Environment and Forest, 1994 (Supreme Court Judgment dated 18.4.1964).

International Seabed Authority. Function for interactive introduction to the Authority, 2002-2006. United Nations Website.

Notification No. S.O. 88(E) of Government of India, Ministry of Environment and Forest, 1997.

Oceans and Laws of the Sea, 2006. Commission on the Limits of the Continental Shelf (CLCS)—Purpose, function and session. United Nations Website.

Press Information Bureau, Government of India, 2008. Website.

United Nations Convention on the Law of Sea. December 1982. Agreement for the implementations of the provisions of UNO Convention on the Law of the Sea. United Nations Website.

Vijay Kumar, 2004, India and common heritage seabed concept in the international seabed area, Current Science, 86(6).

Chapter 55

CONTINENTAL SHELF

India has laws for its territorial water. This law was enacted after the 40th Amendment in the Constitution in 1976. This Act is known as Territorial Waters, Continental Shelf, Exclusive Zone and other Maritime Zones Act, 1976. The limit of the territorial water is the line, every point of which is at a distance of twelve nautical miles from the nearest point of the appropriate base line.

Later, some modifications were made in this Act by notification of the Government of India dated 15 January 1977. According to this amendment, the "exclusive economic zone " of India has been extended up to a distance of 200 nautical miles into the sea from the shore baseline in conformity with to International Laws of the Sea accepted by the United Nations Organization (UNO).

To make the uniformity throughout the world, United Nations Organization has made a commission; the purpose and function of the commission was on the limit of continental shelf (CLCS) and is described as follows:

1. The commission shall make recommendations to coastal states on the matter related to establishment of those limits; its recommendation and actions shall not prejudice matters relating to the delimitation of the boundaries between states with opposite or adjacent coasts.

2. To consider the data and other materials submitted by the coastal states concerning the outer limits of the continental shelf in the areas where those limits extend beyond 200 nautical miles, and to make recommendations in accordance with Article 76 and Statement of Understanding adapted on 29 August 1980 by the Third United Nations Conference on Law of the Sea.

3. To provide scientific advoice, if requested by the coastal states.

In accordance with Article 76(8), the Commission shall make recommendations of the outer limits of their continental shelf. The limits of the shelf established by the coastal states on the basis of the recommendation shall be final and binding.

However, the UN has made a commission for coastal states to increase the outer limit of the shelf from 200 nautical miles to facilitate the implementation of the recommendations of United Nations Convention on the Law of the Sea.

Definition of Continental Shelf :

The continental shelf is defined by geologists as the continental shelf margin which starts between the shoreline and the shelf break or, where there is no noticeable slope, between the shoreline and the point where the depth of the superjacent water is approximately between 100 and 200 metres.

According to the Convention, the continental shelf of a coastal state comprises the submerged prolongation of the land territory of the coastal state—the seabed and subsoils in the marine areas that extend beyond its territorial sea to the outer edge of the continental margin, or to a distance of 200 nautical miles (Nautical is an international unit of distance for sea and air navigation; equal to 6.076.115449 ft. or 1.852 km. = 1.15 meters), where the outer edge of the continental margin does not extend up to that distance. The continental margin consists of the seabed and subsoil of the shelf, the slope and the rise. It does not include the deep ocean floor with its oceanic ridges or the subsoil thereof (Figs. 55.1, 55.2 & 55.3).

Fig. 55.1 : Oceans of the world

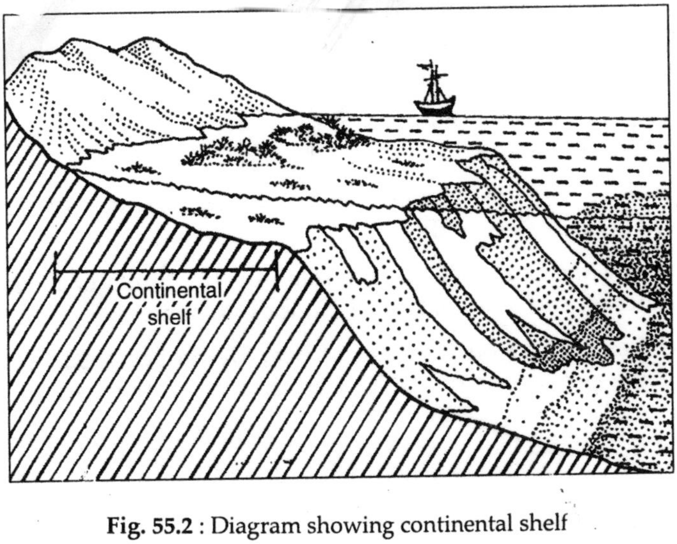

Fig. 55.2 : Diagram showing continental shelf

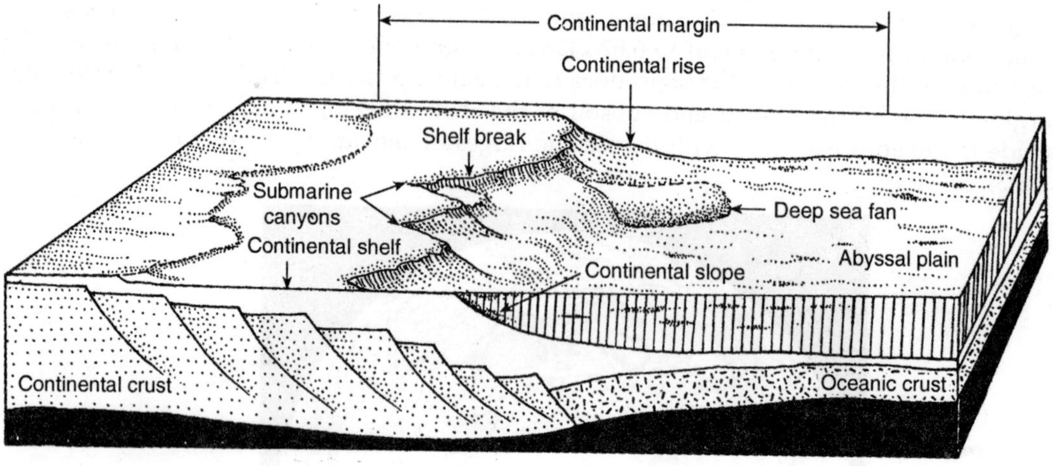

Fig. 55.3 : Continental Shelf showing principle mechanisms of deposition: currents, waves, and animal/sediment interactions

According to Article 76, the coastal state may establish the outer limits of its juridical continental shelf wherever the continental margin extends beyond 200 nautical miles by establishing the foot of the continental slope, by meeting the requirements stated for the thickness of sedimentary rocks, by satisfying geomorphologic requirements and by meeting instance and depth criteria, or by any combination of these methods (Article 76, paragraph 47).

The commission after several meetings appointed members and asked for the funds so that they can help the developing countries in their quest for training and participating in the functions of the commission.

BIBLIOGRAPHY

Amendment Notification of Coastal Regulation Zone, Ministry of Environment and Forest, 1994 (Supreme Court Judgment dated 18.4.1964).

International Seabed Authority. Function for interactive introduction to the Authority, 2002-2006. United Nations Website.

Notification No. S.O. 88(E) of Government of India, Ministry of Environment and Forest, 1997. F.No. L 11011/12/94-1A-III.

Oceans and Laws of the Sea, 2006. Commission on the limits of the continental Shelf (CLCS)—Purpose, function and session. United Nations Website.

United Nations Convention on the Law of Sea, December 1982. Agreement for the implementations of the provisions of UNO Convention on the Law of the Sea, United Nations Website.

AUTHOR INDEX

SUBJECT INDEX

Some exciting gifts are up for grabs with this book!!

FISH AND FISHERIES

Santosh Kumar & Manju Tembhre

FREE Gifts Rs. 1000/-

Dear Readers

By now you must have gone through this book in the entirety. So we invite your assessment on this title with absolute impartiality and candid observations. After all, we publishers blindly depend on your benchmarking. Friends, please fill up the open spaces below with CAPITAL LETTERS. You may add some extra sheets of paper if the space provided is not sufficient for you.

i. What are the important topics of your syllabus which are not adequately covered in this book?

ii. In which chapters and on what topics the treatment of the subject-matter by the author is not satisfactory and why?

iii. Have you come across the typos, factual inaccuracies, inconsistencies, and other forms of mistakes in the book? If yes, then please specify the mistakes and page numbers.

iv. Name top three books on the same subject, which, in your consideration, are better than this book. Could you provide some reasons?

Your 1st Preference: ....

Your 2nd Preference: ....

Your 3rd Preference: ....

v. Any suggestion to improve the quality of illustrations, diagrams, etc.?

vi. How did you like the cover design? Any suggestion on colours?

vii. Suggestions for further improvement of production-standard:

viii. Who recommended you this book:

☐ Teacher ☐ Friend ☐ Parents ☐ Bookseller ☐ Librarian

ix. Name of the person who has recommended:

Designation: ....

Address: ....

Phone No.: ....

e-mail: ....

x. Name and address of the bookseller from whom you have purchased this book:

....

....

....

xi. Name and address of the library where you would like to find the book:

....

....

xii. Name and address of your institution (Please mention the University or Board):

....

....

xiii. Your name, e-mail, and complete postal address:

....

....

....

xiv. A few other titles published by us that you would like to read:

....

....

The best assessment will be rewarded half-yearly. The award will be in the form of our publications, as decided by the Editorial and Marketing Operations Division, amounting to Rs. 1000/- (Rupees one thousand) only.

....

Please mail the filled up coupon at your earliest to :

Editorial & Marketing Operations Division

NEW CENTRAL BOOK AGENCY (P) Ltd

8/1 Chintamoni Das Lane, Kolkata 700 009

West Bengal, INDIA

Phone + Fax: 033 + 2241 8411, 2241 8573, 2241 9094

e-mail: ncbapvtltd@eth.net, ncba@cal2.vsnl.net.in

Website: www.centralgroup.com